Geology, Hydrogeology, and Environmental Remediation:
Idaho National Engineering and Environmental Laboratory, Eastern Snake River Plain, Idaho

Edited by

Paul Karl Link
Department of Geosciences
Idaho State University
Pocatello, Idaho 83209
USA

and

L. L. Mink
Idaho Water Resources Research Institute
205 Morrill Hall
University of Idaho
Moscow, Idaho 83844-3011
USA

SPECIAL PAPER
353

Geological Society of America
3300 Penrose Place
P.O. Box 9140
Boulder, Colorado 80301-9140
2002

Published by The Geological Society of America, Inc.
3300 Penrose Place, P.O. Box 9140, Boulder, Colorado 80301
www.geosociety.org

Printed in U.S.A.

GSA Books Science Editor Abhijit Basu
GSA Books Editor Rebecca Herr
Cover design by Margo Good

Library of Congress Cataloging-in-Publication Data

Geology, hydrogeology, and environmental remediation: Idaho National Engineering and Environmental Laboratory, Eastern Snake River Plain, Idaho / edited by Paul Karl Link and Leland L. Mink.
     p. cm. — (Special Paper ; 353)
     Includes bibliographical references and index.
     ISBN 0-8137-2353-1
     1. Geology—Idaho. 2. Geology—Snake River Plain (Idaho and Or.) 3. Hydrogeology—Idaho. 4. Hydrogeology—Snake River Plain (Idaho and Or.) 5. Idaho Natinal Engineering an Environmental Laboratory—Environmental aspects. I. Link, P. K. II. Mink, Leland L. III. Special papers (Geological Society of America) ; 353.

QE104.S59 G46 2001
557.96—dc21   2001023206

Cover: Landsat Thematic Mapper false color composite image of the eastern Snake River Plain. Area of the Idaho National Engineering and Environmental Laboratory outlined in black. Image processing by John C. Dohrenwend, Southwest Satellite Imaging, 223 South State Street, Teasdale, Utah, 84773-0141; e-mail: dohrenwend@rkymtnhi.com.

10 9 8 7 6 5 4 3 2 1

Contents

GSA Data Repository Item 2002041: Identification of basalt interflow zones with borehole geophysical and video logs at the Idaho National Engineering and Environmental Laboratory, Idaho: John A. Glover, John A. Welhan, and Linda L. Davis.

Geological Society of America
Special Paper 353
2002

Foreword

Robert J. Creed
Trish St. Clair
U.S. Department of Energy, Idaho Operations Office, Idaho Falls, Idaho 83401, USA

The eastern Snake River Plain (ESRP) is a northeast-trending subsiding volcanic trough bounded by northerly trending Basin and Range structures to the north and south. The lithology, trend, and age of the ESRP are consistent with an origin related to the movement of the North American plate in a southwesterly direction over the Yellowstone hotspot. From a distance the ESRP seems to be a flat, featureless basaltic plain. At this perspective one could conclude that the subsurface features that control important geohydrologic properties would be similarly featureless and easily characterized. However, at the meter to tens of meters scales that are important for risk assessment at the Idaho National Engineering and Environmental Laboratory (INEEL) and geohydrological characterization for other water-resource issues, the forces and processes that have shaped the ESRP over the past 4 m.y. have produced a complex system of basalts, sedimentary interbeds, and structural features.

In this Geological Society of America Special Paper, the editors and authors have made significant contributions to our understanding of the composition, geometry, and evolution of the ESRP. Although the U.S. Department of Energy (DOE) and other risk assessment and remediation plans will always require thorough and site-specific investigations, these papers provide a robust and independent context within which various characterization and remediation models can be developed and assessed. This book is of special significance because it will be the first multidisciplinary peer-reviewed regional geoscience summary of any DOE site.

The editors and authors are to be commended in reaching a very productive middle ground between the scientific prerogatives of generating and testing hypotheses (wherever this process might lead) and the need for acceptable characterization data and methods to support risk assessment and resource management needs of the INEEL and the State of Idaho. We hope that the collaboration and multidisciplinary techniques reflected in these papers will serve as an example of how this integrated approach can be applied fruitfully to other sites and regions.

The independent and robust nature of these investigations should give stakeholders and regulators confidence that key characterization issues are or can be well defined. This volume will also give scientists and science and remediation research managers a better understanding of the nature and magnitude of uncertainties and knowledge gaps in important aspects of the ESRP. This knowledge will contribute to the systematic evolution of a productive consensus on how to focus resources and hypotheses to fill these gaps and reduce uncertainties in the assessment of risk associated with remediation strategies and water-resource characterization and development.

Confidence in this process will be essential as organizations with groundwater risk-assessment responsibilities across the country pursue the scientifically and politically defensible characterization of vadose-zone contaminant fate and transport. The INEEL in particular is organizing a systematic scientific effort to understand, coordinate, and integrate vadose-zone characterization efforts across the DOE complex. This book serves as an excellent example of how such issues can be addressed while scientists across the region benefit from pursuing the science needs of the Department of Energy.

A truly outstanding aspect of this volume is the depth, breadth, and coordination of the multidisciplinary techniques used to enhance our understanding of the ESRP and its aquifer. As one studies the systematics of fracture flow and flow in heterogeneous systems, it becomes clear that such a multidisciplinary approach is required. Thus, in this volume there are papers on the microbiology, geochemistry, sedimentology, and hydrology of the ESRP.

We hope that this book will encourage others to support and participate in the evolution of the field of earth sciences as a multidisciplinary problem-solving discipline. A necessary aspect of this evolution is the formulation and pursuit of scientific ideas. Some of these ideas will fail in the validation process and others will lead to concepts and technology that we cannot envision today. The generation and testing of these ideas is a source of endless fascination to earth scientists everywhere.

All involved are indebted to the DOE Environmental Management Office of Technology and Development (EM-50),

Creed, R.J., and St. Clair, T., 2002, Foreword, *in* Link, P.K., and Mink, L.L., eds., Geology, Hydrogeology, and Environmental Remediation: Idaho National Engineering and Environmental Laboratory, Eastern Snake River Plain, Idaho: Boulder, Colorado, Geological Society of America Special Paper 353, p. 1–2.

which provided the money; volume coeditor Roy Mink (Idaho Water Resources Research Institute [IWRRI], Idaho Universities Consortium), who led the research; Clay Nichols, U.S. Department of Energy Idaho Operations Office; Don Maiers (INEEL); and Sarah Bigger and Kathy Owens (IWRRI), who managed Idaho Universities Consortium grants and successfully encouraged collaboration among Idaho university, INEEL laboratory, and U.S. Geological Survey personnel.

MANUSCRIPT ACCEPTED BY THE SOCIETY NOVEMBER 2, 2000

Geological Society of America
Special Paper 353
2002

Introduction to the hydrogeology of the eastern Snake River Plain

Roy C. Bartholomay*
Linda C. Davis
U.S. Geological Survey, Pocatello, Idaho 83209, USA
Paul Karl Link
Department of Geosciences, Idaho State University, Pocatello, Idaho 83209, USA

ABSTRACT

This chapter gives a general overview of the hydrogeology of the eastern Snake River Plain, the Idaho National Engineering and Environmental Laboratory (INEEL), and a description of the INEEL Lithologic Core Storage Library, a source of data for many of the chapters in this volume. It also summarizes definitions and lithostratigraphic terminology for the volume. This volume summarizes geoscience research on the INEEL site in the 1990s. The chapters are written by scientists from many organizations, including INEEL contractors, universities, the U.S. Geological Survey, the state of Idaho, and the Idaho Water Resources Research Institute.

HISTORY

In 1949, the National Reactor Testing Station, a 2300 km² site on the eastern Snake River Plain in southeastern Idaho (Fig. 1), was established by the U.S. Atomic Energy Commission, which later became the U.S. Department of Energy (DOE). The site was used initially for peacetime atomic-energy applications, nuclear safety research, defense programs, and advanced energy concepts. The National Reactor Testing Station was renamed the Idaho National Engineering Laboratory (INEL) in 1974 to reflect the broad scope of engineering activities taking place at various facilities at the site. In 1997, the INEL was renamed the INEEL because of additional emphasis on environmental research. Current INEEL research activities emphasize spent nuclear-fuel management; hazardous and mixed waste management and minimization; cultural resources preservation; and environmental engineering, protection, and remediation (DOE-ID, 1996; Becker et al., 1998).

Various Federal Government contractors operate the INEEL under the supervision of three DOE offices: the Idaho Operations Office (DOE-ID), the Pittsburgh Naval Reactors Office, and the Chicago Operations Office. The government contractors and many other organizations conduct scientific research at the INEEL with authorization by DOE-ID. The State of Idaho INEEL Oversight Program is involved with public education, emergency response, and environmental monitoring. The Idaho Water Resources Research Institute and Inland Northwest Research Alliance (INRA) coordinate university research.

The U.S. Geological Survey has undertaken systematic research at the INEEL since 1949. This work initially characterized water resources before development of nuclear-reactor testing facilities (Walker, 1964; Barraclough et al., 1967, 1976; Nace et al., 1975). Since then, the U.S. Geological Survey has operated a ground-water quality and water-level measurement monitoring network to provide data for research on hydrologic trends and to delineate the movement of facility-related radiochemical and chemical wastes in the Snake River Plain aquifer.

Studies of hydrology, geology, and waste remediation at the INEEL are the focus of this Special Paper, and are needed because of historical waste-disposal practices; however, they also provide better understanding of the regional Snake River Plain-Yellowstone volcanic and hydrologic system. Wastewater containing chemical and radiochemical wastes was discharged

---

*E-mail: rcbarth@usgs.gov

Bartholomay, R.C., Davis, L.C., and Link, P.K., 2002, Introduction to the hydrogeology of the eastern Snake River Plain, *in* Link, P.K., and Mink, L.L., eds., Geology, Hydrogeology, and Environmental Remediation: Idaho National Engineering and Environmental Laboratory, Eastern Snake River Plain, Idaho: Boulder, Colorado, Geological Society of America Special Paper 353, p. 3–9.

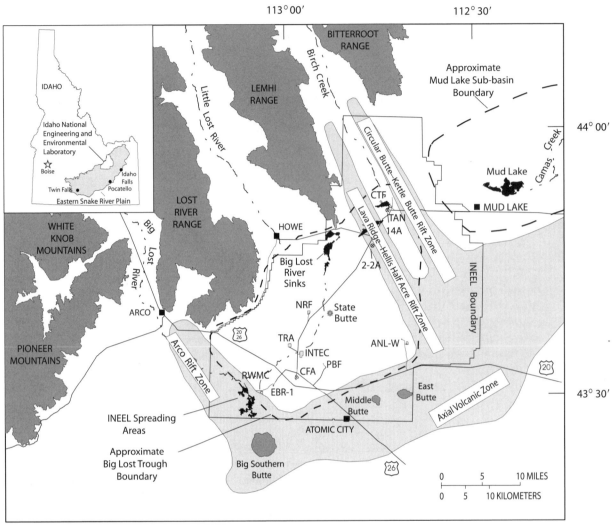

Figure 1. Map showing locations of major geologic features and Department of Energy facilities in the Idaho National Engineering and Environmental Laboratory (INEEL) area, eastern Snake River Plain.

BOUNDARY OF THE IDAHO NATIONAL ENGINEERING AND ENVIRONMENTAL LABORATORY

 APPROXIMATE RIFT ZONE BOUNDARIES

2-2A   LOCAL WELL IDENTIFIER

SELECTED FACILITIES AT THE IDAHO NATIONAL ENGINEERING AND ENVIRONMENTAL LABORATORY

ANL-W	ARGONNE NATIONAL LABORATORY--WEST	NRF	NAVAL REACTORS FACILITY
CFA	CENTRAL FACILITIES AREA	PBF	POWER BURST FACILITY
CTF	CONTAINED TEST FACILITY (Formerly called Loss of Fluid Test Facility--LOFT)	RWMC	RADIOACTIVE WASTE MANAGEMENT COMPLEX
EBR-1	EXPERIMENTAL BREEDER REACTOR NO. 1	TAN	TEST AREA NORTH
INTEC	IDAHO NUCLEAR TECHNOLOGY AND ENGINEERING CENTER	TRA	TEST REACTOR AREA

to ponds and wells on the INEEL until 1983. Since 1983, most aqueous wastes have been discharged to infiltration ponds. Solid and liquid radioactive and chemical wastes have been buried in trenches and pits excavated in surficial sediment at the Radioactive Waste Management Complex. As a result of waste-disposal practices, several monitoring wells at the INEEL contain elevated concentrations of tritium, strontium-90, chromium, sodium, chloride, sulfate, nitrate, and purgeable organic compounds (Bartholomay et al., 1997). Many of the constituents in the wastewater enter the aquifer indirectly by percolation through the unsaturated zone (Pittman et al., 1988).

Geologic hazards and geological, biological, and hydrologic controls on waste management are the financial rationale for much of the research presented in this volume. Ostenaa et al. (this volume) present state-of-the-art flood frequency estimates for the INEEL and suggest a more conservative flood hazard than previous studies have suggested. Watwood et al. (this volume) describe experimental bioremediation techniques in basalt aquifers. Nimmer et al. (this volume) discuss groundtruthing a recirculating tracer test in fractured basalt at the University of Idaho field experiment station.

INEEL Lithologic Core Storage Library

In 1990, the U.S. Geological Survey, in cooperation with the DOE-ID, established the Core Storage Library at the INEEL. The facility was established to consolidate, catalog, and permanently store nonradioactive drill cores and cuttings from investigations of the subsurface conducted at and near the INEEL and to provide a location for researchers to examine, sample, and test these materials.

The Core Storage Library currently stores ~16150 m of drill cores and several suites of drill cuttings. Most of the cores and cuttings were drilled at or near the INEEL for studies of subsurface geohydrologic processes related to waste migration potential, seismic potential, and characterization of the Snake River Plain aquifer. Anderson et al. (1996) provided a partial list of data sources pertaining to many cores stored at the Core Storage Library. A complete description of the facility, procedures for its use, and cores and cuttings available for research can be found in Davis et al. (1997).

Since the 1950s, >500 test holes, auger holes, and wells have been drilled at and near the INEEL to characterize hydrologic and geologic conditions in the subsurface and to supply water to facilities. Data derived from drill cores, such as petrographic analyses, paleomagnetic properties, ^{40}Ar/^{39}Ar age dates, geochemistry, and natural-gamma geophysical logs from boreholes on the INEEL, have been used to determine the subsurface stratigraphy. Many studies of the eastern Snake River Plain aquifer and the INEEL, including several chapters in this volume, are based on core, cuttings, and geophysical borehole data, and reflect ongoing characterization and interpretation of the geology, hydrogeology, and environmental remediation activities at the INEEL and eastern Snake River Plain.

GEOHYDROLOGIC SETTING OF THE E SNAKE RIVER PLAIN

The eastern Snake River Plain is a semiarid l ...u iocss-mantled plain above the subsided late Miocene to recent northeast-striking track of the Yellowstone hotspot (e.g., Nace et al., 1975; Whitehead, 1986; Pierce and Morgan, 1992). This track is perpendicular to northwest-trending basin-and-range faults and volcanic rift zones on the plain (Kuntz et al., 1992; Pierce and Morgan, 1992; Parsons et al., 1998). The INEEL site straddles two topographic depressions, the Big Lost Trough (Bestland et al., this volume; Geslin et al., this volume) and, to the northeast, the Mud Lake subbasin (Fig. 1). The northeast-striking Axial volcanic zone and several diffuse, northwest-striking volcanic rift zones served as volcanic vents for basalt, andesite, and rhyolite (Hackett and Smith, 1992; Hughes et al., 1997, 1999, this volume; Kuntz et al., this volume; Champion et al., this volume).

Pleistocene basaltic vents and their products form topographically elevated, locally diffuse, volcanic constructs, which are concentrated along the Axial volcanic zone and in volcanic rift zones perpendicular to the axis of the plain (Fig. 1; Kuntz et al., 1992, 1994; Anderson and Liszewski, 1997). In places, Quaternary rhyolitic domes stand as high as 610 m above the surface of the plain (Hackett and Smith, 1992; Hughes et al., 1997, 1999; McCurry et al., 1999).

Snake River Plain aquifer

The Snake River Plain aquifer is recharged by seepage from the upper reaches of the Snake River and from tributaries and canals, infiltration from irrigation and precipitation, and underflow from tributary valleys on the perimeter of the plain. Discharge from the aquifer primarily is by pumping for irrigation and flow from springs to the Snake River (Mann and Knobel, 1990). Spring discharge fluctuates as a result of changes in water use, irrigation practices, and precipitation (Kjelstrom, 1992). In 1995, ~4600 × 10^6 m^3 of groundwater was discharged from springs along the Snake River downstream from Twin Falls (Bartholomay et al., 1997).

Recharge to the Snake River Plain aquifer at the INEEL is influenced by local surface drainage. The Big Lost River drains more than 3600 km^2 of mountainous area that includes parts of the Lost River and Pioneer Ranges west of the INEEL (Fig. 1). The average stream flow in the Big Lost River below the Mackay Reservoir for the 79 yr period of record (water years 1905, 1913–1914, and 1920–1995) was ~275 × 10^6 m^3/yr (Brennan et al., 1996, p. 217). Flow in the Big Lost River infiltrates to the Snake River Plain aquifer along its channel and at sinks and playas at the river's terminus in the Big Lost River sinks, in the central Big Lost Trough. To avoid flooding at the INEEL facilities (Ostenaa et al., this volume), excess runoff has been diverted since 1965 to spreading areas in the southwestern part of the INEEL, where much of the water rapidly infiltrates to the aquifer. Measured infiltration losses at various discharges range from ~0.02 to 0.5 m³ s⁻¹ km⁻¹ (Bennett, 1990). Other surface drainages that provide recharge at the INEEL include Birch Creek, Little Lost River, and Camas Creek (Fig. 1).

The Snake River Plain aquifer is one of the most productive in the United States (U.S. Geological Survey, 1985, p. 193). Movement of water is generally from northeast to southwest. The hydraulic gradient is ~0.6–19 m/km and averages ~2.3 m/km (Lindholm et al., 1988). Gradients are smallest in the central part of the plain southwest of the INEEL, where there is a thick section of transmissive basalt (Lindholm, 1991). Water moves horizontally through basalt interflow zones and vertically through joints and interfingering edges of interflow zones. Welhan et al. (this volume, chapter 9), Wylie et al. (this volume), and Gégo et al. (this volume) present hydrologic models that build on previous studies (Welhan and Wylie, 1997). Infiltration of surface water, high rates of pumping, geologic conditions, and seasonal fluxes in recharge and discharge locally affect the movement of groundwater (Garabedian, 1986; Welhan et al., Chapter 15, this volume). McLing et al. (this volume) discuss the composition of shallow and deep thermal parts of the aquifer, and Morse and McCurry (this volume) relate these factors to alteration of basalts.

At the INEEL, depth to water ranges from about 61 m in the northern part to >274 m in the southeastern part. Water levels have fluctuated through time because of wet and dry weather periods. Aquifer water levels generally declined at the INEEL from 1987 to 1995 because of drought conditions, but have increased from 1995 to 2000. Water flows southward and southwestward beneath the INEEL at an average hydraulic gradient of 0.75 m/km. The gradient ranges from ~0.2 to 2.8 m/km (Bartholomay et al., 1997). Estimates of groundwater flow velocities at the INEEL, based on the apparent movement of several tracer constituents in the aquifer system, range from ~1.2 to 6.1 m/day (Mann and Beasley, 1994, p. 24). The range of transmissivity of basalt in the upper part of the aquifer is from 0.1 to 70 600 m²/day (Ackerman, 1991, p. 30). The hydraulic conductivity of underlying rocks is 0.0006–0.01 m/day, several orders of magnitude smaller (Mann, 1986, p. 21). The effective base of the aquifer ranges in depth from ~250 to 520 m below land surface in the western part of the INEEL and may be >580 m below land surface in the eastern part (Anderson and Liszewski, 1997, p. 11; Morse and McCurry, this volume).

Subsurface geology

Miocene and younger rhyolitic lava flows and tuffs are exposed locally at the surface and exist at depth under most of the eastern Snake River Plain. The deepest drill hole in the area is INEL-1, a 3160-m-deep test hole drilled in 1979 northwest of the central facilities area (Doherty et al., 1979; McBroome et al., 1981; Embree et al., 1982; Morgan et al., 1984) that penetrated ~658 m of basalt and sediment and 2501 m of rhyolitic volcanic rocks (tuffs and flows) (Mann, 1986). Hundreds

of basalt flow groups, individual basalt flows, and sedimentary interbeds are present in the subsurface at the INEEL. Basalt flows make up ~85% of the volume of deposits in the unsaturated zone and aquifer (Anderson and Liszewski, 1997, p. 11). The basaltic rocks and sedimentary interbeds combine to form the Snake River Plain aquifer, which is the main source of groundwater in eastern Idaho, and upon which Idaho's agricultural economy is based (Link and Phoenix, 1996).

Flows, flow groups, and members

Stratigraphic subdivision of the basalt lavas at the INEEL is an ongoing research priority. Basalt flow groups and sedimentary interbed units, initially defined in the subsurface by natural-gamma logs, are the fundamental stratigraphic division at the facility scale (e.g., Anderson and Bartholomay, 1995). The basalt flows (as much as 30 m thick) are locally altered (Fromm et al., 1994; Morse and McCurry, 1997, this volume) and consist mainly of medium to dark gray vesicular to dense olivine basalt (Geist et al., this volume, chapter 12; Hughes et al., this volume).

On the eastern Snake River Plain, each defined basaltic shield represents a flow group, or lava field, which is a sequence of flow units erupted within a relatively short period of time from an eruptive center composed of a single vent or series of vents related to a common magma (Kuntz et al., 1980, 1994). When distinction between two or more widely spaced flow groups cannot be made, flow groups may be combined into members, an informal designation that allows further definition when better resolution is available. Welhan et al. (1997) and Wetmore et al. (1997) used the term supergroup for these combinations of flow groups, but the term member is preferable, because supergroup has a specific, and much larger, stratigraphic connotation. These members, composed of flow groups, are coeval but represent multiple, essentially unrelated, eruptive centers. Geochronology, paleomagnetic polarity determinations, gamma-ray log signatures, and geochemistry have been used to determine the genesis and architecture of these basalt lavas and their inclusion within flow groups. Geochemical techniques provide an independent test of flow group and lava flow correlation and are discussed by Geist et al. (this volume, chapter 4) and Hughes et al. (this volume).

Sediments

Modern surficial sediments include lacustrine, fluvial, playa, and eolian facies associations (Scott, 1982; Pierce and Scott, 1982; Garabedian, 1986; Malde, 1991; Kuntz et al., 1994; Gianniny et al., 1997; Mark, 1999; Mark and Thackray, this volume). On the INEEL, sediments have accumulated in two depocenters or sedimentary basins, the Big Lost Trough and the Mud Lake subbasin (Geslin et al., 1999; this volume; Bestland et al., this volume; Gianniny et al., this volume). During wet Pleistocene climate cycles, Lake Terreton expanded to cover both the Big Lost Trough and Mud Lake subbasins, with concomitant expansion in lacustrine facies and increased water flow and sediment transport by the Big Lost River and other streams emptying onto the Snake River Plain (Stearns et al., 1939; Scott, 1982; Geslin et al., 1999; Ostenaa et al., 1999; this volume).

Surface sediments provide analogs to subsurface sediments of the INEEL (Garabedian, 1986; Anderson and Bartholomay, 1995; Mark, 1999; Mark and Thackray, this volume). Core hole 2-2A, drilled to 915 m below land surface in the central part of the Big Lost Trough, contains three thick Quaternary sedimentary intervals, each totaling more than 70 m (Bestland et al., this volume). The total thickness of Neogene and Pleistocene sediment in this hole is >500 m.

Test Area North geology

The Test Area North facility, at the northern end of the Big Lost Trough (Fig. 1), is the focus of several of the papers in this volume. This work was funded by a DOE initiative through the Idaho Water Resources Research Institute to perform cooperative research investigating remediation of a trichloroethylene plume east of the Test Area North facility (Kaminsky et al., 1994; Sorenson et al., 1996). Near the facility, the uppermost basaltic lava flows, located at and near the surface, are part of the ca. 940 ka flow-group N (Lanphere et al., 1994; Geist et al., this volume chapter 4). The area has thus been volcanically quiet throughout much of the Pleistocene (Anderson and Lewis, 1989; Anderson, 1991; Anderson et al., 1996). On the basis of the thin (1–3 m thick) sedimentary units intercalated with these basalts, this area was a paleotopographic high during filling of the Big Lost Trough.

Big Lost Trough

To the southwest, in the north-central part of the Big Lost Trough, near the Big Lost River sinks and core holes 2–2A and Test Area North 14A (Fig. 1), a much thicker sedimentary succession is present (Anderson and Bowers, 1995; Geslin et al., 1997, this volume). Farther to the southwest, the percentage of basalt is greater (Kuntz et al., 1992; Champion et al., this volume). The surface gradient of the Big Lost River flattens north of State Butte as the river passes from a shallow basalt substrate to thicker sediment of the Big Lost Trough (Fig. 1). In the subsurface, the Arco–Big Southern Butte volcanic rift zone defines the southern end of the Big Lost Trough (e.g., Kuntz et al., 1992; Wetmore, 1998). Thick basalt underlies the Central Facilities Area, Idaho Nuclear Technology and Engineering Center, and Radioactive Waste Management Complex.

ACKNOWLEDGMENTS

We gratefully acknowledge reviews by L. Flint Hall of the Idaho National Engineering and Environmental Laboratory (INEEL) Oversight Program and Richard P. Smith of the INEEL.

REFERENCES CITED

Ackerman, D.J., 1991, Transmissivity of the Snake River Plain aquifer at the Idaho National Engineering Laboratory, Idaho: U.S. Geological Survey Water-Resources Investigations Report 91-4058 (DOE/ID-22097), 35 p.

Anderson, S.R., 1991, Stratigraphy of the unsaturated zone and uppermost part of the Snake River Plain aquifer at the Idaho Chemical Processing Plant and Test Reactors Area, Idaho National Engineering Laboratory, Idaho: U.S. Geological Survey Water-Resources Investigations Report 91-4010, 71 p.

Anderson, S.R., Ackerman, D.J., Liszewski, M.J., and Freiburger, R.M., 1996, Stratigraphic data for wells at and near the Idaho National Engineering Laboratory, Idaho: U.S. Geological Survey Open-File Report 96-248 (DOE/ID-22127), 27 p., and diskette.

Anderson, S.R., and Bartholomay, R.C., 1995, Use of natural-gamma logs and cores for determining stratigraphic relations of basalt and sediment at the Radioactive Waste Management Complex, Idaho National Engineering Laboratory, Idaho: Journal of the Idaho Academy of Science, v. 31, no. 1, p. 1–10.

Anderson, S.R., and Bowers, B., 1995, Stratigraphy of the unsaturated zone and uppermost part of the Snake River Plain aquifer at Test Area North, Idaho National Engineering Laboratory, Idaho: U.S. Geological Survey Water-Resources Investigations Report 95-4130, 47 p.

Anderson, S.R., and Lewis, B.D., 1989, Stratigraphy of the unsaturated zone at the Radioactive Waste Management Complex, Idaho National Engineering Laboratory, Idaho: U.S. Geological Survey Water-Resources Investigations Report 89-4065, 54 p.

Anderson, S.R., and Liszewski, M.J., 1997, Stratigraphy of the unsaturated zone and the Snake River Plain aquifer at and near the Idaho National Engineering Laboratory, Idaho: U.S. Geological Survey Water-Resources Investigations Report 97-4183 (DOE/ID-22142), 65 p.

Barraclough, J.T., Robertson, J.B., and Janzer, V.J., 1976, Hydrology of the solid waste burial ground, as related to the potential migration of radionuclides, Idaho National Engineering Laboratory: U.S. Geological Survey Open-File Report 76-471, 183 p.

Barraclough, J.T., Teasdale, W.E., and Jensen, R.G., 1967, Hydrology of the National Reactor Testing Station, Idaho, 1965: U.S. Geological Survey Open-File Report, 107 p.

Bartholomay, R.C., Tucker, B.J., Ackerman, D.J., and Liszewski, M.J., 1997, Hydrologic conditions and distribution of selected radiochemical and chemical constituents in water, Snake River Plain aquifer, Idaho National Engineering Laboratory, Idaho, 1992 through 1995: U.S. Geological Survey Water-Resources Investigations Report 97-4086 (DOE/ID-22137), 57 p.

Becker, B.H., Burgess, J.D., Holdren, K.J., Jorgensen, D.K., Magnuson, S.O., and Sondrup, A.J., 1998, Interim risk assessment and contaminant screening for the Waste Area Group 7 remedial investigation: Idaho Falls, Department of Energy, DOE/ID-10569, variously paginated.

Bennett, C.M., 1990, Streamflow losses and ground-water level changes along the Big Lost River at the Idaho National Engineering Laboratory, Idaho: U.S. Geological Survey Water-Resources Investigations Report 90-4067 (DOE/ID-22091), 49 p.

Brennan, T.S., O'Dell, I., Lehmann, A.K., and Tungate, A.M., 1996, Water resources data, Idaho, water year 1995, Volume 1: Great Basin and Snake River Basin above King Hill: U.S. Geological Survey Water-Data Report ID-95-1, 452 p.

Davis, L.C., Hannula, S.R., and Bowers, B., 1997, Procedures for use of, and drill cores and cuttings available for study at, the Lithologic Core Storage Library, Idaho National Engineering Laboratory, Idaho: U.S. Geological Survey Open-File Report 97-124 (DOE/ID-22135), 31 p.

DOE-ID, 1996, Idaho National Engineering Laboratory comprehensive facility and land use plan: DOE/ID-10514, U.S. Department of Energy, Idaho Operations, variously paginated.

Doherty, D.T., McBroome, L.A., and Kuntz, M.A., 1979, Preliminary geological interpretation and lithologic log of the exploratory geothermal test well (INEL-1), Idaho National Engineering Laboratory, eastern Snake River Plain, Idaho: U.S. Geological Survey Open-File Report 79-1248, 9 p.

Embree, G.F., McBroome, L.A., and Doherty, D.J., 1982, Preliminary stratigraphic framework of the Pliocene and Miocene rhyolite, eastern Snake River Plain, Idaho, in Bonnichsen, B., and Breckenridge, R.M., eds., Cenozoic geology of Idaho: Idaho Bureau of Mines and Geology, Bulletin 26, p. 333–343.

Fromm, J.M., Hackett, W.R., and Stephens, J.D., 1994, Primary mineralogy and alteration of basalts and sediments in drill cores from the Idaho National Engineering Laboratory, eastern Snake River Plain: International Symposium on the Observation of the Continental Crust Through Drilling, 7th, Santa Fe, New Mexico, Abstracts, unpaginated.

Garabedian, S.P., 1986, Application of a parameter estimation technique to modeling the regional aquifer underlying the eastern Snake River Plain, Idaho: U.S. Geological Survey Water-Supply Paper 2278, 60 p.

Geslin, J.K., Gianniny, G.L., Link, P.K., and Riesterer, J.W., 1997, Subsurface sedimentary facies and Pleistocene stratigraphy of the northeastern Idaho National Engineering Laboratory: Controls on hydrogeology, in Sharma, S., and Hardcastle, J.H., eds., Proceedings of the 32nd Symposium on Engineering Geology and Geotechnical Engineering: Boise, Idaho, p. 15–28.

Geslin, J.K., Link, P.K., and Fanning, C.M., 1999, High-precision provenance determination using detrital-zircon ages and petrography of Quaternary sands on the eastern Snake River Plain, Idaho: Geology, v. 27, no. 4, p. 295–298.

Gianniny, G.L., Geslin, J.K., Link, P.K., and Thackray, G.D., 1997, Quaternary surficial sediments near Test Area North (TAN), northeastern Snake River Plain: An actualistic guide to aquifer characterization, in Sharma, S., and Hardcastle, J.H., eds., Proceedings of the 32nd Symposium on Engineering Geology and Geotechnical Engineering: Boise, Idaho, p. 29–44.

Hackett, W.R., and Smith, R.P., 1992, Quaternary volcanism, tectonics, and sedimentation in the Idaho National Engineering Laboratory area, in Wilson, J.R., ed., Field guide to geologic excursions in Utah and adjacent areas of Nevada, Idaho, and Wyoming: Utah Geological Survey Miscellaneous Publication 92–3, p. 1–18.

Hughes, S.S., Smith, R.P., Hackett, W.R., and Anderson, S.R., 1999, Mafic volcanism and environmental geology of the eastern Snake River Plain, Idaho, in Hughes, S.S., and Thackray, G.D., eds., Guidebook to the geology of eastern Idaho: Pocatello, Idaho, Idaho Museum of Natural History, p. 143–168.

Hughes, S.S., Smith, R.P., Hackett, W.R., McCurry, M., Anderson, S.R., and Ferdock, G.C., 1997, Bimodal magmatism, basaltic volcanic styles, tectonics and geomorphic processes of the eastern Snake River Plain, Idaho: Provo, Utah, Brigham Young University Geology Studies, v. 42, part 1, p. 423–458.

Kaminsky, J.F., Keck, K.N., Schafer-Perini, A.L., Hersley, C.F., Smith, R.P., Stormberg, F.J., and Wylie, A.H., 1994, Remedial investigation final report with addenda for the Test Area North groundwater Operable Unit 1-07B at the Idaho National Engineering Laboratory, Volume 1: Idaho Falls, Idaho, EG&G Idaho, Incorporated, Report EGG-ER-10643, variously paginated.

Kjelstrom, L.C., 1992, Assessment of spring discharge to the Snake River, Milner Dam to King Hill, Idaho: Water Fact Sheet, U.S. Geological Survey Open-File Report 92-142, 2 p.

Kuntz, M.A., Covington, H.R., and Schorr, L.J., 1992, An overview of basaltic volcanism of the eastern Snake River Plain, Idaho, in Link, P.K., Kuntz, M.A., and Platt, L.B., eds., Regional geology of eastern Idaho and western Wyoming: Geological Society of America Memoir 179, p. 227–267.

Kuntz, M.A., Dalrymple, G.B., Champion, D.E., and Doherty, D.J., 1980, Petrography, age, and paleomagnetism of volcanic rocks at the Radioactive Waste Management Complex, Idaho National Engineering Laboratory, Idaho, with an evaluation of potential volcanic hazards: U.S. Geological Survey Open-File Report 80-388, 63 p.

Kuntz, M.A., Skipp, B., Lanphere, M.A., Scott, W.E., Pierce, K.L., Dalrymple, G.B., Champion, D.E., Embree, G.F., Page, W.R., Morgan, L.A., Smith, R.P., Hackett, W.R., and Rodgers, D.W., 1994, Geologic map of the Idaho National Engineering Laboratory and adjoining areas, eastern Idaho: U.S. Geological Survey Miscellaneous Investigations Map I-2330, scale 1:100000.

Lanphere, M.A., Kuntz, M.A., and Champion, D.E., 1994, Petrology, age and paleomagnetics of basaltic lava flows in coreholes at Test Area North (TAN), Idaho National Engineering Laboratory: U.S. Geological Survey Open-File Report 94-686, 49 p.

Lindholm, G.F., 1991, Summary of the Snake River Plain regional aquifer-system analysis in Idaho and eastern Oregon: U.S. Geological Survey Open-File Report 91-98, 62 p.

Lindholm, G.F., Garabedian, S.P., Newton, G.D., and Whitehead, R.L., 1988, Configuration of the water table and depth to water, spring 1980, water-level fluctuations, and water movement in the Snake River Plain regional aquifer system, Idaho and eastern Oregon: U.S. Geological Survey Hydrologic Investigations Atlas HA-703, scale 1:500 000, 1 sheet.

Link, P.K., and Phoenix, E.C., 1996, Rocks, rails and trails (second edition): Pocatello, Idaho, Idaho Museum of Natural History, 194 p.

Malde, H.E., 1991, Quaternary geology and structural history of the Snake River Plain, Idaho and Oregon, in Morrison, R.B., ed., Quaternary nonglacial geology, conterminous U. S.: Boulder, Colorado, Geological Society of America, The Geology of North America, v. K-2, p. 251–280.

Mann, L.J., 1986, Hydraulic properties of rock units and chemical quality of water for INEL-1: A 10,365-foot-deep test hole drilled at the Idaho National Engineering Laboratory, Idaho: U.S. Geological Survey Water-Resources Investigations Report 86-4020 (IDO-22070), 23 p.

Mann, L.J., and Beasley, T.M., 1994, Iodine-129 in the Snake River Plain aquifer at and near the Idaho National Engineering Laboratory, Idaho, 1990–91: U.S. Geological Survey Water-Resources Investigations Report 94-4053 (DOE/ID-22115), 27 p.

Mann, L.J., and Knobel, L.L., 1990, Radionuclides, metals, and organic compounds in water, eastern part of A & B Irrigation District, Minidoka County, Idaho: U.S. Geological Survey Open-File Report 90-191 (DOE/ID-22087), 36 p.

Mark, L.E., 1999, Hydrologic and sedimentologic characterization of surficial sedimentary facies, northern Idaho National Engineering and Environmental Laboratory, eastern Idaho[M.S. thesis]: Pocatello, Idaho, Idaho State University, 186 p.

McBroome, L.A., Doherty, D.J., and Embree, G.F., 1981, Correlation of major Pliocene and Miocene ash-flow sheets, eastern Snake River Plain, in Tucker, T.E., ed., Montana Geological Society Field Conference and Symposium Guidebook to Southwest Montana: Billings, Montana, Empire Printers, p. 323–330.

McCurry, M., Hackett, W.R., and Hayden, K., 1999, Cedar Butte and cogenetic Quaternary rhyolite domes of the eastern Snake River Plain, in Hughes, S.S., and Thackray, G.D., eds., Guidebook to the geology of eastern Idaho: Pocatello, Idaho, Idaho Museum of Natural History, p. 169–180.

Morgan, L.A., Doherty, D.J., and Leeman, W.P., 1984, Ignimbrites of the eastern Snake River Plain: Evidence for major caldera-forming eruptions: Journal of Geophysical Research, v. 89, p. 8665–8678.

Morse, L.H., and McCurry, M., 1997, Possible correlations between basalt alteration and the effective base of the Snake River Plain aquifer at the Idaho National Engineering and Environmental Laboratory, in Sharma, S., and Hardcastle, J.H., eds., Proceedings of the 32nd Symposium on Engineering Geology and Geotechnical Engineering: Boise, Idaho, p. 1–14.

Nace, R.L., Voegeli, P.T., Jones, J.R., and Deutsch, M., 1975, Generalized geologic framework of the National Reactor Testing Station, Idaho: U.S. Geological Survey Professional Paper 725-B, 49 p.

Ostenaa, D.A., Levish, D.R., Klinger, R.E., and O'Connell, D.R., 1999, Phase 2 paleohydrologic and geomorphic studies for the assessment of flood risk for the Idaho National Engineering and Environmental Laboratory, Idaho: Denver, Colorado, U.S. Bureau of Reclamation, Geophysics, Paleohydrology and Seismotectonics Group, Report 99-7, 112 p. plus appendices.

Parsons, T., Thompson, G.A., and Smith, R.P., 1998, More than one way to stretch: A tectonic model for extension along the plume track of the Yellowstone hotspot and adjacent Basin and Range province: Tectonics, v. 17, p. 221–234.

Pierce, K.L., and Morgan, L.A., 1992, The track of the Yellowstone hot spot: Volcanism, faulting, and uplift, *in* Link, P.K., Kuntz, M.A., and Platt, L.B., eds., Regional geology of eastern Idaho and western Wyoming: Geological Society of America Memoir 179, p. 1–53.

Pierce, K.L., and Scott, W.E., 1982, Pleistocene episodes of alluvial-gravel deposition, southeastern Idaho, *in* Bonnichsen, B., and Breckenridge, R.M., eds., Cenozoic geology of Idaho: Idaho Bureau of Mines and Geology Bulletin 26, p. 685–702.

Pittman, J.R., Jensen, R.G., and Fischer, P.R., 1988, Hydrologic conditions at the Idaho National Engineering Laboratory, 1982 to 1985: U.S. Geological Survey Water-Resources Investigations Report 89-4008 (DOE/ID-22078), 73 p.

Scott, W.E., 1982, Surficial geologic map of the eastern Snake River Plain and adjacent areas, 111°–115° W., Idaho and Wyoming: U.S. Geological Survey Miscellaneous Investigations Series Map I-1372, scale 1:250 000, 2 sheets.

Sorenson, K.S., Jr., Wylie, A.H., and Wood, T.R., 1996, Test Area North site conceptual model and proposed hydrogeologic studies Operable Unit 1-07B: Idaho Falls, Idaho, Department of Energy, Document INEL-96/0105, Parsons ES-25.9.9.31, variously paginated.

Stearns, H.T., Crandall, L., and Steward, W.G., 1939, Geology and groundwater resources of the Mud Lake region, Idaho, including the Island Park area: U.S. Geological Survey Water-Supply Paper 818, 125 p.

U.S. Geological Survey, 1985, National water summary, 1984: Hydrologic events, selected water-quality trends, and ground-water resources: U.S. Geological Survey Water-Supply Paper 2275, 467 p.

Walker, E.H., 1964, Subsurface geology of the National Reactor Testing Station, Idaho: U.S. Geological Survey Bulletin 1133-E, 22 p.

Welhan, J., Funderberg, T., and Smith, R.P., 1997, Stochastic modeling of hydraulic conductivity in the Snake River Plain aquifer. 1. Geologic constraints and conceptual approach, *in* Sharma, S., and Hardcastle, J.H., eds., Proceedings of the 32nd Symposium on Engineering Geology and Geotechnical Engineering: Boise, Idaho, p. 75–92.

Welhan, J.A., and Wylie, A., 1997, Stochastic modeling of hydraulic conductivity in the Snake River Plain aquifer. 2. Evaluation of lithologic controls at the core and borehole scales, *in* Sharma, S., and Hardcastle, J.H., eds., Proceedings of the 32nd Symposium on Engineering Geology and Geotechnical Engineering: Boise, Idaho, p. 93–107.

Wetmore, P.H., 1998, An assessment of physical volcanology and tectonics of the central eastern Snake River Plain based on the correlation of subsurface basalts at and near the Idaho National Engineering and Environmental Laboratory, Idaho [M.S. thesis]: Pocatello, Idaho, Idaho State University, 118 p.

Wetmore, P.H., Hughes, S.S., and Anderson, S.R., 1997, Model morphologies of subsurface Quaternary basalts as evidence for a decrease in the magnitude of basaltic magmatism at and near the Idaho National Engineering and Environmental Laboratory, Idaho, *in* Sharma, S., and Hardcastle, J.H., eds., Proceedings of the 32nd Symposium on Engineering Geology and Geotechnical Engineering: Boise, Idaho, p. 45–58.

Whitehead, R.L., 1986, Geohydrologic framework of the Snake River Plain, Idaho and eastern Oregon: U.S. Geological Survey Hydrologic Investigations Atlas HA-681, scale 1:1 000 000, 3 sheets.

MANUSCRIPT ACCEPTED BY THE SOCIETY NOVEMBER 2, 2000

Geological Society of America
Special Paper 353
2002

Pliocene and Quaternary stratigraphic architecture and drainage systems of the Big Lost Trough, northeastern Snake River Plain, Idaho

Jeffrey K. Geslin*
Paul Karl Link
James W. Riesterer
Department of Geosciences, Idaho State University, Pocatello, Idaho 83209, USA
Mel A. Kuntz
U.S. Geological Survey, Denver, Colorado 80225, USA
C. Mark Fanning
Research School of Earth Sciences, Australia National University, Canberra, ACT 2601, Australia

ABSTRACT

The geometry, volcanic-sedimentary stratigraphic architecture, and distribution of clastic sedimentary facies reflect a complex tectonic setting and fluctuations in climatic conditions during the past 2.5 m.y. in the Big Lost Trough on the eastern Snake River Plain. Interaction of the migrating Yellowstone hotspot and developing Basin and Range structures controlled the spatial distribution of volcanic rift zones that define the margins of the Big Lost Trough, an arid, underfilled basin. The volcanic-sedimentary stratigraphy of the basin is characterized by basaltic volcanic units that offlap eruptive centers and downlap into the basin, and clastic sedimentary units that onlap adjacent volcanic rift zones. Climatically influenced interactions of a fluvial-playa-eolian depositional system of the Big Lost River and a lacustrine system of Lake Terreton are reflected in the composition and architecture of the sedimentary basin fill.

Petrographic and U/Pb detrital-zircon geochronology analyses of subsurface sands compared with analyses of modern fluvial and eolian sands allow definitive determination of the provenance of the subsurface deposits. Petrographic and detrital-zircon data suggest that the Big Lost River has been the dominant source of sediment for at least the past 1 m.y. Big Lost River deposits found in the middle and northern parts of the basin suggest that the river system prograded northward during lowstands of Lake Terreton. Lowstands of Lake Terreton are also associated with development of an eolian system that reworked the fluvial deposits. The abundance of Big Lost River and eolian sands in the middle of the basin documents the effective damming of sediment by the volcanic rift zone that defines the northern basin margin. X-ray diffraction data suggest that subsurface playa or marginal lacustrine deposits along the northeastern basin margin contain abundant gypsum, indicating that ancient arid climate cycles were drier than the modern arid climate.

*Present address: ExxonMobil Upstream Research Company, P.O. Box 2189, Houston, TX 77252. E-mail: Jeff.K.Geslin@exxonmobil.com

Geslin, J.K., Link, P.K., Riesterer, J.W., Kuntz, M.A., and Fanning, C.M., 2002, Pliocene and Quaternary stratigraphic architecture and drainage systems of the Big Lost Trough, northeastern Snake River Plain, Idaho, *in* Link, P.K., and Mink, L.L., eds., Geology, Hydrogeology, and Environmental Remediation: Idaho National Engineering and Environmental Laboratory, Eastern Snake River Plain, Idaho: Boulder, Colorado, Geological Society of America Special Paper 353, p. 11–26.

INTRODUCTION

The Pliocene to Holocene Big Lost Trough is a small, closed sedimentary basin situated in the middle of the Idaho National Engineering and Environmental Laboratory (INEEL) on the eastern Snake River Plain, Idaho (Fig. 1). The northern INEEL, and the Test Area North (TAN) facility in particular, have been the focus of geological and hydrological investigations related to the remediation of contaminated groundwater in the Snake River Plain aquifer. Ultimately, it is crucial for hydrogeologic models and remediation technologies used in the northern INEEL to incorporate the stratal architecture of the vadose zone and aquifer in the northern end of the Big Lost Trough.

Volcanic rift zones and the Axial volcanic zone of the Snake River Plain (Fig. 1) define the boundaries of the Big Lost Trough on all but its northwestern side, and the stratal architecture under the INEEL reflects the interaction between basaltic volcanism and sedimentary systems operating in the intervening basin. The Lost River Range, Lemhi Range, and Beaverhead Mountains, part of the Basin and Range system, are located along the northwestern basin margin. Streams in valleys between these ranges, including the Big Lost River, Little Lost River, and Birch Creek, are the dominant source of clastic sediments in the Big Lost Trough (e.g., Geslin et al., 1999). The Circular Butte–Kettle Butte volcanic rift zone defines a low-relief ridge on the northeastern end of the Big Lost Trough, and serves as a barrier between the basin and the adjacent Mud Lake subbasin (Fig. 1; e.g., Anderson and Bowers, 1985; Geslin et al., 1997; Mark and Thackray, this volume).

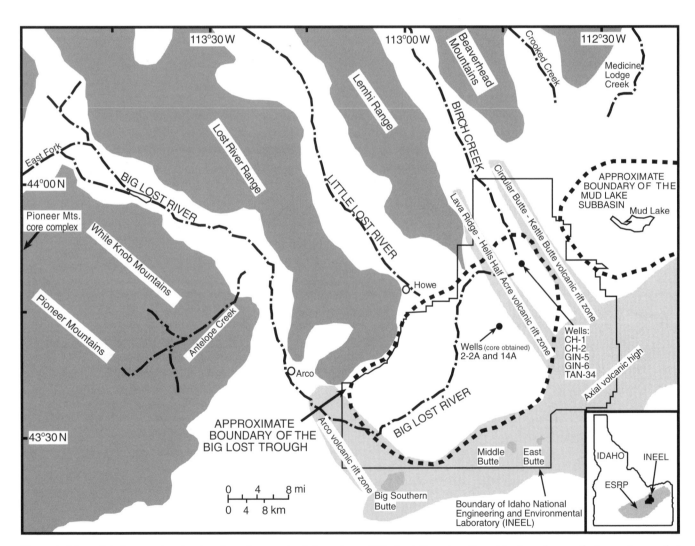

Figure 1. Map of study area showing location of approximate boundary of Big Lost Trough, adjacent basin and range features, streams that flow into basin, approximate locations of volcanic features on Snake River Plain, wells at Test Area North (CH-1, CH-2, GIN-5, GIN-6, TAN-34), and wells in middle of basin (2-2A, 14A).

The elevated topographic axis of the eastern Snake River Plain defines the southeastern side of the Big Lost Trough, and the Arco volcanic rift zone defines its southwestern border.

In this chapter we present the stratigraphic relationships of intercalated basalt flows and clastic sedimentary units in the northern Big Lost Trough, and discuss compositional analyses of sediments on the surface and in the shallow subsurface. We also document, using the results of detailed provenance analysis, the drainage history of the fluvial systems that deliver sand and gravel to the basin. These results predict that, in the upper part of Big Lost Trough strata, there is a progressively increasing abundance of high-permeability sediments (sands and gravels) from the northern end to the middle of the basin. X-ray diffraction analyses suggest that surficial fine-grained lacustrine

and playa sediments, as well as subsurface fine-grained sediments from the middle of the basin, are composed predominantly of smectite, kaolinite, and chlorite; this probably reflects weathering of volcanic and fine-grained sedimentary rocks in fluvial source areas. Fine-grained subsurface sediments found at the northeastern basin margin, in the vicinity of the TAN facilities, contain abundant gypsum, reflecting evaporative depositional conditions.

GEOLOGIC SETTING

The Big Lost Trough is located within the eastern Snake River Plain, a semiarid lava- and loess-mantled plain that cuts across the northwest-trending Basin and Range structure of

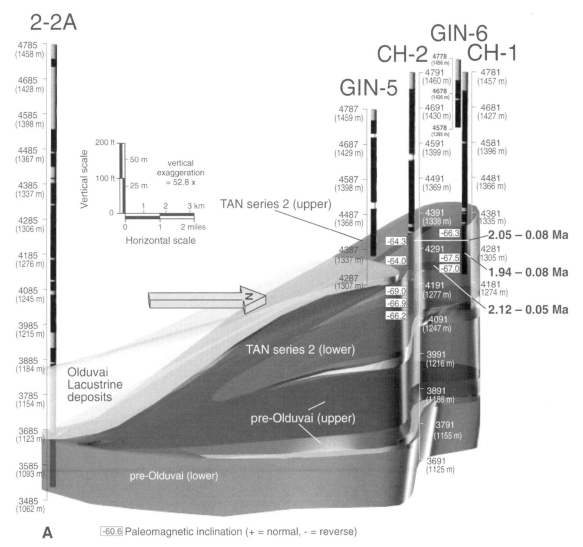

Figure 2. Three-dimensional correlation diagrams showing basalt flow groups and clastic sediments from northeastern basin margin (wells CH-1, CH-2, GIN-5, GIN-6) to middle of basin (well 2-2A). A: 2 Ma. B (page 14): 1 Ma. C (page 15): Modern. Well locations are shown in Figure 1. Basalt is darkly shaded and sedimentary units are lightly shaded. Abbreviations for lava flows Q-R and IJ reflect terminology of Hughes et al. (this volume).

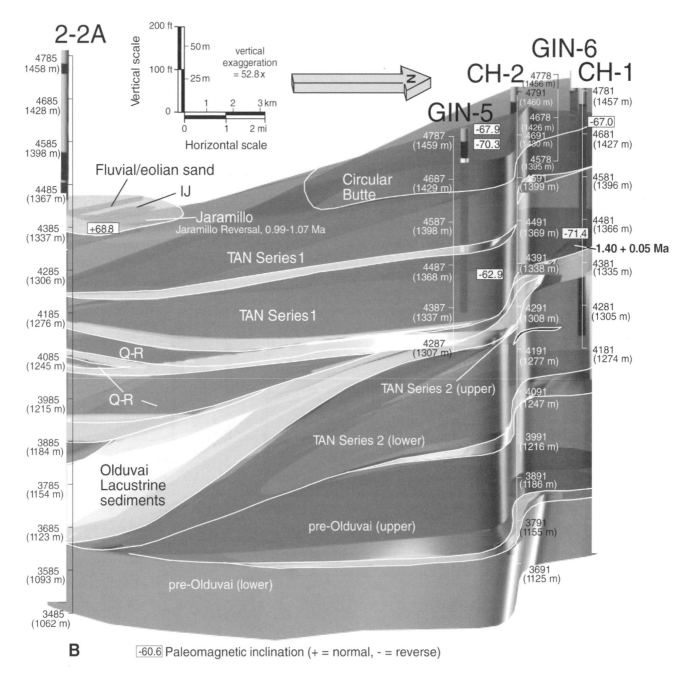

B   -60.6 Paleomagnetic inclination (+ = normal, − = reverse)

Figure 2. (*continued*)

eastern Idaho and is the track of the Yellowstone hotspot (e.g., Nace et al., 1975; Morgan, 1992; Pierce and Morgan, 1992). Topographic depressions that were created on the eastern Snake River Plain by differential subsidence or development of Basin and Range structures define several small sedimentary basins, including the Big Lost Trough (Fig. 1; Geslin et al., 1999) and, to the northeast, the Mud Lake subbasin (Gianniny et al., this volume). Geologic maps of the INEEL area (e.g., Scott, 1982; Kuntz et al., 1994) establish the general sedimentary-volcanic

setting of the Big Lost Trough. Strata of the basin are part of the Pliocene to Holocene Snake River Group, and overlie a series of rhyolitic ignimbrites and minor sedimentary interbeds (Morgan et al., 1984; Champion et al., 1988; Anders et al., 1989; Hackett and Smith, 1992; Kuntz et al., 1992; Morgan, 1992; Pierce and Morgan, 1992).

Basaltic lava fields define many of the landforms on the eastern Snake River Plain and basalt flows dominate the upper part of the subsurface geology of the area. Surficial features

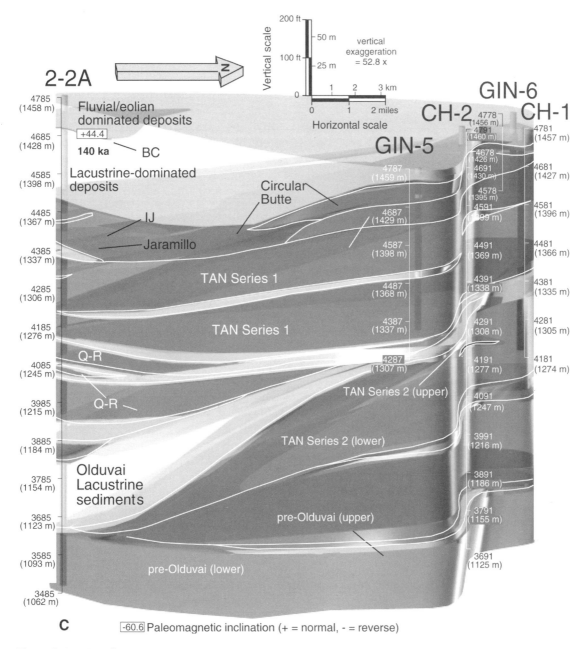

Figure 2. (*continued*)

also include small shield volcanoes and rhyolite domes (e.g., Kuntz et al., 1994). Basalt flows erupted along volcanic rift zones that parallel the northwest-southeast orientation of the adjacent Basin and Range structures (e.g., Kuntz et al., 1992). One of these volcanic rift zones, the Circular Butte–Kettle Butte volcanic rift zone, defines the northeastern end of the Big Lost Trough (Fig. 1). The stratigraphic architecture of the northern end of the basin (see following) reflects the interaction between basaltic volcanism along this volcanic rift zone and sedimentation in the basin.

Sediments are delivered to the Big Lost Trough primarily by three fluvial systems: the Big Lost River, the Little Lost River, and Birch Creek, which drain mountains to the northwest of the basin (Fig. 1; Geslin et al., 1999). Small streams that drain the southern end of the Beaverhead Mountains, north of the basin, and eolian systems operating on the eastern Snake River Plain contribute volumetrically less abundant clastic sediments. Eolian processes primarily winnow and redeposit sediments delivered by the fluvial systems. Examination of the modern configuration of the fluvial systems shows that both the

Little Lost River and Birch Creek have relatively small drainage areas and terminate in playas or sinks at the mouths of their respective valleys. However, the Big Lost River has a relatively large drainage area in the mountains of central Idaho and it traverses nearly the entire length of the Big Lost Trough, terminating in a series of playas near the northern end of the basin (Fig. 1; Kuntz et al., 1994).

The stratigraphic architecture of basalt flows and fluvial, eolian, playa, and lacustrine sediments in the Big Lost Trough controls groundwater flow, both in the vadose zone and the Snake River Plain aquifer (e.g., Mark and Thackray, this volume). In the Snake River Plain aquifer, ~60 m below the surface at the northern end of the Big Lost Trough, flow is to the south and southwest, driven by the regional head of the Mud Lake system (e.g., Whitehead; 1992; Spinazola, 1994). The regional flow in the aquifer suggests that as groundwater contaminants below the TAN facility migrate, their transport rate will be increasingly influenced by the hydrologic properties of sediments in the basin (see Mark and Thackray, this volume).

TECTONIC AND CLIMATIC SETTING

The Big Lost Trough is an arid, underfilled basin (cf. Carroll and Bohacs, 1999). It is underfilled because fluvial and lacustrine systems never breach basin margins defined by the surrounding topographically high volcanic rift zones. Generally, the basin geometry is a byproduct of these volcanic rift zones; however, fluvial systems that transport sediment into the basin are controlled by adjacent Basin and Range structures. The volcanic rift zones that define the northeastern and southwestern basin margins are generally collinear with Basin and Range structures adjacent to the Snake River Plain. Both the Basin and Range structures and volcanic rift zones are related to regional northeast to southwest extension (Kuntz et al.,

1992). Local subsidence of the Big Lost Trough could have resulted from crustal thinning and contraction related to the rhyolite caldera history of the Snake River Plain, specifically the 6.0 Ma Blue Creek caldera that underlies the area (Kuntz et al., 1992; Morgan, 1992; Blair and Link, 2000). Large-scale subsidence of the eastern Snake River Plain is probably due to both thermal contraction and loading by a midcrustal mafic sill (McQuarrie and Rodgers, 1998). Therefore, the tectonic development of the Big Lost Trough represents complex interactions between the migration of the Yellowstone hotspot and Basin and Range extension.

Sedimentation in the Big Lost Trough was dominated by eolian, fluvial, and playa systems during drier climatic conditions and a shallow lacustrine system, Lake Terreton, during wetter climatic conditions (Geslin et al., 1997, 1999; Gianniny et al., 1997; Blair and Link, 2000). Pleistocene Lake Terreton existed intermittently in both the Big Lost Trough and Mud Lake subbasin for the past ~2 m.y. (Stearns et al., 1939; Scott, 1982; Gianniny et al., this volume; Bestland et al., this volume). The low-relief nature of the surface of the eastern Snake River Plain probably resulted in very rapid fluctuations in the lateral extent of Lake Terreton; therefore, lacustrine sediments in the area provide a fairly sensitive indicator of relatively wet climates. Geslin et al. (1999) suggested that fluvial-playa-eolian systems operating today are analogous to depositional systems that operated in the Big Lost Trough during Pleistocene and Holocene dry climates.

VOLCANIC AND SEDIMENTARY STRATIGRAPHY

The stratigraphy of the Big Lost Trough is somewhat complex because basaltic volcanic units offlap eruptive centers and downlap into the basin, while clastic sedimentary units onlap adjacent volcanic rift zones (e.g., Hughes et al., 1998). Three-

TABLE 1. CATEGORIES USED FOR SAND POINT COUNTS OF FRAMEWORK GRAINS AND RECALCULATED PLOTS

Grain category definitions		Categories*	Recalculated parameters
Qp	Aphanitic polycrystalline quartz	Qp	*QFLt:*
Qm	Monocrystalline quartz	Qm	Q = Qm
P	Plagioclase feldspar	P	F = P + K
K	Potassium feldspar	K	Lt = Lv + Lm + Ls + Qp
Lvv	Vitric volcanic lithic fragments	Lv	
Lvf	Felsitic volcanic lithic fragments	Lv	*LmLvLst:*
Lvml	Microlitic volcanic lithic fragments	Lv	Lm = Lm
Lvl	Lathwork volcanic lithic fragments	Lv	Lv = Lv
Lmi	Low-grade metaigneous lithic fragments	Lmv	Lst = Ls + Qp
PolyM	Polycrystalline phyllosilicates	Lms	
QMF(t)	Quartz-mica-feldspar aggregate with tectonite fabric	Lms	*QpLvmLsm:*
			Qp = Qp
QMF(a)	Quartz-mica-feldspar aggregate without tectonite fabric	Lms	Lvm = Lv + Lmv
			Lsm = Ls + Lms
Ls(arg)	Argillaceous sedimentary lithic fragments	Ls	
Ls(cb)	Carbonate sedimentary lithic fragments	Ls	*QmKP:*
Mica	Monocrystalline phyllosilicates	M	Qm = Qm
Dense	Dense mineral grains	D	K = K
C	Carbonate cement (not a grain category)	C	P = P

*See Ingersoll et al. (1984)

dimensional stratigraphic diagrams, constructed from well data, were created to illustrate the development of the stratal architecture in the northern end of the Big Lost Trough (Fig. 2). Figure 2 encompasses a region that includes TAN, the focus of geologic, hydrologic, and environmental studies, and areas hydrologically downgradient to the south. The three-dimensional stratigraphic diagrams correlate basalt flow groups (flow group classification based on Hughes et al. [this volume] as modified from Lanphere et al. [1994] and Anderson and Bowers [1995]) and clastic sedimentary successions from wells CH-1, CH-2, GIN-5, GIN-6, and TAN-34 at TAN to well 2-2A near the middle of the basin (Fig. 1). Geochronology data for basalt flow groups are from whole-rock $^{40}Ar/^{39}Ar$ analyses (Lanphere et al., 1994). A single date of 140 ka in the upper sedimentary succession in well 2-2A is from Geslin et al. (1999) and is based on amino acid racemization in ostracode shells. Paleomagnetic data are from Champion et al. (1988). The stratigraphy shown

in these diagrams represents basaltic lava flows and clastic sediments deposited over the past ~2.5 m.y.

The stratal architecture of the northern end of the basin (Fig. 2) documents the construction of basaltic eruptive centers along the Circular Butte-Kettle Butte volcanic rift zone. The southward downlapping or offlapping of pre-Olduvai, TAN series 1 and 2, and Circular Butte basalt flow groups (Fig. 2), along with surface data (Kuntz et al., 1994), suggests that many of the flows moved generally northeast to southwest, from the volcanic rift zone toward the middle of the basin. Basalt flow group geometry, along with the northward onlap of sediments onto the volcanic rift zone, suggests that the volcanic rift zone was a positive topographic feature throughout basin history. Some of the smaller post-Olduvai flow groups present in well 2-2A (Fig. 2B) are not present in wells along the northern basin margin, and could have erupted from a volcanic center to the south, and then flowed northward into the basin. The strati-

TABLE 2. RECALCULATED MODAL POINT-COUNT DATA FOR SURFICIAL SEDIMENTS OF THE BIG LOST TROUGH AND ADJACENT DRAINAGES

Sample No.	QmFLt			LmLvLst			QpLvmLsm			QmKP		
	Qm	F	Lt	Lm	Lv	Lst	Qp	Lvm	Lsm	Qm	K	P
Big Lost River channel												
1PL96	20.4	14.9	64.7	12.3	69.4	18.3	7.9	69.4	22.7	57.8	0.0	42.2
7JG97	22.4	25.1	52.5	35.3	41.8	22.9	10.4	41.8	47.8	47.1	19.6	33.3
8JG97	12.9	15.8	71.4	32.3	45.3	22.4	13.7	45.3	41.0	44.9	16.7	38.4
9JG96	18.3	14.6	67.1	1.2	67.9	31.0	11.2	67.9	20.9	55.6	17.3	27.2
10JG96	12.5	15.7	71.8	18.0	67.1	15.0	8.7	67.1	24.3	44.3	21.4	34.4
15JG96	27.3	10.6	62.1	9.1	67.1	23.8	5.2	67.1	27.8	72.1	4.5	23.4
mean	19	16	65	18	60	22	10	60	31	54	13	33
σ	6	5	7	13	13	5	3	13	11	11	9	7
Little Lost River channel												
3JG97	36.0	10.5	53.4	29.4	44.0	26.6	13.8	44.0	42.2	77.4	7.9	14.7
4JG97	10.7	2.8	86.5	4.0	5.3	90.7	25.7	5.3	69.0	79.0	6.5	14.5
5JG96	27.6	2.2	70.2	11.5	26.2	62.3	2.6	26.2	71.2	92.6	1.2	6.2
5JG97	37.0	7.4	55.6	18.9	47.2	33.9	15.9	47.2	36.9	83.3	3.8	12.9
6JG97	40.9	4.9	54.2	18.7	45.9	35.4	25.7	45.9	28.4	89.4	0.5	10.1
mean	31	6	64	17	34	50	17	34	49	84	4	12
σ	12	4	14	9	18	27	10	18	19	7	3	4
Birch Creek channel												
1JG97	18.0	6.4	75.6	6.2	3.8	90.0	35.0	3.8	61.2	73.9	21.8	4.2
2JG97	14.6	5.8	79.6	54.7	3.9	41.4	20.7	3.9	75.4	71.4	19.4	9.2
3JG96	15.6	4.1	80.2	0.0	40.3	59.7	10.0	40.3	49.7	79.2	7.3	13.5
11JG96	14.9	9.2	75.8	6.9	13.0	80.1	12.5	13.0	74.5	61.7	20.0	18.3
mean	16	6	78	17	15	68	20	15	65	72	17	11
σ	1	2	2	26	17	22	11	17	12	7	7	6
Small drainages north of Big Lost Trough (Medicine Lodge Creek)												
1JR97	32.6	4.9	62.5	20.8	28.3	50.8	26.1	28.3	45.6	87.0	4.3	8.7
2JR97	65.9	6.3	27.8	29.5	19.1	51.4	31.4	19.0	49.5	91.2	6.6	2.2
4JR97	57.4	6.9	35.7	39.2	16.9	43.8	33.1	16.9	50.0	89.3	6.0	4.7
5JR97	32.4	9.3	58.3	38.5	22.5	39.0	22.5	22.5	55.0	77.6	6.3	16.1
mean	47	7	46	32	22	46	28	22	50	86	6	8
σ	17	2	17	8	5	6	5	5	4	6	1	6
Eolian (sand dunes)												
1JG96	22.8	29.6	47.6	9.2	52.1	38.7	2.8	52.1	45.2	43.5	17.6	38.9
2PL96	28.0	5.4	66.6	1.2	17.6	81.1	13.9	17.6	68.4	84.0	0.0	16.0
7JR96	31.5	16.0	52.6	5.4	43.8	50.8	8.7	43.8	47.5	66.5	5.0	28.4
12JG96	27.8	15.1	57.1	15.6	49.4	34.9	2.2	49.4	48.3	64.9	8.9	26.2
14JG96	30.0	9.8	60.2	8.2	30.9	61.0	1.1	30.9	68.0	75.3	7.9	16.9
mean	28	15	57	8	39	53	6	39	55	67	8	25
σ	3	9	7	6	14	19	6	14	12	15	7	9

graphically highest basalt flow group at the northern end of the Big Lost Trough (Circular Butte flows, Fig. 2C) is ca. 1 Ma, suggesting that the area has been volcanically quiet through much of the Pleistocene (Anderson and Lewis, 1989; Anderson, 1991; Anderson et al., 1996).

A thick sedimentary succession is present, most notably in the upper part of the stratigraphic section near the site of well 2-2A, in the north-central part of the basin (Fig. 2C; Anderson and Bowers, 1995; Geslin et al., 1997; 1999; Blair and Link, 2000). Although there are no data on interwell geometry, these sediments apparently onlap the northern basin margin (Fig. 2C), documenting the closed nature of the basin. Strata in the upper part of well 2-2A record the interaction of eolian, fluvial, playa, and lacustrine depositional systems (e.g., Geslin et al., 1997, 1999; Gianniny et al., 1997). These strata document a general shift from lacustrine-dominated sedimentation to fluvial- and/ or eolian-dominated sedimentation, suggesting an overall drying of the climate (e.g., Bestland et al., this volume). In this volume Geslin et al. and Gianniny et al. argue that modern fluvial, playa, and eolian systems are excellent analogs for ancient depositional systems operating in the basin during relatively dry climatic conditions; lacustrine clays and fine sands deposited in Lake Terreton probably represent relatively wetter climatic conditions. Gianniny et al. (this volume) discuss the timing of multiple recent fluctuations of Lake Terreton. Bestland et al. (this volume) discuss older lacustrine sedimentation documented in the deeper core from well 2-2A (Fig. 2A) and emphasize weak paleosols developed on fluvial, playa, eolian, and lacustrine sediments in the upper part of the 2-2A core.

COMPOSITIONAL ANALYSIS OF SEDIMENTARY UNITS

We examined both modern and shallow subsurface sands from the Big Lost Trough using petrographic analysis. Provenance interpretations of fluvial sands, based on modal recalculation of petrographic data, were supported by U/Pb geochronology of detrital zircons. Provenance results, particularly at the northern end of the basin, are used to interpret the evolution of the fluvial systems and document the role of volcanic eruptive centers in controlling the distribution of depositional systems.

The compositions of modern and shallow subsurface fine-grained sediments were evaluated using semiquantitative X-ray diffraction analysis, and are used to characterize depositional environments. Hughes et al. (this volume) present compositional analyses of the basalts and correlations of the basalt flow groups found in the Big Lost Trough.

Provenance of sands

Petrographic analyses. Sand samples were collected from modern fluvial and eolian deposits, and from core retrieved from wells in the middle and northern parts of the basin (Fig. 1). All

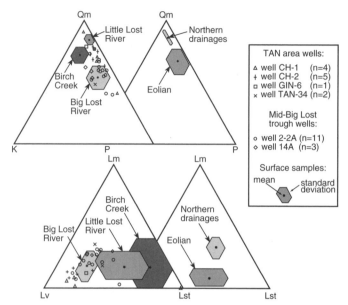

Figure 3. Standard ternary plots of detrital modes from petrographic analysis of sand samples. Qm—monocrystalline quartz, K—potassium feldspar, P—plagioclase feldspar, Lm—metamorphic lithic grains, Lv—volcanic lithic grains, Lst—total sedimentary lithic grains. Polygons represent mean and standard deviation for samples collected from modern drainages (Big Lost River, *n* = 6; Little Lost River, *n* = 5; Birch Creek, *n* = 4; northern drainages, *n* = 4; eolian deposits, *n* = 5). Symbols represent individual samples collected from wells (locations in Fig. 1) in central and northern parts of Big Lost Trough (*n* = 29).

modern fluvial systems that flow into the Big Lost Trough, including the Big Lost River, Little Lost River, and Birch Creek, were sampled, as were several small streams that flow out of the southern Beaverhead Mountains and terminate near the northeastern margin of the basin. Sand samples were examined petrographically; compositions were determined by the Gazzi-Dickinson point-count method (Ingersoll et al., 1984); raw point-count data were recalculated into detrital modes (Table 1). Samples from modern fluvial and eolian deposits, summarized in Table 2, were first compared to each other by plotting detrital mode data on ternary diagrams (Fig. 3). The source areas of the modern streams are generally well differentiated using fields defined by the statistical average and standard deviation for the samples (Fig. 3). Big Lost River sands are distinguished by abundant volcanic lithic material, probably reflecting both the abundance of Challis volcanic rocks in the source area and the abundant Snake River Plain basalts over which the stream flows in its lower reach. Modern eolian sands have an average composition that overlaps several of the fluvial fields (Fig. 3), indicating mixing and redeposition.

To evaluate the provenance of shallow subsurface sands, their recalculated modal compositions (Table 3) were compared to modern sands by plotting subsurface samples individually on the ternary diagrams (Fig. 3). The majority of the shallow subsurface sand samples have modal compositions that are similar

TABLE 3. RECALCULATED MODAL POINT-COUNT DATA FOR SUBSURFACE SEDIMENTS OF THE BIG LOST TROUGH

Sample No.	Depth (m)	QmFLt			LmLvLst			QpLvmLsm			QmKP		
		Qm	F	Lt	Lm	Lv	Lst	Qp	Lvm	Lsm	Qm	K	P
Corehole 2-2A													
18JG96	19.8	26.7	17.4	55.9	21.0	58.0	21.0	17.8	58.0	24.2	60.5	9.7	29.8
19JG96	21.3	24.4	14.8	60.8	21.8	48.2	29.9	25.4	48.2	26.4	62.3	8.2	29.5
20JG96	36.6	26.7	17.9	55.3	31.9	44.1	23.9	16.0	44.1	39.9	59.9	4.1	36.0
21JG96	41.2	20.9	30.4	48.9	4.4	44.4	51.2	0.0	44.4	55.6	40.7	13.9	45.4
22JG96	42.4	30.9	20.9	48.2	22.8	60.8	16.4	4.2	60.8	34.9	59.6	7.4	33.0
23JG96	44.8	15.1	22.1	62.8	13.1	73.5	13.5	2.9	73.5	23.6	40.5	9.8	49.7
24JG96	50.6	23.4	11.5	65.1	23.9	39.1	37.0	25.0	39.1	35.9	67.1	5.9	27.0
25JG96	55.2	15.7	13.0	71.3	27.9	49.1	22.9	15.6	49.1	35.3	54.7	9.5	35.8
26JG96	58.6	27.2	12.7	60.2	28.1	56.1	15.8	10.1	56.1	33.8	68.2	5.3	26.5
27JG96	72.4	30.2	9.3	60.5	20.4	64.6	15.0	10.2	64.6	25.2	76.5	4.5	19.0
28JG96	75.3	25.5	17.4	57.1	29.3	55.0	15.7	12.4	55.0	32.5	59.4	8.6	32.1
29JG96	113.6	19.4	18.0	62.7	19.9	56.0	24.1	17.3	56.0	26.7	51.9	19.1	29.0
30JG96	118.8	21.6	17.2	61.2	32.2	46.4	21.4	11.2	46.4	42.4	55.6	7.0	37.4
31JG96	191.9	41.2	15.2	43.8	7.7	74.0	18.4	16.3	74.0	9.7	73.0	10.3	16.7
Corehole TAN CH1													
34JG96	122.5	44.7	4.7	50.6	0.0	1.6	98.4	25.4	1.6	73.0	90.5	5.4	4.1
36JG97	125.4	17.4	25.2	57.4	1.9	77.4	20.6	7.1	77.4	15.5	40.9	6.1	53.0
37JG96	136.0	33.4	16.2	50.3	6.0	81.2	12.8	7.4	81.2	11.4	67.3	6.8	25.9
38JG96	136.2	31.1	12.7	56.1	8.8	74.5	16.8	11.7	74.5	13.9	71.0	4.7	24.3
Corehole TAN CH2													
39JG96	73.0	35.4	12.9	51.7	12.5	70.1	17.4	15.8	70.1	14.1	73.3	7.2	18.0
41JG96	143.1	34.2	9.0	56.8	14.2	76.5	9.3	9.3	76.5	14.2	79.1	4.3	16.6
42JG96	224.2	35.8	16.2	47.5	9.3	77.6	13.1	9.8	77.6	12.6	68.3	9.4	22.3
43JG96	224.8	29.1	17.4	53.5	14.6	58.7	26.7	14.1	58.7	27.2	62.6	4.5	33.0
44JG96	225.5	32.2	10.6	57.1	17.2	70.6	12.3	10.8	70.6	18.6	75.2	3.3	21.6
Corehole GIN 6													
33JG96	35.7	32.8	8.7	58.5	13.2	64.0	22.8	12.7	64.0	23.2	79.0	9.3	11.7
Borehole 14A													
9JG97	4.2	26.0	18.0	56.0	29.9	35.6	34.4	7.3	35.6	57.1	59.0	15.1	25.9
10JG97	8.8	28.8	18.0	53.2	29.9	45.0	25.1	6.4	45.0	48.6	61.5	18.1	20.4
11JG97	15.0	17.4	15.3	67.2	27.2	51.9	20.9	13.9	51.9	34.2	53.2	16.9	29.9
Borehole TAN 34 (cuttings)													
15-20	4.6–6.1	36.7	11.2	52.1	19.0	48.6	32.4	22.4	48.6	29.0	76.7	4.1	19.2
25–30	7.6–9.1	19.5	21.2	59.3	35.7	46.2	18.1	11.6	46.2	42.2	47.9	18.4	33.7

to those of modern sands from the Big Lost River; a few samples are compositionally similar to modern eolian sands. These data suggest that the Big Lost River has been the dominant source of fluvial sands in the basin since at least the Pliocene.

The abundance of Big Lost River and eolian sands in the middle of the basin documents the effective damming of sediment by the volcanic rift zone that defines the northern basin margin (Fig. 2). These results also suggest that deposition by the Big Lost River in the middle of the basin is commonly associated with dry climatic conditions favorable for the development of an eolian system, similar to present-day conditions.

Detrital-zircon U/Pb geochronology. Sand samples for detrital-zircon analysis were collected from modern fluvial deposits, from shallow core retrieved from well 2-2A in the southern Big Lost Trough, and wells TAN-34, CH-1, and CH-2 at TAN. Detrital zircons were separated from the samples and ~50 zircon grains from each sample were analyzed on SHRIMP I at the Australian National University (Table 4). The ages of individual zircon grains were derived from the weighted mean of ^{206}Pb/^{238}U-, ^{207}Pb/^{235}U- and ^{207}Pb/^{206}Pb-corrected ages, and relative-probability spectra were created by assigning

Gaussian distributions to individual ages and errors and then summing them together in 1 m.y. bins.

It is easy to differentiate the detrital-zircon age spectra for the fluvial systems that provide sediment to the Big Lost Trough (Fig. 4). There are distinctive populations of detrital zircons in the sands from Birch Creek and the small northern drainages (Fig. 4, C–F): (1) several groups of <20 Ma zircons derived from Snake River Plain volcanic rocks; (2) 700-100 Ma zircons derived from the Cretaceous Idaho batholith (Worl et al., 1995); (3) ca. 500 Ma zircons derived from the Ordovician Beaverhead pluton exposed in the Beaverhead Mountains (Evans and Zartman, 1988); and (4) a large group of 1000–2000 Ma zircons and a small group of 2500–3000 Ma zircons likely recycled from Mesoproterozoic to Ordovician strata exposed in the Beaverhead Mountains and Lemhi Range (Oaks et al., 1977; Skipp and Link, 1992; Winston and Link, 1993). Detrital zircons from the Little Lost River (Fig. 4B) also contain a large Paleoproterozoic zircon population (1500–2000 Ma), but this sample can be differentiated from other samples by the presence of a large ca. 50 Ma zircon population, derived from the Eocene Challis Volcanic Group, and by the lack of 500 Ma and <20 Ma zircon populations. The detrital-zircon population in the Big Lost

TABLE 4. DETRITAL-ZIRCON AGE DATA FOR SURFACE AND SUBSURFACE SANDS OF THE BIG LOST TROUGH AND ADJACENT DRAINAGES

Sample 9JG96 Big Lost River			Sample 5JG96 Little Lost River			Sample 11JG96 Birch Creek			Sample 1JR96 Beaver Creek			Sample 2JR96 Crooked Creek			Sample 2JR97 Medicine Lodge Creek			Sample 50JG96 Well 2-2A 55.4 m		
Grain spot	Age ^{206}Pb/^{238}U	±	Grain spot	Age ^{206}Pb/^{238}U	±	Grain spot	Age ^{206}Pb/^{238}U	±	Grain spot	Age ^{206}Pb/^{238}U	±	Grain spot	Age ^{206}Pb/^{238}U	±	Grain spot	Age ^{206}Pb/^{238}U	±	Grain spot	Age ^{206}Pb/^{238}U	±
1.1	2628	57	1.1	1841	43	1.1	503	22	1.1	1851	40	1.1	1646	34	1.1	1669	47	1.1	46.8	0.7
2.1	46.9	0.9	2.1	1614	70	2.1	493	32	2.1	50.2	1.2	3.1	2.1	0.1	2.1	2669	111	2.1	2644	29
3.1	45.4	1.5	3.1	1774.4	36.1	3.1	1055	27	3.1	51.0	1.2	4.1	10.7	1.0	3.1	1730	44	3.1	56.1	1.3
4.1	707	21	4.1	49.9	1.4	4.1	1859	44	4.1	52.1	1.2	5.1	7.7	0.7	4.1	1650	51	4.1	47.0	1.3
5.1	611	40	5.1	1728	38	5.1	33.0	0.9	5.1	52.9	1.3	6.1	1817	50	5.1	1103	59	5.1	46.8	1.0
6.1	46.6	1.9	6.1	49	2	6.1	1289	22	6.1	47.1	2.3	7.1	1399	32	6.1	1429	43	6.1	971	22
7.1	46.5	1.5	7.1	1275.7	47.4	7.1	482	29	7.1	51.2	1.2	8.1	1775	34	7.1	1488	58	7.1	1266	115
8.1	1441	59	8.1	2254	65	8.1	1.1	0.3	8.1	50.9	2.1	9.1	1048	44	8.1	1345	49	8.1	1760	33
9.1	46.4	1.4	9.1	1666	35	9.1	1810	29	9.1	75.3	3.1	10.1	48.9	2.1	9.1	1627	32	9.1	48.6	1.6
10.1	47.3	1.1	10.1	1728	35	10.1	496	13	10.1	91.9	2.0	11.1	1028	18	10.1	1835	46	10.1	1760	50
11.1	49.6	1.5	11.1	1787	41	11.1	44.3	1.2	11.1	50.3	1.4	12.1	1548	64	11.1	1352	32	11.1	48.6	1.0
12.1	48.6	2.0	12.1	1687.4	27.0	12.1	2574	148	12.1	50.9	1.3	13.1	6.5	0.4	12.1	1656	27	12.1	51.5	1.3
13.1	52.9	2.4	13.1	2728	72	13.1	491	29	13.1	64.8	3.9	14.1	32.4	1.5	13.1	6.4	0.6	13.1	1767	23
14.1	50.9	4.7	14.1	51	2	14.1	485	13	14.1	49.2	1.1	15.1	11.9	0.7	14.1	536	26	14.1	1118	32
15.1	45.7	2.1	15.1	668	13	15.1	1665	36	15.1	2.7	0.3	16.1	26.9	7.6	15.1	1703	38	15.1	88.0	2.8
16.1	49.9	2.0	16.1	50.6	1.8	16.1	46.8	1.6	16.1	1791	41	17.1	1598	44	16.1	1693	59	16.1	83.3	2.2
17.1	49.1	1.5	17.1	1667	44	17.1	33.1	2.2	17.1	227	10	18.1	8.4	0.5	17.1	91.5	4.2	17.1	1922	50
18.1	48.3	2.3	18.1	1830	38	18.1	1408	25	18.1	50.4	1.7	19.1	1406	49	18.1	93.9	3.6	18.1	49.5	1.6
19.1	47.4	1.1	19.1	1744	50	19.1	1808	38	19.1	16.6	0.5	20.1	1610	29	19.1	1530	43	19.1	49.2	1.2
20.1	52.4	2.2	20.1	1716.2	46.1	20.1	485	24	20.1	50.4	1.4	21.1	1738	37	20.1	18.4	9.5	20.1	11.3	1.3
21.1	689	39	21.1	1713.0	41.0	21.1	38.2	1.5	21.1	1083	22	22.1	1829	55	21.1	6.3	0.5	21.1	1690	28
22.1	47.9	1.3	22.1	1633	38	22.1	495	13	22.1	1406	52	23.1	1571	95	22.1	1588	105	22.1	50.1	1.7
23.1	51.2	1.9	23.1	1707	44	23.1	1698	39	23.1	47.4	2.4	24.1	2079	38	23.1	92	2	23.1	46.0	1.3
24.1	42.8	1.0	24.1	1812	36	24.1	1809	46	24.1	49.9	1.5	25.1	1594	35	24.1	1072	25	24.1	10.4	0.4
25.1	49.6	1.7	25.1	48	1	25.1	1570	45	25.1	1420	59	26.1	484	22	25.1	1750	72	25.1	2047	44
26.1	49.0	1.2	26.1	1713.8	69.8	26.1	488	13	26.1	49.5	1.2	27.1	484	18	26.1	1750	37	26.1	50.5	1.1
27.1	46.0	0.9	27.1	1402	74	27.1	462	13	27.1	7.6	1.4	28.1	1594	36	27.1	63.5	4.1	27.1	1768	39
28.1	674	16	28.1	48	3	28.1	1857	33	28.1	13.5	0.9	29.1	2682	60	28.1	1765	46	28.1	46.8	1.3
29.1	1449	24	29.1	2546	49	29.1	1454	35	29.1	509	10	30.1	10.3	0.3	29.1	419	12	29.1	93.6	2.1
30.1	49.6	2.3	30.1	1815	51	30.1	1066	25	30.1	1729	40	31.1	1678	36	30.1	1205	44	30.1	2262	32
31.1	570	15	31.1	1661	40	31.1	1418	29	31.1	2785	55	32.1	1777	40	31.1	1782	49	31.1	1058	13
32.1	54.1	1.0	32.1	2615	62	32.1	266	7	32.1	48.8	2.7	33.1	2633	60	32.1	9.8	1.8	32.1	50.2	1.2
33.1	45.4	1.0	33.1	48	1	33.1	1661	38	33.1	48.0	1.7	34.1	47.9	1.1	33.1	101	2	33.1	46.5	1.4
34.1	48.0	1.2	34.1	501	13	34.1	498	28	34.1	1419	38	35.1	49.7	1.9	34.1	1770	27	34.1	1658	23
35.1	48.1	1.5	35.1	2813	76	35.1	452	34	35.1	8.2	5.0	36.1	67.7	1.9	35.1	1774	82	35.1	47.5	1.1
36.1	1248	32	36.1	1840	56	36.1	1081	25	36.1	11.0	7.3	37.1	1686	32	36.1	1715	52	36.1	48.9	1.0
37.1	46.2	1.2	37.1	1355	51	37.1	1255	39	37.1	1275	43	38.1	1160	31	37.1	1789	40	37.1	2180	64
38.1	43.8	2.2	38.1	49.5	1.7	38.1	463	16	38.1	208	8	39.1	1639	37	38.1	1730	39	38.1	49.4	1.1
39.1	1245	32	39.1	1721	39	39.1	489	12	39.1	1288	46	40.1	474	18	39.1	2115	54	39.1	49.4	1.5
40.1	53.8	2.9	40.1	1732	31	40.1	17.2	1.6	40.1	52.3	4.5	41.1	6.7	0.8	40.1	1453	27	40.1	47.2	0.8
41.1	44.0	1.1	41.1	2025.0	45.7	41.1	46.2	1.9	41.1	97.3	3.6	42.1	3.5	0.8	41.1	1144	16	41.1	45.5	1.1
42.1	49.5	2.0	42.1	1817	56	42.1	8.9	1.0	42.1	101	7	43.1	7.7	1.2	42.1	1007	21	42.1	438	10
43.1	44.1	1.5	43.1	2694	45	43.1	1451	28	43.1	96.3	4.5	44.1	1743	38	43.1	1710	42	43.1	45.3	0.9
44.1	46.3	2.0	44.1	1839	37	44.1	637	23	44.1	153	3	45.1	83	29	44.1	1010	51	44.1	48.9	1.5
15.2	47.8	1.1	45.1	1959	77	45.1	8.4	0.8	45.1	47.6	1.4	46.1	929	30	45.1	1729	44	45.1	48.4	1.0
19.2	44.6	1.7	46.1	1	0	46.1	6.6	1.4	46.1	97.6	2.7	47.1	1054	21	46.1	98.3	1.3	46.1	11.0	0.5
28.2	706	21	47.1	1761.8	28.6	47.1	106	6	47.1	77.3	5.7	48.1	13.9	0.5	47.1	1016	26	47.1	613	27
20.2	49.9	1.8	48.1	2677.8	56.2	48.1	1431	54	48.1	48.5	1.2	49.1	1549	50	48.1	1694	34	48.1	1379	55
12.2	50.8	2.1	49.1	1753	29	49.1	1787	37	49.1	1372	26	50.1	1708	35	49.1	395	7	49.1	47.9	1.2
13.2	50.2	1.6	50.1	1521	46	50.1	962	27	50.1	517	18				50.1	1069	65	50.1	46.8	1.0

TABLE 4. DETRITAL-ZIRCON AGE DATA (continued)

Sample 51JG96 Well 2-2A 113.7 m			Sample 52JG96 Well 2-2A 191.9 m			Sample 3JG96 Well TAN 34 ~5.8 m			Sample 47JR96 Well CH 2 136.8 m			Sample 48JR96 Well CH 2 224.4 m			Sample 49JR96 Well CH 1 136.0 m		
Grain spot	Age $^{206}Pb/^{238}U$	±	Grain spot	Age $^{206}Pb/^{238}U$	±	Grain spot	Age $^{206}Pb/^{238}U$	±	Grain spot	Age $^{206}Pb/^{238}U$	±	Grain spot	Age $^{206}Pb/^{238}U$	±	Grain spot	Age $^{206}Pb/^{238}U$	±
1.1	51	1	1.1	49.7	1.1	1.1	741	24	1.1	525	20	1.1	1718	27	1.1	48.5	1.3
2.1	49	1	2.1	51.8	1.1	1.2	734	17	2.1	1849	30	2.1	52.9	2.5	2.1	1.9	0.1
3.1	1853	29	3.1	46.6	1.7	2.1	48.9	1.5	3.1	50.6	1.0	3.1	2674	91	3.1	9.3	0.5
4.1	49	1	4.1	1544	39	3.1	65.0	1.6	4.1	52.2	0.9	4.1	1217	39	4.1	1795	39
5.1	696	14	5.1	1228	74	4.1	1559	68	5.1	1765	34	5.1	1.7	0.2	5.1	1441	21
6.1	49	2	6.1	680	12	5.1	47.6	2.8	6.1	1622	114	6.1	1055	25	6.1	1821	39
7.1	47	1	7.1	49.0	1.1	6.1	47.9	1.4	7.1	8.9	0.6	7.1	34.1	1.9	7.1	1865	32
8.1	48	1	8.1	50.8	1.3	7.1	446	15	8.1	1049	23	8.1	2524	93	8.1	1888	58
9.1	1300	19	9.1	1861	91	8.1	837	23	9.1	1889	28	9.1	49.0	2.0	9.1	57.1	2.1
10.1	1408	25	10.1	545	19	9.1	47.5	1.5	10.1	1109	14	10.1	50.9	1.2	10.1	2030	30
11.1	2105	24	11.1	1751	28	10.1	85.6	3.9	11.1	1249	46	11.1	6.3	0.4	11.1	1119	31
12.1	49	1	12.1	48.9	1.5	11.1	49.1	2.8	12.1	50.4	0.8	12.1	1181	30	12.1	1644	59
13.1	696	14	13.1	47.9	2.6	12.1	1135	101	13.1	1285	14	13.1	1342	35	13.1	1630	26
14.1	1505	27	14.1	47.2	2.3	13.1	80.3	2.8	14.1	1227	18	14.1	1542	37	14.1	1679	32
15.1	48	1	15.1	48.2	2.7	14.1	1747	48	15.1	1279	15	15.1	32.1	2.0	15.1	1638	35
16.1	655	18	16.1	47.8	1.5	15.1	47.0	1.9	16.1	49.3	1.1	16.1	1590	32	16.1	48.2	1.5
17.1	49	1	17.1	573	21	16.1	45.4	2.4	17.1	525	24	17.1	452	7	17.1	1597	37
18.1	91	2	18.1	1836	51	17.1	44.8	1.6	18.1	1204	15	18.1	2700	84	18.1	1651	33
19.1	2445	54	19.1	50.2	2.2	18.1	48.1	1.5	19.1	506	8	19.1	1633	76	19.1	1665	39
20.1	691	9	20.1	49.3	1.2	19.1	46.9	1.7	20.1	47.7	1.3	20.1	7.1	0.3	20.1	2624	59
21.1	49	1	21.1	1185	19	20.1	27.4	0.9	21.1	49.8	1.3	21.1	49.5	1.2	21.1	51.1	1.8
22.1	47	1	22.1	2407	68	21.1	338	13	22.1	149	8	22.1	1693	24	22.1	1740	21
23.1	48	1	23.1	1614	47	22.1	48.0	1.3	23.1	1523	28	23.1	1.9	0.4	23.1	1.6	0.7
24.1	2062	28	24.1	49.1	3.7	23.1	620	14	24.1	509	11	24.1	5.9	0.4	24.1	1915	28
25.1	46	2	25.1	50.5	2.1	24.1	680	18	25.1	963	14	25.1	1949	24	25.1	648	7
26.1	107	5	26.1	48.5	2.9	25.1	757	32	26.1	50.0	0.8	26.1	1379	15	26.1	1418	16
27.1	1392	16	27.1	648	21	26.1	1717	36	27.1	1773	30	27.1	1666	31	27.1	48.7	0.7
28.1	207	4	28.1	729	14	27.1	82.1	3.5	28.1	1578	25	28.1	2664	61	28.1	1035	27
29.1	1479	23	29.1	46.3	4.5	28.1	45.6	1.8	29.1	1890	31	29.1	1727	18	29.1	2536	35
30.1	1492	28	30.1	45.6	3.0	29.1	47.8	1.4	30.1	1750	29	30.1	1666	30	30.1	1351	20
31.1	1707	23	31.1	1125	32	30.1	46.5	1.2	31.1	1538	41	31.1	1466	25	31.1	1925	30
32.1	50.5	0.8	32.1	48.9	1.9	31.1	48.1	1.4	32.1	1508	27	32.1	47.9	1.0	32.1	48.2	1.0
33.1	48.8	1.2	33.1	1753	48	32.1	653	15	33.1	496	11	33.1	77.2	1.4	33.1	49.9	1.2
34.1	710	42	34.1	44.1	2.6	33.1	1006	42	34.1	1159	31	34.1	1865	33	34.1	47.8	0.7
35.1	1505	23	35.1	51.2	2.7	34.1	50.9	1.8	35.1	1838	30	35.1	676	9	35.1	1.9	0.1
36.1	1804	27	36.1	91.2	7.2	35.1	45.5	1.5	36.1	2344	57	36.1	1666	32	36.1	50.4	0.9
37.1	48.5	1.6	37.1	41.8	1.9	36.1	42.5	1.6	37.1	896	32	37.1	48.6	0.8	37.1	48.6	2.3
38.1	696	11	38.1	679	22	37.1	1817	156	38.1	10.4	0.7	38.1	1981	23	38.1	1785	51
39.1	49.6	0.7	39.1	45.3	1.1	38.1	47.1	1.2	39.1	534	12	39.1	1531	15	39.1	1548	46
40.1	1030	18	40.1	595	14	39.1	47.1	1.5	40.1	1617	27	40.1	1405	16	40.1	1544	38
41.1	48.4	1.0	41.1	46.5	1.0	40.1	372	13	41.1	9.8	1.1	41.1	36.0	0.5	41.1	1756	52
42.1	1764	26	42.1	1300	20	41.1	45.4	1.4	42.1	858	35	42.1	1526	27	42.1	1699	24
43.1	46.1	1.0	43.1	43.3	1.5	42.1	48.1	2.9	43.1	475	12	43.1	2827	58	43.1	1210	30
44.1	48.8	1.5	44.1	1511	51	43.1	1855	44	44.1	2553	78	44.1	2322	26	44.1	1721	100
45.1	2890	54	45.1	89.9	1.7	44.1	57.9	1.5	45.1	3359	229	45.1	1780	25	45.1	983	13
46.1	46.9	0.8	46.1	48.4	2.3	45.1	46.9	2.3	46.1	1019	16	46.1	43.2	2.5	46.1	1840	22
47.1	1631	37	47.1	1296	37	46.1	138	10	47.1	48.6	0.9	47.1	1260	21	47.1	1398	32
48.1	2231	40	48.1	645	32	47.1	56.5	3.1	48.1	50.8	2.3	48.1	1327	22	48.1	2.3	0.4
49.1	49.5	0.9	49.1	49.3	2.4	48.1	46.5	1.1	49.1	1839	27	49.1	1473	26	49.1	49.6	0.8
50.1	49.1	1.2	50.1	1348	27	49.1	49.7	3.7	50.1	1223	30	50.1	757	10	50.1	1860	35
						50.1	112	7									

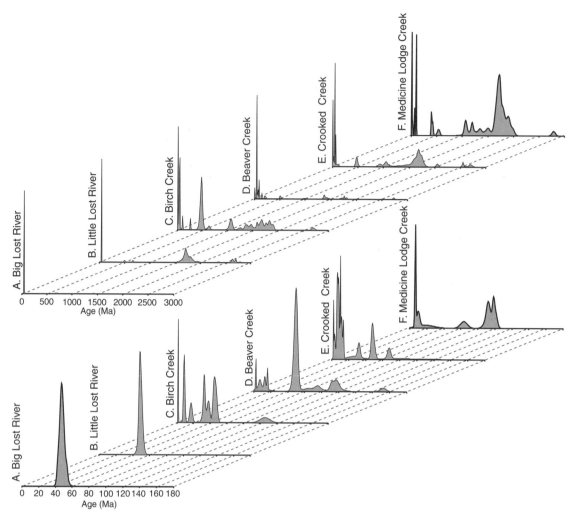

Figure 4. Relative probability plots for detrital-zircon ages determined for modern fluvial sands. Big Lost River, Little Lost River, and Birch Creek (A–C) are primary fluvial systems flowing into Big Lost Trough; smaller streams (D–F) are north of basin. Upper graphs contain results of all analyses (0–3000 Ma zircons); lower graphs are for young (<180 Ma) zircon populations.

River sample (Fig. 4A) is characterized by an abundance of ca. 50 Ma zircon grains and by a relative lack of older zircons.

Subsurface sands from well 2-2A in the central Big Lost Trough have detrital-zircon age spectra (Fig. 5) that are characterized by an abundance of ca. 50 Ma zircons and very few Paleozoic and older zircon grains. Small populations of 70–100 Ma zircons probably indicate derivation from the Cretaceous Idaho batholith. On the basis of visual comparison, detrital-zircon age spectra for these samples are most similar to the age spectra for the modern Big Lost River (Fig. 4A). These data suggest that the Big Lost River drainage was the dominant source, and are in agreement with modal compositional data (Fig. 3).

Subsurface samples from core collected in wells at TAN, near the northern margin of the basin, have detrital-zircon age spectra (Fig. 6) that suggest changing source areas through time. The detrital-zircon age spectrum of a near-surface sand

sample (depth of 6.1–7.6 m in TAN-34; Fig. 6A) is most similar to the signature of the Big Lost River (Fig. 4A). However, deeper samples (Fig. 6, B–D) have detrital-zircon signatures similar to Birch Creek (Fig. 4C) or Crooked Creek (Fig. 4E).

Mineralogy of fine-grained sediments

The compositions of fine-grained sediments from surficial lacustrine and playa deposits were evaluated and compared to fine-grained deposits from the shallow subsurface (Figs. 7 and 8). X-ray diffraction analysis was used to determine the semi-quantitative mineralogy of the <4 μm fraction of the samples, typically clay and evaporite minerals. Semiquantitative mineral abundance was based on the relative intensity of the diffraction peak for the clay or evaporite mineral. Material comprising the remainder of the samples, including quartz silt to very fine sand, organic matter, and detrital or authigenic carbonate, was not

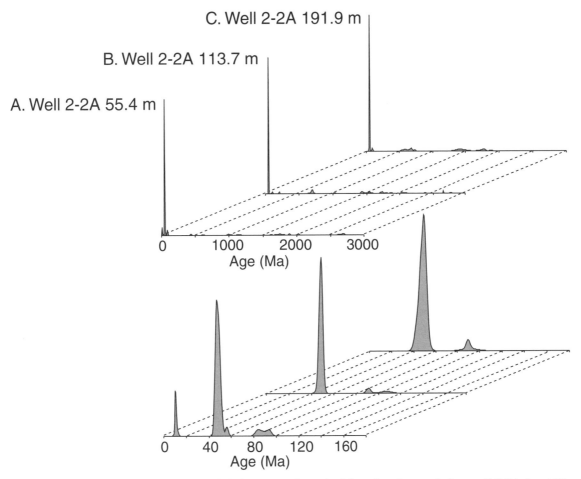

Figure 5. Relative probability plots for detrital-zircon ages determined for subsurface sands from well 2-2A, in middle of Big Lost Trough (Fig. 1). Upper graphs contain results of all analyses (0–3000 Ma zircons); lower graphs are for young (<180 Ma) zircon populations.

analyzed. Analysis of the samples included glycolization, to determine the presence of expanding clays (e.g., smectite), and 550°C heat treatment to differentiate chlorite and kaolinite.

X-ray diffraction data for shallow subsurface samples from well 2-2A (Fig. 8A) are generally similar to the surficial lacustrine and playa deposits (Fig. 7) in that they contain primarily smectite, kaolinite, and chlorite. This mineral assemblage probably reflects weathering of abundant volcanic and sedimentary rocks in fluvial source areas. However, several samples from the TAN wells (Fig. 8B) are distinctly different in that they contain gypsum. These data suggest that ancient evaporative playas or evaporative marginal lacustrine environments existed along the margin of the basin and that climatic conditions during deposition were probably drier than modern conditions.

DISCUSSION

The geometry and stratigraphic architecture of the Big Lost Trough was strongly influenced by the volcanic rift zones that define most of the basin margins. The locations of these vol-

canic rift zones, as well as local and regional subsidence patterns, reflect the interaction between basaltic volcanism and rhyolite caldera formation related to the Yellowstone hotspot and development of Basin and Range structures. Offlapping basalt flow groups and onlapping clastic sedimentary units along the northern margin of the basin (e.g., Fig. 2) represent growth of the volcanic rift zones and simultaneous clastic sedimentation in the basin. A general volcanic quiescence over the past ~1 m.y. has allowed fluvial, playa, eolian, and lacustrine systems to infill the basin to its modern configuration (Fig. 2C).

Sediments deposited in the middle of the basin over the past ~1 m.y. contain fluvial, eolian, and minor playa deposits interbedded with lacustrine sediment; this could represent alternating relatively wetter and relatively drier climatic conditions. This argument is based on documented fluctuations of Lake Terreton over the past 2 m.y. (e.g., Scott, 1982; Gianniny et al., this volume) and the presence of fluvial sands from the Big Lost River in the middle of the basin (Figs. 3–6).

Provenance studies in the Big Lost Trough that involve a comparison of modern fluvial and eolian sands with sands from

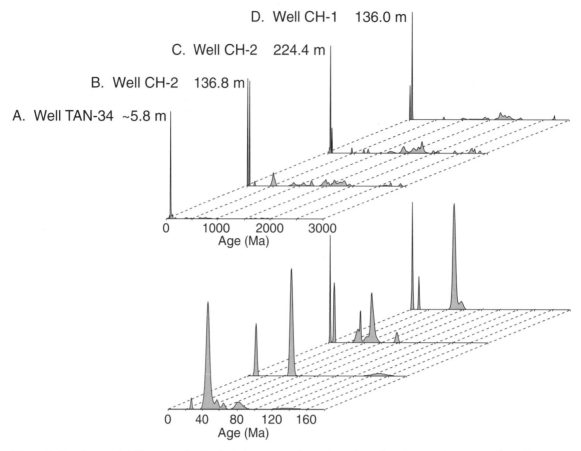

Figure 6. Relative probability plots for detrital-zircon ages determined for subsurface sands from wells at Test Area North, along northeastern margin of Big Lost Trough (Fig. 1). Upper graphs contain results of all analyses (0–3000 Ma zircons); lower graphs are for young (<180 Ma) zircon populations.

the subsurface using both petrography and U/Pb geochronology of detrital zircons provide very robust correlations (Figs. 4–6; Geslin et al., 1999). However, relating the detrital-zircon age spectra from modern Big Lost River sand and correlative subsurface sands to those of bedrock exposures in the source area is somewhat difficult. It is surprising that the Proterozoic to Archean detrital-zircon signature in these samples is faint despite the fact that the upper reaches of the Big Lost River drain the Pioneer Mountains core complex (Fig. 1), where abundant Paleoproterozoic metamorphic rocks crop out (Worl et al., 1995). Possible explanations for this observation include: (1) the great abundance of Eocene Challis volcanics cropping out in the middle reaches of the river have diluted the detrital zircon population; or (2) smaller streams that drain the Pioneer Mountains core complex were captured by the Big Lost River relatively recently, and Proterozoic to Archean zircons have yet to be transported to the lower reaches of the river. These data suggest that provenance determination, even in a well-documented setting like the Big Lost Trough, is a complex task, and that the zircon signature of the Big Lost River requires further study (e.g., Link et al., 1999, 2000).

Subsurface sands in the middle of the Big Lost Trough have provenance signatures indicating deposition by the Big

Lost River, suggesting that the proportion of Big Lost River fluvial sand in the subsurface increases from northeast to southwest, toward the point where the river enters the basin (Fig. 1). The southwestward increase in abundance of subsurface sand-rich sediments, the increase in the thickness of strata from the northern basin margin to basin center (Fig. 2C), and the regional northeast to southwest groundwater gradient all suggest that water flowing in the upper part of the Snake River Plain aquifer will encounter progressively more abundant and more conductive sediments as it migrates down gradient from the TAN facilities toward the middle of the basin. This information should be incorporated into regional hydrologic models of the unsaturated zone and upper part of the Snake River Plain aquifer.

CONCLUSIONS

The Big Lost Trough is an arid underfilled basin, the geometry and volcanic-sedimentary stratigraphic architecture of which reflect the development of surrounding volcanic rift zones. The pattern of volcanic rift zones that define the basin margins, subsidence of the area, and the fluvial systems that provide clastic sediments to the basin reflect the interaction of the migrating Yellowstone hotspot and developing Basin and

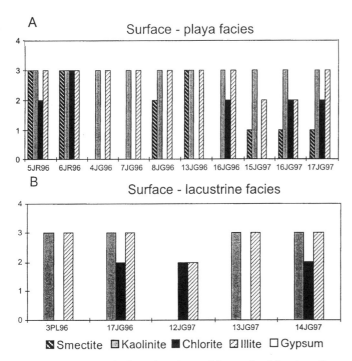

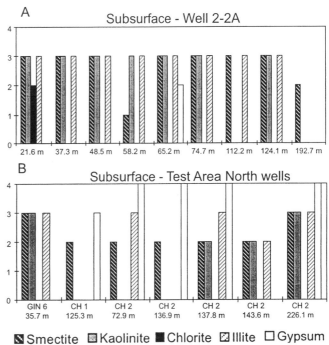

Figure 7. Semiquantitative mineralogy of fine-grained fraction of surficial playa and lacustrine sediments determined using X-ray diffraction analyses. Ranking from 0 to 4 indicates relative abundance of mineral: 4 = mineral dominant in sample, 3 = mineral present in sample, 2 = trace amounts of mineral in sample, 1 = mineral possibly present in sample.

Figure 8. Semiquantitative mineralogy of fine-grained fraction of subsurface deposits determined using X-ray diffraction analyses for samples from well 2-2A, in middle of basin, and from wells at Test Area North, along northeastern margin of basin. Ranking from 0 to 4 indicates relative abundance of mineral: 4 = mineral dominant in sample, 3 = mineral present in sample, 2 = trace amounts of mineral in sample, 1 = mineral possibly present in sample.

Range structures. Correlations of basalt flow groups and sedimentary successions from the basin margin at TAN to the middle of the basin at well 2-2A show the complex stratigraphic relationships created by basalt flows that downlap into the basin and sediments that onlap onto the adjacent volcanic high. Architecture within the clastic sedimentary fill of the basin reflects deposition during climatically influenced progradation and retrogradation of a fluvial-playa-eolian depositional system and transgressions and regressions of Lake Terreton.

Because the Big Lost Trough is still an active basin with easily identifiable sediment source areas, it provides an excellent natural laboratory for conducting provenance studies. By conducting both petrographic and U/Pb detrital-zircon geochronology analyses on subsurface sands, as well as modern fluvial and eolian deposits, the provenance of the subsurface sands has been very tightly defined. These analyses suggest that the Big Lost River has been the dominant source of sediment to the basin for at least the past 1 m.y., and that the river system frequently prograded across the basin during lowstands of Lake Terreton. Fluvial progradation during lowstands of Lake Terreton was associated with eolian reworking of fluvial deposits.

Semiquantitative compositional analyses, using X-ray diffraction, indicate that subsurface playa or marginal lacustrine deposits along the basin margin (at TAN) contain abundant gypsum, suggesting that ancient dry climate cycles were drier than the modern dry climate.

ACKNOWLEDGMENTS

This study was funded by grant DE-FG07-96ID13420 from the Department of Energy to the Idaho Water Resources Research Institute and Idaho Universities Consortium. Mary Kauffman and Savona Anderson assisted with sample collection and analysis. Fruitful discussions of sedimentary systems on the Snake River Plain were held with Gary Gianniny and Glenn Thackray. Access to core was provided by Linda Davis (U.S. Geological Survey). The manuscript was improved by comments from reviewers Larry Middleton, Dick Smith, and Brad Ritts.

REFERENCES CITED

Anders, M.H., Geissman, J.W., Piety, L., and Sullivan, T., 1989, Parabolic distribution of circum-eastern Snake River Plain seismicity and latest Quaternary faulting: mMigratory pattern and association with the Yellowstone hotspot: Journal of Geophysical Research, v. 94, p. 1589–1621.

Anderson, S.R., 1991, Stratigraphy of the unsaturated zone and uppermost part of the Snake River Plain aquifer at the Idaho Chemical Processing Plant and Test Reactors Area, Idaho National Engineering Laboratory, Idaho: U.S. Geological Survey Water-Resources Investigations Report 91-4010, 71 p.

Anderson, S.R., and Bowers, B., 1995, Stratigraphy of the unsaturated zone and uppermost part of the Snake River Plain aquifer at Test Area North, Idaho National Engineering Laboratory, Idaho: U.S. Geological Survey Water-Resources Investigations Report 95-4130, 47 p.

Anderson, S.R., and Lewis, B.D., 1989, Stratigraphy of the unsaturated zone at Radioactive Waste Management Complex, Idaho National Engineering Laboratory, Idaho: U.S. Geological Survey Water-Resources Investigations Report 89-4065, 54 p.

Anderson, S.R., Liszewski, M.J., and Ackerman, D.J., 1996, Thickness of surficial sediment at and near the Idaho National Engineering Laboratory, Idaho: U.S. Geological Survey Open-File Report 96-330, 16 p.

Blair, J.J., and Link, P.K., 2000, Pliocene and Quaternary sedimentation and stratigraphy of the Big Lost Trough from coreholes at the Idaho National Engineering and Environmental Laboratory, Idaho: Evidence for a regional Pliocene lake during the Olduvai normal polarity subchron, *in* Robinson, L., ed., Proceedings of the 35th Symposium on Engineering Geology and Geotechnical Engineering: Pocatello, Idaho, Idaho State University, College of Engineering, p. 163–179.

Carroll, A.R., and Bohacs, K.M., 1999, Stratigraphic classification of ancient lakes: Balancing tectonic and climatic controls: Geology, v. 27, p. 99–102.

Champion, D.E., Lanphere, M.A., and Kuntz, M.A., 1988, Evidence for a new geomagnetic reversal from lava flows in Idaho: Discussion of short polarity reversals in the Brunhes and late Matuyama polarity chrons: Journal of Geophysical Research, v. 93, p. 11667–11680.

Evans, K.V., and Zartman, R.E., 1988, Early Paleozoic alkalic plutonism in east-central Idaho: Geological Society of America Bulletin, v. 100, p. 1981–1987.

Geslin, J.K., Gianniny, G.L., Link, P.K., and Riesterer, J.W., 1997, Subsurface sedimentary facies and Pleistocene stratigraphy of the northern Idaho National Engineering Laboratory: Controls on hydrogeology, *in* Sharma, S., and Hardcastle, J. H., eds., Proceedings of the 32nd Annual Symposium on Engineering Geology and Geotechnical Engineering: Moscow, Idaho, University of Idaho, p. 15–28.

Geslin, J.K., Link, P.K., and Fanning, C.M., 1999, High-precision provenance determination using detrital-zircon ages and petrography of Quaternary sands on the eastern Snake River Plain, Idaho: Geology, v. 27, p. 295–298.

Gianniny, G.L., Geslin, J.K., Riesterer, J.W., Link, P.K., and Thackray, G.D., 1997, Quaternary surficial sediments near Test Area North (TAN), northeastern Snake River Plain: An actualistic guide to aquifer characterization, *in* Sharma, S., and Hardcastle, J.H., eds., Proceedings of the 32nd Annual Symposium on Engineering Geology and Geotechnical Engineering: Moscow, Idaho, University of Idaho, p. 29–44.

Hackett, W.R., and Smith, R.P., 1992, Quaternary volcanism, tectonics, and sedimentation in the Idaho National Engineering Laboratory area: *in* Wilson J.R., ed., Field guide to geologic excursions in Utah and adjacent areas of Nevada, Idaho, and Wyoming: Utah Geological Survey Miscellaneous Publication 92-3, p. 1–18.

Ingersoll, R.V., Bullard, R.R., Ford, R.L., Grimm, J.P., Pickel, J.P., and Sares, S.W., 1984, The effect of grain size on detrital modes: A test of the Gazzi-Dickinson point-counting method: Journal of Sedimentary Petrology, v. 54, p. 103–116.

Kuntz, M.A., Covington, H.R., and Schorr, L.J., 1992, An overview of basaltic volcanism of the eastern Snake River Plain, Idaho, *in* Link, P.K., Kuntz, M.A., and Platt, L.B., eds., Regional geology of eastern Idaho and western Wyoming: Geological Society of America Memoir 179, p. 227–267.

Kuntz, M.A., Skipp, B., Lanphere, M.A., Scott, W.E., Pierce, K.L., Dalrymple, G.B., Champion, D.E., Embree, G.F., Page, W.R., Morgan, L.A., Smith, R.P., Hackett, W.R., and Rodgers, D.W., 1994, Geological map of the Idaho National Engineering Laboratory and adjoining areas, eastern Idaho: U.S. Geological Survey Miscellaneous Investigations Series Map I-2330, scale 1:100 000.

Lanphere, M.A., Kuntz, M.A., and Champion, D.E., 1994, Petrography, age, and paleomagnetism of basaltic lava flows in coreholes at the Test Area North (TAN), Idaho National Engineering Laboratory: U.S. Geological Survey Open-File Report 94-686, 49 p.

Link, P.K., Geslin, J.K., Thackray, G.T., Gianniny, G.L., 1999, Basin architecture of the Pleistocene Bit Lost Trough, eastern Snake River Plain, Idaho: Geological Society of America Abstracts with Programs, v. 31, no. 4, p. A22.

Link, P.K., Geslin, J.K., and Fanning, C.M., 2000, Detrital zircon 'barcodes' from modern and Neogene sands of the Snake River Plain, Idaho: Defining a provenance area requires several grains and single grains mean little: Geological Society of America Abstracts with Programs, v. 32, no. 6, p. A25.

Morgan, L. A., 1992, Stratigraphic relations and paleomagnetic and geochemical correlations of ignimbrites of the Heise Volcanic Field, Eastern Snake River Plain, eastern Idaho and western Wyoming, *in* Link, P.K., Kuntz, M.A., and Platt, L.B., eds., Regional geology of eastern Idaho and western Wyoming, Geological Society of America Memoir 179, p. 215–226.

Morgan, L.A., Doherty, D.J., and Leeman, W.P., 1984, Ignimbrites of the eastern Snake River Plain: Evidence for major caldera forming eruptions: Journal of Geophysical Research, v. 89, p. 8665–8678.

McQuarrie, N., and Rodgers, D.W., 1998, Subsidence of a volcanic basin by flexure and lower crustal flow: The eastern Snake River Plain, Idaho: Tectonics, v. 17, p. 203–220.

Nace, R.L., Voegeli, P.T., Jones, J.R., and Deutsch, M., 1975, Generalized geologic framework of the National Reactor Testing Station, Idaho: U.S. Geological Survey Professional Paper 725-B, 49 p.

Oaks, R.Q., Jr., James, W.C., Francis, G.G., and Schulingkamp, W.J., 1977, Summary of Middle Ordovician stratigraphy and tectonics, northern Utah, southern and central Idaho, *in* Heisey, E.L., Lawson, D.E., Norwood, E.R., Wach, P.H., and Hale, L.A., eds., Rocky Mountain thrust belt geology and resources: Casper, Wyoming, Wyoming Geological Association Guidebook, p. 101–118.

Pierce, K.L., and Morgan, L.A., 1992, The track of the Yellowstone hot spot, *in* Link, P.K., Kuntz, M.A., and Platt, L.B., eds., Regional geology of eastern Idaho and western Wyoming: Geological Society of America Memoir 179, p. 1–53.

Scott, W.E., 1982, Surficial geologic map of the eastern Snake River Plain and adjacent areas, 111°–115° W, Idaho and Wyoming: U.S. Geological Survey Miscellaneous Investigations Series Map I-1372, scale 1:250 000, 2 sheets.

Skipp, B., and Link, P.K., 1992, Middle and Late Proterozoic rocks and Late Proterozoic tectonics in the southern Beaverhead Mountains, Idaho and Montana, a preliminary report, *in* Link, P.K., Kuntz, M.A., and Platt, L.B., eds., Regional geology of eastern Idaho and western Wyoming: Geological Society of America Memoir 179, p. 141–153.

Spinazola, J.M., 1994, Geohydrology and simulation of flow and water levels in the aquifer system in the Mud Lake area of the eastern Snake River Plain, eastern Idaho: U.S. Geological Survey Water-Resources Investigations Report 93–4227, 78 p.

Stearns, H.T., Crandall, L., and Steward, W.G., 1939, Geology and groundwater resources of the Mud Lake region, Idaho, including the Island Park area: U.S. Geological Survey Water-Supply Paper 818, 125 p.

Whitehead, R.L., 1992, Geohydrologic framework of the Snake River Plain regional aquifer system, Idaho and eastern Oregon: U.S. Geological Survey Professional Paper 1409-B, 32 p., 6 plates.

Winston, D., and Link, P.K., 1993, Middle Proterozoic rocks of Montana, Idaho, and Washington: The Belt Supergroup, *in* Reed, J., Simms, P., Houston, R., Rankin, D., Link, P., Van Schmus, R., and Bickford, P., eds., Precambrian of the conterminous United States: Boulder, Colorado, Geological Society of America, Geology of North America, v. C-3, p. 487–521.

Worl, R.G., Link, P.K., Winkler, G.R., and Johnson, K.M., editors., 1995, Geology and mineral resources of the Hailey 1° × 2° quadrangle and the western part of the Idaho Falls 1° × 2° quadrangle, Idaho: Washington, D.C., U.S. Geological Survey Bulletin 2064, v. 1, chapters A-R, separate pagination.

MANUSCRIPT ACCEPTED BY THE SOCIETY NOVEMBER 2, 2000

Geological Society of America
Special Paper 353
2002

Paleoenvironments of sedimentary interbeds in the Pliocene and Quaternary Big Lost Trough, eastern Snake River Plain, Idaho

Erick A. Bestland*
Flinders University, GPO Box 2100, Adelaide 5001, South Australia, Australia
Paul Karl Link
Department of Geosciences, Idaho State University, Pocatello, Idaho 83209, USA
Marvin A. Lanphere
U.S. Geological Survey, Mail Stop 937, Menlo Park, California 94025, USA
Duane E. Champion
U.S. Geological Survey, Mail Stop 910, Menlo Park, California 94025, USA

ABSTRACT

Thick successions of Pliocene and Quaternary sediment recovered from core 2-2A in the central part of the Big Lost Trough on the Idaho National Engineering and Environmental Laboratory (INEEL) record deposition in two climatically controlled depositional systems. During latest Pliocene and early Pleistocene time (2.5–1.5 Ma), wetter climates plus a higher regional water table produced lakes and lake deltas. The eastern Snake River Plain groundwater level was adjusted to the level of Lake Idaho to the west, which drained by middle Pleistocene time. During relatively dryer late Pleistocene time, sediment was deposited in cyclically aggrading playa-lunette systems with a fluvial and loess sediment source.

The central Big Lost Trough was a persistent low area that trapped and preserved sediment. Thus, unlike cores and wells in the northern INEEL around Test Area North, core 2-2A contains a substantial sedimentary record including four upper Pliocene to upper Pleistocene stratigraphic intervals. A thick interbed at depth 292–350 m (958–1148 ft) is dated by a combination of magnetostratigraphy and radiometric ages as ca. 1.7–1.9 Ma, and records infilling of a lake that was probably ice free. In contrast, an overlying interval of earliest Pleistocene age (estimated as ca. 1.5–1.6 Ma) contains both laminated silts and clays (lake-bottom, prodelta facies) and massive silt with abundant pebble- to cobble-sized clasts. This diamictite facies is interpreted as the result of sedimentation during cooler times when profundal suspension-fallout deposits were mixed with clasts of ice-rafted origin (shore and river ice). Furthermore, three distinct sequences of seasonal-ice and ice-free conditions are recognized in this interval and could correspond with the 40 ka glacial-interglacial cycles recognized from marine records. An upper sedimentary interbed is separated from the diamictite interval by a thick succession of basalts. The Matuyama-Brunhes reversal (0.78 Ma) probably occurs near the base of this interbed, making the strata middle and late Pleistocene in age. These strata contain playa deposits, lunette dunes, fluvial and eolian sands and silts, paludal-lacustrine deposits, and weakly developed paleosols. Taken together, the change from lacustrine-dominated deposition during

*E-mail: erick@es.flinders.edu.au

Bestland, E.A., Link, P.K., Lanphere, M.A., and Champion, D.E., 2002, Paleoenvironments of sedimentary interbeds in the Pliocene and Quaternary Big Lost Trough, eastern Snake River Plain, Idaho, *in* Link, P.K., and Mink, L.L., eds., Geology, Hydrogeology, and Environmental Remediation: Idaho National Engineering and Environmental Laboratory, Eastern Snake River Plain, Idaho: Boulder, Colorado, Geological Society of America Special Paper 353, p. 27–44.

the late Pliocene and early Pleistocene to eolian-, playa-, and soil-dominated deposits in the late Pleistocene indicates drying, plus lowering of regional hydrologic base level.

Carbon and oxygen isotope analysis of a well-developed, stage III calcareous paleosol, as well as isotope data from weakly developed paleosols and silty lake beds indicates that the dominant source of calcareous loess, ubiquitous in these deposits, was Paleozoic and Mesozoic carbonate units.

INTRODUCTION

The eastern Snake River Plain (Fig. 1) is an unusual sedimentary-volcanic basin. There are no unifying models for the development of such intracratonic rhyolite-basalt downwarps (e.g., Busby and Ingersoll, 1995). The eastern Snake River Plain is not a fault-bounded rift such as nearby Basin and Range basins, nor does it appear to be solely a volcano-tectonic subsidence feature above a series of calderas. It has subsided as a downwarp, likely loaded by a mid-crustal basaltic sill, along the track of the Yellowstone hotspot (Pierce and Morgan, 1992; McQuarrie and Rodgers, 1998). The modern surface is composed largely of Pleistocene basalt flows with a variable veneer of loess, eolian sand, and fluvial, playa, and lacustrine deposits (Scott, 1982; Hackett and Smith, 1992; Kuntz et al., 1994,

2002). Most of these deposits are pedogenically modified (Ostenaa et al., 1999a). Fresh, Holocene lava flows such as Craters of the Moon and Hells Half Acre make up a significant percentage of the modern surface.

The eastern Snake River Plain is little dissected by erosion; thus the underlying basin fill is largely inaccessible to standard field methods, except for wells, well cores, and geophysical surveys. The basin fill is dominated by basaltic lava flows of local origin. Sedimentary interbeds are subordinate to basalt.

In this chapter we describe sedimentary interbed stratigraphy that is defined by paleomagnetic data, $^{40}Ar/^{39}Ar$ age determinations, and geochemical correlations. Reconstructing the stratigraphy of the INEEL subsurface is a first-order condition for understanding the contaminant plume present near the Test Area North (TAN) facility (stratigraphic location shown in Fig. 2) as well as other INEEL-related plumes.

Big Lost Trough

The Big and Little Lost Rivers flow into the Big Lost Trough, an underfilled, volcanically silled basin north of the Axial Volcanic Zone of the eastern Snake River Plain and south of Basin and Range fault blocks and basins that trend perpendicular to the plain (Gianniny et al., 1997; Geslin et al., 1999) (Fig. 1). From ~lat 113°W, near State Butte, south of Howe, northeastward, the modern basin floor is shallowly incised to form Pleistocene fluvial terraces. Surficial sediments include Holocene and late Pleistocene fluvial, eolian, and playa deposits that largely lack gravel-sized detritus. The age and sedimentology of Holocene deposits indicate a lack of large fluvial floods or lacustrine highstands during the past 10 k.y. (Ostenaa et al., 1999a, 1999b, this volume, chapter 7; Levish et al., 1999).

Examination of the sediment in core 2-2A demonstrates that pebble to cobble detritus has been largely absent from fluvial and lacustrine systems over the central and northern parts of the Big Lost Trough for the past 2 m.y. This suggests that coarse Pleistocene alluvial fan material from the Lemhi and Lost River Ranges did not prograde into the Big Lost Trough to the area represented by core 2-2A. The Big Lost Trough is thus underfilled and stratal geometry is aggradational (Geslin et al., this volume, chapter 2; cf. Carroll and Bohacs, 1999).

STRATIGRAPHIC FRAMEWORK OF CENTRAL AND NORTHERN BIG LOST TROUGH

The physical stratigraphy of sedimentary interbeds, basalt lava flows, and weathered-oxidized rubble zones between lava

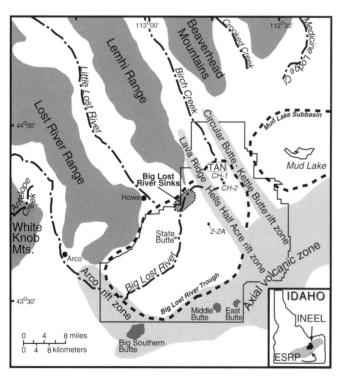

Figure 1. Location map of northern Idaho National Engineering and Environmental Laboratory (INEEL) area on eastern Snake River Plain (ESRP), showing Big Lost Trough, Big Lost River sinks, Axial volcanic zone, and northwest-striking rift zones. Also shown are locations of cores 2-2A, and CH-1 and CH-2 near Test Area North (TAN) (modified from Geslin et al., 1999).

29

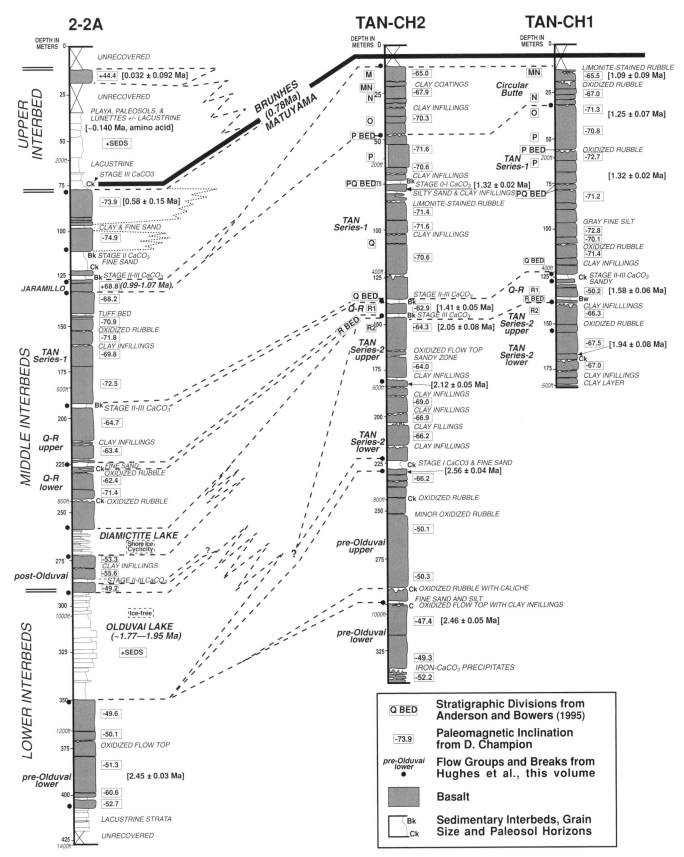

Figure 2. Fence diagram of cores 2-2A, Test Area North (TAN) CH-1, and CH-2 showing sedimentary interbeds, lava flow group correlations (Hughes et al., this volume), stratigraphic divisions, paleomagnetic data, and ^{40}Ar/^{39}Ar age determinations (Lanphere et al., 1994, Hughes et al., this volume, Table 4). Selected stratigraphic designations (e.g., P flow, PQ interbed) of Anderson and Bowers (1995) are also shown.

flows is the first-order control on stochastic modeling of hydrologic properties (Welhan et al., this volume, chapter 15). The stratigraphy, correlation, and age of sedimentary interbeds and lava flow groups in core 2-2A and in two wells (CH-1 and CH-2) near TAN are summarized in Figure 2. The ^{40}Ar/^{39}Ar age determinations are from whole-rock plateaus (Lanphere et al., 1994). Paleomagnetic reversals in stratigraphic sequences of lava flows and sediment (Fig. 2) approximate the Brunhes-Matuyama boundary (0.78 Ma) and the Olduvai reversal (1.77–1.95 Ma; ages from Berggren et al., 1995a, 1995b) as modified from Cande and Kent (1992). Whole-rock geochemical trends and correlation of lava flows and flow groups (Hughes et al., this volume, chapter 10; Geist et al., this volume, chapter 4) are also shown in Figure 2, as are selected stratigraphic designations (e.g., P flow, PQ interbed) of Anderson and Bowers (1995).

Of particular importance in the evaluation and correlation of interbeds are calcareous paleosols, which, by analogy to calcareous soils from dated geomorphic surfaces (Gile et al., 1966; Machette, 1985; Simpson et al., 1999), represent tens of thousands to hundreds of thousands of years of soil formation. Much of the geologic time represented in these stratigraphic sections is contained in calcareous paleosol horizons.

Basalt ash or tuff beds are rare in TAN area cores and 2-2A, indicating a lack of significant nearby basaltic pyroclastic eruptions. This is consistent with the location of wells CH-1 and CH-2 more than 5 km from the Circular Butte vent area and well 2-2A more than 4 km south of the Lava Ridge–Hells Half Acre rift zone (Fig. 1). Thick accumulations of basaltic pyroclastic deposits are generally confined to within a few kilometers of a vent (Cas and Wright, 1988, their Fig. 14.7).

SEDIMENTARY INTERBEDS OF CORE 2-2A

Core 2-2A contains the thickest sedimentary interbeds of the INEEL cores, and as such contains evidence of past environmental conditions on this part of the Snake River Plain (Forester, personal commun., 1979, 1987, 1991; Thompson, 1991). The stratigraphy and sedimentology of the upper half of core 2-2A (427 m) are described in this chapter following standard field methods for paleosols (Retallack, 1988) and sedimentary deposits (Figs. 3–7). In general, the sedimentary interbeds comprise four lithofacies (Table 1): (1) lacustrine, (2) calcareous paleosols, (3) clay-sand facies of aggrading loess-lunette-playa systems with weak pedogenic modification, and (4) sand and silt facies of fluvial-eolian origin (fluveolian sensu Fryberger, 1990, his Table 1).

Lacustrine lithofacies

Deposits of lacustrine origin are present in most of the sedimentary interbeds that are thicker than a few meters (Fig. 2). They consist of fine-grained clay, both massive and laminated, interbedded sand and silt, and, in one interbed, diamictite

layers. Five divisions of lacustrine facies are recognized (Table 1); four of the facies relate to classic Gilbert-type deltas (Gilbert, 1885). Definitive characteristics of lacustrine deposition are indicated by the presence of well-sorted clay and mudstone intervals that are relatively thick (tens of centimeters to a few meters) and that have no indications of pedogenic modification. These mudstones are thought to represent suspension fallout of fine-grained detritus in open quiet water.

Lacustrine conditions are indicated for the following four sedimentary sections of core 2-2A: (1) within the lowest sedimentary section studied with a top at 418.7 m (1340 ft); (2) the Olduvai lake interbed from 292.1 to 349.4 m (958–1146 ft); (3) the middle, diamictite lake interbed from 259.1 to 272.9 m (850–895 ft); and (4) the interval from 57.9 to 68.0 m (190–223 ft) in the upper sedimentary interbed.

The Olduvai lake interval from 292.1 to 349.4 m (958–1146 ft) (Fig. 3) is the thickest and most sedimentologically coherent lacustrine sequence. That is, it contains a classic sequence of coarsening-upward strata interpreted as prodelta, foredelta, and topset beds. It also contains abundant ostracodes (Forester, personal communication 1979). By analogy with similar modern ostracodes, Forester interpreted *Candona rawsoni* and plant remains of *Charophytes* and *Nitellopsis* to indicate small, shallow (~3–15 m deep), freshwater lakes and ponds with pH of ~8–10. These earliest Pleistocene lacustrine conditions at the site of hole 2-2A ended with emplacement of the overlying basalt lava flow (post-Olduvai flow of Hughes et al., this volume, chapter 10, their Fig. 8; below the R sedimentary bed of Anderson and Bowers, 1995).

A significant amount of time elapsed between deposition of the Olduvai lake and the diamictite lake sequence, as indicated by a well-developed calcareous paleosol situated between these two lacustrine sections at 286.0 m (938 ft) depth (Fig. 4). Calcareous soil stages from the American southwest with stage II to III morphologies, similar to the previously mentioned paleosol, range in age from 50 to 500 ka (Machette, 1985). Above this paleosol and lava flows, laminated clay and silt of diamictite lake directly overlie lava at a depth of 272.9 m (895 ft). No paleosol features were observed at the sediment-basalt contact, indicating that this surface was inundated by water not long after the emplacement of the flow.

Deposits of the middle lacustrine interval (diamictite lake) differ from those of Olduvai lake by the common occurrence of diamictite layers. These layers consist of pink, silty clay with pebble- to cobble-size fragments of basalt and rip-up clasts of silty clay (Fig. 8). Layers are both massive and laminated; soft-sediment deformation is visible around some clasts. The clasts are interpreted as dropstones. Clasts are most abundant in the massive beds but are also common in the laminated beds. The clasts have a local origin, and were likely incorporated into the shore ice of lakes or from ice-covered streams flowing into the lakes.

Within the diamictite lake, at least three fining-upward stratigraphic sequences can be defined on the basis of repeating

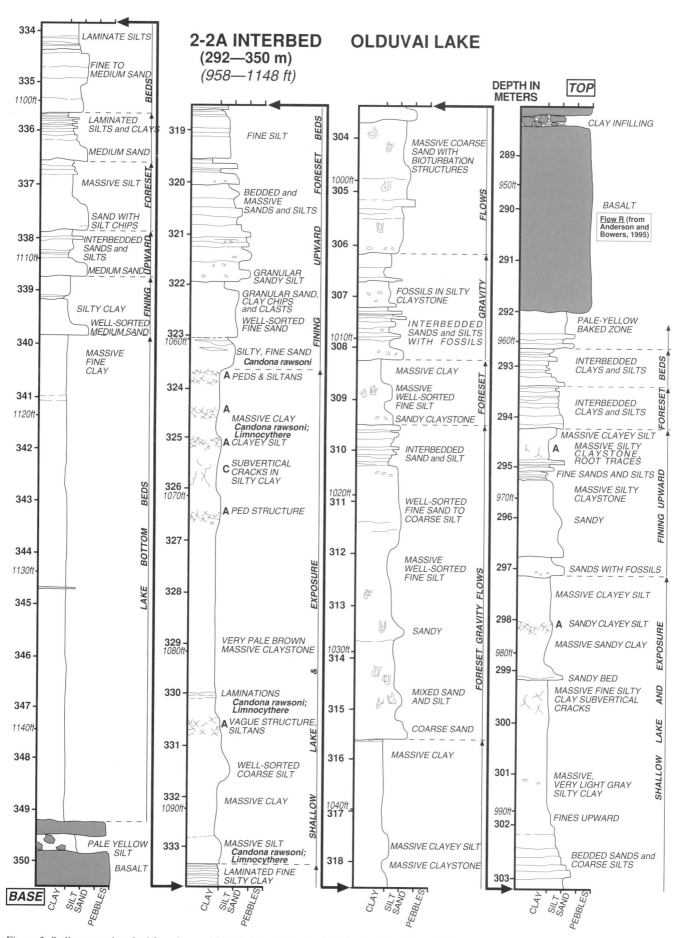

Figure 3. Sedimentary interbed from interval 292–350 m (Olduvai lake). Explanation of symbols is shown in Figure 4. Fining-upward sequences with sedimentological interpretations are indicated by arrows and corresponding text.

32

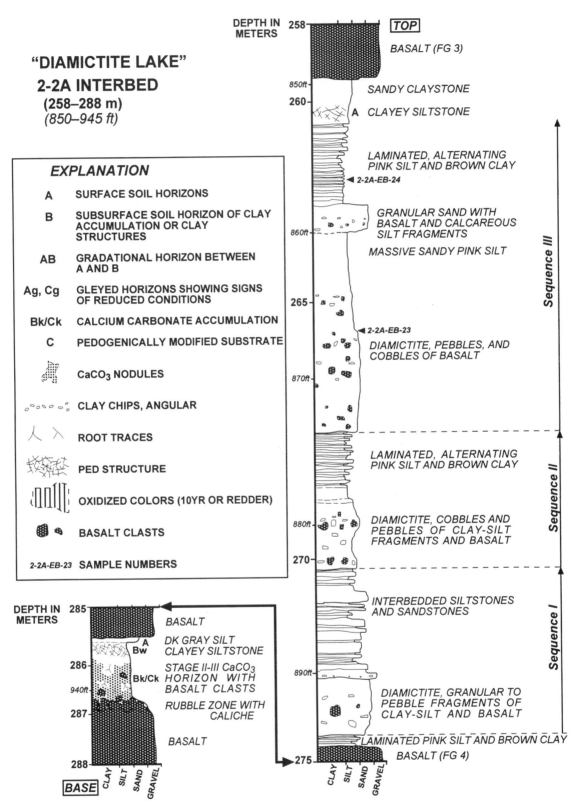

Figure 4. Sedimentary interbed from interval 258–288 m (diamictite lake). Sequences I-III are interpreted as representing deposition during glacial (diamictite layers) and interglacial times (laminated strata) possibly following ca. 40 ka climate cyclicity of late Pliocene–early Pleistocene (Raymo et al., 1990).

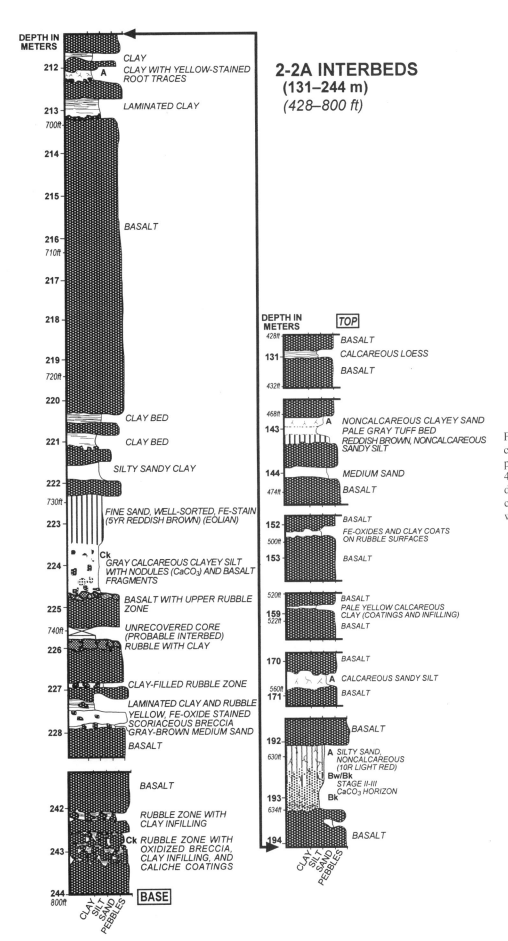

2-2A INTERBEDS
(131–244 m)
(428–800 ft)

DEPTH IN METERS

212 — CLAY
— A — CLAY WITH YELLOW-STAINED ROOT TRACES
213 — LAMINATED CLAY
700ft
214
215
216 — BASALT
710ft
217
218
219
720ft
220 — CLAY BED
221 — CLAY BED
— SILTY SANDY CLAY
222
730ft
— FINE SAND, WELL-SORTED, FE-STAIN (5YR REDDISH BROWN) (EOLIAN)
223
— Ck — GRAY CALCAREOUS CLAYEY SILT WITH NODULES (CaCO₃) AND BASALT FRAGMENTS
224
225 — BASALT WITH UPPER RUBBLE ZONE
740ft — UNRECOVERED CORE (PROBABLE INTERBED)
226 — RUBBLE WITH CLAY
227 — CLAY-FILLED RUBBLE ZONE
— LAMINATED CLAY AND RUBBLE
— YELLOW, FE-OXIDE STAINED
— SCORIACEOUS BRECCIA
228 — GRAY-BROWN MEDIUM SAND
— BASALT
— BASALT
242 — RUBBLE ZONE WITH CLAY INFILLING
— Ck — RUBBLE ZONE WITH OXIDIZED BRECCIA, CLAY INFILLING, AND CALICHE COATINGS
243
244
800ft — CLAY SILT SAND PEBBLES — BASE

DEPTH IN METERS — TOP
428ft — BASALT
131 — CALCAREOUS LOESS
— BASALT
432ft
468ft — A — NONCALCAREOUS CLAYEY SAND
143 — PALE GRAY TUFF BED
— REDDISH BROWN, NONCALCAREOUS SANDY SILT
144 — MEDIUM SAND
474ft — BASALT
152 — BASALT
500ft — FE-OXIDES AND CLAY COATS ON RUBBLE SURFACES
153 — BASALT
520ft — BASALT
159 — PALE YELLOW CALCAREOUS CLAY (COATINGS AND INFILLING)
522ft — BASALT
170 — BASALT
— A — CALCAREOUS SANDY SILT
560ft
171 — BASALT
192 — BASALT
630ft — A — SILTY SAND, NONCALCAREOUS (10R LIGHT RED)
— Bw/Bk — STAGE II-III CaCO₃ HORIZON
193 — Bk
634ft
— BASALT
194 — CLAY SILT SAND PEBBLES

Figure 5. Sedimentary interbeds from core 2-2A from interval 131–244 m. Explanation of symbols is shown in Figure 4. Thin interbeds between lava flows are dominated by very calcareous, white clay coatings and infillings. Bw refers to weakly weathered B horizons in soils.

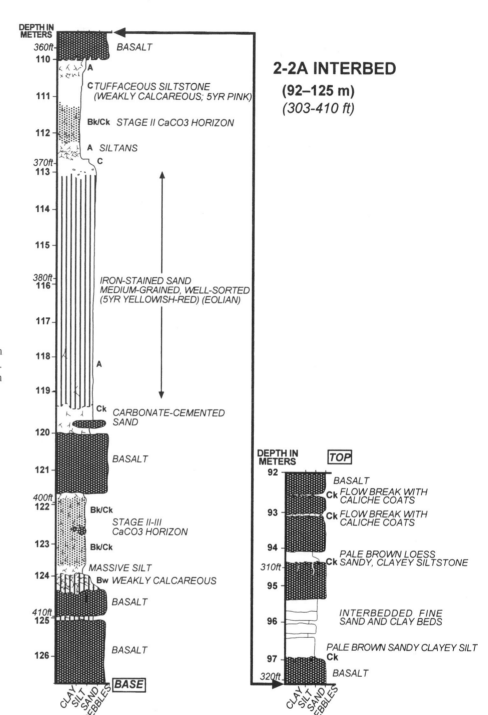

Figure 6. Sedimentary interbeds from core 2-2A from interval 92–125 m. Explanation of symbols is shown in Figure 4.

diamictite-laminated lithofacies (Fig. 4). Bases of sequences are defined at the contact between laminated silts and clays (below) and diamictites (above). Sequence III could be divided into two parts at the break between massive silt and granular sand at 262.2 m (860 ft) (Fig. 4), giving four sequences.

The upper sedimentary interbed contains three stratigraphic intervals interpreted as lacustrine-paludal (Fig. 7) (72.5–70 m, 238–230 ft; 67.5–63 m, 222–208 ft; and 61–59.5 m, 201–196 ft). These intervals have less obvious lacustrine features than the three underlying lacustrine sections because they lack laminated or thinly bedded strata and lack zones with abundant microfossils. Nonetheless, massive light gray clay with little silt content argues strongly for suspension fallout from quiet waters. It is envisioned that these intervals in the upper sedimentary interbed record deposition in small lakes that had low sediment input and significant bioturbation; hence the

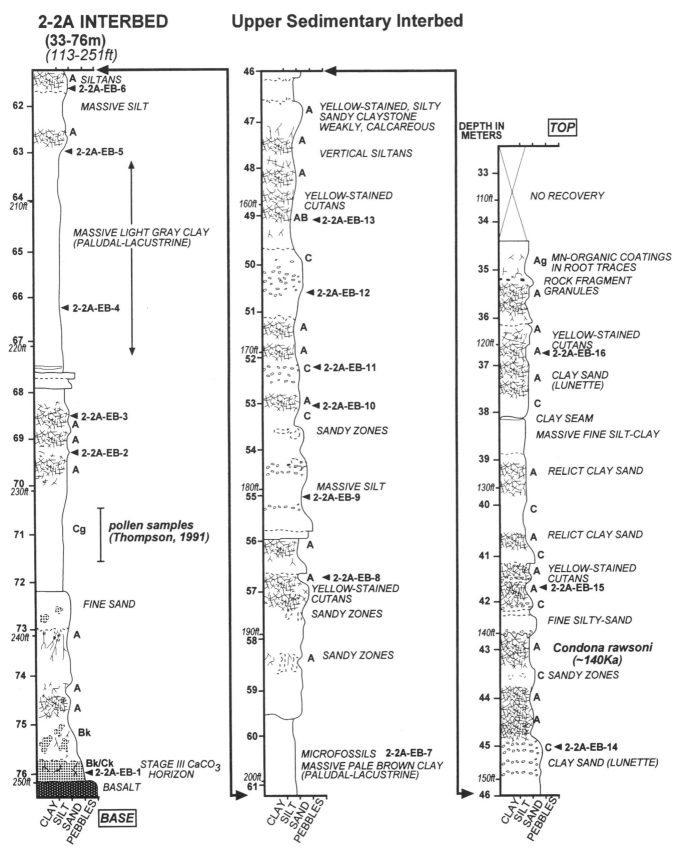

Figure 7. Upper sedimentary interbed from core 2-2A from interval 33–76 m. Explanation of symbols is shown in Figure 4. Sample localities for photographs of Figures 9, 10, and 11 are shown.

TABLE 1. DESCRIPTION AND INTERPRETATION OF SEDIMENTARY AND PEDOGENIC LITHOFACIES IN INTERBEDS FROM CORE 2-2A

Lithofacies	Description	Interpretation	Examples (Core 2-2A)
Lacustrine facies			
Massive to laminated clay	Clay and silty clay, very pale gray, very calcareous, mostly massive, some laminations.	Lake bottom bioturbated suspension fallout, prodelta, and distal prodelta	210–220 ft (64–67 m) 1120–1140 ft (341–347 m)
Laminated clay and silt	Laminated to thinly bedded clay, silty clay, pale gray, pale brown, very calcareous.	Prodelta suspension fallout	1094–1096 ft (333–334 m) 874–877 ft (226–267 m)
Interbedded massive sand	Bedded, structureless fine- to coarse-grained sand, some normal grading, bioturbation structures common.	Delta-front, proximal prodelta, sediment gravity flows	1030–1040 ft (314–317 m) 1050–1069 ft (320–326 m)
Thinly bedded sand and silt	Fine- to coarse-grained sand, interbedded with coarse silt, thin to medium bedded.	Delta-plain, distributary channel, and interchannel deposition	760–766 ft (232–233 m)
Diamictite	Massive, pink to pale gray silty clay with sand, granules, pebbles, and cobbles of basalt and rip-up clasts of silty clay.	Lake bottom suspension fallout, bioturbated with ice-rafted dropstones	860–870 ft (262–265 m)
Calcareous paleosol facies			
Fine-grained calcium carbonate	Silty claystone to sandy siltstone with displacive micrite carbonate cement, indurated, very pale yellow to very pale gray, open root traces common and coated with sparry calcite or yellow-brown clay film, admixtures of sand and gravel and basalt clasts.	Ck soil horizons of stage II and III development	940 ft (287 m) 400 ft (122 m) 249 ft (76 m)
Brown silty-clay	Silty, sandy claystone, 10YR brown to 5YR reddish brown, weakly calcareous to noncalcareous, ped structures and cutans weak to moderately developed, occasional calcium carbonate nodules.	Bw to Bt soil horizons	408 ft (124 m) 631 ft (192 m) 940 ft (287 m)
Clay sand facies (Playa-Lunette)			
Clay sand and silt	Granular to sand-sized subrounded to angular and tabular (rip-ups) clasts of mud-clay, very pale brown-gray, very calcareous.	Reworked playa bottom clays by aeolian processes to form lunettes (clay sand dunes)	120 ft (37 m) 148–151 ft (45–46 m)
Mixed mud and silt	Silty, sandy claystone, very pale brown-gray, very calcareous, mm- to cm-scale ped structure, cutans common of yellow-stained clay, common siltans (silty cutans).	Soil A horizons developed on lunette and playa muds and silts	115–123 ft (35–37 m) 155–160 ft (47–49 m)
Sand and silt facies (Fluvial-Eolian)	Poorly sorted sand.	Fluvial channels	(maybe non in 2-2A)
Sorted sand and silt	Fine to medium sand, most well sorted, common reddish-brown iron staining in thick sand beds, weakly to non-calcareous.	Aeolian sand and silt from proximal fluvial system	239 ft (73 m) 470 ft (143 m) 730 ft (223 m)
Coarse sand with clay chips	Poorly sorted, fine to coarse sand with clay chips to small pebble size, chip very calcareous.	Mixed aeolian, fluvial along playa margins and fluvial channels	164–183 ft (50–56 m)

Note: CK refers to calcareous soil horizons, Bt refers to a clayey soil horizon, and Bw refers to weakly weathered soil horizons.

lack of delta facies and general lack of bedding. In addition, these intervals have associated A horizons, indicating subaerial exposure and colonization by plants.

Calcareous paleosol lithofacies

Soils of arid to subhumid regions are well known for their distinctive morphologies of calcium carbonate accumulation (Gile et al., 1966). Stages of accumulation have been recog-nized and dated approximately by age estimates of geomorphic surfaces in the American southwest on which these soils formed (Machette, 1982, 1985; McFadden, 1988). The time needed for the formation of calcium carbonate stages has been shown to vary with precipitation levels and dust flux rates (Machette, 1985; Marion et al., 1985). Nevertheless, calcium carbonate stages can serve as approximate indicators of the duration of formation of calcareous paleosols. As such, they are widely used in Quaternary geology studies (Birkeland et al., 1991).

Stages I–III of calcium carbonate accumulation as defined by Gile et al. (1966) and refined by Machette (1985) are recognized in paleosols from INEEL cores. According to these studies, stage I consists of scattered calcium carbonate nodules and is thought to develop in 10–80 k.y. Stages II and III consist of coalesced nodules, stage III having >50% coalesced nodules. Stage II requires from 50 k.y. to more than 200 k.y., and stage III requires from 100 to 500 k.y.

Ongoing studies of Quaternary surficial deposits and soils in the Big Lost Trough have identified calcium carbonate stages I–III and estimated the amount of time necessary to develop them (Simpson et al., 1999). Based on radiocarbon dates, soils as young as 2–2.5 ka have stage I morphologies. Gravelly, latest Pleistocene (12–50 ka) soils have stage II morphology. Older soils (50–90 ka) have stage I–III morphologies, depending on gravel content; horizons rich in gravel appear to accumulate calcium carbonate more rapidly (Simpson et al., 1999). Thus, the Simpson study is in general agreement with studies from the American southwest whereby stage II morphology represents >10 k.y. and soils with stage III morphology represent >50 k.y. The time of formation for stage I morphology appears to be more rapid on the Snake River Plain than in the deserts of the southwest.

In the subsurface of the northern INEEL, paleosols with calcareous horizons occur almost exclusively in stratigraphic positions directly above lava flows. These calcareous horizons are moderately indurated (Kauffman and Geslin, 1998) and therefore have much higher preservation potential than the corresponding less indurated B and A horizons, which are commonly eroded prior to burial. Displacive micrite is the dominant carbonate precipitate and occurs both mixed with clay and as relatively pure nodules that coalesce to form a mottled microscopic texture (Fig. 9).

In several stratigraphic intervals in core 2-2A, clayey and silty horizons interpreted as soil B horizons are present above indurated calcium carbonate zones. The best two examples occur at 192.1 m (630 ft) and 286.0 m (938 ft) depth (Figs. 4 and 5). Both consist of weakly calcareous to noncalcareous silty clay or clayey silt and contain abundant root traces. The noncalcareous or weakly calcareous character of these horizons is significant because of the ubiquitous presence of calcareous silt and clay in the Big Lost Trough depositional system. The lack of calcareous material in these horizons most likely indicates

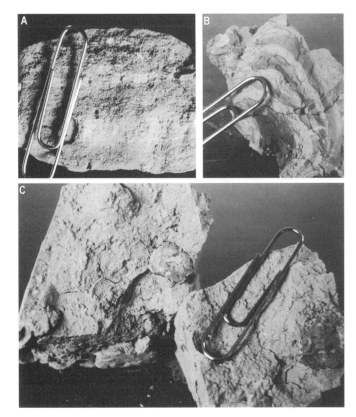

Figure 8. Photographs of laminated beds and diamictite fabric from diamictite lake interbed. A: Laminated sand, silt, and clay from 269.5 m depth of Figure 4. B: Laminated clay and silt from 266.8 m depth, Figure 4. C: Diamictite with pebble-size basalt clasts interpreted as dropstones (from 265.2 m, Fig. 4). Paper clip is 1 cm wide. See Figure 4 for sample locations.

Figure 9. Photographs of fabric and structures of clay sand and paleosol facies in upper sedimentary interbed. A: Granular-sized, subangular to subrounded silty clay clasts, interpreted as lunette dune deposits (sample 2-2A-EB-14, Fig. 7). B: Close-up of lunette clay-sand showing coarse silt between granules of clay clasts (same sample as A). C: Transition from clay-sand texture to ped structure (sample 2-2A-EB-15, Fig. 7). D: Ped structure defined by yellow clay coating root traces and cutan surfaces (sample 2-2A-EB-16, Fig. 7). Scale is in millimeters.

sufficient weathering and leaching of calcium carbonate from the B horizon during times of relatively low dust flux. This is further interpreted to represent wetter times, i.e., Pleistocene glacial intervals. The majority of calcareous loess deposition occurred just after these wetter maxima, during waning glacial stages.

Clay sand and silt lithofacies (playa-lunette)

Playa basins are a common depositional-geomorphic environment in subhumid to semiarid climates on gently sloping surfaces subject to ephemeral flooding. These playas are relatively small, internally drained depressions that are thought to be important sites of groundwater recharge. Their origin has been a topic of discussion for some time (Gilbert, 1895). Ideas of their origin fall into two general camps: (1) dissolution and transport of clastic material by infiltrating groundwater (Osterkamp and Wood, 1987), and (2) eolian removal of clastic material from the playa floor (Gilbert, 1895; Holliday et al., 1996). In addition, many of the eastern Snake River Plain playas are due in part to damning by lava flows or represent closed, low areas on lava flow surfaces.

Accretion of eolian sediment adjacent to playa margins commonly creates crescent-shaped, dune-like features composed of clay pellets. These clay-sand dunes are termed lunettes. Deflation of clay aggregates from desiccated playa-floor deposits has been recognized as both a mechanism for playa maintenance and a source of detritus for the associated lunettes (Sabin and Holliday, 1995).

Playas and lunettes in the modern Big Lost River sinks and near TAN contain two types of clay aggregates, which are in the sand to small pebble grain size (clay sand grains) (Levish et al., 1999; Mark and Thackray, this volume). The most conspicuous type consists of angular, light colored clay intraclasts of granule to small pebble size that probably originated as deflation chips (Fig. 9B). They occur in sandy as well as clayey beds. The second type is more typical of lunette clay sand reported elsewhere, and consists of subrounded, sorted, sand-sized grains made of aggregates of clay (Fig. 9A). Both types of aggregates are present as clasts throughout the upper interbed (33.5–57.9 m; 110–190 ft) section of core 2-2A (Fig. 7).

Much of the upper interbed of core 2-2A (34.5–57.9 m, 113–190 ft) consists of pedogenically modified silty clay (Figs. 7 and 10). We interpret this sediment to be loess that was reworked in playa-lunette systems and underwent varying degrees of pedogenic modification. Such a grain-size distribution, dominated by clay but with significant percentages of silt and sand, is typical of loess deposits on this part of the Snake River Plain (Forman et al., 1993). Textural transitions, in which granule to sand-sized clay clasts of primary eolian detritus grade into larger clay aggregates that consist of pedogenically homogenized clay with ped structures (Fig. 11), are common. Silt and fine sand are common constituents in the primary clay sand

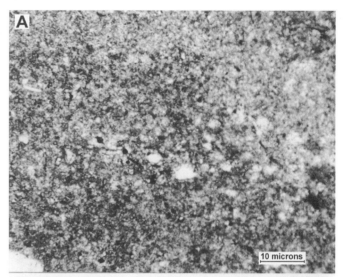

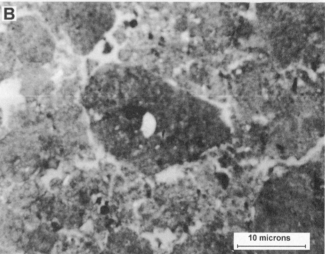

Figure 10. Photomicrographs of (A) primary, massive silty claystone (sample 2-2A-EB-5, Fig. 7), and (B) ped structure in A horizon of paleosol from upper sedimentary interbed (sample 2-2A-EB-15, Fig. 7).

deposits (Fig. 10A). In some paleosol horizons, these coarser constituents have concentrated along ped boundaries, forming siltans (cutans defined by silt).

Ostracodes of the species *Candona rawsoni* occur at the ~44.2 m (145 ft) level of core 2-2A. On the basis of amino-acid racemization, these fossils are thought to be ca. 140 ka and correlative with the Little Valley unit of Lake Bonneville sequence (Gianniny et al., 1999; this volume). This ostracode species occurs in both permanent and ephemeral freshwater lakes and ponds that have seasonal salinity variations (Forester, 1987). This lake may have formed as an ephemeral lake during the Bull Lake glaciation, even though the sandy sediments that host the ostracodes are sandwiched by soil-modified playa deposits (Fig. 7).

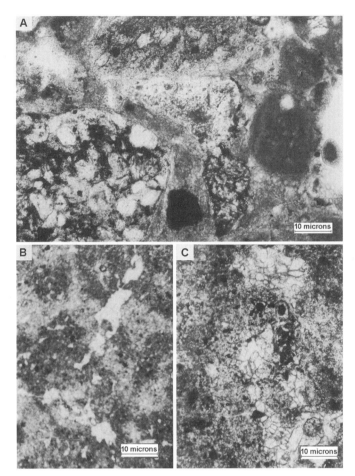

Figure 11. Photomicrographs of calcium carbonate textures from paleosol at base of upper sedimentary interbed (sample 2-2A-EB-1, Fig. 7). A: Carbonate micrite precipitate enclosing pebble and sand clasts. B: Mottled texture of displacive micrite precipitate. Darker areas consist of mix of siliciclastic clay and micrite. Bright areas are sparry calcite cement filling root traces and other primary soil voids and concentrations of clay.

Sand and silt lithofacies (fluvial-eolian)

Sandy beds occur as two basic types. The first is well-sorted, massive to thick-bedded, fine to medium sand, interpreted as eolian dunes. Two iron-stained, relatively thick, seemingly massive sand interbeds are present in core 2-2A, at 222.6 m (730 ft) (Fig. 5) and between 112.8 and 118.9 m (370–390 ft) (Fig. 6). In addition, thin beds of fine- to medium-grained eolian sand are common in many of the nonlacustrine interbeds. The sand probably originated in fluvial channels or playas, and has been reworked locally from these sources by wind. Fine sand is common in some terrace soils proximal to the Big Lost Trough today.

The second type of sandy lithofacies is poorly sorted fine-grained to granule sand, containing small-pebble-sized clay chips and silt, and is present throughout the paleosol-dominated part of the upper interbed (30–60 m, 100–200 ft) (Fig. 7). These poorly sorted sands are interpreted as fluvially derived sediment with varying amounts of eolian reworking and deposition (fluveolian, sensu Fryberger, 1990).

INTERFLOW RUBBLE ZONES

A common feature in many of the lava flow rubble zones is the presence of very fine grained coatings and infillings of calcareous clay (Figs. 5 and 6). These features are mostly not calcium carbonate precipitates, although in some cases a mix of detrital clay and caliche occurs (Fig. 5, 243 m, 796 ft). Instead, the coated, laminated character of this clay argues for it being washed in or illuviated by drip water into void spaces. In this interpretation, the coatings and infillings would accumulate in the vadose zone. Another possibility is that they were deposited by groundwaters in the saturated zone. However, such coatings now occur above the water table. It is envisioned that dust accumulating on mostly barren lava flow surfaces is washed down into near-surface void spaces to form these coatings and infillings.

CARBON AND OXYGEN ISOTOPES

Analyses of stable isotopes of carbon and oxygen in carbonate from bulk sediment samples of loess, paleosol horizons, lacustrine silt and clay intervals, and shell fragments of gastropods and ostracodes in lacustrine beds were made to assess whether a paleoclimatic signal was obtainable from these strata. The isotopic identification of pedogenic carbonate was also a goal of this analysis, because these strata contain abundant, largely calcareous, loess. Three groupings of the data are recognized (Fig. 12; Table 2): (1) calcareous silt and clay from lacustrine strata, (2) pedogenic carbonate from paleosols, and (3) shells of calcium carbonate from lacustrine settings. The stratigraphic occurrence of fossil shell fragments in lacustrine beds was not sufficient to determine whether the isotopic variation that exists in the shell samples contains a paleoclimatic-signal.

One important factor when analyzing carbonate in these deposits is the large flux of calcareous loess associated with the ends of glacial maxima (Forman et al., 1993). Much of the loess originates from fluvial sediment eroded from Paleozoic carbonate strata of east-central Idaho (Fig. 1). This calcareous loess composes ~10–30 wt% of the finer textured soils exposed on the plain today (Scott, 1982; Ostenaa et al., 1999a). Thus, the question that carbon and oxygen isotopes can help to answer is how much pedogenic overprint of the loess has occurred in the various types of deposits and paleosols.

Relatively unaltered loess was analyzed in order to establish a baseline isotopic composition. The best candidates for nonpedogenically altered loess were thought to be the very calcareous, thinly bedded to laminated silt layers in the lacustrine intervals. These strata have a $\delta^{13}C$ (Peedee belemnite, PDB) near 0.0‰, and thus are thought to retain most of their original

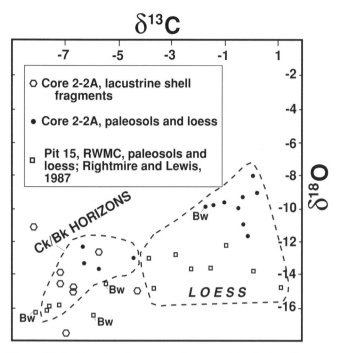

Figure 12. Scatter plot of carbon and oxygen isotopic data from paleosols, lacustrine beds, and shell fragments from upper half of core 2-2A. Symbol Bw refers to B horizons of paleosols. See Table 2 for specific data. Data from Rightmire and Lewis (1987) are plotted for reference.

marine carbonate carbon signature. Recrystallization of fine-grained marine limestone detritus during pedogenesis in Quaternary alluvial-soil systems would impart, to varying degrees, a pedogenic signal to carbon and oxygen isotopic values. Marine limestones, by definition, are close to zero on the carbon and oxygen isotopic scale (PDB) and have not varied widely during the Phanerozoic. Greater degrees of pedogenic modification and less loess deposition would produce a signal closer to the standard C_3 vegetation value for pedogenic carbonate (between $-9‰$ and $-12‰$; Cerling and Quade, 1993). This assumes that at the latitude of southern Idaho there would be little contribution of C_4 vegetation, which is reasonable given the modern distribution of C_3 and C_4 vegetation in North America (Terri and Stowe, 1976).

Two samples from the diamictite lake and one sample of lacustrine clay from the upper sedimentary interbed have carbon isotopic values are very close to zero (PDB standard). However, their oxygen isotopic values range from $-9‰$ to $-12‰$, considerably more negative than marine carbonates. Oxygen exchange in calcium carbonate minerals occurs much more readily than does carbon exchange (Anderson and Arthur, 1983). Thus, these $\delta^{18}O$ values are interpreted as reflecting oxygen exchange in the calcareous detritus with oxygen from the terrestrial environment, which has much more negative values. This exchange probably occurred during weathering and erosion of carbonate rock units prior to loess deposition.

Carbon and oxygen isotopic values from paleosol horizons that lacked visible and microscopic pedogenic carbonate features cluster close to the loess values (Bw samples in Fig. 12). The $\delta^{13}C$ is slightly more negative than the loess. This is interpreted as reflecting increased exchange of carbon and oxygen between loess grains and soil-derived carbon and oxygen with greater degrees of pedogenic modification. This exchange probably does not include pedogenic crystallization of carbonate, because no such features were observed megascopically or microscopically.

The gap in isotopic values between paleosol horizons with pedogenic carbonate and paleosol horizons without pedogenic carbonate (Bw samples) is thought to reflect the addition of pedogenic carbonate that crystallized in equilibrium with soil-derived CO_2 and O_2 as well as destruction (dissolution) of calcareous loess. This gap is seen in both the scatter plot of carbon versus oxygen (Fig. 12) and in the depth profile of a stage III calcareous paleosol at the base of the upper sedimentary interval (Fig. 13).

Rightmire and Lewis (1987) documented a similar spread of carbon and oxygen isotopic values from soils and near-surface paleosols in trenches near the Radioactive Waste Management Complex near the southwestern border of the INEEL. Their data are plotted in Figure 12 for reference and, by comparison with data collected here, lack the gap between the soil and loess signature. Rightmire and Lewis (1987) interpreted the spread in isotopic data to be due to glacial-interglacial climate change. It is unlikely that the carbon isotopic composition of soils would change so radically during climate swings. A more likely interpretation is that pedogenesis has overprinted an original marine limestone signature. Similar findings of pedogenic carbonate overprinting calcareous parent material of marine origin have been documented in Quaternary soils of central Texas (Nordt et al., 1998).

An alternative interpretation of the shift in carbon isotopes from near 0‰ to $-5‰$ and $-6‰$ would involve a vegetation shift between C_4 grasses, giving near zero values, to C_3, which would be much more negative (Monger et al., 1998). However, the scarcity of C_4 grasses at the latitude of the eastern Snake River Plain makes this interpretation unlikely (Terri and Stowe, 1976).

DISCUSSION

Synthesis of the paleoenvironmental record and chronology of the four sedimentary intervals in core 2-2A in the central part of the Big Lost Trough allows for a comparison of local conditions with the global climatic record (Fig. 14). From a depositional perspective, in which lake sediments are interpreted as indicators of wetter periods, the four interbeds, which span the time from 2 Ma to the past few hundred thousand years, indicate a general trend toward drier and probably cooler conditions. This scenario is based on the transition from a lacustrine-dominated system between 2 Ma and 1.5 Ma to a

TABLE 2. STABLE ISOTOPE RESULTS OF OXYGEN AND CARBON IN CARBONATE FROM BULK SAMPLES AND FOSSIL SHELLS OF GASTROPODS AND OSTRACODES FROM THE UPPER HALF OF CORE 2-2A

Sample depth		Type of sample	$\delta^{13}C_{PDB}$	$\delta^{18}O_{PDB}$	Comments
(feet)	(meters)				
121	36.9	Bulk sediment	−0.98	−9.75	Loess-paleosol
137	41.8	Bulk sediment	−0.41	−10.05	Loess-paleosol
187	57.0	Bulk sediment	−0.26	−9.37	Loess-paleosol
203	61.9	Bulk sediment	0.11	−8.11	Loess-paleosol
217	66.1	Bulk sediment	0.27	−9.16	Lacustrine clay
247	75.3	Silt clasts	−1.41	−9.89	Calcareous rip-up clasts
248	75.6	Bulk sediment	−1.69	−10.01	Bw paleosol horizon
249	75.9	Bulk sediment	−5.73	−13.77	Bk/Ck paleosol horizon
249.5	76.0	Bulk sediment	−4.37	−13.15	Bk/Ck paleosol horizon
253	77.1	Vesicular fill	−6.28	−13.35	Calcareous clay vesicle fill
262	79.9	Vesicular fill	−6.34	−12.37	Calcareous clay vesicle fill
856	261.0	Bulk sediment	−0.09	−11.74	Shell fragments
868	264.6	Bulk sediment	−0.22	−11.11	Shell fragments
1016	309.7	Shell separates	−8.20	−11.03	Shell fragments
1049	319.7	Shell separates	−5.78	−12.59	Shell fragments
1050-1	320.0	Shell separates	−7.20	−14.44	Shell fragments
1050-2	320.0	Shell separates	−7.21	−13.81	Shell fragments
1052-1	320.6	Shell separates	−6.74	−14.81	Shell fragments
1052-2	320.6	Shell separates	−6.75	−14.69	Shell fragments
1057	322.2	Shell separates	−4.34	−14.81	Shell fragments
1365	416.1	Shell separates	−7.02	−17.49	Shell fragments

Note: Analyses were done at the Illinois State Geological Survey, Isotope Geochemistry Laboratory. PDB is Peedee belemnite. Bk/Ck refers to calcareous soil horizons.

playa-, loess-, and soil-dominated system with periodic ephemeral lakes during the past few hundred thousand years.

The indication of a change across the Pliocene-Pleistocene boundary, from lakes that were ice free to lakes with seasonal ice, suggests a cooling trend, following the overall global Quaternary climate. This finding, if true, is in concert with the record from the North Pacific, which shows the interval from 2.0 to 1.6 Ma being a relatively warm interval, followed by cooling (Sancetta and Silvestri, 1986).

However, the pollen record from core 2-2A is not directly compatible with the interpretation of a simple cooling and drying trend during the Pleistocene (Thompson, 1991). In the Olduvai lake beds, the abundance of steppe vegetation types, i.e., *Artemisia* and *Chenopodiineae* (sagebrush, greasewood, and saltbrush), is much higher than in the lacustrine beds of the playa-loess facies in the upper sedimentary interbed (Thompson, 1991). Likewise, *Pinus* and *Picea* (pine and spruce) are much more abundant in the lacustrine deposits in the upper sedimentary interval than in the Olduvai lake. Thus the pollen record suggests that lacustrine deposition during the late Pleistocene is associated with cool and/or moist conditions, whereas latest Pliocene lacustrine deposition occurred during drier and/or warmer conditions.

We suggest that ephemeral late Pleistocene lakes were associated with glacial periods that were cool and relatively moist, within the Pleistocene trend toward aridity, whereas the late Pliocene lakes persisted during both glacial and interglacial periods, and were surrounded by steppe vegetation. The pollen record in the Olduvai lake may be dominated by vegetation species that existed during interglacial periods of longer duration.

In addition to climatic indicators, the evolution of the Snake River Plain hydrologic system had a first-order control on lakes of the Big Lost Trough. During the Pliocene and early Pleistocene the western Snake River Plain underwent extensive lacustrine deposition (Glenns Ferry Formation; Swirydczuk et al., 1979; Kimmel, 1982; Middleton et al., 1985; Lee et al., 1995; McDonald et al., 1996). The stratigraphy of the Glenns Ferry Formation demonstrates a 240 km westward regression of the eastern lake margin from Hagerman on the east to Froman Ferry on the west from 3.5 to 1.5 Ma (Repenning et al., 1995; Link et al., 2002). This westward lake-margin regression likely contributed to a lowering of the groundwater table upstream. By this reasoning, lakes in the eastern Snake River Plain, upstream from the Glenns Ferry lake, may have drained as Lake Idaho retreated to the west and became restricted in size.

CONCLUSIONS

1. The four, thick sedimentary interbeds in the upper half of core 2-2A include lacustrine-dominated intervals of late Pliocene and early Pleistocene age, and an upper, late Pleistocene sedimentary interbed that is dominated by playa, loess, and paleosol deposits with lesser amounts of lacustrine strata.

2. An age of 1.9–1.8 Ma (Olduvai reversal) is assigned to the lacustrine strata at 292.0–350.0 m (958–1148 ft) (Olduvai lake). Early Pleistocene lacustrine strata between 259.1 and

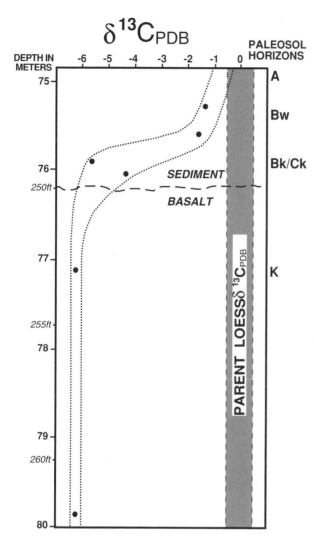

Figure 13. Depth profile of well-developed calcareous paleosol at base of upper sedimentary sequence showing carbon isotope values of various paleosol horizons. Vertical profile is interpreted to represent increased pedogenic modification with depth, producing more negative $\delta^{13}C$ values reflecting overprinting of marine signature in loess with pedogenic carbonate signature.

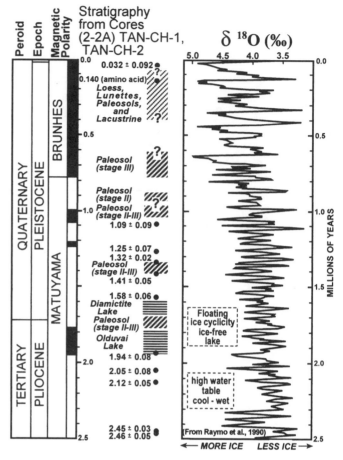

Figure 14. Composite stratigraphy from core 2-2A and age dates from cores (Test Area North, TAN) CH-1 and CH-2 compared to marine ^{18}O record (from Raymo et al., 1990). Time span interpreted for diamictite lake corresponds to approximately three glacial-interglacial cycles. Upper sedimentary interbed is entirely within time period of large-scale glacial-interglacial.

272.9 m (850–895 ft) (diamictite lake) depth contain prominent diamictite facies interpreted as the product of ice-rafted sedimentation and are estimated to span the interval from 1.6 to 1.5 Ma.

3. Strata in the upper sedimentary interbed (34.5–76.5 m, 113–251 ft) contain textures and structures indicative of deposition in aggrading playa systems with significant loess input. Eolian reworking of playa deposits in lunette-like dunes is indicated by abundant granule to coarse-sand-sized clay pellets in these deposits. Most of the deposits in the upper sedimentary interbed were affected by weak pedogenic modification.

4. Isotopic analysis of paleosols and loess indicates an original Paleozoic carbon isotopic signature in the loess; however, the oxygen has been reset. Weak pedogenesis of loess has only limited affects on carbon and oxygen isotopic values. However, in paleosols with pedogenic carbonate, carbon values approach levels expected for soils under C_3 vegetation.

ACKNOWLEDGMENTS

This work is funded by U.S. Department of Energy–Idaho Water Resources Research Institute (DOE-IWRRI) grant DE-FG07-96ID13420. This paper was reviewed internally by L. Mark, J. Blair, and S.R. Anderson, and externally by W.E. Scott, D.R. Lemon, S.L. Forman, J.K. Geslin, G.I. Smith, and J.W. Hillhouse. Access to the U.S. Geological Survey Core Library at the Idaho National Engineering and Environmental Laboratory was kindly provided by L. Davis.

REFERENCES CITED

Anderson, S.R., and Bowers, B., 1995, Stratigraphy of the unsaturated zone and uppermost part of the Snake River Plain aquifer at Test Area North, Idaho National Engineering Laboratory, Idaho: U.S. Geological Survey Water-Resources Investigations Report 95-4130, 47 p.

Anderson, T.F., and Arthur, M.A., 1983, Stable isotopes of oxygen and carbon and their application to sedimentologic and paleoenvironmental problems, *in* Arthur, M.A., Anderson, T.F., Kaplan, I.R., Veizer, J., and Land, L.S., eds., Stable isotopes in sedimentary geology: Tulsa, Oklahoma, Society of Economic Paleontologists and Mineralogists short course 10, p. 1–151.

Berggren, W.A., Hilgren, F.J., Langerers, C.G., Kent, D.V., Obradovich, J.D., Raffi, I., Raymo, M.E., and Shackleton, N.J., 1995a, Late Neogene chronology: New perspectives in high-resolution stratigraphy: Geological Society of America Bulletin., v. 107, p. 1272–1287.

Berggren, W.A., Kent, D.V., Swisher, C.C., and Aubry, M-P., 1995b, A revised Cenozoic geochronology and chronostratigraphy, *in* Berggren, W.A., et al., eds., Geochronology, time scales and global stratigraphic correlation: SEPM (Society for Sedimentary Geology), Special Volume 54, p. 129–212.

Birkeland, P.W., Machette, M.N., Haller, K.M., 1991, Soils as a tool for applied Quaternary geology: Utah Geological and Mineral Survey Miscellaneous Publication 91-3, 67 p.

Blair, J.J., and Link, P.K., 2000, Pliocene and Quaternary sedimentation and stratigraphy of the Big Lost Trough from coreholes at the Idaho National Engineering and Environmental Laboratory, Idaho: Evidence for a regional Pliocene lake during the Olduvai normal polarity subchron, *in* Robinson, L., ed., Proceedings of the 35th Symposium on Engineering Geology and Geotechnical Engineering: Pocatello, Idaho, Idaho State University, College of Engineering, p. 163–179.

Busby, C.J., and Ingersoll, R.V., 1995, Tectonics of sedimentary basins: Cambridge, Massachusetts, Blackwell Science, 579 p.

Cande, S.C., and Kent, D.V., 1992, A new geomagnetic polarity time scale for the late Cretaceous and Cenozoic: Journal of Geophysical Research, v. 97, no. B10, p. 13917–13951.

Carroll, A.R., and Bohacs, K.M., 1999, Stratigraphic classification of ancient lakes: Balancing tectonic and climatic controls: Geology, v. 27, no. 2, p. 99–102.

Cas, R.A.F., and Wright, J.V., 1988, Volcanic successions: Cambridge, England, Chapman and Hall, 528 p.

Cerling, T.E., and Quade, J., 1993, Stable carbon and oxygen isotopes in soil carbonates, *in* Swart, P.K., Lohmann, K.C., McKenzie, J., Savin, S., eds., Climate change in continental isotopic records: American Geophysical Union Geophysical Monograph 78, p. 217–231.

Forester, R.M., 1987, Late Quaternary paleoclimate records from lacustrine ostracodes, *in* Ruddiman, W.F., and Wright, H.E. Jr., eds., North American and adjacent oceans during the last deglaciation: Boulder, Colorado, Geological Society of America, The Geology of North America, v. K-3, p. 261–276.

Forester, R.M., 1991, Pliocene-climate history of the western United States derived from lacustrine ostracodes: Quaternary Science Review, v. 10, p. 133–146.

Forman, S.L., Smith, R.P., Hackett, W.R., Tullis, J.A., McDaniel, P.A., 1993, Timing of Late Quaternary glaciations in the western United States based on the age of loess on the Eastern Snake River Plain, Idaho: Quaternary Research, v. 40, p. 30–37.

Fryberger, S.G., 1990, Great Sand Dunes depositional system: An overview, *in* Fryberger, S.G., Krystinik, L.F., and Schenk, C.J., eds., Modern and ancient eolian deposits: Petroleum exploration and production: Denver, Colorado, Rocky Mountain Section, Society for Sedimentary Geology (SEPM), p. 1.1–1.9.

Geslin, J.K., Link, P.K., and Fanning, C.M., 1999, High-precision provenance determination using detrital-zircon ages and petrography of Quaternary sands on the eastern Snake River Plain, Idaho: Geology, v. 27, no. 4, p. 295–298.

Gianniny, G.L, Geslin, J.K., Riesterer, J.W., Link, P.K., and Thackray, G.D., 1997, Quaternary surficial sediments near Test Area North (TAN), northeastern Snake River Plain: An actualistic guide to aquifer characterization, *in* Sharma, S., and Hardcastle, J.H., eds., Proceedings of the 32nd Symposium on Engineering Geology and Geotechnical Engineering: Boise, Idaho, p. 29–44.

Gianniny, G.L., Thackray, G.D., and Kaufman, D.S., 1999, Late Quaternary highstands of Lake Terreton, Idaho: Geological Society of America Abstracts with Programs, v. 31, no. 4, p. A-14.

Gilbert, G.K., 1885, The topographic features of lake shores: U.S. Geological Survey 5th Annual Report, p. 69–123.

Gilbert, G.K., 1895, Lake basins created by wind erosion: Journal of Geology, v. 3, p. 47–49.

Gile, L.H., Peterson, F.F., and Grossman, R.B., 1966, Morphology and genetic sequence of carbonate accumulation in desert soils: Soil Science, v. 101, p. 347–360.

Hackett, W.R., and Smith, R.P., 1992, Quaternary volcanism, tectonics, and sedimentation in the Idaho National Engineering Laboratory area, *in* Wilson, J.R., ed., Field guide to geologic excursions in Utah and adjacent areas of Nevada, Idaho, and Wyoming: Utah Geological Survey Miscellaneous Publication 92-3, p. 1–18.

Holliday, V.T., Hovorka, S.D., Gustavson, T.C., 1996, Lithostratigraphy and geochronology of fills in small playa basins on the southern High Plains, United States: Geological Society of America Bulletin, v. 108, p. 953–965.

Kauffman, M.E. and Geslin, J.K., 1998, Impermeable horizons in Neogene sediments on the eastern Snake River Plain, Idaho: Calcretes in fluvial and eolian deposits and fine-grained lacustrine and playa deposits: Geological Society of America Abstracts with Programs, v. 30, no. 6, p. 12.

Kimmel, P.G., 1982, Stratigraphy, age, and tectonic setting of the Miocene-Pliocene lacustrine sediments of the western Snake River Plain, Oregon and Idaho, *in* Bonnichsen, B., and Breckenridge, R.M., eds., Cenozoic geology of Idaho: Idaho Geological Survey Bulletin 26, p. 559–578.

Kuntz, M.A., Skipp, B., Lanphere, M.A., Scott, W.B., Pierce, K.L., Dalrymple, G.B., Champion, D.E., Embree, G.F., Page, W.R., Morgan, L.A., Smith, R.P., Hackett, W.R., and Rodgers, D.W., 1994, Geological map of the Idaho National Engineering Laboratory and adjoining areas, eastern Idaho: U.S. Geological Survey Miscellaneous Investigations Series Map I-2330, scale 1:100 000.

Kuntz, M.A., Geslin, J.K., Rodgers, D.W., Hodges, M.K.V., Mark, L.E., and Link, P.K., 2002, Geologic map of the northern part of the Idaho National Engineering and Environmental Laboratory Around Test Area North and the Big Lost River Sinks, eastern Idaho: Idaho Geological Survey Technical Report, scale 1:100 000, (in press).

Lanphere, M.A., Kuntz, M.A., and Champion, D.E., 1994, Petrology, age and paleomagnetics of basaltic lava flows in coreholes at Test Area North (TAN), Idaho National Engineering Laboratory: U.S. Geological Survey Open-File Report 94-686, p. 490.

Lee, D.E., Link, P.K., and Ore, H.T., 1995, Characterization of the Glenns Ferry Formation in the Fossil Gulch area, Hagerman Fossil Beds National Monument: Pocatello, Idaho, Idaho Museum of Natural History, Geology Report 1, Completion Report for National Park Service Subagreement 3 to Contract CA-9000-0-0013, 59 p.

Levish, D.R., Ostenaa, D.A., Klinger, R.E., Simpson, D.T., 1999, Holocene lunettes on the Birch Creek playa, Butte County, Idaho: Implications for a post-glacial Lake Terreton: Geological Society of America Abstracts with Programs, v. 31, p. A-21.

Link, P.K., Fanning, C.M., and Godfrey, A.E., 2002, Detrital zircon evidence for Pliocene southward drainage of the Snake River Plain, Idaho, *in* Bonnichsen, B., White, C.M., and McCurry, M., eds., Tectonic and Magmatic

Evolution of the Snake River Plain Volcanic Province: Idaho Geological Survey Bulletin 30, p. 30.

Machette, M.N., 1982, Quaternary and Pliocene faults in the La Jencia and southern parts of the Albuquerque-Belen basins, New Mexico: Evidence of fault history from fault-scarp morphology and Quaternary geology, *in* Grambling, J.A., and Wells, S.G., eds., Albuquerque Country II: New Mexico Geological Society Guidebook, 33rd Field Conference, p. 161–169.

Machette, M.N., 1985, Calcic soils of the southwestern United States, *in* Weide, D.L., ed., Soils and Quaternary geology of the southwestern United States: Geological Society of America Special Paper 203, p. 1–21.

Marion, G.M., Schlesinger, W.H., and Fonteyn, P.J., 1985, CALDEP: A regional model for soil $CaCO_3$ (caliche) deposition in southwestern deserts: Soil Science, v. 139, p. 468–481.

McDonald, G.H., Link, P.K., and Lee, D.E., 1996, An overview of the geology and paleontology of the Pliocene Glenns Ferry Formation, Hagerman Fossil Beds National Monument: Northwest Geology, v. 26, p. 16–45.

McFadden, L.D., 1988, Climatic influences on rates and processes of soil development in Quaternary deposits of southern California, *in* Reinhardt, J., and Sigleo, W.R., eds., Paleosols and weathering through geologic time: Techniques and applications: Geological Society of America Special Paper 216, p. 153–177.

McQuarrie, N., and Rodgers, D.W., 1998, Subsidence of a volcanic basin by flexure and lower crustal flow: The eastern Snake River Plain, Idaho: Tectonics, v. 17, p. 203–220.

Middleton, L.T., Porter, M.L., and Kimmel, P.G., 1985, Depositional settings of the Chalk Hills and Glenns Ferry Formations west of Bruneau, Idaho, *in* Flores, R.M., and Kaplan, S.S., eds., Cenozoic paleogeography of the West-Central United States: Denver, Colorado, Rocky Mountain Section of the Society of Economic Paleontologists and Mineralogists, p. 37–53.

Monger, H.C., Cole, D.R., Gish, J.W., and Giordano, T.H., 1998, Stable carbon and oxygen isotopes in Quaternary soil carbonates as indicators of ecogeomorphic changes in northern Chihuahuan Desert, USA: Geoderma, v. 82, p. 137–172.

Nordt, L.C., Hallmark, C.T., Wilding, L.P., and Boutton, T.W., 1998, Quantifying pedogenic carbonate accumulations using stable carbon isotopes: Geoderma, v. 82, p. 115–136.

Ostenaa, D.A., Levish, D.R., Klinger, R.E., and O'Connell, D.R.H., 1999a, Phase 2 paleohydrologic and geomorphic studies for the assessment of flood risk for the Idaho National Engineering and Environmental Laboratory, Idaho: Denver, Colorado, U.S. Bureau of Reclamation, Geophysics, Paleohydrology and Seismotectonics Group, Report 99-7, 112 p. plus appendices.

Ostenaa, D.A., Levish, D.R., and Klinger, R.E., 1999b, Holocene paleofloods on the Big Lost River, Idaho National Engineering and Environmental Laboratory, Idaho: Limits on extreme flood magnitude for long time periods: Geological Society of America Abstracts with Programs, v. 31, p. A51.

Osterkamp, W.R., and Wood, W.W., 1987, Playa-lake basins on the southern High Plains of Texas and New Mexico. 1. Hydrologic, geomorphic, and geologic evidence for their development: Geological Society of America Bulletin, v. 99, p. 215–223.

Pierce, K.L., and Morgan, L.A., 1992, The track of the Yellowstone hot spot: Volcanism, faulting, and uplift, *in* Link, P. K., Kuntz, M.A., and Platt, L.B., eds., Regional geology of eastern Idaho and western Wyoming: Geological Society of America Memoir 179, p. 1–53.

Raymo, M.E., Ruddiman, W.F., Shackleton, N.J., Oppo, D.W., 1990, Evolution of Atlantic-Pacific ^{13}C gradients over the last 2.5 million years: Earth and Planetary Science Letters, v. 97, p. 353–368.

Repenning, C.A., Weasma, T.R., Scott, G.R., 1995, The Early Pleistocene (Latest Blancan-Earliest Irvingtonian) Froman Ferry Fauna and history of the Glenns Ferry Formation, Southwestern Idaho: U.S. Geological Survey Bulletin 2105, 86 p.

Retallack, G.J., 1988, Field recognition of paleosols, *in* Reinhardt, J., and Sigleo, W.R., eds., Paleosols and weathering through geologic time: Techniques and applications: Geological Society of America Special Paper 216, p. 1–20.

Rightmire, C.T., and Lewis, B.C., 1987, Hydrogeology and geochemistry of the unsaturated zone, Radioactive Waste Management Complex, Idaho National Engineering Laboratory, Idaho, U.S. Geological Survey Water Resources Investigations Report 87-4198, 89 p.

Sabin, T.J., and Holliday, V.T., 1995, Morphometric and spatial relationships of playas and lunettes on the Southern High Plains: Association of American Geographers Annals, v. 85, p. 286–305.

Sancetta, C., and Silvestri, S., 1986, Pliocene-Pleistocene evolution of the North Pacific ocean-atmosphere system, interpreted from fossil diatoms: Paleoceanography, v. 1, p. 163–180.

Scott, W.E., 1982, Surficial geologic map of the eastern Snake River Plain and adjacent areas, 111°–115° W, Idaho and Wyoming; U.S. Geological Survey Miscellaneous Investigations Series Map I-1372, scale 1:250 000, 2 plates.

Simpson, D.T., Kolbe, T.E., Ostenaa, D.A., Levish, D.R., Klinger, R.E., 1999, Lower Big Lost River chronosequence: Implications for glacial outburst flooding: Geological Society of America Abstracts with Programs, v. 31, p. A56.

Swirydczuk, K., Wilkinson, B.H., and Smith, G.R., 1979, The Pliocene Glenns Ferry oolite: Lake margin carbonate deposition in the southwestern Snake River Plain: Journal of Sedimentary Petrology, v. 49, p. 995–1004.

Terri, J.A., and Stowe, L.G., 1976, Climatic patterns and distribution of C_4 grasses in North America: Oecologia, v. 23, p. 1–12.

Thompson, R.S., 1991, Pliocene environments and climates in the western United States: Quaternary Science Review, v. 10, p. 115–132.

MANUSCRIPT ACCEPTED BY THE SOCIETY NOVEMBER 2, 2000

Geological Society of America
Special Paper 353
2002

Subsurface volcanology at Test Area North and controls on groundwater flow

Dennis J. Geist*
Rachel A. Ellisor
Elisa N. Sims
Department of Geological Sciences, University of Idaho, Moscow, Idaho 83844, USA
Scott S. Hughes
Department of Geology, Idaho State University, Pocatello, Idaho 83209, USA

ABSTRACT

Detailed core logging, paleomagnetic measurements, and an internally consistent geochemical database are used to develop a picture of the subsurface geology above the Q-R interbed at Test Area North in the northern part of the Snake River Plain. Our argon dating and minor changes in paleomagnetic inclination match perfectly those of Lanphere et al., indicating that the 120 m sequence erupted in a short time interval (<1 k.y.) over the entire area, ca. 1.13 ± 0.24 Ma. The flows are inflated pahoehoe lavas, and their brecciated and weathered flow tops provide preferential flow paths that dominate groundwater flow. The compositional variation of the basalts is relatively large, the compositional evolution is markedly consistent, and internal differentiation of individual cooling units is only locally important. Thus, careful logging coupled with chemical analysis is the ideal correlation tool. Our cross sections are substantially different from those previously constructed because we can demonstrate that most of the flows are much smaller than previously supposed and disprove certain correlations. At least nine flow groups are identifiable and only two lava flows extend across the area. Several of the flows have widths <100 m in the line of section. The most important contact for groundwater flow pinches out to the northwest and thins dramatically to the east, the direction that the contaminant plume is oriented. Anisotropy of magnetic susceptibility measurements were made to permit estimates of lava flow directions. These data cluster about N70W and N25E, which we believe are the directions to two vents that erupted these lava flows. Our interpretation is that this terrain was much different than that currently exposed at the surface of the Snake River Plain, and involved many long, narrow flows emplaced on an irregular surface at a very rapid accumulation rate.

INTRODUCTION

Bedrock constituted of lava flows has extraordinarily heterogeneous hydraulic conductivity, ranging over a million-fold in different parts of single flow units of Snake River Plain basalts (e.g., Welhan et al., 2001). The most significant variations of hydraulic conductivity in lava flows are related to structures that develop as the flows are either actively moving or those that form shortly after stagnation and solidification. Although weathering and sediment deposition amplify the heterogeneity

---

*E-mail: dgeist@uidaho.edu

Geist, D.J., Ellisor, R.A., Sims, E.N., and Hughes, S.S., 2002, Subsurface volcanology at Test Area North and controls on groundwater flow, *in* Link, P.K., and Mink, L.L., eds., Geology, Hydrogeology, and Environmental Remediation: Idaho National Engineering and Environmental Laboratory, Eastern Snake River Plain, Idaho: Boulder, Colorado, Geological Society of America Special Paper 353, p. 45–59.

of hydraulic conductivity, those processes merely augment the primary features imparted by lava flow emplacement.

Groundwater flow at the Test Area North (TAN) site (Fig. 1) of the Idaho National Engineering and Environmental Laboratory is counter to that predicted by simple hydrologic models. Contours of the water table are oriented nearly east-west in the TAN vicinity (Sorenson et al., 1996), indicating that groundwater flow should be southward. However, the direction of flow, as indicated by the delineation of a contaminant plume, is east for ~1 km from the site, before it curves to the south. One hypothesis is that the direction of flow is not solely controlled by the potential head, but also by orientation of heterogeneity in hydraulic conductivity. For example, one possibility is that a strongly permeable, narrow lava flow is oriented east-west through the area, and channelizes flow obliquely to the potentiometric surface.

In order to evaluate this hypothesis, we have undertaken a detailed study of the lava flows beneath the TAN site that were sampled by core drilling. We have used geochemical, petrological, and geophysical tools to decipher the subsurface geology at TAN, augmented by our experience in lava flow emplacement at active basaltic volcanoes on ocean islands. Our study adds to those already performed by U.S. Geological Survey geologists on drill core that existed prior to 1995 (Lanphere et al., 1994; Anderson and Bowers, 1995; Knobel et al., 1995). Our work involves detailed logging, focusing on structures related to the emplacement of the flow, geochemical analysis, age determinations, and determination of the paleomagnetic decli-

nation. In addition, we performed studies of the anisotropy of magnetic susceptibility of the flows, in the hope of ascertaining flow directions, and applied statistical tools to the data sets. The ultimate goal is to use information from the cores to evaluate the continuity of individual flows and thus infer controls on groundwater flow.

Because our interest is in the controls of lava flow structure on the travel of the contaminant plume, we have focused on the geology below the lavas of Circular Butte (Casper, 1999), because the water table is at ~62 m (Sorenson et al., 1996). We have also only considered lavas above the so-called Q-R sedimentary interbed (~130 m depth; Anderson and Bowers, 1995), which is an effective aquitard that restricts vertical transfer of groundwater at the site (Sorenson et al., 1996).

FLOW STRUCTURE AND EMPLACEMENT DURATION

Welhan et al. (2001) proposed a hierarchy of terms for geologic units within a section of basaltic lava flows; i.e., "lobe" for the individual units that inflate and cool, "unit" for lobes that erupted from the same vent and advanced together, and "group" for a series of units erupted from the same volcano. This terminology is difficult to apply with confidence to the one-dimensional sampling that drill cores provide. Individual lobes are straightforward to identify in core (when recovery is complete), but it is impossible to relate these lobes to units in the absence of two-dimensional exposure. On the basis of our interpretation of stratigraphic correlation and flow direction, we believe the entire section between the Q-R interbed and the Circular Butte flows to be one or possibly two flow groups, as defined here. In this work, we refer to "lobes" whenever we observe flow boundaries, and "units" when sets of lobes have statistically distinctive compositions. Unfortunately, the paleomagnetic measurements reported here and by Lanphere et al. (1994) do not define these designations beyond that of the "flow group."

Detailed logging of more than 400 m of lava drill cores from six holes (Fig. 2) has allowed us to identify the flow unit boundaries and the internal vertical divisions within those flows based on the variations of crystal and vesicle populations and groundmass textural variations. With these data a stratigraphic section of buried Snake River Plain basalts was constructed for each of the five core holes described (Ellisor et al., 1998). All of the flows are buried beneath between 10–60 m of sediment, and none is exposed at the surface near the TAN site.

Flow structures and fluid transport

Primary lava flow structures form large-scale preferential pathways through which water and contaminants flow (for a thorough discussion, see Welhan et al., 2001). The most permeable constituents of the lavas are likely to include open spaces within rubbly upper crusts and through the scoriaceous

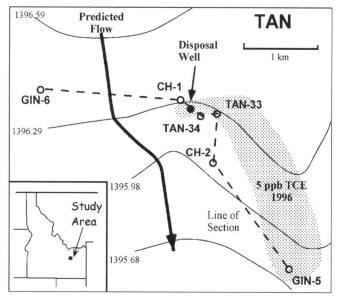

Figure 1. Study area is located in northern part of eastern Snake River Plain. Map shows locations of six holes we are studying and line along which cross sections were made. Approximate limits of contaminant plume are shown as shaded area. Contours show elevation of water table (in meters) (from Sorenson et al., 1996); thick line perpendicular to contours is predicted direction of flow. TCE is trichloroethylene.

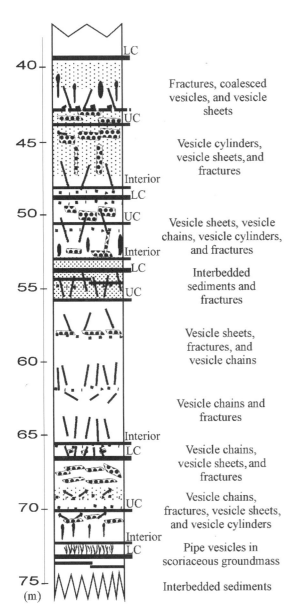

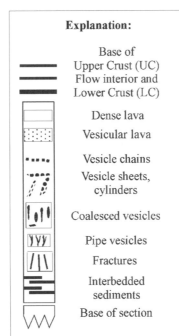

Figure 2. Representative stratigraphic section of logged sections from (Test Area North) TAN CH-2, meant to illustrate geometric relationships of flow features discussed in text.

lower crusts of some flows (Fig. 2). Vesicular, scoriaceous, and rubbly, broken zones of each of the five drill cores described make up 25%–39% of all core described. This provides a substantial proportion of the total core thickness through which fluids could migrate relatively rapidly. The finely porous texture of the remaining dense and finely vesicular zones provides very limited porosity and permeability. The vertically oriented vesicles or vesicle pipes that terminate in larger horizontal cavities (vesicle sheets) at flow breaks or internal boundaries (e.g., at 48–54 m; Fig. 2) may permit fluids to pool in the horizontal vesicle sheets then trickle further down through the lava flow via a vertical vesicle pipe or fracture.

Near-vent pahoehoe lavas tend to be strongly vesicular (>30% by volume), to occur in thin ("shelly"; <20 cm thick) lobes, to retain pyroclastic textures, and to be interlayered with

spatter and basaltic tephra (bombs, Pele's tears, Pele's hair, and ash) (Kuntz et al., 1992). None of these features are apparent in any of the investigated core from the TAN site. The flows therefore are interpreted to be in a medial to distal facies with respect to their vents. Thus, it is unlikely that there are any dikes (which form relatively impermeable barriers to horizontal flow) or wide vertical fissures in the TAN area.

Inflation mechanism of Snake River Plain lava flows

All of the identified flows in the drill core are pahoehoe flows: rubbly aa flow tops are not observed, and the characteristic vesicle shapes found in aa are absent. The 22 flow units are identified as inflated pahoehoe flows and are 3–30 m thick. Inflation of basalt flows occurs during emplacement when flows increase in thickness from initial 10–100 cm thicknesses to tens of meters by intrusion of fresh magma into the core of the flow (e.g., Hon et al., 1994; Self et al., 1997).

Most inflated flow lobes have three identifiable zones (Table 1; Fig. 2); where the upper crust and dense interiors of the flow are each ~40%–60% of the total flow thickness, the remaining 10% of the flow thickness is the lower crust (as defined by Self et al., 1998). Upper crusts of the flows observed in the Snake River Plain core were identified on the basis of coarse vesicles (3–7 mm) in a glassy groundmass. The groundmass of many of the upper crusts is discolored, oxidized to purple or brown, indicating incipient weathering. Broken or rubbly segments of the core and sediment (presumably accumulated in joints connected to the surface) are common constituents of the upper crust. Vesicles decrease in number and increase in size downward through the upper crust.

Flow interiors have a dense groundmass and few vesicles (generally ≤3 mm and <1% by volume) except in concentrated zones such as vesicle sheets (subhorizontal swarms of vesicles, e.g., Self et al., 1998), vesicle cylinders (Goff, 1996), and vesicle chains (aligned individual vesicles, oriented from horizontal to vertical). Vesicles in the flow interior are commonly coalesced, some forming large cavities (to 3 cm). Coalesced vesicles have teardrop or pear shapes and are commonly surrounded by smaller vesicles that are hosted in a denser groundmass than that found farther from the coalesced vesicle. Flow interiors are commonly crosscut by both steeply and shallowly dipping fractures, which obviously form after partial solidification and stagnation, following emplacement. Several cases exist in which fractures crosscut otherwise uniform vesicle sheets, indicating that fracturing occurred after the latest stages

of vesicle segregation and that the vesicle-rich zones do not significantly affect the fracture pattern. Rare vesicle-lined fractures suggest that some fractures developed while the lava interior was still partly molten.

The basal crust is characterized by glassy groundmass and is more vesicular than flow interiors, but less vesicular than upper crusts. Basal crusts are often discolored and pipe vesicles occasionally occur (Self et al., 1997). The occurrence of pipe vesicles indicates emplacement on a very shallow (<2°) slope (Walker, 1987), consistent with our cross section. Rubbly breccia only rarely exists at the bases of the flow units.

Estimation of emplacement duration

The solidification of inflated lavas was described empirically by Hon et al. (1994), where crustal cooling rates are determined using the thickness of the upper crust (C, in meters) and the square-root of time (t, in hours):

$$C = 0.0779\ t^{0.5} \qquad (1)$$

The coefficient for equation 1 was determined by a least-squares fit of thermal and crustal thickness data for sheet flows on Kilauea and the Makaopuhi lava lake (Hon et al., 1994), so is specific to crustal growth rates under those conditions of emplacement. The application of the time parameter in equation 1 to inflated Snake River Plain flows is reasonable as a first-order estimate, because the physical emplacement process of inflation dictates the square root of time with crustal thickness relationship. The consistent growth rate of upper crust along the lengths of the Kilauean flows (Hon et al., 1994) demonstrates that solidification of the crust occurs by conduction. Likewise, the coefficient in equation 1 is specific to Kilauean flows, but it should be similar for most tholeiitic basalts, the physical properties of which are not considerably different. The rate of thermal conduction (which governs the coefficient of equation 1) is governed by thermal diffusivity of the magma, specific heat of the magma, latent heat of crystallization, and the temperature gradient. Of these, the first three factors should not differ significantly between Hawaiian and Snake River Plain lavas. Differences in the thermal gradient imposed by the molten core of the flow and its exterior may differ significantly, owing to the compositional controls on the various liquidus and solidus values of lavas.

The initial melt and solidification temperatures of the Kilauean flows are ~1142 °C and 1070 °C (Hon et al., 1994). Estimation of the temperature of emplacement of the Snake River Plain basalts, based on thermodynamic calculations using the MELTS algorithm (Ghiorso and Sack, 1995), are 1230 °C for purely molten magma and 1110 °C when the flow is 60% crystalline (the point at which the flow will become essentially rigid). Thus, the thermal gradients are comparable, and the empirical parameters determined for Hawaiian tholeiite should be applicable to Snake River Plain basalts. Flow emplacement

TABLE 1. REPRESENTATIVE DIMENSIONS OF COOLING UNITS

Core	Flow#	Upper crust (m)	Flow interior (m)	Lower crust (m)	Total thickness (m)	Time of upper crust formation (h)*
GIN-6	2	1.2	3.0	0.3	4.5	242
TAN-CH-1	2	10.9	8.5	2.4	21.8	19611
	3	7.9	4.5	1.5	13.9	10229
	5	5.8	3.6	0.3	9.7	5463
	6	1.2	7.3	2.4	10.9	242
	7	3.9	4.5	3.0	11.5	2557
GIN-5	2	1.2	5.8	1.5	8.5	242
	3	6.4	4.8	0.6	11.8	6673
	4	9.1	3.9	4.2	17.3	13619
	5	4.2	3.3	0.6	8.2	2966
TAN-CH-2	1	1.8	2.1	0.9	4.8	545
	2	0.6	3.9	0.6	5.2	61
	3	0.9	1.2	0.8	2.9	136
	4	3.8	3.6	0.6	8.0	2364
	5	1.5	4.5	0.9	7.0	378
	6	1.2	2.1	0.9	4.2	242
	7	2.1	9.7	2.4	14.2	741
	8	2.0	2.2	0.9	5.2	689
TAN-34	2	3.9	16.4	2.4	22.7	2557

*Cooling times calculated according to the formula of Hon et al. (1994)

times range from 61 h (2.5 days) to 19 600 h (2.2 yr) for flows that are ~1–2 km long (Fig. 3). By comparison, the longest duration of crustal growth of Kilauean flows reported have emplacement times of 350 h (14.5 days) and maximum thickness of nearly 4 m (Hon et al., 1994). These durations also compare to estimates of eruption duration of younger Snake River Plain lava flows on the basis of calculations of the head required to drive the eruptions, which range from 1 day to 2 months (Kuntz, 1992). Some of the flows sampled by the TAN cores are nearly as thick as some Columbia River flood basalts, the thicknesses of which average ~30 m (Reidel et al., 1989), but range from 3 to 100 m (Reidel et al., 1989; Self et al., 1997).

$^{40}Ar/^{39}Ar$ AND K-Ar AGE DATA

A study by Lanphere et al. (1994) placed strong constraints on the ages of the flows within the interval above the Q-R interbed. Duplicate K-Ar measurements on a sample from within the interval, at 54 m depth in CH-1, yielded an age of 1.25 ± 0.14 Ma (reported uncertainties are twice the standard deviation of multiple analyses). A triplicate measurement on a sample from 27 m depth yielded 1.04 ± 0.08 Ma, although this flow underlies a surficial flow from Circular Butte that has a $^{40}Ar/^{39}Ar$ plateau age of 1.09 ± 0.02 Ma (Lanphere et al., 1994). This contradiction, along with the reversed polarity of this flow, led Lanphere et al. (1994) to suggest that the lower flow is older than 1.07 Ma. Samples from flows directly below the Q-R interbed yielded K-Ar ages of 1.58 ± 0.12 in CH-1 and 1.41 ± 0.10 Ma in CH-2 (Lanphere et al., 1994).

To complement these age determinations, $^{40}Ar/^{39}Ar$ plateau ages were measured for samples from 67 m and 128 m depth in TAN-34. The measurements were made in the laboratory of W. McIntosh at New Mexico Institute of Mining and Technology according to methods described by Hughes et al. (this volume). The upper sample yields an age of 1.01 ± 0.10 Ma and the lower one is 1.20 ± 0.14 Ma (Fig. 4).

PALEOMAGNETIC INCLINATION STUDIES

Analytical techniques

Because all samples are from diamond drill core that was not oriented when it was extracted, only paleomagnetic inclinations (and thus polarity) are measurable. The 2.5 cm long by 2.5 cm diameter subcores were drilled from diamond drill cores of holes TAN-33 and TAN-34; the smaller cores were drilled perpendicular to the axis of the diamond drill core (thus the

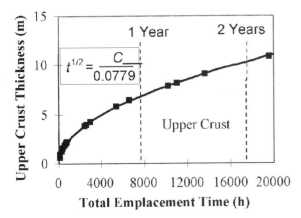

Figure 3. Calculated emplacement times of Test Area North flows; values are from Table 1.

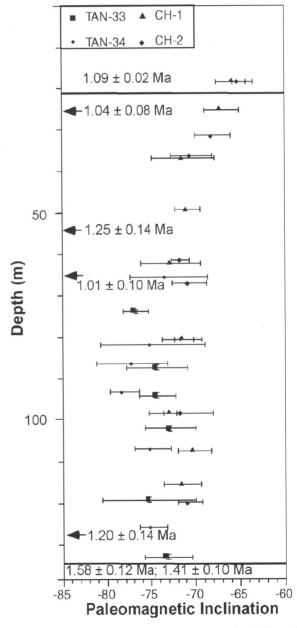

Figure 4. Paleomagnetic inclination measurements in 12 flows from TAN-33 and TAN-34 compared to paleomagnetic and age data obtained by Lanphere et al. (1994) and $^{39}Ar/^{40}Ar$ age data from this study.

axes of the paleomagnetic samples are horizontal). From 3 to 10 samples were drilled from a total of 12 flow lobes, 77 samples in all. One sample from each lobe was sequentially demagnetized in an alternating field demagnetizer to 800 oersteds. Every sample checked had a similar pattern of demagnetization. At low levels of demagnetization (<150 oersteds) samples had an overprint, but very consistent directions thereafter. Such a signature is common for reversely polarized basalts with a weathering overprint imparted during times of normal polarization. Consequently, all samples were demagnetized to 150 oersteds before paleomagnetic directions were measured.

Results

All measured samples are reversely polarized and, given their ages, must have been emplaced during the Matayuma reversed polarity epoch (Table 2). We have combined our measurements from holes TAN-33 and TAN-34 with those of Lanphere et al. (1994) from TAN CH-1 and CH-2 (Fig. 4). Between the Q-R interbed and the flow break at ~34 m in CH-1, all of the inclinations are within the uncertainty of the measurements (typically ± 2°). The inclination of this entire group of flows from TAN-33 and TAN-34 is −74.6° ± 3.2° and −75.2° ± 3.9°, compared to −71.5° ± 0.6° and −71.0° ± 0.7° for the same interval in TAN-CH1 and CH-2 (Lanphere et al., 1994). Thus, all paleomagnetic inclinations are within measurement uncertainty through the depth interval that is the focus of this study.

GEOCHEMICAL CORRELATION

Although the lavas were logged in detail and petrography has been performed on many of them (Lanphere et al., 1994, and our data), even to the experienced eye they are petrographically almost identical. Thus, we have compiled an internally consistent geochemical database from the literature (Knobel et

al., 1995) and our own analyses, and applied a simple algorithm to form a permissible cross section through the TAN area.

Analytical techniques

Samples were taken from cores TAN-33 and TAN-34. Although rotary cuttings were also analyzed from the upper 60 m of these two holes and TAN-35 and TAN-39, some of the cuttings were found to be mixtures from two or more flows. Thus, we only report samples taken from cores (although the other data are available upon request from D.J. Geist). Samples were selected from parts of the cores that lack visible alteration or segregation veins. They were all prepared in facilities at the University of Idaho and first crushed in a steel jawcrusher, which reduces sample size to ~3 mm in diameter. Chunks that were not exposed to the drill bit were then hand-picked and ground to a very fine powder in a tungsten-carbide shatterbox. Fused beads of the rocks and Li-tetraborate were made according to the specifications of Johnson et al. (1999).

Analyses reported by both us and Knobel et al. (1995) were performed by X-ray fluorescence in the Geoanalytical Laboratory of Washington State University according to techniques described by Johnson et al. (1999), thus providing an internally consistent geochemical database that minimizes analytical bias and whose analytes have similar uncertainties (Table 3). Estimates of precision and accuracy were given by Johnson et al. (1999), but for the purposes of this study, another estimation of precision was made by analyzing in triplicate a basalt from the Galápagos Islands that is similar in composition to the lavas analyzed here. The three analyses are on separate beads that were made on separate occasions in the University of Idaho facility and analyzed in separate batches at Washington State University. Thus we believe that this is the best estimate of the internal precision of that laboratory. All of the analytes reported here, with the exception of Rb, have relative standard deviations (RSDs) of <5%, and most are <1% (Table 3). For this reason, we do not consider the Rb concentrations for correlation purposes. Likewise, we do not consider Ba, because the correction procedure for an interfering peak changed between the study of Knobel et al. (1995) and our analyses (Diane Johnson, 1999, personal commun.).

Correlation techniques and results

A number of sophisticated statistical tools exist for correlating lavas on the basis of their chemical composition (e.g., Albarede, 1995). These are purely mathematical tests, however, and do not necessarily consider fundamental geologic relationships such as superposition, petrography, paleomagnetic and age limitations, and flow structure.

Correlation of basaltic lavas on the basis of geochemistry has the following four potential problems.

1. As discussed by Welhan et al. (2001), lavas from individual eruptions can form innumerable overlapping, anasto-

TABLE 2. PALEOMAGNETIC INCLINATION MEASUREMENTS

Depth (m)	n*	Mean inclination (°)	±
TAN-33			
74	7	−76.7	1.5
87	3	−74.3	3.5
94	5	−74.3	2.1
102	5	−72.8	2.9
119	5	−75.2	5.3
133	5	−73.1	2.7
TAN-34			
66	10	−73.0	4.4
82	8	−74.8	5.9
86	8	−77.1	4.0
93	7	−78.0	1.6
107	7	−74.8	2.1
126	8	−74.8	1.6

*n = number of measurements.

TABLE 3. ANALYSES OF CORE SAMPLES FROM TAN SITE

Sample	SiO$_2$	Al$_2$O$_3$	TiO$_2$	FeO*	MnO	CaO	MgO	K$_2$O	Na$_2$O	P$_2$O$_5$	Ni	Cr	Sc	V	Sr	Zr	Y	Nb
TAN-34																		
14	46.86	14.05	3.31	14.62	0.216	9.96	6.72	0.62	2.64	1.01	57	186	33	326	309	334	50	40.2
54	46.88	13.82	3.40	14.87	0.218	9.67	6.99	0.70	2.58	0.86	83	194	34	337	318	321	48	35.3
61	48.22	15.16	2.84	13.13	0.197	9.13	7.30	0.66	2.68	0.67	69	207	29	303	325	258	38	33.6
69	47.62	14.61	2.92	13.99	0.207	9.28	7.39	0.69	2.58	0.69	61	197	26	300	323	268	38	35.1
81	47.74	15.42	2.31	13.01	0.195	9.61	8.03	0.54	2.60	0.54	80	245	27	259	334	201	33	23.2
87	47.63	14.68	2.54	13.56	0.202	9.64	8.15	0.58	2.50	0.52	83	234	30	285	311	199	33	22.2
93	48.00	14.77	2.61	13.42	0.201	9.58	7.70	0.60	2.58	0.55	86	240	28	280	312	207	35	23.8
98	47.74	14.67	2.76	13.63	0.204	9.45	7.63	0.65	2.61	0.65	79	233	27	280	325	230	37	27.3
103	47.88	15.04	2.49	13.04	0.193	9.62	8.15	0.57	2.52	0.50	103	262	30	275	291	207	33	21.6
128	47.43	14.97	2.11	12.53	0.191	10.15	9.28	0.44	2.49	0.40	116	383	34	279	239	160	30	15.5
TAN-33																		
63	46.93	14.51	3.19	13.98	0.22	9.68	7.66	0.57	2.46	0.79	105	247	30	319	328	287	42	32.1
74	48.47	13.49	3.60	14.16	0.22	9.52	6.01	0.91	2.69	0.95	37	151	29	350	304	346	53	45.2
80	48.13	16.07	2.45	12.07	0.20	9.76	7.61	0.50	2.62	0.59	78	222	27	269	347	212	34	25.4
88	47.89	14.77	2.62	13.27	0.20	9.40	7.99	0.62	2.62	0.61	86	241	32	285	318	221	37	27.8
97	48.12	15.09	2.53	12.99	0.20	9.45	7.88	0.54	2.64	0.56	91	236	31	270	315	204	34	23.7
102	48.16	14.18	3.26	13.65	0.21	9.67	6.76	0.74	2.64	0.73	54	195	30	335	313	263	41	32.7
109	47.79	15.22	2.50	12.79	0.19	9.82	7.98	0.59	2.59	0.52	104	259	29	279	294	209	35	22.0
110	47.87	15.64	2.38	12.21	0.19	9.85	8.18	0.60	2.56	0.51	99	269	29	267	299	202	33	21.6
121	47.80	15.25	2.42	12.68	0.19	9.78	8.30	0.53	2.56	0.48	114	274	31	292	277	191	33	19.9
122	47.82	15.11	2.98	13.10	0.21	9.80	7.10	0.64	2.54	0.70	69	226	31	308	328	241	39	29.5
132	47.52	15.16	2.05	12.11	0.19	10.05	9.60	0.42	2.50	0.40	113	392	24	268	238	155	29	15.7
GIN-5																		
9	46.51	13.81	3.43	15.22	0.23	9.77	6.55	0.65	2.58	1.24	52	144	34	339	313	431	60	53.8
13	46.94	14.75	3.22	14.05	0.21	9.91	6.68	0.54	2.66	1.02	65	163	29	340	323	345	47	42.8
19	46.67	14.41	2.99	14.50	0.22	9.55	7.61	0.54	2.52	0.99	96	198	28	327	312	330	46	42.6
22	46.55	13.88	3.22	14.94	0.22	9.56	7.55	0.55	2.46	1.07	72	176	31	352	310	359	51	44.6
25	46.70	14.13	3.34	14.47	0.22	9.88	7.21	0.55	2.49	1.00	73	198	29	356	309	331	50	40.6
29	47.01	14.53	2.96	14.14	0.22	9.59	7.57	0.52	2.50	0.96	77	181	31	332	311	319	47	40.4
37	46.97	14.05	3.23	14.43	0.22	9.62	7.35	0.66	2.56	0.92	70	206	31	340	305	315	48	39.0
51	47.15	13.95	3.24	14.74	0.22	9.33	7.44	0.67	2.47	0.78	89	217	32	354	314	303	44	33.8
57	46.93	13.98	3.56	14.83	0.22	9.32	7.40	0.55	2.40	0.82	76	223	34	376	317	305	44	33.2
60	46.99	14.13	3.31	14.51	0.22	9.42	7.44	0.68	2.52	0.79	86	203	29	342	320	301	44	33.7
65	47.48	14.06	3.32	14.37	0.21	9.27	7.31	0.66	2.52	0.80	94	222	32	353	312	304	43	33.9
74	46.71	14.62	3.15	14.28	0.21	9.43	8.08	0.50	2.41	0.61	106	263	34	357	321	249	38	25.4
81	46.76	13.73	3.83	15.01	0.22	9.48	7.08	0.58	2.56	0.73	93	207	27	409	315	304	45	32.9
93	47.22	14.76	2.98	14.33	0.21	9.34	7.30	0.60	2.60	0.66	92	199	30	325	337	260	40	27.6
101	48.36	14.92	2.76	13.34	0.20	9.07	7.41	0.67	2.60	0.67	75	216	30	318	327	253	37	31.2
106	49.02	14.86	2.82	13.16	0.20	8.96	7.02	0.67	2.55	0.74	66	191	27	297	328	275	42	36.2
110	48.06	14.89	2.72	13.49	0.20	9.22	7.39	0.70	2.65	0.69	66	195	32	295	330	259	39	33.6
118	47.87	14.57	2.76	13.57	0.21	9.39	7.91	0.62	2.53	0.56	78	235	32	293	309	216	35	25.1
124	47.98	14.87	2.60	13.06	0.20	9.56	8.08	0.57	2.55	0.53	91	268	31	304	277	217	35	22.0
GIN-6																		
22	47.34	14.01	3.23	14.24	0.22	9.55	7.09	0.65	2.74	0.93	68	210	32	342	300	323	47	39.7
27	47.16	13.99	3.33	14.36	0.22	9.54	7.06	0.67	2.70	0.96	65	202	33	359	303	327	49	38.1
34	46.77	13.83	3.51	15.03	0.22	9.56	6.87	0.62	2.63	0.96	101	191	29	361	315	342	50	38.8
41	48.09	14.95	2.74	13.51	0.20	9.26	7.15	0.71	2.68	0.70	66	195	32	269	331	264	39	33.7
49	48.26	14.98	2.68	13.30	0.20	9.30	7.12	0.68	2.81	0.66	71	183	28	293	336	252	38	32.4
52	48.13	15.47	2.43	12.96	0.19	9.09	7.63	0.62	2.87	0.62	79	199	28	267	338	231	36	29.1
58	48.10	15.29	2.60	12.98	0.19	9.16	7.35	0.63	3.04	0.64	68	192	26	274	335	244	36	29.8
61	48.30	14.86	2.70	13.11	0.20	9.31	7.25	0.66	2.94	0.66	105	220	32	287	326	242	38	30.1
Precision	0.1%	0.4%	0.5%	1.4%	0.8%	0.6%	1.6%	0.5%	0.5%	1.0%	4.1%	2.3%	3.1%	1.8%	0.1%	0.3%	2.0%	4.4%

mosing lobes (a lobe being defined as having mostly solidified before being overlain by the next youngest flow). Hence, determining different time-stratigraphic units from field (or drill core) relations is not always possible, especially from a quickly erupted sequence where there are neither paleomagnetic variation nor disconformities. Hence, flow unit boundaries can be impossible to identify and differentiate.

2. Lavas from a single eruption (hence forming a single flow unit of many thousands of flow lobes) may have a range of compositions. The best-documented example of this is the Pu'u O'o eruption at Kilauea; there, the secular variation in major and trace element and isotopic composition exceeds the analytical uncertainty by more than ten fold (e.g., Garcia et al., 1996). Thus single flow units may be misinterpreted as having been from several different flow groups if the flow groups are defined as being compositionally distinct.

3. Lavas from some volcanoes have almost no compositional variation over eruptive cycles spanning many millennia from dozens of vents, thus many flow groups may be equivalent compositionally (e.g., Mauna Loa, Hawaii, Rhodes, 1982; Sierra Negra, Galapagos, Reynolds et al., 1995). In this case, geochemical correlation is impossible, much less the assignment of flow groups. Using compositional criteria alone, most of the flows from two of the world's largest volcanoes would be assigned to the same flow group.

4. Differentiated liquids can be redistributed within a flow due to crystal fractionation (e.g., Goff, 1996) or vapor-phase transfer, and weathering can alter compositions. This may be avoided by careful sampling in the interior of flows, avoiding segregation veins (which are vesicular) and weathered zones. The very limited range in the concentrations of elements that are included in the observed phenocryst phases (e.g., Ni, Cr, Sr, Mg, Al) and excellent correlations between compatible and incompatible element concentrations lead us to conclude that these processes did not strongly affect the samples reported here.

A simple, straightforward series of geological observations and statistical techniques were applied to assess correlation at the TAN site. First, cores from each drill hole were logged, special attention being placed on identifying the margins of individual flow lobes. These were identified by glassy margins, vesicle distribution, rubble zones, weathered zones, and thin sedimentary interbeds. Although individual eruptions may produce many of these intervals, each flow unit must also be marked by a flow lobe contact.

Second, the geochemical database was screened so we would not consider elements that have little total variation. Elements with a relative standard deviation of <5% among the entire database were not considered, nor were elements such as Rb and Ba with suspect imprecision. The petrologic processes that caused compositional variations of these lavas simply did not change Si, Al, Mn, Ca, and Na enough for them to contribute to the compositional fingerprint used to correlate flow units. This left Ti, Fe, Mg, Sc, K, P, Ni, Cr, Zr, Nb, Sr, Y, and V.

Third, the analyses from each sample j were compared to a sample from each adjacent lobe $(j - 1)$ by the formula

$$D = \left[\sum_{i=1}^{n} |(C_i^j - C_i^{j-1})|/C_i^j \right]/n \qquad (2)$$

for each element i (n, number of total elements), where C is concentration and D is the average deviation, a measure of the average relative difference of each analyte for every sample. An arbitrary cutoff of $D = 7\%$ (a value much greater than simple analytical precision) was applied: analyses with $D < 7\%$ are deemed to come from the same flow unit (or the same phase of an eruption that experienced secular compositional variation), and those with $D > 7\%$ are deemed to come from a different flow unit. These downhole correlations are illustrated in Figure 5 by the shaded vertical arrows. Compositionally similar flows range in thickness from ~3 m to ~40 m, and large compositional discontinuities are generally at places

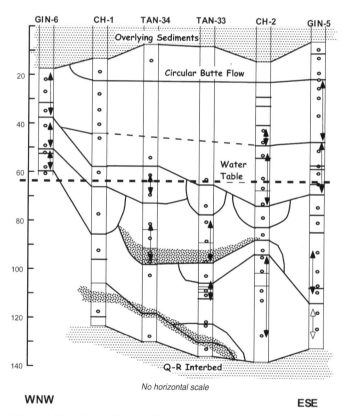

Figure 5. Correlation diagram linking lava flows with statistically similar compositions. Horizontal lines in columns indicate identified flow lobe boundaries and circles indicate sample locations. Vertical arrows connect lobes interpreted to be of same flow unit. Water table is shown. Stippled zones show interpretation of geometries of high hydraulic conductivity zones. Note that this is not cross section, as there is no horizontal scale.

where excessive weathering or sediment deposition were noticed during visual inspection of the core.

Fourth, flows grouped downhole were correlated to flow groups in adjacent drill holes. The criterion was based on the average difference. A limit in the average deviation of 10% was used. In addition, a time-stratigraphic break is assumed to correspond to any clear compositional break in any single hole, thus flows in the adjacent holes could not cross such a break.

The cross section we have generated is not proven but is permissible given the data. The details of most previous cross sections of the site, which are not based on compositional and paleomagnetic constraints, are either inconsistent with the geochemical data (Anderson and Bowers, 1995; Sorenson et al., 1996) or are much less detailed (Lanphere et al., 1994) than necessary to address the issue of lithologic controls on groundwater flow. These previous studies essentially lump all of the units that we have identified into two flow groups.

DIRECTION OF LAVA FLOWS

Anisotropy of magnetic susceptibility (AMS) is a technique that has been applied to determine the direction of flow of basaltic lavas in several settings, with mixed success (Elwood, 1978; Canon-Tapia et al., 1995, 1996, 1997). Interpretation of AMS data is not always straightforward, especially for inflated pahoehoe flows, because flow lobes may be protruded in directions different than that of the bulk flow and late movement of the lava after stagnation due to cooling and crystallization (Canon-Tapia et al., 1995). Nevertheless, in some settings AMS has been shown to be a useful indicator of flow direction in different types of lava (Canon-Tapia et al., 1995, 1996, 1997). Thus, samples from six flows each in TAN-33 and TAN-34 were measured for AMS in the paleomagnetism laboratory at the University of New Mexico. Representative stereograms showing our AMS results are depicted in Figure 6. Because the core is not oriented in the horizontal direction, the subcores were rotated until the thermoremanent magnetic north was oriented due geographic south (the flows are reversely polarized). This introduces extra uncertainty into the AMS directions, because we do not know how close to geographic south the magnetic pole was at the time of magnetization, and there may be unaccounted-for variation in the paleomagnetic declination through the section. The direction of maximum intensity is very irregular in multiple samples from the same flows (Fig. 7).

These samples do not have strongly developed anisotropy (Table 4). Although there are several parameters that can be used to signify the extent of anisotropy, we have chosen the A parameter of Canon-Tapia (1994), where:

$$A = 100* [1 - (k_3 + k_2)/2k_1] \quad (3)$$

where k_3, k_2, k_1 are the maximum, intermediate, and minimum magnetic susceptibilities. The A values range from 0 (isotropic;

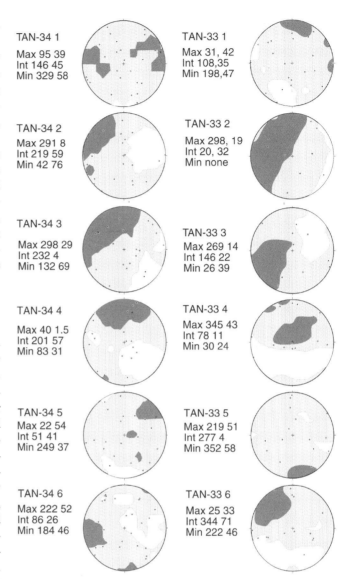

Figure 6. Stereograms of anisotropy of magnetic susceptibility. Diagrams are contoured for maximum intensity, showing calculated mean vector according to Fisher distribution. Contours are by Kamb method. Lighter shaded region is for 2–4 sigma and darker region is for 4–6 sigma from random distribution for maximum intensity.

a sphere) to 100% (maximum anisotropy; a line). The A values from the TAN flows average 1.3% ± 0.4%. This is significantly smaller anisotropy than that of near-vent aa flows from Xitle, Mexico, and Mauna Kea, Hawaii, which have A values mostly 2%–3% (Canon-Tapia et al., 1996). The A values from TAN flows are greater than those from Hawaiian S- and P-type pahoehoe (Wilmoth and Walker, 1993), which are typically <1% (Canon-Tapia et al., 1997). It is interesting to note that the A values from these flows match almost precisely those from Hawaiian toothpaste lava (Rowland and Walker, 1987), a variety of lava that is intermediate between aa and pahoehoe.

The shape of the magnetic parameter can be defined by:

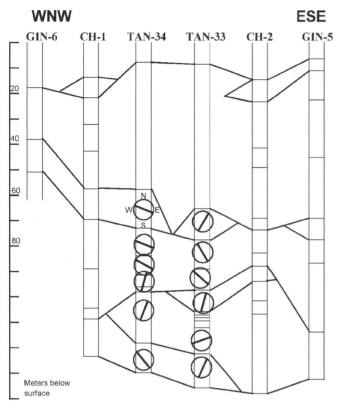

Figure 7. Cross section showing declination of Fisher average of maximum intensity from TAN-33 and TAN-34. These may correspond to flow directions.

$$B = 100* (k_3 - 2k_2)/(k_1 + 1), \qquad (4)$$

which ranges from -100 to $+100$. $B < 0$ indicates a foliation without well-developed lineation, whereas $B > 0$ indicates lineation. A B value close to 0 can indicate either random orientation or a combination of lineation and foliation (Canon-Tapia, 1994). The B values from all flows are not significantly different from zero, although we note that the averages from each individual flow are <0 (Table 4). Canon-Tapia et al. (1997) found that B does not differ significantly between different flow types and is typically near zero.

INTERPRETATIONS

Time span of the eruptive sequence, accumulation rates, and analogue models

The K-Ar and ^{39}Ar-^{40}Ar dating techniques yield four ages from the sequence that overlap within the uncertainties of the measurement. As stated by Lanphere et al. (1994), the 1.04 ± 0.08 Ma age from near the top of CH-1 is inconsistent with its stratigraphic position and paleomagnetism. As well, the CH-1 lava with a 1.25 ± 0.14 age (Lanphere et al., 1994) overlies

the lava with an age of 1.01 ± 0.10 from TAN-34 (Fig. 5). Hence, the data are consistent with the entire sequence between ~135 m (above the Q-R interbed) and 25 m (below the Circular Butte flows) depth as having erupted over a very short interval ca. 1.13 ± 0.24 Ma (the uncertainty here being the propagated uncertainty of the individual measurements; i.e., uncertainty assuming all the ages are the same).

Secular variation of the Earth's magnetic field takes place at variable rates at different times and places; thus, an accurate assessment of the time it took for emplacement of this 95 m sequence of lavas cannot be accurately estimated solely on the basis of the lack of change in the paleomagnetic inclination. However, studies of well-dated basaltic lava flows from Hawaii indicate rates of change in inclination of ~10° per millennium (Holcomb et al., 1986), an amount much greater than the observed variation in this sequence of flows. Thus, the sequence of flows between the Q-R interbed and the Circular Butte flows probably erupted in a very short interval of time, probably <200 yr. They must, therefore, be from a single eruptive episode and constitute a single flow group. These data, however, do not indicate whether more than one volcano might be involved.

The most conservative estimate of average accumulation rate of these lavas is 0.3 mm/yr, which assumes the maximum uncertainties in the argon ages determined for the flow at 127 m in TAN-34 and at 28 m in CH-1 (which, according to Lanphere et al. [1994], is demonstrably incorrect). A more realistic estimate for the average accumulation rate of lavas from the Q-R interbed to the top of the flow at 60 m depth in TAN-34 is 13 cm/yr. This estimate presumes an interval of 500 yr, which is consistent with the both the paleomagnetism and age measurements.

An average accumulation rate of 13 cm/yr is extraordinary, even for a volcanic terrain. The rapid accumulation may be due to proximity to a vent, an episode with large eruption rates, or ponding of lava in a topographic low. As discussed herein, we interpret the lavas underlying TAN as being in a medial to distal facies with respect to their vents. Neither our cross sections nor those of Anderson and Bowers (1995) show evidence of a local basin centered on TAN. Thus, we believe that the large accumulation rate of lava at TAN ca. 1.1 Ma results from an episode of exceptional volcanic activity. If this interpretation is correct, then there may not be a good analogue for the emplacement of the subsurface flows at TAN anywhere on the surface of the Snake River Plain. This is an important observation, because flow morphology depends strongly on the rate of emplacement of lava flows (Walker, 1975; Rowland and Walker, 1990), which in turn depends on the evolutionary stage of any given volcano. Although long-term averages may have little relation to the short-term emplacement rates and flow morphology, the extraordinary difference in rate of magma supply certainly indicates that the style of volcanism may have been different during the TAN event. Some oceanic volcanoes have accumulation

TABLE 4. ANISOTROPY OF MAGNETIC SUSCEPTIBILITY DATA

Depth	K Min	D Min	I Min	K Int	D Int	I Int	K Max	D Max	I Max	A*	B† shape
TAN-33											
74.5	0.9847	116	−47.9	0.9889	201.7	3.7	0.9985	108	41.9	1.17	0.54
74.8	1	14	−61.1	1.006	299.2	8.2	1.009	33	27.5	0.59	−0.30
74.8	1.134	86	−76.4	1.1	34.9	8.6	1.158	307	−10.4		
72.6	0.9296	316	28.9	0.9365	301.3	−60.4	0.9378	42	−6.1	0.51	−0.60
72.6	0.9579	310	26.1	0.9617	249.0	−44.9	0.9643	201	33.8	0.47	−0.12
73.5	1.041	2	−47.2	1.127	272.4	0.8	1.208	183	−42.8	10.26	−0.41
73.5	0.9382	90	0.9	0.9449	1.9	−58.7	0.9457	360	31.3	0.44	−0.62
87.1	0.9158	8	−14.7	0.935	80.7	48.8	0.9417	290	37.5	1.73	−1.33
87.1	0.5565	276	76.1	0.5574	208.5	−5.5	0.5604	300	−12.8	0.62	0.37
88.3	0.5984	43	13.3	0.6044	332.8	−55.3	0.6095	305	31.4	1.33	−0.15
93.2	1.188	22	−33.1	1.193	0.6	55.0	1.2	105	10.0	0.79	0.17
93.2	0.6227	45	45.9	0.6253	312.1	2.8	0.6268	39	−43.9	0.45	−0.18
93.8	1.046	9	39.0	1.05	332.0	−44.5	1.063	82	−19.5	1.41	0.85
93.8	1.097	203	12.3	1.109	191.0	−77.4	1.13	113	2.6	2.39	0.80
96.9	1.536		15.2	1.543		72.4	1.5		−8.6		−3.33
101.2	0.5956	58	−70.5	0.6046	93.3	16.1	0.6081	180	−10.7	1.32	−0.90
101.8	0.6198	201	86.7	0.6265	44.0	3.1	0.6288	134	−1.3	0.90	−0.70
101.8	0.6579	340	13.4	0.6581	69.6	−1.3	0.6634	334	−76.5	0.81	0.77
102.9	0.477	180	76.7	0.483	98.0	−1.9	0.488	188	−13.1	1.64	−0.20
102.9	0.4531	81	−52.1	0.4546	86.2	37.8	0.4558	354	2.3	0.43	−0.07
117.8	0.6418	248	67.9	0.6514	272.6	−20.4	0.6543	179	−8.0	1.18	−1.02
117.8	0.5901	12	58.5	0.5962	292.3	−6.1	0.5979	26	−30.8	0.79	−0.74
120.0	0.7391	104	54.5	0.7501	356.8	12.2	0.7528	79	−32.7	1.09	−1.10
120.0	0.6103	328	−13.9	0.614	238.0	−0.3	0.6165	328	76.1	0.71	−0.19
121.2	0.4194	355	35.5	0.4247	240.9	30.0	0.4302	302	−40.0	1.89	0.05
131.7	1.363	59	−71.0	1.379	63.4	19.0	1.395	333	1.3	1.72	0.00
132.0	0.2335	119	44.2	0.2357	72.8	−35.5	0.2415	2	27.8	2.86	1.49
132.0	0.2429	150	82.5	0.2448	177.4	−6.6	0.2464	87	−3.4	1.03	−0.12
135.4	0.3676	286	−53.4	0.3696	255.9	15.5	0.3726	156	32.8	1.07	0.27
135.4	0.2839	250	37.0	0.2862	293.4	−43.8	0.2878	359	23.6	0.96	−0.24
TAN-34											
62.5	0.699	193	61.9	0.702	156.2	−23.1	0.7081	73	15.0	1.07	0.44
62.5	0.5251	307	27.4	0.5275	23.4	−25.1	0.5324	258	−51.2	1.15	0.47
63.1	0.8		73.3	0.8086		−14.3	0.8191		−8.4	1.81	0.23
63.1	0.7843	224	70.8	0.7854	247.0	−17.6	0.7885	155	−7.1	0.46	0.25
63.4	0.7	348	64.3	0.7768	334.1	−25.1	0.7796	67	−5.4	5.28	−9.49
63.4	0.6714	289	49.7	0.6776	216.3	−14.7	0.6844	137	36.8	1.45	0.09
67.1	0.6857	136	69.7	0.6915	31.6	5.4	0.6967	120	−19.5	1.16	−0.09
67.1	0.658	324	−43.8	0.6607	243.2	9.0	0.6685	342	44.8	1.37	0.76
71.7	2.061	345	−25.0	2.07	72.1	6.5	2.086	329	64.1	0.98	0.34
71.7	2.572	71	−4.0	2.587	340.2	−6.3	2.6	13	82.5	0.79	−0.08

(Table 4 continued on next page)

rates more comparable to those at TAN and may be more appropriate analogues than Holocene Snake River Plain flows (see Welhan et al., 2000, for a different opinion). The accumulation rate at the coast of Mauna Loa, Hawaii, which is the largest volcano on Earth, has been ~2 mm/yr for the past several millennia (Lipman, 1995). Kilauea, which is in a more juvenile stage of growth, has an accumulation rate of ~6.4 cm/yr low on its flanks. The flanks of Sierra Negra, the largest volcano in the Galápagos archipelago, have an accumulation rate of ~0.5 mm/yr (Reynolds et al., 1995). Thus, the dimensions and morphologies of flows erupted from those volcanoes may be more appropriate analogues for the TAN flows than young Snake River Plain flows, the total accumulation rate of which is smaller than one-hundredth that of the TAN event, when measured over a similar time scale.

Development of the volcanic landscape

Although the directions of lineation of the maximum principal susceptibility are scattered, one hypothesis is that the statistical averages reflect the direction of transport of the lavas (Canon-Tapia et al., 1995, 1997). In TAN-34, the maximum susceptibility is oriented west-northwest in the youngest three flows and north-northeast in the two underlying flows (Figs. 6 and 7). This is consistent with the apparent dips of the tops of these units: down to the east in the younger flows and roughly horizontal in the older ones. In TAN-33, the maximum susceptibility is more irregular, but is mostly in the same two directions (Fig. 7). The lack of correlation between the AMS-determined directions in some of the correlated flow units indicates either that AMS is not always a reliable indicator of

TABLE 4. ANISOTROPY OF MAGNETIC SUSCEPTIBILITY DATA (continued)

Depth	K Min	D Min	I Min	K Int	D Int	I Int	K Max	D Max	I Max	A*	B† shape
TAN-34											
80.9	0.6253	182	73.5	0.6354	178.2	−16.5	0.6399	269	−1.0	1.49	−0.88
80.9	0.4578	355	65.9	0.4628	54.2	−12.8	0.466	319	−20.0	1.22	−0.39
81.5	0.5955	5	59.1	0.5977	78.0	−10.1	0.6004	162	28.9	0.63	0.08
81.5	0.5	22	41.8	0.5386	338.8	−39.3	0.5422	269	23.3	4.22	−6.46
82.2	0.8476	99	15.3	0.8486	353.0	45.9	0.8512	23	−40.1	0.36	0.19
82.2	0.7967	240	−87.1	0.7986	213.6	2.6	0.7999	124	−1.3	0.28	−0.08
82.8	0.8817	228	2.1	0.8883	313.2	−65.5	0.8899	319	24.3	0.55	−0.56
85.2	0.6221	156	61.2	0.6303	124.1	−24.9	0.639	41	13.5	2.00	0.08
85.8	0.5239	297	66.0	0.5286	229.3	−9.7	0.5341	143	21.7	1.47	0.15
85.8	0.5261	117	62.8	0.5314	30.7	−2.2	0.5343	122	−27.3	1.04	−0.45
86.2	0.5969	94	59.2	0.6014	6.5	−1.8	0.6038	98	−30.8	0.77	−0.35
86.2	0.5654	351	−52.6	0.5707	62.9	13.5	0.5734	143	−34.1	0.93	−0.45
86.5	1.96	90	65.5	1.975	23.2	−10.1	1.982	117	−22.1	0.73	−0.40
86.5	2.083	178	−54.2	2.098	237.4	20.3	2.106	316	−28.1	0.74	−0.33
88.6	0.4708	182	60.7	0.4731	83.3	4.7	0.4761	171	−28.9	0.87	0.15
92.0	0.981	134	−0.8	0.9867	50.9	83.5	0.9922	44	−6.4	0.84	−0.02
92.0	0.9288	119	69.9	0.9401	106.3	−20.0	0.9473	18	4.0	1.36	−0.43
92.9	0.9192	358	71.1	0.928	264.4	1.2	0.9337	354	−18.9	1.08	−0.33
93.5	1.058	32	−11.2	1.061	74.5	75.0	1.065	124	−9.9	0.52	0.09
93.5	1.054	334	−56.1	1.065	268.2	15.6	1.067	7	29.2	0.70	−0.84
94.5	0.9054	97	−27.8	0.9095	126.0	58.9	0.9156	14	12.9	0.89	0.22
94.8	4.8	189	62.9	4.886	195.7	−27.0	4.904	104	−2.8	1.24	−1.39
99.4	1.171	202	73.0	1.176	240.2	−13.4	1.182	148	−10.2	0.72	0.08
101.5	0.6271	207	44.2	0.6345	146.5	−26.9	0.6384	256	−33.8	1.19	−0.55
103.7	0.3105	259	5.9	0.3124	344.9	−35.0	0.3137	177	−54.3	0.72	−0.19
109.5	0.7288	279	−63.1	0.731	174.5	−7.5	0.7378	261	25.6	1.07	0.62
109.8	0.7635	71	−50.0	0.7686	150.6	8.8	0.7714	54	38.6	0.69	−0.30
109.8	0.6059	238	40.0	0.654	316.1	−14.3	0.6582	210	−46.5	4.29	−6.67
115.1	0.4471	346	63.4	0.4505	290.5	−15.7	0.4531	27	−20.8	0.95	−0.18
122.2	0.2957	306	−72.6	0.2986	319.8	16.9	0.2996	49	−3.8	0.82	−0.63
124.9	0.2452	190	63.8	0.2457	144.2	−19.0	0.2476	240	−17.4	0.87	0.57
124.9	0.2357	255	−55.3	0.236	328.4	10.9	0.2367	51	−32.4	0.36	0.17
125.8	0.2328	45	−51.9	0.2345	67.2	35.9	0.2354	149	−11.1	0.74	−0.34
125.8	0.1992	282	35.2	0.2007	208.2	−21.9	0.2015	323	−46.7	0.77	−0.35
126.8	0.1914	31	59.6	0.1926	342.4	26.7	0.1939	325	−13.4	0.98	0.05
126.8	0.1662	133	30.8	0.1671	206.7	−25.2	0.1694	85	−48.2	1.62	0.83
128.0	0.486	142	17.4	0.4876	188.9	−65.2	0.4936	57	−17.1	1.38	0.89

Note: Values (K), declination (D), and inclination (I) of three principal susceptibilities are reported. Samples have been rotated about a vertical axis such that TRM north points to geographic south.
*A is a measure of the degree of anisotropy, from Canon-Tapia (1994).
†B is a measure of the shape of the magnetic fabric, from Canon-Tapia (1994).

flow direction in these lavas, or that in some situations flow direction was highly irregular over fairly short length scales (scale of the flow thickness; Canon-Tapia et al., 1995). Although the individual lava flows show much scatter in the direction of the maximum susceptibility, the data from all of the flows cluster in two principal directions: N20E and N60W (Fig. 8). On the basis of these measurements and the regional topography, we suggest that two vents erupted the magmas that underlie TAN, one north-northeast of the site and the other west-northwest of the site.

According to our cross section (Fig. 9), most of the flow units are discontinuous over the 4 km line of section, and some are limited to <~400 m wide in the line of section. Stratigraphic models for this part of the Snake River Plain that correlate flow units or even flow groups over a distance of >10 km may be based on gross compositional similarities and thus are not time-equivalent units. In addition, the limited aerial extent of individual flow units renders conclusions about tilting and subsidence questionable (Anderson et al., 1995). Our opinion is that the evidence for localized subsidence or tilting is very weak, and the consistent paleomagnetic inclinations across the TAN site preclude tilting of >~2°.

Paleotopography as measured from the base of the surficial sediments (the top of Circular Butte lava) indicates gradients of ~0.9° between TAN-34 and CH-1. Likewise, the top of the Q-R interbed has essentially the same gradient. These are apparent gradients, because the drill hole coverage is along a single one-dimensional section and the true gradient could be larger. Overall, there appears to be a consistent topographic gradient down to east between holes, and the slopes were slight. The shallowness of the slopes is also consistent with the widespread appearance of P-type pahoehoe in the drill core samples (Walker, 1987) and the inflationary mechanisms for their emplacement (Hon et al., 1994).

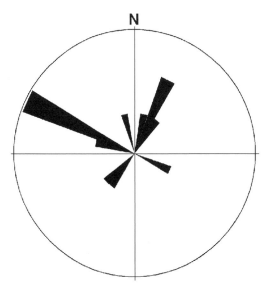

Figure 8. Rose diagram of Fisher average of maximum intensity. Note well-defined clusters at N20E and N70W.

Lithologic controls on preferential flow paths

Wylie et al. (this volume) identified a zone of exceptionally high hydraulic conductivity (~900 m/day) at 94–98 m depth in hole TAN-34. This zone is at the bottom of a sequence of thin flow lobes that are interbedded with rubble, all of which are from the same flow unit. These thin flows are just above a thick, dense lava. According to our cross section (Figs. 5 and 9), the high-conductivity flow unit pinches out to the northwest before reaching CH-1, where no such high-conductivity zone appears at equivalent depth (A. Wylie, 2000, personal commun.). This independently supports our interpretative cross section and exemplifies the lack of continuity of the individual high-conductivity zones. Moreover, the high-conductivity zone is only a couple of meters thick in CH-2, which may inhibit large-scale movement of groundwater to the south.

The second-highest conductivity measured in TAN-34 is at the interval 115–120 m and is ~700 m/day (Wylie et al., this volume). The highest conductivity noted in CH-1 is at 110–114 m depth and the two zones are part of the same flow unit on our section (Figs. 5 and 9). Note that this is also a primary breccia (or rubble) zone at a flow top, and it immediately overlies an impermeable layer, the Q-R interbed (Anderson and Bowers, 1995). This flow unit is not widespread; it pinches out to the east and does not appear in CH-2.

These observations support the idea that the lava flows in this region are not laterally extensive, hence the aquifer cannot be thought of as a layered system, even on a scale of hundreds of meters. The zones of high hydraulic conductivity must be series of flow units that are intermittently interconnected along vertical pathways such as the sides or toes of lava flows or vertical fractures. There are no aquifers that are laterally continuous beyond several hundred meters at shallow levels beneath the TAN site. These interpretations are consistent with those of Welhan et al. (2001), who studied surficial Snake River Plain lavas and presented stochastic models of local- and regional-scale conductivity heterogeneity in the Snake River Plain aquifer (Welhan and Reed, 1997).

We suggest, therefore, that the high-flow zones that control the dispersal of the contaminant plume channel groundwater flow to the east for ~200 m, then pinch out, causing groundwater flow to veer south. The southward change in the groundwater flow direction may be facilitated by a high-conductivity zone associated with a flow from the predicted vent to the north-northeast of the TAN site. In any event, the sides or toe of a lava flow must abut the side or top of another flow, providing a network of rubbly, high-flow zones in the subsurface. Future core sampling in the vicinity of the southward bend of the contaminant plume would provide a rigorous test of this hypothesis. Likewise, we emphasize that the locations of the available drill cores provide a two-dimensional view of the geometrical relationships between these lava flows; further drilling to the northeast and southwest of the disposal well would provide the three-dimensional view that is necessary for understanding these issues.

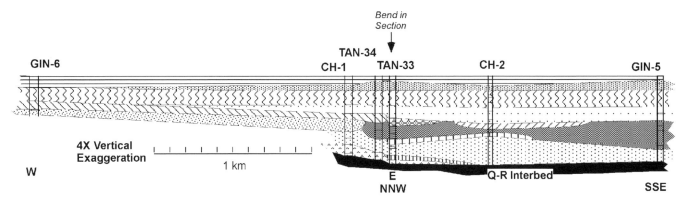

Figure 9. Cross section, to scale, of Test Area North site. Note that 4× vertical exaggeration strongly distorts dips between units, making them appear much steeper than they actually are.

CONCLUSIONS

Attempts to correlate basaltic lava flows at contaminated sites using petrography, paleomagnetism, and downhole geophysical methods have severe shortcomings. Even to the experienced eye, basaltic lava flows from the same volcanic system can be petrographically identical. Gamma-ray logging only distinguishes gross changes in the volumetric concentration of several naturally occurring radioactive isotopes (Anderson and Bowers, 1995), and it misses significant compositional variations inherent in an evolving volcanic system. Paleomagnetic measurements are limited to inclination, due to core rotation and the rate of secular variation. However, modern analytical techniques can reliably determine compositions for more than 20 analytes at modest cost. Given the great power and ease of geochemical analysis, geochemical correlation tools should be a routine practice in all drill holes at contaminated sites.

We have used the full combination of these techniques to understand the subsurface geometry and morphology of lava flows beneath the TAN site. Unfortunately, the existing drill-hole pattern gives only a two-dimensional slice through the site; our interpretations are readily testable by further drilling to the north and southwest of the TAN site, which would provide a more three-dimensional view of the subsurface geology of this small part of the Snake River Plain. Nevertheless, there is little doubt that groundwater flow is strongly channelized in this basalt-hosted aquifer. Our interpretation is that the channelization may be caused by the direction of original flow of lavas and the locations of the boundaries of the lavas.

ACKNOWLEDGMENTS

This work was funded by U.S. Department of Energy grant DE-FG07-96ID13420 through the Idaho Water Resources Research Institute. We thank Linda Davis and Travis McCling for help with access to core curated by the U.S. Geological Survey and Idaho National Engineering and Environmental Laboratory. Diane Johnson of the Washington State University geoanalytical facility provided competent analytical acumen. We thank John Geissman for opening his anisotropy of magnetic susceptibility lab to us. Reviews by Steve Reidel, Paul Link, and Richard Smith improved our presentation and rationale. Dale Ralston is thanked for bringing us into this project, providing his hydrologic know-how, and reviewing an early draft of this paper. This work is dedicated to him at the time of his premature retirement from academia.

REFERENCES CITED

Albarede, F, 1995, An introduction to geochemical modeling: Cambridge, Cambridge University Press, 564 p.

Anderson, S.R., and Bowers, B., 1995, Stratigraphy of the unsaturated zone and uppermost part of the Snake River Plain aquifer at Test Area North, Idaho National Engineering Laboratory, Idaho: U.S. Geological Survey Water-Resources Investigation Report 95-4130, 47 p.

Canon-Tapia, E., 1994, AMS parameters: Guidelines for their rational selection: Pure and Applied Geophysics, v. 142, p. 365–382.

Canon-Tapia, E., Walker, G.P.L., and Herrero-Bervera, E., 1995, Magnetic fabric and flow direction in basaltic pahoehoe lava of Xitle Volcano, Mexico: Journal of Volcanology and Geothermal Research, v. 65, p. 249–263.

Canon-Tapia, E., Walker, G.P.L., and Herrero-Bervera, E., 1996, The internal structure of lava flows: Insights from AMS measurements 1: Near-vent a'a: Journal of Volcanology and Geothermal Research, v. 70, p. 21–36.

Canon-Tapia, E., Walker, G.P.L., and Herrero-Bervera, E., 1997, The internal structure of lava flows: Insights from AMS measurements 2: Hawaiian pahoehoe, toothpaste lava and a'a: Journal of Volcanology and Geothermal Research, v. 76, p. 19–46.

Cashman, K., Thornber, C.R., and Kauahikaua, J.P., 1999, Cooling and Crystallization of lava in open channels, and the transition of pahoehoe lava to a'a: Bulletin of Volcanology, v. 61, p. 306–323.

Casper, J.L., 1999, The volcanic evolution of Circular Butte [M.S. thesis]: Pocatello, Idaho, Idaho State University, 113 p.

Ellisor, R., Sims, E., Geist, D., and Ralston, D., 1998, Controls of basaltic lava flow structures on groundwater flow, Snake River Plain, Idaho: Eos (Transactions, American Geophysical Union), v. 79, p. F959.

Elwood, B.B., 1978, Flow and emplacement direction determined for selected basaltic bodies using magnetic anisotropy measurements: Earth and Planetary Science Letters, v. 41, p. 254–264.

Garcia, M.O., Rhodes, J.M., Trusdell, F.A., and Pietruszka, A.J., 1996, Petrology of lavas from the Puu Oo eruption of Kilauea Volcano. 3. The Kupaianaha pisode (1986–1992): Bulletin of Volcanology, v. 58, p. 359–379.

Ghiroso, M.S., and Sack, R.O., 1995, Chemical mass transfer in magmatic processes. 4. A revised and internally consistent thermodynamic model for the interpolation and extrapolation of liquid-solid equilibria in magmatic systems at elevated temperatures and pressures: Contributions to Mineralogy and Petrology, v. 119, p. 197–212.

Goff, F., 1996, Vesicle cylinders in vapor-differentiated basalt flows: Journal of Volcanology and Geothermal Research, v. 71, p. 167–185.

Holcomb, R., Champion, D., and McWilliams, M., 1986, Dating recent Hawaiian lava flows using paleomagnetic secular variation: Geological Society of America Bulletin, v. 97, p. 829–839.

Hon, K., Kauahikaua, J., Denlinger, R., and Mackay, K., 1994 Emplacement and inflation of pahoehoe sheet flows: Observations and measurements of active lava flows on Kilauea Volcano, Hawaii: Geological Society of America Bulletin, v. 106, p. 351–370.

Johnson, D M., Hooper, P.R., and Conrey, R.M., 1999, XRF analysis of rocks and minerals for major and trace elements on a ingle low dilution Litetraborate fused bead: Advances in X-ray Analysis, v. 41, p. 843–867.

Knobel, L.L., Cecil, L.D., and Wood, T.R., 1995, Chemical composition of selected core samples, Idaho National Engineering Laboratory, Idaho: U.S. Geological Survey Open-File Report 95-748, 59 p.

Kuntz, M., 1992, A model-based perspective of basaltic volcanism, eastern Snake River Plain, Idaho: Geological Society of America Memoir, v. 179, p. 289–304.

Kuntz, M., Covington, H.R., and Schorr, L.J., 1992, An overview of basaltic volcanism of the eastern Snake River Plain, Idaho, *in* Link, P.K., Kuntz, M.A., and Platt, L.B., eds., Regional geology of eastern Idaho and western Wyoming: Geological Society of America Memoir 179, p. 227–267.

Lanphere, M.A., Kuntz, M.A., and Champion, D.E., 1994, Petrography, age, and paleomagnetism of basaltic lava flows in coreholes at Test Area North (TAN), Idaho National Engineering Laboratory: U.S. Geological Survey Open-File Report 94-686, 47 p.

Lipman, P.W., 1995, Declining growth of Mauna Loa volcano during the last 10,000 years: Rates of lava accumulation versus gravitational subsidence, *in* Rhodes, J.M., and Lockwood, J.P., eds., Mauna Loa revealed: American Geophysical Union Geophysical Monograph 92, p. 45–80.

Reidel, S.P., Tolan, T.L., Hooper, P.R., Beeson, M.H., Fecht, K.R., Bentley, R.D., and Anderson, J.L., 1989, The Grande Ronde Basalt, Columbia River Basalt Group: Stratigraphic descriptions and correlations in Washington, Oregon, and Idaho, *in* Reidel, S.P., and Hooper, P.R., eds., Volcanism and tectonism in the Columbia River Flood-Basalt Province, Geological Society of America Special Paper 239, p. 21–53.

Reynolds, R., Geist, D., and Kurz, M., 1995, Physical volcanology and structural development of Sierra Negra volcano, Galápagos Archipelago: Geological Society of America Bulletin, v. 107, p. 1398–1410.

Rhodes, J.M., 1982, Homogeneity of lava flows: Chemical data for historic Mauna Loa eruptions: Journal of Geophysical Research, v. 88, p. 869–879.

Rowland, S., and Walker, G.P.L., 1987, Toothpaste lava: Characteristics and origin of a lava structural type transitional between pahoehoe and a'a,: Bulletin of Volcanology, v. 49, p. 631–641.

Rowland, S.K., and Walker, G.P.L., 1990, Pahoehoe and a'a in Hawaii: Volumetric flow rate controls the lava structure: Journal of Volcanology and Geothermal Research, v. 52, p. 615–628.

Self, S., Thordarson, T., and Kesztheyli, L., 1997, Emplacement of continental flood basalt lava flows, *in* Mahoney, J.J., and Coffin, M.F., eds., Large igneous provinces: Continental, oceanic, and planetary flood volcanism: American Geophysical Union Monograph 100, p. 381–410.

Self, S., Kesztheyli, L., and Thordarson, T., 1998, The importance of pahoehoe: Annual Reviews of Earth and Planetary Sciences, v. 26, p. 81–110.

Sorenson, K.S., Wylie, A.H., and Wood, T.R., 1996, Test Area North site conceptual model and proposed hydrogeologic studies operable unit 1-07B: Idaho Falls, Idaho, Report submitted to Lockheed Idaho Technologies Company and U.S. Dept. of Energy by Parsons Engineering Science, Inc., 196 p.

Walker, G.P.L., 1975, The lengths of lava flows: Royal Society of London Philosophical Transactions, v. 274, p. 107–118.

Walker, G.P.L., 1987, Pipe vesicles in Hawaiian basaltic lavas: Their origin and potential as a paleoslope indicator: Geology, v. 15, p. 84–87.

Welhan, J.A., and Reed, M.F., 1997, Geostatistical analysis of regional hydraulic conductivity variations in the Snake River Plain aquifer, eastern Idaho: Geological Society of America Bulletin, v. 109, p. 855–868.

Welhan, J.A., Johannesen, C.M., Glover, J.A., Davis, L.L., and Reeves, K.S., 2002, Overview and synthesis of lithologic controls on aquifer heterogeneity in the eastern Snake River Plain, Idaho, *in* Bonnichsen, B., White, C.M., and McCurry, M., eds., Tectonic and magmatic evolution of the Snake River Plain Volcanic Province: Idaho Geological Survey Bulletin 30 (in press).

Wilmoth, R.A., and Walker, G.P.L., 1993, P-type and S-type pahoehoe: A study of vesicle distribution patterns in Hawaiian lava flows: Journal of Volcanology and Geothermal Research, v. 55, p. 129–142.

MANUSCRIPT ACCEPTED BY THE SOCIETY NOVEMBER 2, 2000

Geological Society of America
Special Paper 353
2002

Sedimentologic and hydrologic characterization of surficial sedimentary facies in the Big Lost Trough, Idaho National Engineering and Environmental Laboratory, eastern Idaho

Linda E. Mark*
Glenn D. Thackray
Department of Geosciences, Idaho State University, Pocatello, Idaho 83209, USA

ABSTRACT

Hydrologic and sedimentologic properties of surficial sedimentary facies in the Big Lost Trough bear upon the hydrologic characteristics of subsurface sedimentary interbeds within the Snake River Plain aquifer. Surficial sedimentary facies include pedogenically modified lake-floor sediments covered in many areas by longitudinal dunes, pedogenically modified loess deposits, deposits of relatively small, historically active playas, and interfluve sediments cut by coarse-grained fluvial channel deposits. During interglacial time, playas become seasonally active, dunes are active, and the interfluve facies is dissected by isolated fluvial channels. During glacial periods, when lake levels are high, lake-floor and loess sediments are deposited and the fluvial braid-plain aggrades thick sheets of gravel in the southern Big Lost Trough. The surficial sedimentary facies are inferred to be analogous to the sedimentary interbeds, the depositional history and geometry of which are fundamentally climatically controlled.

Graphical and statistical analyses of grain-size data suggest that sedimentary facies can be partially or completely identified by grain-size distribution. Fluvial channel sediments exhibit a coarser, statistically distinctive grain-size distribution. Most dune samples can be distinguished by their cumulative frequency curves, or statistically using their d_{84} grain-size parameters, and lake-floor sediments can be distinguished using their d_{16} grain-size parameters. The loess, playa-bottom, and interfluve sediments cannot be completely distinguished through grain-size characteristics.

The channel sedimentary facies forms a distinct hydrologic facies, and the non-channel sediments form two overlapping hydrologic facies. The interfluve, playa-bottom, and dune facies form one hydrologic facies, and the lake-floor, loess, interfluve, and playa-bottom facies form a second nonchannel hydrologic facies. The hydraulic conductivity (K) of the channel facies ranges from $10^{0.9}$ cm/s to $10^{-1.7}$ cm/s, and the field-saturated hydraulic conductivities (K_{fs}) of the dune, playa-bottom, interfluve, loess, and lake-floor facies are in a broad continuum from $10^{-1.8}$ cm/s to $10^{-4.7}$ cm/s. Channel sediments likely form laterally continuous ribbons in the subsurface, as on the surface, cutting through fine-grained areas with low K, and thus may provide pathways for rapid groundwater flow.

*Current address: U.S. Geological Survey, Cascades Volcano Observatory,
1300 SE Cardinal Court, Building 10, Suite 100, Vancouver, WA 98683.
E-mail: lemark@usgs.gov

Mark, L.E., and Thackray, G.D., 2002, Sedimentologic and hydrologic characterization of surficial sedimentary facies in the Big Lost Trough, Idaho National Engineering and Environmental Laboratory, eastern Idaho, *in* Link, P.K., and Mink, L.L., eds., Geology, Hydrogeology, and Environmental Remediation: Idaho National Engineering and Environmental Laboratory, Eastern Snake River Plain, Idaho: Boulder, Colorado, Geological Society of America Special Paper 353, p. 61–75.

INTRODUCTION

Sedimentary interbeds are a key component of the Snake River Plain aquifer. Although occupying far less volume than basalt, the sedimentary interbeds play a significant role in controlling groundwater flow. The distribution and characteristics of the sediments, influenced by climatically controlled lake-level fluctuations, determine the degree to which they retard or facilitate movement of water and contaminants.

In particular, modeling flow and transport of a plume of trichloroethylene and other pollutants within the Snake River Plain aquifer, below the northern portion of the Idaho National Engineering and Environmental Laboratory (INEEL), necessitates quantification of the hydrologic and sedimentologic characteristics of the sedimentary interbeds. Identification of depositional facies within the sedimentary interbeds is important both for understanding the hydrologic role of the sedimentary interbeds and for deciphering the sedimentary and climatic history of the Pliocene to Holocene Big Lost Trough (e.g., Geslin et al., 1999).

This chapter describes research intended to quantify the hydrologic and sedimentologic properties of surficial sedimentary facies, which are analogous to the sedimentary interbeds (e.g., Bartholomay, 1990; Kaminsky et al., 1994; Geslin et al., 1999). In addition, this research aims to understand the degree to which pedogenic carbonate buildup (caliche development) affects the hydraulic conductivity of the sediments. Greater understanding of the characteristics of and interrelationships between hydraulic conductivity, sedimentary facies, and climate in the Big Lost Trough should aid in the understanding of similar sequences of surficial sediments and sedimentary interbeds elsewhere.

GEOLOGIC AND PALEOCLIMATIC SETTING

Big Lost Trough

The study area is located largely within the Big Lost Trough and entirely within the INEEL. It extends southwest from Circular Butte to the Naval Reactor Facility (Fig. 1). The Big Lost Trough is an underfilled sedimentary and volcanic basin surrounded on the west and northwest by the Lost River, Lemhi, and Beaverhead mountain ranges and separated from the rest of the eastern Snake River Plain by volcanic constructs (Gianniny et al., 1997). It receives fluvial sediment from rivers draining these mountain ranges. The Big Lost River currently flows into the Big Lost Trough, terminating in small, infiltrative playas. The Little Lost River and Birch Creek were water sources for the Big Lost Trough in the past, but are currently diverted for irrigation purposes. The northeast-trending Axial volcanic zone is a topographic high that forms both the southeastern boundary of the Big Lost Trough and the central axis of the eastern Snake River Plain (Hackett and Smith, 1992).

The Circular Butte–Kettle Butte volcanic rift zone composes the northeastern boundary of the Big Lost Trough and separates the Big Lost Trough from the Mud Lake subbasin of Lake Terreton (Gianniny et al., 1997). The Arco volcanic zone forms the southwestern boundary of the Big Lost Trough. These basaltic topographic highs cause eolian sediment to be partially trapped, and fluvial and lacustrine sediment to be wholly trapped, within the closed basin. Fluvial and lacustrine sediment is redistributed as loess and wind-blown sand.

Lake-level history

Lake-level variations play a prominent role in determining facies distribution. Pleistocene Lake Terreton had water levels high enough to inundate both the Mud Lake subbasin and much of the Big Lost Trough (e.g., Scott, 1982) (Fig. 1). Because the Big Lost Trough exhibits very low relief, small changes in the level of Lake Terreton produced large changes in its geographic extent (Geslin et al., 1999).

Recent research has elucidated lake-level histories on the northeastern Snake River Plain, illuminating the effect of climate on the sedimentary system. The inferred cool, wet periods have alternated with dry, possibly interglacial (or interstadial) periods, characterized by an absence of Lake Terreton in the Big Lost Trough and a considerably reduced Mud Lake. The oldest dated highstand is represented by lacustrine sediment in cores from the Mud Lake subbasin and the Big Lost Trough and is dated as between 120 and 160 ka; a subsequent highstand has been dated as 88 ± 8 ka (Gianniny et al., this volume). The Mud Lake subbasin also underwent a highstand around the time of the last glacial maximum. Forman and Kaufman (1997) obtained a radiocarbon age of 21.2 ka on Mud Lake sediments several meters above current lake level. A radiocarbon age of 11.6 ka was obtained on a gastropod shell from the North Lake embayment within the Mud Lake subbasin, suggesting an elevated lake level during latest Pleistocene time (Gianniny et al., this volume). Forman and Kaufman (1997) obtained radiocarbon ages on aquatic snails and charcoal indicating a Mud Lake highstand less than 1000 years B.P.

Subsurface geology and hydrogeology

Subsurface sedimentary interbeds separate basalt flow units in the Big Lost Trough (Anderson and Bowers, 1995). The sedimentary interbeds in the northern portion of the Big Lost Trough are as much as 12 m thick (Kaminsky et al., 1994). They are much thinner than sedimentary interbeds in the middle of the Big Lost Trough (Anderson and Bowers, 1995; Geslin et al., 1997), which are as thick as 58 m in the upper half of well 2-2A (Fig. 1) (Blair and Link, 2000; Bestland et al., this volume).

The top 60 m of the volcanic-sedimentary sequence beneath the northern Big Lost Trough is unsaturated; below this

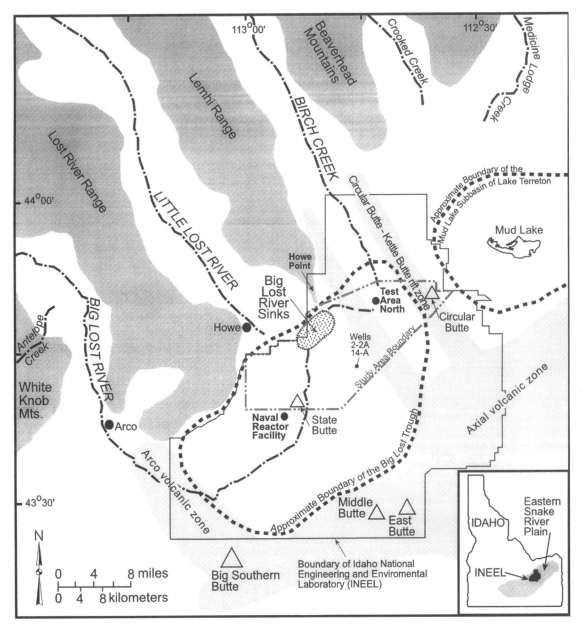

Figure 1. Map of Idaho National Engineering and Environmental Laboratory (INEEL) showing outline of Big Lost Trough. Modified from Geslin et al. (1999), after Gianniny et al. (1997). Test Area North and Naval Reactor Facility are INEEL facilities.

depth is the Snake River Plain aquifer, which extends from ~60 m to 270 m below the surface (Anderson and Bowers, 1995). Although unfractured basalt is relatively impermeable, the subsurface basalt in the Snake River Plain aquifer contains numerous fractures and rubble zones that promote very high hydraulic conductivity. Highly fractured basalt may provide unrestricted avenues for vertical and horizontal flow (Anderson, 1991). Within this hydrogeologic setting, sedimentary interbeds are typically assumed to act as aquitards. Their sedimentologic

and hydrologic properties, however, have not been previously investigated in detail.

Surficial sediments and sedimentary interbeds

Surface features that make up the sedimentary system in the northern portion of the Big Lost Trough consist of lake-floor deposits from Pleistocene Lake Terreton, modern playa deposits, eolian dunes, loess deposits, fine-grained interfluve deposits (various depositional environments), and alluvial channels of the Big Lost River and Birch Creek (e.g., Scott, 1982; Kuntz et al., 1994). The sedimentary interbeds contain deposits of fluvial, eolian, and lacustrine origin (Nace et al., 1975; Reed and Bartholomay, 1994; Anderson and Bowers, 1995). Geslin et al. (1997) interpreted sedimentary interbeds in core obtained from 73 to 226 m below the northern Big Lost Trough to be of playa-bottom, playa-margin, lake-margin, lake-floor, and fluvial origin. Sand-rich subsurface deposits, dominated by Big Lost River and eolian sediments, suggest periods of dry climate similar to that of today (Geslin et al., 1999). These deposits alternate with clay-rich deposits that suggest periods of wetter climate marked by Lake Terreton highstands. The sedimentary interbeds represent about the same range of grain sizes as the surficial sediments (Nace et al., 1975).

Mineralogic characteristics of subsurface sedimentary interbeds correlate closely with those of surficial deposits. The mineralogy of sedimentary interbeds below the southern and central portions of the Big Lost Trough is similar to that of the modern Big Lost River (Bartholomay, 1990), and detrital-zircon age spectra from subsurface sands within the Big Lost Trough also suggest deposition by the Big Lost River (Geslin et al., 1999). Furthermore, the sedimentary interbeds below the northern Big Lost Trough are mineralogically similar to those of the modern Birch Creek drainage (Bartholomay, 1990). Therefore, the sedimentary interbeds were likely deposited in a basin very similar to the modern one (Bartholomay, 1990; Geslin et al., 1999).

METHODS

Investigative methods, which were described fully in Mark (1999), included field observation, Guelph permeameter tests, grain-size analysis, and carbonate dissolution analysis. Field observations documented the geomorphic and sedimentologic setting of each study site, and shallow test pits (<1.5 m) permitted observation of near-surface stratigraphy and soil carbonate stages (e.g., Birkeland, 1999). The field-saturated hydraulic conductivity (K_{fs}) of each of the sedimentary facies, with the exception of the channel facies, was determined in situ at each of the selected field sampling locations using a Guelph permeameter (as discussed in Reynolds and Elrick, 1986). Approximately 600–1000 g of sediment was collected with a hand trowel at the depth of each Guelph permeameter test. Hydraulic conductivity of the channel sediments is too high to be measured with the Guelph permeameter; therefore, larger samples were collected and hydraulic conductivity was determined solely through grain-size analysis. An equation derived by Shepherd (1989), specifically for channel sediments, was used to estimate the K of the channel sediments:

$$K = (.1652)d^{1.65} \qquad (1)$$

where K = hydraulic conductivity (cm/s) and $d = d_{50}$ (mm) grain size.

Grain-size analysis, using American Society for Testing and Materials method D 422-63 (ASTM, 1986), was performed on clastic sediments of the Big Lost Trough. These sediments include clay-sized particles (clay minerals and pedogenic carbonate), silt-sized particles (pedogenic carbonate and clastic mineral grains), and sand-sized particles (mineral and lithic grains as well as grain aggregates). The Udden-Wentworth grain-size scale is used throughout this chapter, and nomenclature of sediments is based on Pettijohn et al. (1972, Fig. 1-2A). Mineralogic analyses were presented in Bartholomay et al. (1989), and modal compositional analyses of sands can be found in Gianniny et al. (1997) and Geslin et al. (1999; this volume). Carbonate-cemented grain aggregates are highly durable and likely act hydrologically as sand grains. These aggregates are generally sturdy enough to withstand grain-size analysis and are therefore recorded as sand-sized material. Clay-cemented grain aggregates are considerably less durable than carbonate-cemented aggregates and are unlikely to act hydrologically as sand grains. They were disaggregated during grain-size analysis and are recorded as the size of their constituent particles. Both types of aggregates are common to the dune facies and are thought to be the product of playa deflation.

Carbonate dissolution (Mark, 1999; adapted from Pudney, 1994) was performed on 15 samples in order to determine their mass percent of calcium carbonate. While pedogenic carbonate content is of greatest interest, the results of this method include any lithogenic carbonate that may be present as sedimentary particles.

Statistical analyses were performed on the log K_{fs} data (field-saturated hydraulic conductivity determined by the Guelph permeameter tests) and log K data (hydraulic conductivity, determined from grain-size analysis) in order to determine the degree to which the sedimentary facies exhibit similar hydrologic properties. Similar tests were performed on the grain-size data in order to test for differences in grain size between the sedimentary facies. Where possible, the statistical software package Statistical Product and Service Solutions (SPSS for Windows, release 8.0.0, 1997) was used. An analysis of variance (ANOVA) test, a parametric test discussed by Zar (1999) that assumes equal variances and normality, and Tukey's (1953) Honestly Significant Difference (HSD) post hoc (follow-up) test, as cited in Zar (1999), were used to investigate differences in hydraulic conductivity between the sedimentary facies. Once the differences in hydraulic conductivity between the sedimentary facies were established, the sedimentary facies could be grouped into hydrologic facies. A one-way multivariate analysis of variance (MANOVA) test, as discussed in Rencher (1995), and Tukey's (1953) HSD post hoc test were used to test for differences in grain size between the sedimentary facies. A 95% confidence level was used for all statistical tests.

SEDIMENTARY SYSTEM

Sedimentary facies

This study recognizes six surficial sedimentary facies in the northern Big Lost Trough. The surficial sediments were originally divided by Gianniny et al. (1997) into seven surficial sedimentary facies, following the work of Nace et al. (1975) and Scott (1982). These are the dune, loess, fluvial channel–braid-plain, fluvial overbank, playa–Lake Terreton margin, playa (sinks) bottom, and Lake Terreton bottom facies. These were modified slightly for the purpose of this study. The fluvial channel–braid-plain facies was renamed the channel facies. Instead of using the overbank facies designation, the interfluve designation was used for the large areas of fine-grained sediments between fluvial channels south of Howe Point; some of these are overbank sediments, but others are of loess, playa-bottom, and/or lake-floor origin, simply cut by fluvial channels. We did not study the playa–Lake Terreton margin facies, because it covers only small areas and typically consists of dunes.

Figure 2 depicts the surficial geology of the northern INEEL. The sedimentary characteristics of the surficial sedimentary facies, including mean grain size, clay percentage, sorting, and skewness, were determined from grain-size analysis (Table 1). Facies were described in greater detail in Mark (1999).

Lake-floor facies. The lake-floor facies consists of Pleistocene Lake Terreton sediments and is exposed near the northeastern boundary of the INEEL near Mud Lake and Circular Butte and southwest of Test Area North (TAN) to the border of the INEEL, abutting the Little Lost River Valley near Howe (Figs. 1 and 2). In many areas (e.g., south and southeast of Circular Butte and southwest of TAN), longitudinal dunes overlie the lake-floor facies. South of the Big Lost River sinks, fine-grained sediments of possible Lake Terreton nearshore origin are overlain by loess and/or overbank deposits and are cut by fluvial channels.

The lake-floor facies is generally massive and primarily consists of silty clay; it is the finest grained of the sedimentary facies. Clay minerals within the lake-floor facies include kaolinite, illite, and smectite, and minor amounts of chlorite (Kauffman and Geslin, 1998). Within the top meter, lake-floor sediment is characterized by soil structure (blocky peds) or is massive. It lacks laminations, likely because of bioturbation and pedogenesis.

Carbonate buildup within the lake-floor facies ranges from none to stage II (cf. Birkeland, 1999). Stage II to II+ carbonate buildup in fine-grained alluvium at the INEEL is estimated to be 10 k.y. old or older (Simpson et al., 1999). By analogy, the lake-floor sediments with stage II carbonate buildup have likely maintained stable surfaces for 10 k.y. or longer.

Playa-bottom facies. The playa-bottom facies includes modern and historically active infiltrative playa-bottom deposits. The playa-bottom facies consists of locally structureless sediment ranging from silty clay to silty clay sand. Playa-bottom sediments locally contain a layer of fine- to medium-grained sand to ~6 cm thick. The playa-bottom deposits are coarser than lake-floor sediments, exhibit considerably less pedogenic alteration than do the lake-floor deposits, and generally do not exhibit visible pedogenic carbonate buildup within the top meter. This lack of pedogenic carbonate horizons is likely a result of very recent activity of the playas.

Dune facies. The dune facies is composed largely of northeast-trending longitudinal dunes. Parabolic dunes have formed downwind of playas 1 and 2 (Fig. 2). Scott (1982) also noted the presence of transverse dunes. The activity level of the dunes varies; some dune fields are inactive, supporting mature sagebrush and forming subdued topography. Others are vegetated more sparsely, have a well-defined shape, and may still be active.

The dune facies consists of cross-bedded, locally laminated, sand to silty clay sand, except where adjacent to playas. Sand grains are typically angular to subrounded. The presence of clay-cemented aggregates—recorded as the size of their constituent particles because of the wet-sieving procedure—causes the dune facies to be recorded as poorly sorted to very poorly sorted. Clay content typically ranges from 7% to 19% (Mark, 1999). Some dune sites display a thin (0.5 cm), friable surface crust that may result from loess deposition. Dune sediments do not contain visible pedogenic carbonate buildup.

In addition to sand, silt, and clay-cemented aggregates, the INEEL dunes contain limited quantities of durable, rounded, sand-sized, carbonate-cemented aggregates that locally compose more than 5% of individual samples. The percentage of aggregates, both carbonate and clay-cemented, varies partly with location of the dune: dunes adjacent to playas generally have a higher percentage of aggregates than do dunes farther from playas.

Grain-size characteristics of individual dunes similarly vary with distance from local sediment sources. Three dune sites adjacent to playa 2 coarsen on a local (meter) scale as their distance from the playa increases. This coarsening reflects a relative decrease in fine-grained (finer than 3.5 phi) material. The relative decrease in fine-grained material may be caused by the destruction of clay aggregates during transport away from the playa and the subsequent eolian removal of the disaggregated clay.

In general, dunes farther from fine-grained lacustrine and playa sediment sources are characterized by a coarser grain size than are dunes closer to lacustrine-playa sediment sources. The coarser dunes likely receive the majority of their sediment from coarser grained, alluvium-dominated source areas rather than from finer grained, lacustrine-dominated source areas.

The generally fine mean grain size of Big Lost Trough dunes could have several explanations. These dunes receive large amounts of fine-grained sediment from local lacustrine sources, giving the dunes a finer than normal grain size. Clay aggregates, which are recorded as the size of their constituent

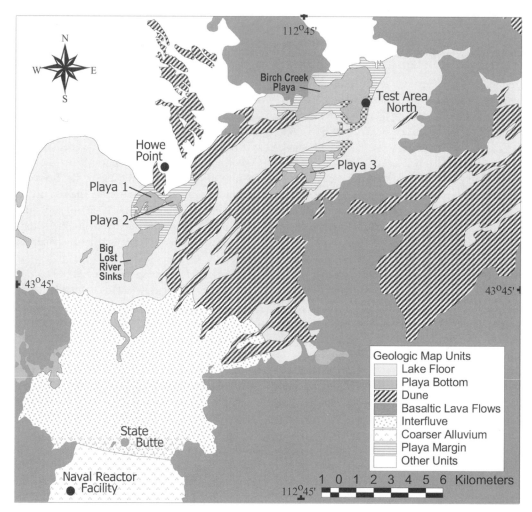

Figure 2. Geologic map of northern Idaho National Engineering and Environmental Laboratory (INEEL). Modified from Kuntz et al. (1994). Test Area North and Naval Reactor Facility are INEEL facilities. Note that interfluve area shown on map was mapped as fine-grained alluvial deposits (Qmtf) by Kuntz et al. (1994).

particles, certainly make the grain-size distributions finer than they otherwise would be. The INEEL dunes may also be over-printed with loess. Vegetation on the dunes has helped to sta-bilize the dunes to varying degrees and may act to trap fine-grained eolian sediment and prevent its removal from the dune system. Pedogenesis may also increase the percentage of fine-grained sediment. Note that the dune grain-size data in this study are all from the top 1 m of dune sediment. The cores of dunes may be coarser than the top 1 m, which is more likely to have been affected by loess accumulation and pedogenesis.

Loess facies. The loess facies chiefly consists of wind-transported silt with lesser percentages of sand and clay. Loess at the INEEL generally ranges in thickness from <10 cm to 3 m (Nace et al., 1975). Although loess is widely distributed at the INEEL on basaltic lava flows and interfluve areas, the loess

sites that were sampled were all located on lower to middle Pleistocene basalt flows (Kuntz et al., 1994). These locations were chosen because of their lack of alluvial sediment contam-ination and ample thickness of loess (at least 0.7 m) needed in order to perform Guelph permeameter tests; younger basalt flows do not have an adequate thickness of loess.

The loess facies is generally massive and is primarily sandy clay silt. Silt content ranges from 57% to 20% and averages 41%. Much of the loess has an eolian sand component, likely derived from nearby alluvium. The loess is typically massive and does not exhibit primary sedimentary structures, although at some locations relatively sand-rich layers are present.

Pedogenic carbonate buildup in the loess ranges from none to stage III (cf. Birkeland, 1999). Forman et al. (1993) con-cluded, on the basis of thermoluminescence dates on loess at

TABLE 1. RANGE OF MEAN GRAIN SIZES, SORTING, AND SKEWNESS FOR EACH OF THE SIX SEDIMENTARY FACIES

Facies	Mean grain size	Clay content	Sorting	Skewness
Lake floor	Finer than 6.1 phi (fine silt) to finer than 7.8 phi (very fine silt); averages finer than 7.1 phi (very fine silt)	24%–60%	Poorly sorted*	Nearly symmetrical to strongly coarse skewed[†]
Playa bottom	4.6 phi (coarse silt) to finer than 7.3 phi (very fine silt); averages finer than 6.2 phi (fine silt)	18%–60%	Poorly sorted to very poorly sorted*	Strongly fine skewed to strongly coarse skewed[†]
Dune	2.7 phi (fine sand) to finer than 5.9 phi (medium silt); averages finer than 4.1 phi (coarse silt)	Usually 7%–19%	Poorly sorted to very poorly sorted	Typically strongly fine skewed to nearly symmetrical[†]
Loess	4.9 phi (coarse silt) to finer than 6.0 phi (medium silt); averages finer than 5.5 phi (medium silt)	15%–40%	Poorly sorted to very poorly sorted*	Typically strongly fine skewed to nearly symmetrical[†]
Channel	−2.4 phi (pebble) to 1.7 phi (medium sand); averages 0.2 phi (coarse sand)	less than 3% combined clay and silt	Ranges from moderately sorted to very poorly sorted	Strongly fine skewed to strongly coarse skewed
Interfluve	4.5 phi (coarse silt) to finer than 7.0 phi (fine silt); averages finer than 5.7 phi (medium silt)	9%–34%	Poorly sorted to very poorly sorted*	Strongly fine skewed to strongly coarse skewed[†]

Note: Data from Mark (1999). Sorting and skewness were calculated according to Folk and Ward (1957).
* These are maximum estimates of sorting due to a lack of grain-size data in the finer than ~8.2 phi range. The sediment may be more poorly sorted than indicated.
[†] These are maximum estimates of skewness due to a lack of grain-size data in the finer than ~8.2 phi range. The sediment may be less skewed than indicated.

the INEEL, that there were at least two periods of loess deposition within the past 100 k.y. on the Snake River Plain: 80–60 ka and 40–10 ka. The long-term stability of the loess-covered surfaces has favored the formation of considerable pedogenic carbonate buildup.

Channel facies. The channel facies contains channel deposits of the modern Big Lost River, in addition to continuous channel deposits from a dozen or more abandoned channels that cut the landscape. In some areas between Howe Point and Naval Reactor Facility (Fig. 1), abandoned channels are located within 0.1 km of each other. Some of the abandoned channels are at least as large as the modern Big Lost River channel.

The channel facies typically contains layered or graded deposits of sand and pebbles. The grain size of the channel facies decreases from pebble and granule dominated in the southern Big Lost Trough (south of State Butte) to sand dominated with lesser pebbles and granules in the northern Big Lost Trough (north of State Butte) (Mark, 1999). Channel sand grains are frequently angular to subrounded, whereas channel pebbles are frequently subrounded to well rounded. Minor pedogenic carbonate buildup (sub-stage I) is present within the channel facies at some locations, but it does not appear to cement the sediments.

The skewness and sorting of the channel sediment are dependent on the relative location downstream (Mark, 1999). The skewness and sorting range from strongly fine skewed and very poorly sorted at the three southern (upstream) sampling locations to strongly coarse skewed and moderately sorted at one of the northernmost (downstream) sampling locations.

Interfluve facies. The interfluve facies includes interchannel areas between Howe Point and the Naval Reactor Facility (Figs. 1 and 2). Interfluve areas have undergone multiple depositional processes and contain loess, playa-bottom sediments, overbank deposits, lake-floor sediments, or a combination thereof, depending on location. Because of their hybrid nature, the interfluve sediments are indistinguishable from the loess and playa-bottom sediments. In general, the interfluve facies has undergone less pedogenic alteration and carbonate buildup than have the lake-floor and loess facies. Visible carbonate buildup is generally negligible.

Summary of the sedimentary facies interrelationships

The distribution of sedimentary facies and pedogenic carbonate buildup in the Big Lost Trough was controlled in large part by climatic fluctuations, as reflected in the lake-level history (Table 2). It is generally inferred that, during glacial climates, the precipitation/evaporation ratio was higher and Lake Terreton rose. Under these conditions, the lake-floor facies would undergo active deposition, and there would be few or no playa deposits or dune deposits. Loess deposition is generally believed to increase during glacial periods, while decreasing during interglacials when pedogenesis increases (Forman et al., 1993). Coarser alluvial deposits would likely be deposited at an increased rate during glacial periods. These deposits are seen in the southern Big Lost Trough, where laterally extensive Pleistocene alluvial deposits containing pebbles and granules imply the former presence of extensive, aggrading braid-plains. Glacial episodes are likely represented in the subsurface by laterally extensive lake-floor deposits with very little pedogenic alteration, widespread loess with limited pedogenic alteration, and coarser, more extensive channel–braid-plain deposits.

Interglacial conditions, which generally have resulted in low lake levels, lead to increased facies complexity due to

TABLE 2. CLIMATE DEPENDENT FACIES RELATIONSHIPS

Lake-level conditions	Lake floor	Playa bottom	Dune	Loess	Channel northern INEEL	Interfluve dissection	Braid-plain southern INEEL
Modern lowstand		X	X		X	X	
11.6–21.2 ka highstand	X			X			X

Note: An X indicates active deposition in a facies, except in the case of interfluve sediments, where it indicates facies dissection. Ages are approximate ages of the most recent Lake Terreton highstand and lowstand; see text. We expect these lake-level dependent facies sequences to be repeated in older highstand-lowstand climate cycles. INEEL—Idaho National Engineering and Environmental Laboratory.

eolian and fluvial redistribution of fluvial and lacustrine sediments, the development of playas, and pedogenic alteration of stable sediments. Modern surficial sediments resulting from a dry climate include pedogenically modified lake sediments covered in many areas by dunes, pedogenically modified loess deposits, active or recently active playa deposits of relatively limited extent, and interfluve sediments cut by active or recently active channels. In the subsurface, periods of dry, interglacial climate are likely reflected by deposits similar to the current surficial deposits.

ANALYSIS OF SEDIMENTOLOGIC AND HYDROLOGIC DATA

In order to assess the relationships within and between surficial facies, we performed qualitative and statistical analyses on grain size and hydraulic conductivity data. These analyses permit us to evaluate the similarity of the facies to one another and to determine the degree to which distinct sedimentary facies act as distinct hydrologic facies. These analyses are particularly useful for analyzing the sedimentology and hydrology of subsurface sediments.

Grain size

Cumulative frequency curves: Results. Standard grain-size analysis was performed on 74 sediment samples from the six sedimentary facies (Fig. 3). Visual analysis of Figure 3 indicates that some of the sedimentary facies are considerably more distinct than others. The channel facies is distinct. The dune facies is largely distinct, with the exception of two cumulative frequency curves that indicate a finer grain-size distribution, overlapping those of other facies. Both of these anomalous samples come from the edge or side of dunes that are immediately adjacent to playa-bottom sediments (see discussion in previous section). The loess facies appears to overlap considerably the interfluve and playa-bottom facies, but appears to be largely distinct from the lake-floor facies. The loess facies is also largely distinct from the dune facies, with the exception of the two finest-grained dune samples. The interfluve facies overlaps with the loess and playa-bottom facies but is largely distinct from the lake-floor and dune facies. The playa-bottom facies appears to have the largest range in grain size of any facies except for the channel facies. It commonly overlaps the loess,

interfluve, and lake-floor facies, but is distinct from the dune and channel facies. The lake-floor facies is the finest-grained of the six sedimentary facies. It overlaps the playa-bottom facies but is otherwise largely distinguishable from the other facies.

Cumulative frequency curves: Discussion. While few of the sedimentary facies can be completely distinguished from others on the basis of their cumulative frequency curves, visual analysis of grain-size curves narrows the range of possible facies. Using this grain-size distribution information, the facies or group of facies to which a sample belongs can be determined. The channel facies has a distinct grain-size distribution, and thus should be identifiable with confidence in the subsurface. Dune sediments from areas that are not immediately adjacent to playas can also be identified readily on the basis of grain size. Loess, interfluve, and playa-bottom sediments cannot be distinguished from each other on the basis of cumulative frequency curves, but both loess and interfluve sediments can be distinguished from lake-floor sediments.

Playa bottom and lake-floor sediments cannot be distinguished using cumulative frequency curves. Biota, particularly ostracodes, and sedimentary structures may be useful in differentiating between these two facies in the subsurface. Differentiation between playa-bottom and lake-floor sediments has important hydrologic implications, because playas are relatively small, ephemeral bodies of water, whereas lake sediments represent large, longer lived bodies of water. Thus, lakes should be represented in the subsurface by extensive, relatively continuous layers of low hydraulic conductivity, whereas playa sediments should form much more discontinuous sediment layers of somewhat higher hydraulic conductivity.

Statistical analyses performed on grain-size data

Statistical analyses of grain size: Results. Statistical analyses were performed on the grain-size data in order to determine quantitatively the similarity or dissimilarity of the grain-size distributions of the sedimentary facies. A MANOVA parametric test, as discussed by Rencher (1995), was conducted in order to determine the effect of the five nonchannel sedimentary facies on the six dependent variables. Those variables are cumulative percent coarser grain-size parameters (d_5, d_{16}, d_{50}, d_{84}, d_{90}, and d_{95}). These grain-size parameters were chosen because of their importance in describing grain-size distributions and, potentially, in controlling hydraulic conductivity. A MANOVA

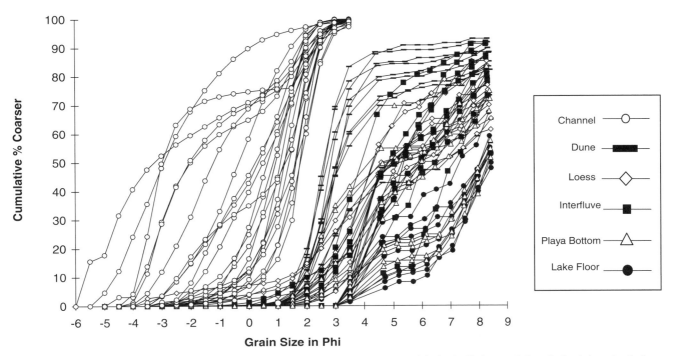

Figure 3. Cumulative frequency curves from six sedimentary facies. Note that channel facies is distinct, and dune facies is largely distinct. Loess, interfluve, and playa-bottom facies overlap considerably. Lake-floor facies is somewhat distinct but overlaps playa-bottom facies.

test on all six facies is not presented here because the sample sizes would have been unequal; thus a MANOVA test on all six facies could have yielded invalid results.

Statistical hypotheses (in each case, H_0 represents the null hypothesis and H_a represents the alternative hypothesis) comprise the following: H_0, the population means for the five nonchannel facies are equal (i.e., the grain-size means for all of the facies are the same); and H_a, the population means of at least two of the facies are different (i.e., at least two of the facies have different grain size means).

The p-value for Pillai's Trace (the MANOVA test statistic used to determine whether or not to reject H_0) is <0.001. Therefore, H_0 is rejected. In other words, the test shows that the grain-size means of at least two of the facies are different. Therefore, follow-up tests were run in order to test for effects between the facies (i.e., differences in grain size for each tested grain-size parameter).

When effects between the five nonchannel facies are tested, p is ≤0.001 for d_5, d_{16}, d_{50}, d_{84}, and d_{95} (phi). Therefore, H_0 is rejected, indicating that at least two of the facies are distinct for the d_5, d_{16}, d_{50}, d_{84}, and d_{95} grain-size parameters. However, $p = 0.163$ for d_{90}, indicating that the five nonchannel facies are not significantly different for the d_{90} grain-size parameter.

Tukey's (1953) HSD test was conducted in order to determine which of the facies are distinct with respect to each tested grain-size parameter (Fig. 4). As indicated by the Tukey's

(1953) HSD test (Fig. 4), the dune facies appears to be distinct on the basis of the d_{84} grain-size parameter. Likewise, the lake-floor facies appears to be distinct on the basis of the d_{16} grain-size parameter. Otherwise, facies are indistinct with respect to individual grain-size parameters.

Statistical analyses of grain size: Discussion. The MANOVA and Tukey's (1953) HSD test results indicate that there is overlap in grain-size parameters between the sedimentary facies. The individual sedimentary facies, with the possible exception of the channel, dune, and lake-floor facies, are not completely distinguishable on the basis of their grain-size distributions. The channel facies is clearly distinct from the other sedimentary facies based on its overall coarser grain size, as seen in Figure 3. The dune and lake-floor facies are distinct based on single grain-size parameters (d_{84} and d_{16}, respectively). However, given that neither of these facies is distinct for any of the other grain-size parameters, their classification as statistically distinct facies is tentative. Visual analysis of the cumulative frequency curves (Fig. 3) suggests that the dune facies is largely distinct, but the lake-floor facies is not; the grain-size curves of the lake-floor facies overlap considerably with those of the playa-bottom facies. The median grain size (d_{50}) appears to be the optimum grain-size parameter, of the six tested, for distinguishing statistically between the sedimentary facies; it produces four overlapping groups, each containing two facies.

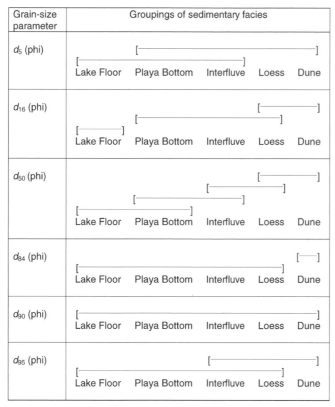

Figure 4. Results of Tukey's (1953) Honestly Significant Difference (HSD) tests performed on the five, non-channel sedimentary facies and six grain-size parameters. Horizontal lines group facies that are statistically similar based on the listed grain-size parameter. For example, for the d_{16} (phi) grain-size parameter, the loess and dune facies are statistically similar, so they are grouped together; the playa bottom, interfluve, and loess facies are statistically similar, so they are grouped together; and the lake floor facies is statistically distinct. Note that there is overlap between the first and second lines (groups). Also note that, for example, the dune and interfluve facies are not grouped together for the d_{16} (phi) grain-size parameter. Therefore, the dune and interfluve facies' d_{16} (phi) grain-size parameter is statistically different.

Guelph permeameter and grain-size estimates of hydraulic conductivity

Hydraulic conductivity results. There were 80 Guelph permeameter tests completed at unique locations and depths within the lake floor, loess, interfluve, playa-bottom, and dune facies. The error on Guelph permeameter tests in this study is $\leq 10^{-0.2}$ cm/s (Mark, 1999). The hydraulic conductivities (K) of 24 sites within the channel facies were determined using grain-size analyses and the Shepherd (1989) method.

Table 3 displays descriptive log K_{fs} statistics for each of the six facies. As seen in Table 3, the mean log K_{fs} of the lake floor, loess, interfluve, and playa-bottom facies are very similar, within 0.5 of each other. The hydraulic conductivity of the dune facies is slightly higher. Overall, the fine-grained sedimentary facies have low hydraulic conductivities relative to the channel facies.

Hydraulic conductivity: Discussion. Quantification and comparison of the hydraulic conductivity of the sedimentary facies and basalt are important for understanding water flow and pollutant migration within the Snake River Plain aquifer. Sedimentary interbeds, with which the surficial sediments are analogous, are typically assumed to represent aquitards, but their hydraulic conductivities actually span a broad range. As shown in Table 4, the K_{fs} of the nonchannel surficial sedimentary facies partly overlaps that of core-scale and aquifer-scale massive basalt hydraulic conductivity (see Table 4 for basalt data references). The K of the channel sediments in the northern Big Lost Trough, based on 21 sediment samples used to estimate hydraulic conductivity with the Shepherd (1989) method, is significantly higher than that of the finer grained facies. The hydraulic conductivity values of the channel sediments are similar to estimates of both core-scale and aquifer-scale basalt rubble and interflow-zone hydraulic conductivity. Estimates of K from three channel samples collected from the southern Big Lost Trough, which is upstream within the Big Lost River fluvial system, exhibit an even higher K, exceeding the K estimates of the basalt rubble and interflow zones.

Hydrologic facies

Statistical analyses of facies K_{fs} *data: Results.* Statistical analyses were performed on the log K_{fs} and log K data in order to determine the degree to which individual sedimentary facies act as statistically distinct hydrologic facies. An ANOVA test, as discussed in Zar (1984), was performed on the log K_{fs} data from the lake-floor, loess, interfluve, playa-bottom, and dune facies as well as on the log K data from the channel facies. The three coarsest channel sites, consisting of channel gravel from the southern Big Lost Trough, were omitted from this statistical test because their presence would have violated the ANOVA test's assumption of equal variances and normality. The ANOVA test indicates at least one significant difference in mean log hydraulic conductivity between the six facies ($F_{5,\,95}$ = 52.027, $p < 0.001$; i.e., degrees of freedom between groups = 5, degrees of freedom within groups = 95, F value = 52.027, and asymptotic significance is <0.001). Having at least one significant difference means that the log K_{fs} and log K values are significantly different for at least one sedimentary facies, and that a post hoc test is needed in order to determine how the sedimentary facies are grouped into hydrologic facies.

Based on log K_{fs} and log K, Tukey's (1953) HSD post hoc test demonstrates that the sedimentary facies may be grouped together into the hydrologic facies shown in Figure 5. The first hydrologic facies is distinct and composed of the channel sedimentary facies; the second hydrologic facies is composed of the interfluve, playa-bottom, and dune sedimentary facies and overlaps the third hydrologic facies, which is composed of the lake-floor, loess, interfluve, and playa-bottom sedimentary facies.

TABLE 3. DESCRIPTIVE STATISTICS FOR LOG K_{fs} AND LOG K DATA FOR EACH FACIES

Facies	Mean	Median	Standard deviation	Variance	Skewness	Range	Minimum	Maximum	n
Lake floor	−3.3	−3.2	0.6	0.3	−1.0	2.2	−4.7	−2.5	21.0
Playa bottom	−2.8	−2.9	0.4	0.1	1.1	1.2	−3.3	−2.1	10.0
Dune*	−2.5	−2.4	0.4	0.2	0.1	1.3	−3.1	−1.8	17.0
Loess	−3.1	−3.0	0.5	0.2	−0.3	1.7	−4.0	−2.2	20.0
Channel	−0.8	−1.2	0.9	0.7	0.9	2.5	−1.7	0.9	24.0
Interfluve	−3.0	−3.1	0.4	0.2	0.1	1.4	−3.7	−2.3	12.0

Note: K_{fs}—field-saturated hydraulic conductivity; K—hydraulic conductivity. Pre-log units are cm/s.
* Three of the log K_{fs} values in this data set are slightly outside of the stated operating range of the Guelph permeameter. However, because they are within $10^{-0.2}$ cm/s of the stated operating range, it was decided that they should be included within the data set. Their exclusion would have skewed the results.

TABLE 4. K_{fs} AND K OF SURFICIAL SEDIMENTS COMPARED WITH K OF BASALT

Surficial sediments (cm/s)	Basalt: core scale (cm/s)	Basalt: aquifer scale (cm/s)
Nonchannel facies: $10^{-4.7}$ to $10^{-1.8}$	Massive basalt*: 10^{-9} to 10^{-3}	Massive basalt*: $<10^{-3}$
Channel facies (northern INEEL only): $10^{-1.7}$ to $10^{+0.5}$	Basalt rubble*: $10^{-0.8}$ (estimate)	Interflow zone[†]: $>10^{-0.8}$ (estimate)
Average: $10^{-1.0}$		
Channel facies (southern INEEL only)[§]: $10^{+0.7}$ to $10^{+0.9}$	Basalt rubble*: $10^{-0.8}$ (estimate)	Interflow zone[†]: $>10^{-0.8}$ (estimate)

Note: K_{fs}—field-saturated hydraulic conductivity; K—hydraulic conductivity. Nonchannel facies K_{fs} values are from Guelph permeameter tests; channel facies' K are estimated from grain-size analysis using the Shepherd (1989) method. Basalt core-scale data are from direct measurements but at a scale that does not pertain directly to well-testing, whereas basalt aquifer-scale estimates are indirect but have been shown to be reasonable for interpreting well-test data and for aquifer modeling (Welhan et al., this volume, chapter 15).
* Knutson et al. (1990).
[†] Knutson et al. (1992, 1993).
[§] Range is based on only three samples.

Hydrologic facies discussion. When the results of Tukey's (1953) HSD tests performed on the grain-size parameters of the sedimentary facies (Fig. 4) are compared with the groupings of sedimentary facies into hydrologic facies (Fig. 5), it can be seen that none of the statistical groupings of the sedimentary facies based on individual grain-size parameters correlate exactly to the statistical groupings of the sedimentary facies into hydrologic facies. Mark (1999) showed that, based on a limited sample size, there is a moderate to good correlation between d_{90} (phi) and log K_{fs} and d_{50}–d_{90} (phi) and log K_{fs}, whereas there is moderate to no correlation between the other tested grain-size parameters and log K_{fs}. These correlations indicate that the fine end of the grain-size distribution is likely to have a more significant effect on the K_{fs} than is the coarser end. Although grain size plays an important role in governing K_{fs}, it appears that the hydrologic facies are not solely dependent on the individual grain-size parameters used to determine the statistical groupings of the sedimentary facies; there is variability in K_{fs} beyond that which can be attributed to grain size alone. The hydraulic conductivities of the sedimentary facies, and thus the hydrologic facies, are likely governed by their range of particle sizes (or a portion thereof), sorting, packing, porosity, particle shape, and fracture flow. For clay-rich sediments, fracture flow, structure, carbonate buildup, swelling of clays, and particle charge are likely to play an increasingly important role in controlling K_{fs}. Nonetheless, the groupings of the sedimentary and hydrologic facies are still helpful for understanding groundwater flow in relation to grain size.

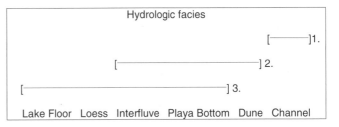

Figure 5. Groupings of sedimentary facies into hydrologic facies. Horizontal lines indicate hydrolic facies, which include the corresponding sedimentary facies shown at the bottom of the figure.

Carbonate buildup in the surface and subsurface

Carbonate buildup within some surficial sedimentary facies is similar to that found in the subsurface. This finding is important when using the surficial sediments as hydrologic analogues. Kauffman and Geslin (1998) found that some sandy subsurface samples from wells 2-2A and 14-A (Fig. 1) contain more than 60% calcium carbonate, as measured by petrographic analysis. Calcium carbonate buildup we measured within surficial dune, lake-floor, and loess sediments ranges from 12% to 57% by mass, as determined by dissolution analysis on a limited number (n = 15) of samples (Mark, 1999). Bestland et al. (this volume) have documented the presence of sub-stage I to stage III carbonate buildup in paleosols in core from well 2-2A (Fig. 1); the higher stage calcareous horizons are typically in the form of displacive micrite and are found almost exclusively directly above basalt flows. We observed visible carbonate buildup ranging from none to stage III within surficial loess sediments, also located directly above basalt flows (Mark, 1999). Surficial visible carbonate buildup in lake-floor sediments ranges from none to stage II, whereas visible carbonate buildup in dune, channel, playa-bottom, and interfluve sediments ranges from none to sub-stage I. More stable facies, namely the lake-floor and loess facies, have more pronounced carbonate buildup than do more active facies, such as the dune and playa-bottom facies. Thus, the range of carbonate buildup in subsurface sediments, as well as the proximity of carbonate buildup horizons to basalt flows, appears to be similar to that of pedogenic carbonate buildup found within the top meter in surficial loess and lake-floor sediments (Mark, 1999). However, carbonate buildup in sediments of other facies may be less pronounced than that found in the subsurface.

Stage carbonate and calcium carbonate percentage versus log K_{fs}

In order to determine the degree to which carbonate content influences hydraulic conductivity, we examined trends in the percent calcium carbonate data (Table 5) and visual stage of carbonate buildup (sensu Birkeland, 1999).

Calcium carbonate percentage and stage carbonate versus log K_{fs}: Results. The mass percent of calcium carbonate (determined by dissolution analysis) for three facies was plotted against the log K_{fs} in order to examine relationships between these two variables (Fig. 6, A, B, and C). On the basis of the limited data ($n = 5$ for each of three facies), there is a weak to nonexistent correlation ($r = -0.081$) between percent carbonate and log K_{fs} for the loess facies (Fig. 6A; Table 5). However, there is a weak to moderate negative correlation ($r = -0.389$) between percent carbonate and log K_{fs} for the dune facies, and a moderate to good negative correlation ($r = -0.675$) between percent carbonate and log K_{fs} for the lake-floor facies.

Relationships between visual carbonate stage and K_{fs} vary between facies. Facies that typically have low visual carbonate stages, such as the interfluve, playa-bottom, and dune facies, do not appear to have a correspondence between observed stage carbonate and log K_{fs} within an individual facies. In contrast, facies that generally have higher stages of visible carbonate buildup, such as the lake-floor and loess facies, generally exhibit a positive correspondence between observed stage car-

bonate and log K_{fs}. That is, higher carbonate stages correspond with higher log K_{fs}.

Calcium carbonate percentage and stage carbonate versus log K_{fs}: Discussion. Overall, there is a weak to moderate negative correlation between calcium carbonate percentage and log K_{fs}. Of the lake-floor, loess, and dune facies, the lake-floor facies has the most pronounced negative correlation between mass percent of calcium carbonate and log K_{fs}. Low visual carbonate stages do not appear to have an effect on K_{fs}, whereas higher visual carbonate stages appear to correspond more strongly with log K_{fs}, at least in fine-grained sediments.

It was expected that the calcium carbonate percentage and observed stage of carbonate buildup versus log K_{fs} would have the same effect on log K_{fs} (either both positive or both negative). The correlation between calcium carbonate percentage and log K_{fs} is negative, whereas the correspondence between observed carbonate stage and log K_{fs} is generally positive, at least for the lake-floor and loess facies. These results suggest that the mass percent of calcium carbonate may have a different effect on K_{fs} than does the distribution of pedogenic carbonate buildup. One possible explanation for this phenomenon is that fine-grained pedogenic carbonate, which may make up a significant portion of the mass, may act to clog pores more effectively than do calcium carbonate filaments or nodules, which are more responsible for the observed stage designation.

CONCLUSIONS AND IMPLICATIONS FOR HYDROLOGY OF THE SNAKE RIVER PLAIN AQUIFER

The distribution of surficial sedimentary facies is in part controlled by climate, as reflected in Lake Terreton levels; these same factors have likely influenced the distribution of subsurface facies in the Big Lost Trough. During inferred cool and wet glacial climates, the lake-floor facies undergoes active deposition, there are few to no playas or dunes, and loess and alluvial deposition occurs at an increased rate. During interglacial periods, facies complexity increases as fluvial and eolian redistribution of lacustrine and fluvial sediments occurs, playas and dunes develop, and pedogenic alteration of stable sediments occurs.

Climate has a pronounced effect on the hydrology of the sedimentary system by controlling the distribution of the sedimentary facies. For example, playa sediments represent short-lived, ephemeral lakes of relatively small geographic extent that

TABLE 5. CALCIUM CARBONATE PERCENTAGE VERSUS LOG K_{fs}

Facies	r	Correlation	r^2	Mean CaCO$_3$%	Standard deviation CaCO$_3$%	Mean log K_{fs}	Standard deviation log K_{fs}
Lake floor	−0.675	moderate to good	0.455	32.0	13.0	−3.3	0.55
Dune	−0.389	weak to moderate	0.151	14.8	2.8	−2.5	0.44
Loess	−0.081	none to weak	0.00651	36.2	12.8	−3.1	0.46
All 3 facies	−0.465	weak to moderate	0.216	27.7	13.8	−3.0	0.55

Note: K_{fs}—field-saturated hydraulic conductivity; r—Pearson's correlation coefficient. Pre-log units on mean log K_{fs} are cm/s.

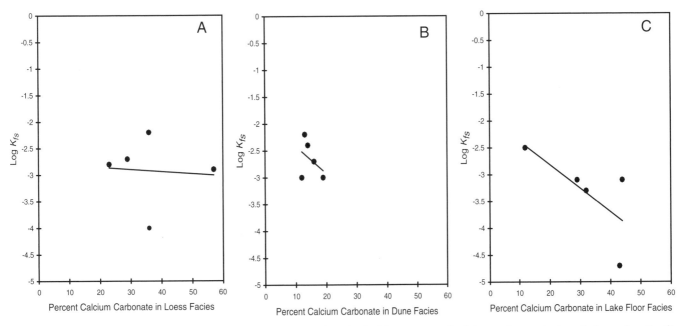

Figure 6. A: Mass percent of calcium carbonate in loess facies vs. log K_{fs}. B: Mass percent of calcium carbonate in dune facies vs. log K_{fs}. C: Mass percent of calcium carbonate in lake-floor facies vs. log K_{fs}. K_{fs} = field-saturated hydraulic conductivity. Pre-log units are cm/s.

exist during interglacials. Therefore, they are represented as relatively thin, discontinuous lenses of fine-grained sediments of low hydraulic conductivity that likely impede or may possibly deflect groundwater flow. In contrast, lake-floor sediments represent longer lived lakes of large geographic extent that existed during glacial time. They are represented by thicker, laterally extensive, clay-rich sediments that act as aquitards, impeding groundwater flow and contaminant transport.

The sedimentary facies to which a sample belongs can be identified partially or totally by its grain-size distribution. Visual and statistical analyses of cumulative frequency curves indicate that the channel facies is distinct based on grain size alone. Dune sediments (excepting those from dunes immediately adjacent to playas) typically can be identified on the basis of their overall grain-size distribution, as well as statistically using the d_{84} parameter. Samples from the loess, playa-bottom, and interfluve facies cannot be completely identified by grain-size analysis, but the range of facies to which they may belong can be significantly narrowed to the point where other identification tools, such as carbonate buildup patterns, sedimentary structure, and biota may be useful in positively identifying them. Using statistical analysis, the lake-floor facies appears to be distinct based on the d_{16} grain-size parameter, although it is not statistically distinct using other grain-size parameters. Biota (especially ostracodes) and structure may be very useful in helping to make the critical distinction between the lake-floor and playa-bottom facies.

The K_{fs} and K of surficial sediments should be representative of the range of K values of noncemented and nonlithified sediments observed in subsurface cores. Thus, the sediments of

the northern Big Lost Trough should act largely to impede groundwater flow, with the notable exception of the channel facies. The nonchannel sedimentary facies combine to form two overlapping hydrologic facies. The channel sedimentary facies forms its own hydrologic facies. The K_{fs} for the nonchannel surficial sediments is in a broad continuum between $10^{-1.8}$ cm/s and $10^{-4.7}$ cm/s. The K of the channel facies ranges from $10^{0.9}$ to $10^{-1.7}$ cm/s.

The nonchannel facies should largely act as aquitards, whereas the channel facies will readily transmit water. The K_{fs} values of the nonchannel facies are similar to that of massive basalt found within the aquifer. However, the K of the channel facies is similar to that of the basalt interflow zones. Therefore, the channel facies can be expected to facilitate rather than to impede groundwater flow.

Pedogenic carbonate buildup, particularly within the lake-floor and loess facies, is similar to that found within the sedimentary interbeds. Although small amounts of visible carbonate buildup do not appear to affect K_{fs}, significant amounts of visible carbonate, like that seen in some loess and lake-floor sediments, are likely to affect K_{fs} in fine-grained sediments. The distribution of pedogenic carbonate buildup appears to have a different effect on K_{fs} than does the mass percent of calcium carbonate. The surficial K_{fs} and K results presented here should not be used to estimate the K of core that is very heavily cemented or excessively compacted.

The hydrologic importance of the individual sedimentary facies in the subsurface depends in large part on the three-dimensional architecture of the facies. The surficial channel sediments in the northern Big Lost Trough are continuous ribbons of high hydraulic conductivity. Subsurface channel sediments can be expected to be similar. While channel sediments only make up a small fraction of the surface area, their implications for groundwater flow may be significant. On the surface, they typically cut through regions characterized by low K_{fs}. Therefore, in the subsurface they may represent avenues for rapid groundwater flow in areas otherwise characterized by low K_{fs}. In the southern Big Lost Trough, surficial gravels form a laterally extensive layer of high hydraulic conductivity, and similar layers may be present in the subsurface. Ultimately, the channel facies may represent a rough sedimentary analogue of a basalt rubble zone.

Dune sediments, which have the second highest hydraulic conductivity of the facies studied, are commonly in the form of longitudinal dunes overlying lake-floor sediments. Although the hydraulic conductivity of the dune sediments is lower than that of the channel facies, the coarser grained dune sediments may also facilitate, or at least fail to impede, groundwater flow in some regions. It is possible that the dune sediments that overlie lake-floor sediments would aid in directing horizontal groundwater flow parallel to the dunes, across the lake-floor sediments.

The most extensive loess deposits are located on top of basalt flows. Due to the low hydraulic conductivity of the loess, there is a large contrast between the hydraulic conductivity of the loess and that of underlying basalt rubble. Because loess mantles the basalt topography by filling in depressions and cracks, it may also reduce the hydraulic conductivity of the basalt with which it has contact by filling in void spaces.

Thick, clay-rich lake-floor successions are expected in the subsurface in the same areas that were covered by Lake Terreton during late Pleistocene time, i.e., from north of State Butte to the Mud Lake area (Figs. 1 and 2). These thick, broad, laterally extensive, low-hydraulic-conductivity deposits can be expected to act as regional aquitards, which may result in perched water or extensive horizontal flow, at least where they are not dissected by channels. A long-lived lake, as suggested from deep subsurface core data by Bestland et al. (this volume), would produce very thick, laterally extensive, clay-rich lake deposits. Multiple wet/dry cycles, as suggested by Geslin et al. (1999) and Gianniny et al. (this volume), would lead to multiple, somewhat thinner lake deposits in the subsurface interspersed with somewhat coarser grained material. Although playa-bottom sediments can also be expected to act as aquitards, they differ from the lake-floor sediments in that they are small, localized, and not laterally extensive. Therefore, it is possible that they may deflect groundwater flow around them, diverting it to higher hydraulic conductivity regions, or that they may act as small aquitards.

Thus, it is clear that the identification of sedimentary facies and evaluation of their hydrologic properties bear significantly on understanding groundwater flow in the Snake River Plain aquifer. The broad-scale hydrologic properties of the sediments are governed not only by their specific grain-size distributions

and hydraulic conductivities, but also by their lateral and vertical distribution. Therefore, understanding flow in this aquifer—or any other that includes packages of highly variable terrestrial sediments—hinges on thorough understanding of spatial and temporal variability in the sedimentary system.

ACKNOWLEDGMENTS

We thank Paul Link for helpful comments throughout the course of this research, Allan Wylie for suggestions regarding the hydrologic aspects of this project, and Jeffrey Geslin, J. Matthew Davis, and Scott Burns for thoughtful reviews that led to substantial refinement of this paper. Statistical guidance from Teri Peterson and Jane Chang was very helpful. We also thank Diana Boyack for redrafting the maps and field assistants Jonathan Walters, Michelle Byrd, Gaylon Lords, Jordan Vaughn, and Will Glasgow. U.S. Department of Energy grant DE-FG07-96ID13420 to the Idaho Water Resources Research Institute–Idaho Universities Consortium supported this research.

REFERENCES CITED

American Society for Testing and Materials (ASTM), 1986, Standard method for particle-size analysis of soils: Annual Book of ASTM Standards, v. 4.08, p. 116–126.

Anderson, S.R., 1991, Stratigraphy of the unsaturated zone and uppermost part of the Snake River Plain aquifer at the Idaho Chemical Processing Plant and Test Reactors Area, Idaho National Engineering Laboratory, Idaho: U.S. Geological Survey Water-Resources Investigations Report 91-4010, 71 p.

Anderson, S.R., and Bowers, B., 1995, Stratigraphy of the unsaturated zone and uppermost part of the Snake River Plain aquifer at Test Area North, Idaho National Engineering Laboratory, Idaho: U.S. Geological Survey Water-Resources Investigations Report 95-4130 (DOE/ID-22122), 47 p.

Bartholomay, R.C., 1990, Mineralogical correlation of surficial sediment from area drainages with selected sedimentary interbeds at the Idaho National Engineering Laboratory, Idaho: U.S. Geological Survey Water-Resources Investigations Report 90-4147, 18 p.

Bartholomay, R.C., Knobel, L.L., and Davis, L.C., 1989, Mineralogy and grain size of surficial sediment from the Big Lost River drainage and vicinity, with chemical and physical characteristics of geologic materials from selected sites at the Idaho National Engineering Laboratory, Idaho: U.S. Geological Survey Open-File Report 89-384, 74 p.

Birkeland, P.W., 1999, Soils and geomorphology: New York, Oxford University Press, 430 p.

Blair, J.J., and Link, P.K., 2000, Pliocene and Quaternary sedimentation and stratigraphy of the Big Lost Trough from coreholes at the Idaho National Engineering and Environmental Laboratory, Idaho: Evidence for a regional Pliocene lake during the Olduvai normal polarity subchron, *in* Robinson, L., ed., Proceedings of the 35th Symposium on Engineering Geology and Geotechnical Engineering: Pocatello, Idaho, Idaho State University, College of Engineering, p. 163–179.

Folk, R.L., and Ward, W., 1957, Brazos River bar: A study in the significance of grain size parameters: Journal of Sedimentary Petrology, v. 27, p. 3–26.

Forman, S.L., and Kaufman, D., 1997, Late Quaternary oscillations of Lake Terreton, eastern Snake River Plain, Idaho: Geological Society of America Abstracts with Programs, v. 29, no. 6, p. A253.

Forman, S.L., Smith, R.P., Hackett, W.R., Tullis, J.A., McDaniel, P.A., 1993, Timing of late Quaternary glaciations in the western United States based on the age of loess on the eastern Snake River Plain, Idaho: Quaternary Research, v. 40, p. 30–37.

Geslin, J.K., Gianniny, G.L., Link, P.K., and Riesterer, J.W., 1997, Subsurface sedimentary facies and Pleistocene stratigraphy of the northern Idaho National Engineering Laboratory: Controls on hydrogeology, *in* Sharma, S. and Hardcastle, J.H., eds., Proceedings of the 32nd Symposium on Engineering Geology and Geotechnical Engineering: Moscow, Idaho, University of Idaho, Department of Civil Engineering, p. 15–28.

Geslin, J.K., Link, P.K., and Fanning, C.M., 1999, High-precision provenance determination using detrital-zircon ages and petrography of Quaternary sands on the eastern Snake River Plain, Idaho: Geology, v. 27, no. 4, p. 295–298.

Gianniny, G.L., Geslin, J.K., Riesterer, J.W., Link, P.K., and Thackray, G.D., 1997, Quaternary surficial sediments near Test Area North (TAN), northeastern Snake River Plain: An actualistic guide to aquifer characterization, *in* Sharma, S., and Hardcastle, J.H., eds., Proceedings of the 32nd Symposium on Engineering, Geology and Geotechnical Engineering: Moscow, Idaho, University of Idaho, Department of Civil Engineering, p. 29–44.

Hackett, W.R., and Smith, R.P., 1992, Quaternary volcanism, tectonics, and sedimentation in the Idaho National Engineering Laboratory area, *in* Wilson, J.R., ed., Field guide to geologic excursions in Utah and adjacent areas of Nevada, Idaho, and Wyoming: Utah Geological Survey Miscellaneous Publication 92-3, p. 1–18.

Kaminsky, J.F., Keck, K.N., Schafer-Perini, A.L., Hersley, C.F., Smith, R.P., Stormberg, F.J., and Wylie, A.H., 1994, Remedial investigation final report with addenda for the Test Area North groundwater operable unit 1-07B at the Idaho National Engineering Laboratory: Idaho Falls, Idaho, EG&G Idaho, Inc. Report EGG-ER-10643, variously paged.

Kauffman, M.E., and Geslin, J.K., 1998, Impermeable horizons in Neogene sediments on the eastern Snake River Plain, Idaho: Calcretes in fluvial and eolian deposits and fine-grained lacustrine and playa deposits: Geological Society of America Abstracts with Programs, v. 30, no. 6, p. 12.

Knutson, C.F., McCormick, K.A., Smith, R.P., Hackett, W.R., O'Brien, J.P., and Crocker, J.C., 1990, Radioactive Waste Management Complex vadose zone basalt characterization: Idaho Falls, Idaho, EG&G Idaho, Inc., Report EGG-WM-8949, 126 p.

Knutson, C.F., McCormick, K.A., Crocker, J.C., Glenn, M.A., and Fishel, M.L., 1992, 3D vadose zone modeling: Idaho Falls, Idaho, EG&G Idaho, Inc., Report EGG-ERD-10246, variously paged.

Knutson, C.F., Cox, D.O., Dooley, K.J., and Sisson, J.B., 1993, Characterization of low-permeability media using outcrop measurements *in* Proceedings of the 68th Annual Technical Conference and Exhibition of the Society of Petroleum Engineers: Houston, Texas, p. 729–739.

Kuntz, M.A., Skipp, B., Lanphere, M.A., Scott, W.E., Pierce, K.L., Dalrymple, G.B., Champion, D.E., Embree, G.F., Page, W.R., Morgan, L.A., Smith, R.P., Hackett, W.R., and Rodgers, D.W., 1994, Geologic map of the Idaho National Engineering Laboratory and adjoining areas, eastern Idaho: U.S. Geological Survey Miscellaneous Investigations Series Map I-2330, scale 1:100 000, 1 sheet.

Mark, L.E., 1999, Hydrologic and sedimentologic characterization of surficial sedimentary facies, northern Idaho National Engineering and Environmental Laboratory, eastern Idaho [M.S. thesis]: Pocatello, Idaho, Idaho State University, 186 p.

Nace, R.L., Voegeli, P.T., Jones, J.R., and Deutsch, M., 1975, Generalized geologic framework of the National Reactor Testing Station, Idaho: U.S. Geological Survey Professional Paper 725-B, 49 p.

Pettijohn, F.J., Potter, P.E., and Siever, R., 1972, Sand and sandstone: New York, Springer-Verlag, 618 p.

Pudney, W., 1994, Physical properties of sediments affecting saturated vertical water flow at the Idaho National Engineering Laboratory [M.S. Thesis]: Pocatello, Idaho, Idaho State University, 92 p.

Reed, M.F., and Bartholomay, R.C., 1994, Mineralogy of selected sedimentary interbeds at or near the Idaho National Engineering Laboratory, Idaho: U.S. Geological Survey Open-File Report 94-374, 19 p.

Rencher, A.C., 1995, Methods of multivariate analysis: New York City, John Wiley and Sons, 627 p.

Reynolds, W.D., and Elrick, D.E., 1986, A method for simultaneous in situ measurement in the vadose zone of field-saturated hydraulic conductivity, sorptivity, and the conductivity-pressure head relationship: Ground Water Monitoring Review, v. 6, no. 1, p. 84–95.

Scott, W.E., 1982, Surficial geologic map of the eastern Snake River Plain and adjacent areas, 111° to 115° W., Idaho and Wyoming: U.S. Geological Survey Miscellaneous Investigations Series Map I-1372, scale 1:250 000, 2 sheets.

Shepherd, R.G., 1989, Correlations of permeability and grain size: Ground Water, v. 27, p. 633–638.

Simpson, D.T., Kolbe, T.E., Ostenaa, D.A., Levish, D.R., and Klinger, R.E., 1999, Lower Big Lost River chronosequence: Implications for glacial outburst flooding: Geological Society of America Abstracts with Programs, v. 31, no. 4, p. A56.

Tukey, J.W., 1953, The problem of multiple comparisons: Princeton, New Jersey, Princeton University, Department of Statistics, 396 p.

Zar, J.H., 1999, Biostatistical analysis (fourth edition): Upper Saddle River, New Jersey, Prentice-Hall, Inc., variously paged.

MANUSCRIPT ACCEPTED BY THE SOCIETY NOVEMBER 2, 2000

Geological Society of America
Special Paper 353
2002

Late Quaternary highstands in the Mud Lake and Big Lost Trough subbasins of Lake Terreton, Idaho

Gary L. Gianniny*
Department of Geosciences, Fort Lewis College, Durango, Colorado 81301, USA
Glenn D. Thackray
Department of Geosciences, Idaho State University, Pocatello, Idaho 83209, USA
Darrell S. Kaufman
*Department of Geology and Department of Environmental Sciences, Northern Arizona University,
Flagstaff, Arizona 86011, USA*
Steven L. Forman
Department of Earth and Environmental Sciences, University of Illinois, Chicago, Illinois 60607-7059, USA
Michael J. Sherbondy
R.R. 1, Box 1096, Ruffsdale, Pennsylvania 15679, USA
Delda Findeisen
Department of Geosciences, Idaho State University, Pocatello, Idaho 83209, USA

ABSTRACT

Pleistocene Lake Terreton on the northeastern Snake River Plain, Idaho, contains a record of five episodes of higher effective moisture during the past 160 k.y. Subsurface lacustrine sediments were analyzed from three cores (CB-20, CB-21, CB-23) from the Mud Lake subbasin of Lake Terreton and one (2-2A) from the Big Lost Trough subbasin. Surface sediments were examined in shallow excavations in the North Lake embayment of the Mud Lake basin. The chronology of the sediments is defined by thermoluminescence (TL) and infrared stimulated luminescence (IRSL) age estimates from lacustrine sediment, amino acid racemization age estimates from ostracodes and gastropods, and accelerator mass spectroscopy radiocarbon dates from lacustrine gastropods.

Mud Lake cores contain evidence of three highstands during the past 160 k.y. Evidence of highstands includes laminated lacustrine mud, ostracodes, and gastropods. The oldest highstand occurred between ca. 120 and 160 ka, documented by TL and IRSL age estimates ranging from 130 ± 12 ka to 152 ± 13. This highstand correlates broadly with lacustrine sediment in core 2-2A in the Big Lost Trough subbasin, which yielded ostracodes with amino acid ratios consistent with the Little Valley cycle in the Bonneville Basin (ca. 150 ka). Lacustrine sediment from a second highstand yielded luminescence age estimates of 88 ± 7 ka and 84 ± 6 ka. A younger, undated interval of lacustrine sediment may correlate with a ca. 22–11 ka highstand.

Surficial lacustrine deposits in the Mud Lake and Big Lost Trough subbasins of Lake Terreton, described in this and previous studies, document two additional highstands (ca. 22–11 ka and after 1 ka).

*E-mail: gianniny_g@fortlewis.org

Gianniny, G.L., Thackray, G.D., Kaufman, D.S., Forman, S.L., Sherbondy, M.J., and Findeisen, D., 2002, Late Quaternary highstands in the Mud Lake and Big Lost Trough subbasins of Lake Terreton, Idaho, *in* Link, P.K., and Mink, L.L., eds., Geology, Hydrogeology, and Environmental Remediation: Idaho National Engineering and Environmental Laboratory, Eastern Snake River Plain, Idaho: Boulder, Colorado, Geological Society of America Special Paper 353, p. 77–90.

INTRODUCTION

Lake-level fluctuations are one of the hallmarks of Pleistocene climatic fluctuations in the western United States. Dramatic fluctuations of lake level and extent are well known from the Bonneville and Lahontan basins (e.g., Scott et al., 1983; Benson and Thompson, 1987; Oviatt et al., 1992; Oviatt, 1997; Oviatt and Miller, 1997) as well as smaller basins, and those fluctuations have proven to be useful indicators of past climatic conditions. In particular, the lake-level records for closed basins are useful because they reflect marked changes in the ratio of precipitation to evaporation.

The eastern Snake River Plain also hosted a sizeable lake or lake complex that appears to have fluctuated in response to late Quaternary climatic fluctuations. Lake Terreton, first described from the presence of emerged lake sediments by Stearns et al. (1939), occupied several hundred square kilometers of the northeastern Snake River Plain in the area of Mud Lake and the Big Lost Trough (Fig. 1), the latter being a terminal basin now occupied by the Idaho National Engineering and Environmental Laboratory. Whereas Lake Terreton has long been assumed to have fluctuated in response to Pleistocene glacial-interglacial climatic shifts, neither the detailed record of lake fluctuations nor the relationship of the fluctuations to the underlying climatic factors have been studied.

The links between climate and Lake Terreton levels are complex; Lake Terreton has occupied two hydrologically distinct basins. Lake Terreton fills the Big Lost Trough basin largely in response to increased surface water discharge in the Big Lost River system (and secondarily the Little Lost River and Birch Creek drainages; Geslin et al., 1999). With current water-table conditions, the Big Lost Trough basin is a groundwater recharge area, several hundred meters above the eastern Snake River Plain aquifer. In contrast, the Mud Lake basin, which contains a small (~20 km²) perennial lake, is largely a groundwater discharge area (Spinazola, 1994), fed at the surface only by small creeks. Its fluctuations are driven in large part by groundwater recharge in the northeastern Snake River Plain and Yellowstone plateau. Despite the hydrographic and hydrologic complexity, precipitation/evaporation ratios on the eastern Snake River Plain and adjacent uplands have likely been the dominant control on fluctuations in lake level. Thus, Lake Terreton fluctuations likely reflect first-order changes of regional climate.

The intent of this paper is to document the timing of major Lake Terreton highstands during the past 160 k.y. This study focuses on first-order lake-level fluctuations in the Mud Lake subbasin of Lake Terreton. The highstands we document herein reflect lake levels at least ~5 m above the modern elevation of Mud Lake. The study is based principally upon sediment records in cores retrieved from the western portion of the Mud Lake subbasin, as well as sediment records from shallow ex-

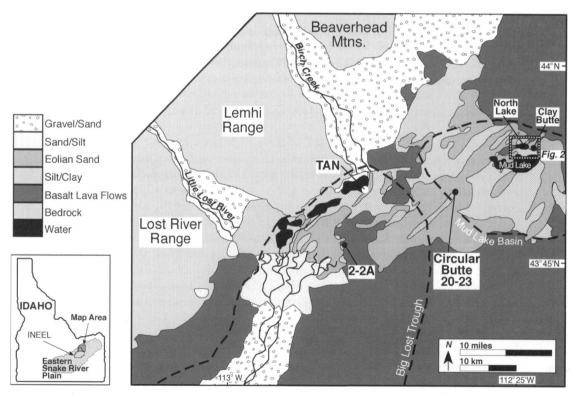

Figure 1. Map showing generalized sediment patterns in the Mud Lake and Big Lost Trough subbasins of Lake Terreton. INEEL is Idaho National Engineering and Environmental Laboratory.

cavations near the center of that subbasin. We suggest correlation, based on chronologic data, of sediments from one Mud Lake highstand to lacustrine sediment in core 2-2A from the Big Lost Trough basin.

Beyond their implications for paleoclimatic reconstruction, lake-level fluctuations in the Lake Terreton basin are of particular interest because they provide a means to predict spatial continuity of aquifer characteristics. In particular, laterally extensive lacustrine muds likely represent prominent aquicludes in this portion of the eastern Snake River Plain aquifer (see Mark and Thackray, this volume).

GEOLOGIC SETTING AND PREVIOUS INVESTIGATIONS

The eastern Snake River Plain is a broad, elongate, low-relief topographic depression, representing the major single geomorphic feature of eastern Idaho. The depression is underlain by a thick (>1000 m) sequence of Tertiary and Quaternary rhyolitic and basaltic volcanic rocks with intercalated terrestrial sediments (e.g., Hackett and Smith, 1992). Outcrops of strata are rare; the surficial sediment-volcanic facies distributions serve as an analogue for the subsurface sequence, which is revealed in numerous drill cores from the Idaho National Engineering and Environmental Laboratory and surrounding areas.

Subsidence of this portion of the eastern Snake River Plain initiated ca. 10 Ma (e.g., McQuarrie and Rodgers, 1998; Humphreys et al., 1999). More recent, asymmetric subsidence in the Big Lost Trough subbasin is suggested by the apparent displacement of 0.5 Ma basalt flows to depths of 200 m in the northern end of the basin (Wetmore, 1998), as well as similar displacement of ca. 1.7 Ma lacustrine sediment (Blair and Link, 2000). Topography and hydrography of both Lake Terreton subbasins are also controlled by the volcanic edifices of the axial volcanic zone and volcanic rift zones (e.g., Hackett and Smith, 1992), which define the sill heights and drainage divides across much of the eastern Snake River Plain.

Sediments and the lake-level fluctuations that they record have been studied by several workers. Stearns et al. (1939) first described sediments of Pleistocene Lake Terreton, although only in reconnaissance fashion. Spinazola (1994) compiled water well logs across the Mud Lake subbasin for the purposes of hydrologic assessment. He constructed a cross section suggesting a sedimentary sequence as much as 150 m thick beneath the center of the basin. Forman and Kaufman (1997) described a preliminary chronology of latest Pleistocene and Holocene lake-level fluctuations at Mud Lake, suggesting that the late Pleistocene expansion was underway by ca. 21 200 ^{14}C yr B.P. and reached as much as 10 m above modern lake level. They also concluded that a lake highstand 2–4 m above modern lake level occurred after 1000 ^{14}C yr B.P. In the Big Lost Trough subbasin, Bright and Davis (1982) obtained a radiocarbon date on lacustrine gastropods suggestive of a lake highstand in the northernmost portion of the Big Lost Trough ca. 700 ^{14}C yr

B.P. Levish et al. (1999) studied in detail the evidence for Holocene lake highstands in that area and concluded that positive evidence for late Holocene lacustrine conditions is restricted to the ~10 km^2 Birch Creek playa (Fig. 1). They further concluded that the extensive lake floor surrounding the playa contains evidence of a late Pleistocene highstand, but lacks evidence for a laterally extensive Holocene lake.

The surficial sedimentary system in the study area reflects the interaction of fluvial, lacustrine, and eolian processes. Surficial sediments on the northern Idaho National Engineering and Environmental Laboratory and surroundings were described by Nace et al. (1975), Scott (1982), Bartholomay and Knoble (1989), Gianniny et al. (1997), and Mark and Thackray (this volume). Modern fluvial input to the subbasins terminates in isolated playas in the Big Lost Trough subbasin and in Mud Lake. Eolian activity has had a strong effect on sedimentation and geomorphic development at various times during the Holocene, and likely during previous relatively warm, dry periods, redistributing sediment originally deposited in fluvial and lacustrine settings. Gianniny et al. (1997), Bestland et al. (this volume), and Mark and Thackray (this volume) describe seven distinct sedimentary facies in the northern Big Lost Troughs. We use that facies framework (Table 1) for the study of Mud Lake sediments.

METHODS AND RESULTS

We have described sediments and associated fauna in two areas of the Mud Lake subbasin (Fig. 1; Table 2). Cores from three wells (CB-20, CB-21, CB-23), drilled in 1994 for the environmental assessment of the Jefferson County Landfill site in the southwestern portion of the basin, near Circular Butte, were described in detail (e.g., grain size, color, sedimentary structures, mineral composition). The geochronology of the cores is established by luminescence methods (on sediments) and Ar/Ar methods (on basalt). In addition, we use amino acid racemization (AAR) ratios to correlate lacustrine sediments between the two basins. These cores are hereafter referred to as the Circular Butte (CB) cores.

We also described sediments in test pits excavated in the North Lake embayment, a dry lake bed near the center of the Mud Lake basin. The floor of the embayment is ~4–5 m above the modern surface of Mud Lake. We dug 10 pits to 2.5 m deep by 2 m wide along a north-south transect across the dry bed of North Lake (Fig. 2). Four of these pits were dug on swales where the lake sediments are directly exposed; the other six were dug on dune surfaces. Sediment characterization methods included analysis of sedimentary structures, facies contacts, grain size, color, and mineral composition (Sherbondy and Gianniny, 1999). The North Lake test pits yielded 94 sediment samples taken at 10 cm intervals to depths of nearly 2 m. Grain-size distributions were determined for 57 samples through sieve and hydrometer analysis. Ages of gastropods in these sediments

TABLE 1. SUMMARY OF DEPOSITIONAL FACIES CHARACTERISTICS

Eolian Depositional Facies
  Dunes
    Fine sand to very fine silt, poorly sorted*
    Typically strongly fine-skewed to near-symmetrical*
    Planar cross-bedding, asymmetrical ripples with high
      ripple index, avalanche deposits, graded beds capped
      with silt laminae
    Composition: volcanic lithics, very altered
  Loess
    Coarse to medium silt, poorly sorted to very poorly
      sorted, typically strongly fine-skewed to near-
      symmetrical*
    Structureless
Fluvial Depositional Facies
  Fluvial Channel
    Poor to moderate sorting, pebble to medium sand
    Strongly fine-skewed to strongly coarse-skewed
    Trough, planar cross-bedding, asymmetrical ripples
    Distal fining within systems (e.g. Big Lost River)
    Distinctive source area signature (Bartholomay, 1990;
      Geslin et al., 1999)
  Interfluve
    Moderate to very poorly sorted, coarse silt to fine silt
    Strongly fine-skewed to strongly coarse-skewed*
    Trough, planar cross-bedding, asymmetrical ripples
    Distal fining within systems
    Bioturbation by vegetation, vertebrates, and arthropods
Lacustrine Depositional Facies
  Playa/Lake Terreton Margin
    Sand-rich, clay to coarse sand, very poorly sorted
    Oxidized or calcified root molds, plant fragments
    Abundant ostracodes, and *Chara* oogonia
    Oscillation ripples (silt and sand)
  Playa Bottom
    Clay rich to fine sand, very poorly sorted, near-
      symmetrical to strongly coarse-skewed*
    Rare ostracodes, gastropods, copepods and *Chara*
      oogonia
    Mud cracks and intraclasts
    Oscillation ripples (silt and sand)
  Lake Terreton Floor:
    Fine silt to very fine silt, poorly sorted, near-symmetrical
      to strongly coarse-skewed*
    Ostracodes, mollusks and *Chara* oogonia
    Well-laminated mud-silt couplets

Note: Modified from Gianniny et al., 1997, and Mark and Thackray, this volume. Sorting and skewness were calculated according to Folk and Ward (1957).
*These are maximum estimates of sorting and skewness due to a lack of grain-size data in the finer than approximately 8.2 phi range. The sediment may be more poorly sorted and/or less skewed than indicated.

were determined through AAR and accelerator mass spectrometry (AMS) radiocarbon methods.

All cores and test pits were also sampled for ostracodes and mollusks. Ostracodes were separated by soaking sediment samples in a deionized water-sodium hexametaphospate solution for 2–7 days and sieving at 100 mesh, following the methods of R.M. Forester (personal commun., 1998). Ostracodes were identified using conventional criteria such as muscle scars, size, shape, ornamentation, and inner wall construction.

Chronologic data were obtained through several methods. Amino acid racemization age estimates (and associated ostra-

code identifications) were determined by the Amino Acid Geochronology Laboratory, Northern Arizona University, according to analytical procedures detailed in Kaufman and Manley (1998); ostracodes were identified according to Kaufman (2000). Luminescence age estimates were obtained by the Luminescence Dating Laboratory at the University of Illinois, Chicago. The optical and thermoluminescence dating procedures are described in the following. Radiocarbon dates were determined by the University of Colorado AMS Radiocarbon Research Laboratory. The Ar/Ar dating on basalt in the Mud Lake cores was completed by Bill McIntosh of the New Mexico Institute of Mining and Technology and by Duane Champion of the U.S. Geological Survey Ar-Ar laboratory, and are reported in detail in Hughes et al. (this volume).

Optical and thermoluminescence dating

Luminescence geochronology is based on the time-dependent dosimetric properties of quartz and feldspar. The technique has been used previously to determine ages of 100 ka and older for sediments that received sufficient solar radiation prior to deposition (e.g., Forman et al., 2000). Usually, exposure of sediment to sunlight for a few minutes to a few hours eliminates most of the stored luminescence signal from mineral grains, thereby resetting the age at the time of burial. The technique is especially effective for sediments that have been baked by overriding lava flows (Forman et al., 1994a). Exposure of sediment to heating by lava eliminates most, if not all, of the acquired luminescence. Exposure of these minerals to ionizing radiation from the decay of naturally occurring radioisotopes in sediments results in the trapping of electrons that accumulate following burial, after the sediment is shielded from further light or heat exposure. Excitation of sediments by heat or light releases the time-stored electrons as luminescence emissions. The intensity of the luminescence can be used as a measure of sample age by dividing the laboratory-determined paleodose (radiation dose [in Grays] estimated to yield the natural luminescence emissions) by an estimate of the rate at which the sample was irradiated following burial and light closure (dose rate).

Luminescence dating of water-laid sediment can be problematic because of the potential for insufficient solar resetting prior to deposition. Water-laid sediment often receives limited light exposure during deposition because of the light-filtering effects of water and turbidity (Jerlov, 1976). Thus, water-laid sediment may have a higher residual luminescence level than subaerial deposits that have received extensive light exposure. Traditional thermoluminescence (TL) techniques, such as the total-bleach and the regeneration techniques, assume full solar resetting of the TL signal, and can yield overestimates in the age of water-laid sediments. The recent advent of optical dating, including infrared stimulated luminescence (IRSL) (Aitken and Xie, 1992; Lang, 1994), provides a more sensitive tool for discriminating inherited from postdepositional luminescence emissions, ultimately providing a better geochronometer for water-

TABLE 2. TAXA AND ENVIRONMENTAL OCCURRENCES FOR OSTRACODES RECOVERED FROM CORES FROM WELLS CB-20, CB-22, AND CB-23

Ostracode taxa	Well	Depth (m [ft])	Environmental range (from literature)
Candona bretzi	CB-20	18.3 [60]	Pleistocene sediments of the Yukon (Delorme, 1970b)
Candona caudata	CB-20	48.6 [159.6]	"Canals, lakes, ponds, among grass and weeds. June. Ill., Mass., Wash., Mont." (Tressler, 1959: p. 686) .".commonly occurs in streams and considerable depth in lakes. On the Canadian Prairies the species is concentrated along the southern edge of the boreal forest and in the mixed woods zone as well as along the Rocky Mountain foothills and the Cypress Hills. Worldwide distribution" (Delorme, 1970b)
Candona elliptica	CB-20	12.3 [40.5], 16–16.8 [53–54], 47 [154], 48.5 [159.6]	"Muddy, weedy bottoms of lakes and ponds" (Tressler, 1959 p. 680) " . . . occurs rarely in the southeastern portion of the Canadian Prairies. Restricted to North America" (Delorme, 1970b)
	CB-23	12.3 [40.5]	
Candona ohioensis	CB-21	153.0	"Weedy margins of lakes. Nov. Ohio" (Tressler, p.678). " . . . occurs most commonly in the mixed-woods zone and southern fringes of the boreal forest of the Canadian Prairies, in lakes. North American distribution" (Delorme, 1970b)
	CB-23	53.0, 162.5	
Candona rawsoni	CB-20	32.9 [108.0]	"*Candona rawsoni* commonly occurs in permanent and ephemeral freshwater and saline lakes and ponds on the prairies and forest-prairies transition in the United States and Canada. . . . It does not live in freshwater lakes having little seasonal salinity variation" (Forester et al., 1987) *C. rawsoni* is "often common in environments characterized by high variability and low predictability" and is a "euryhaline species whose members live primarily in prairie lakes, including saline ones." (Forester et al., 1994, p. 98)
	CB-23	12.3 [40.5], 16.2 [53.0], 49.5 [162.5]	
	2-2A	44 [144]	
Cytherissa cf. lacustris	CB-20	16.5–16.9 [54–55.5]	" . . . common in deep lakes (greater than 10 ft or 3 m) in the boreal forests of Canada" (Delorme, 1970c). " . . . lives in dilute, cold, stenotopic, boreal forest lakes" (Forester, 1991), 20°>water temperature<?15°C (Forester, 1991)
	CB-23	16.2 [53.0]	
Eucypris serrata	CB-23	12.3 [40.5]	"Restricted [in Canada?] to the southern portion of the interior plains, particularly near the Alberta-Saskatchewan boundary" (Delorme, 1970a)
Limnocythere friabilis	CB-20	16–16.9, [53–55.5], 60.5 [18.4]	" . . . has only been encountered occasionally in lakes in the interior plains [of Canada?]" (Delorme, 1971). *L. friabilis* is known living "from only in a few dilute lakes in Oregon." In cores from Lake Michigan, "*L. friabilis* was found only in the littoral and sublittoral cores. In these cores its abundance increases toward the shore suggesting its abundance might provide an estimate of shore-line proximity" (Forester et al., 1994)
	CB-21	15.4 [50.5]	
	CB-23	16.2 [53.0]	
	2-2A	44 [144]	
Limnocythere sp.	CB-20	18.4 [60.5]	Limnocythere lives mostly in saline lakes that exhibit seasonal variability in temperature and chemistry (Forester et al., 1987)

Note: Species identifications by Delda Findeisen and Gary Gianniny except those identified by Jordon Bright (Northern Arizona University) that are marked by an asterisk.

laid sediment (e.g., Smith et al., 1986; Hütt et al., 1988; Forman, 1999). If TL and IRSL ages agree, as in this study, there is little to no inherited luminescence and the ages are considered finite estimates.

The IRSL geochronometer, similar to TL, is reset by exposure of sediment to sunlight prior to deposition. However, TL analysis uses heat to release photons within a mineral; IRSL dating uses infrared light (880 ± 80 nm) to rapidly release from the most light-sensitive traps in the mineral lattice. The decrease in the IRSL signal after exposure to sunlight for as little as ~20 s is similar to the reduction in the TL signal after a 20 h light exposure (Forman 1998); therefore, IRSL is useful for dating

sediments (especially water-laid sediments) that might have received brief (<4 h) and wavelength-restricted light exposure.

TL methods. TL emissions were measured on the fine-grained (4–11 μm), polymineral fraction of sediments. All samples were analyzed by the total-bleach method, with the residual level measured on four aliquots of each sample. The residual level was obtained by exposure to 8 h of a UV-dominated spectra from a 275 W sunlamp, which results in near-total resetting of the TL signal. A more realistic solar-resetting level was measured after 16 h of natural sunlight exposure in Columbus, Ohio. The partial-bleach technique was also performed on one sample. Dose-rate estimates were derived from U and Th con-

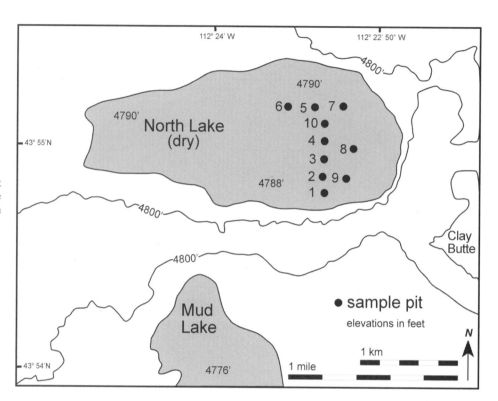

Figure 2. Map showing locations of test pits in the bed of North Lake, and the relationship of North Lake to modern Mud Lake.

centrations, inferred from thick-source alpha counting and the ^{40}K content, calculated from the total K concentration (Table 3). Blue TL emissions were selected for measurement using a Corning 5–58 filter (<2% transmission below 320 nm) (Balescu and Lamothe, 1992). All samples were preheated at 124 °C for at least 48 h prior to analysis to remove potential instability (anomalous fading) in the laboratory-induced TL signal (e.g., Forman et al., 1994b). After heating, each sample was tested for anomalous fading by storing irradiated (100–450 Gy) material for at least 32 days and comparing to the TL signal of an aliquot that was analyzed <1 h after irradiation (Table 3). The anomalous fading tests reveal no significant instability in the TL emissions of the preheated sample.

The rate of TL ingrowth was evaluated by applying additive beta doses to the natural TL signal by a series of irradiations with a calibrated ^{90}Sr/^{90}Y source. The highest radiation dose added to the natural TL signal was at least six times the calculated equivalent dose, which is sufficient for accurate extrapolation. The natural and additive-dose data were fitted by a saturating exponential function (cf. Huntley et al., 1988) over the range of temperatures, usually between 250 and 350 °C, that encompasses at least 90% of the measured TL signal and exhibits a plateau in equivalent dose values.

IRSL methods. Optical stimulation of sediments was accomplished using infrared emissions (880 ± 80 nm) from a ring of 30 diodes (Spooner et al., 1990) with an estimated energy delivery of 17 mW cm^{-2}. The resultant blue (Schott BG-39; <5% transmission below 360 nm) emissions for polymineral 4–11 μm fraction was measured from the sediments. The

background count rate for measuring blue emissions was low at 80 counts/s, with a signal to noise ratio of >20. Samples were excited for 90 s, and the resulting IRSL signal was recorded in 1 s increments. Similar additive beta-dose, normalization, and equivalent-dose computational procedures used in TL analysis (e.g., Forman et al., 1994b) were also employed to determine IRSL ages. One difference between TL and IRSL procedures is the shorter and higher temperature preheat, at 160 °C for 5 h, for IRSL analysis (cf. Aitken and Xie, 1992). Measurement of the IRSL signal was also delayed at least 1 day after preheating. Tests for anomalous fading of the laboratory-induced and preheated IRSL signal, after >32 days storage, revealed insignificant (≤8%) reduction in signal, indicating stability of the laboratory and natural infrared emissions.

RESULTS

Mud Lake subbasin: Circular Butte cores

Four sediment cores were retrieved near the western edge of the Mud Lake basin (T. 6 N., R. 33 W., Section 13) in 1994 by Holladay Engineering, Inc. Since their retrieval and analysis for the purposes of landfill placement and design, the cores have been housed in the U.S. Geological Survey core facility at the Idaho National Engineering and Environmental Laboratory. We have described the cores in detail in order to examine the evidence for lacustrine highstands. The log of core CB-20 (Fig. 3) is representative of the subsurface sequence in this area.

TABLE 3. THERMOLUMINESCENCE AND INFRARED STIMULATED LUMINESCENCE DATA AND AGES FOR SEDIMENTS FROM MUD LAKE CORES, TERRETON, IDAHO

Laboratory number	Sediment type	Core and depth (m)	a count (ks/cm2)*	Th (ppm)*	U (ppm)*	Unsealed/ sealed†	K₂O (%)§	TL a-value*	IRSL a-value*	TL dose rate (Gy/ka)**	IRSL dose rate**	TL residual level‡	IRSL residual level‡	TL De temperature range (°C)§§	IRSL De time range (s)§§	TL De (grays)##	IRSL De (grays)##	TL age estimate (ka)***	IRSL age estimate (ka)***
UIC-612	Baked	CB-23/48.3	0.63 ± 0.03	9.3 ± 1.2	2.4 ± 0.5	1.01 ± 0.03	2.24 ± 0.02	0.06 ± 0.01		3.34 ± 0.13		Thermal		250–400		538.8 ± 13.6		161 ± 13	
UIC-613	Baked	CB-20/48.1	0.73 ± 0.04	12.5 ± 1.5	2.3 ± 0.5	0.95 ± 0.03	2.86 ± 0.03	0.12 ± 0.01		4.76 ± 0.18		Thermal		250–400		>1487 ± 41		>313 ± 25	
UIC-614	Baked	CB-23/48.1	0.77 ± 0.04	12.9 ± 1.7	2.4 ± 0.6	1.00 ± 0.02	2.53 ± 0.03	0.05 ± 0.01		3.83 ± 0.15		Thermal		250–400		516.7 ± 34.7		135 ± 13	
UIC-623	Lacustrine	CB-21/30.8	0.65 ± 0.03	8.7 ± 1.2	2.7 ± 0.4	0.94 ± 0.06	2.83 ± 0.03	0.04 ± 0.01	0.04 ± 0.01	3.58 ± 0.14	3.58 ± 0.14	8h UV	1h SL	250–350	2–90	546.5 ± 10.2	532.1 ± 0.9	152 ± 13	149 ± 11
UIC-624	Lacustrine	CB-20/31.3	0.67 ± 0.04	9.6 ± 1.3	2.6 ± 0.5	0.95 ± 0.06	2.49 ± 0.02	0.04 ± 0.01	0.05 ± 0.01	3.38 ± 0.13	3.48 ± 0.13	8h UV	1h SL	250–400	2–90	437.3 ± 15.4	508.7 ± 5.7	130 ± 12	148 ± 12
UIC-632	Lacustrine	CB-21/18.3	0.76	8.2 ± 1.3	3.7 ± 0.5	0.98 ± 0.07	2.71 ± 0.03	0.07 ± 0.01	0.04 ± 0.01	4.11 ± 0.16	3.74 ± 0.15	8h UV	1h SL	250–400	2–90	362.5 ± 12.1	312.2 ± 0.5	88 ± 7	84 ± 6

*U and Th ppm values calculated from alpha count rate, assuming secular equilibrium.
†Ratio of bulk alpha count rate under unsealed and sealed counting conditions. A ratio of >0.94 indicates little or no radon loss.
§Percent potassium determined on homogenized sediment sample by Activation Laboratory Ltd. Ontario, Canada.
#The measured alpha efficiency factor as defined by Aitken and Bowman (1975).
**Dose rate value includes a 0.12 ± 0.02 Gy/k.y. contribution from cosmic radiation (Prescott and Hutton, 1988) with an assumed moisture content of 25% ± 5% by weight.
‡For lava-baked sediments residual level defined by TL remaining after heating to 500°C. For lacustrine sediments residual for TL calculations is residual level after 8 h light exposure to a UV dominated sunlamp; IRSL analysis is the residual after 1 h sunlight exposure in Chicago, Illinois.
§§Temperature range or seconds since light exposure used to calculate equivalent dose (De).
##All thermoluminescence De (equivalent dose) measurements were made with a Corning 5/58 and HA-3 filters in front of the photomultiplier tube. TL samples were preheated to 124°C for 48 h prior to analysis. Infrared stimulated luminescence (IRSL) measurements were made with BG-39 filters in front of the photomultiplier tube. IRSL samples were preheated to 140°C for 10 h, and subsequently stored for 24 h, prior to analysis.
***Errors are at 1σ and calculated by averaging the errors across the temperature or time range. All samples were tested for anomalous fading by storing irradiated (450 Gy) samples for at least 32 days and comparing the luminescence signal to an unstored aliquot. The anomalous fading ratios are between 1.00 and 0.92 indicate little or no fading, within analytical resolution.

Sediment data, facies assignments, and documentation of lacustrine fluctuations

The cores reveal a variety of sediments, ranging from laminated mud to thin layers of very fine gravel (Fig. 3), as well as 11 m of basalt near the base of the core. Laminated clay and fine to medium sand dominate the cores. Most of the mud-rich portions of the core contain minor fine to very fine sand. Although mold impressions of ostracodes occur within these laminated clays, most ostracode shells appear to have been dissolved in situ. Rounded clay intraclasts (1–4 mm in diameter), and disrupted clay laminae are common in most of the sand-bearing portions of the cores. The very poorly sorted sand and pebble sediments are completely unconsolidated; recovery was poor, and sedimentary structures are not preserved. Micritic pedogenic carbonate cements are uncommon and are best developed in sediments overlying the basalt near the base of the core.

Using sedimentary structures and grain-size characteristics and the criteria in Table 1, we have assigned facies designations to the sediments at each level (Fig. 3). Facies shifts—from sub-wave-base lacustrine, to lake margin, to fluvial and eolian—record the regression of the lake margin. It is possible that some of the sandy sediments interpreted as fluvial were deposited by eolian processes, and vice versa. While this is an important paleoenvironmental distinction, it has little effect on our conclusions because both facies represent low lake levels.

The cores contain four intervals of laminated clay and silty clay with lacustrine ostracodes (see following discussion). In core CB-20, these intervals are found at depths of ~47–52 m, 28–31 m, 12–18 m, and 8 m. Ostracodes are sparse or moldic in the deposits at 28–31 m and 8 m depths. We have focused our attention on the second (28–31 m) and third (12–18 m) of these lacustrine intervals due to unresolved difficulties in defining the age of lowest interval and the lack of datable sediment in the uppermost interval.

The sediment core was sampled for biotic analysis, luminescence age estimation, and amino acid age estimation. Because the wells are closely spaced with respect to the breadth of the basin and reveal a similar stratigraphic sequence, we consider samples taken from a particular depth in one core to be representative of that depth within our master log of core CB-20.

Ostracode species data

Ostracode species data provide specific paleoenvironmental information. Data on ostracode species occurrences and environmental tolerances in cores CB-20, CB-21, CB-23, and 2–2A are summarized in Table 2 and Figure 4. Ostracode species data for the laminated mud intervals indicate well-circulated lakes with seasonal variation in salinity. Of particular interest is the presence at a depth of 16.5 m of *Cytherissa* cf. *lacustris* in CB-20, which is only known from cold boreal forest lakes with water depths >3 m (Delorme, 1970c; Forester, 1991).

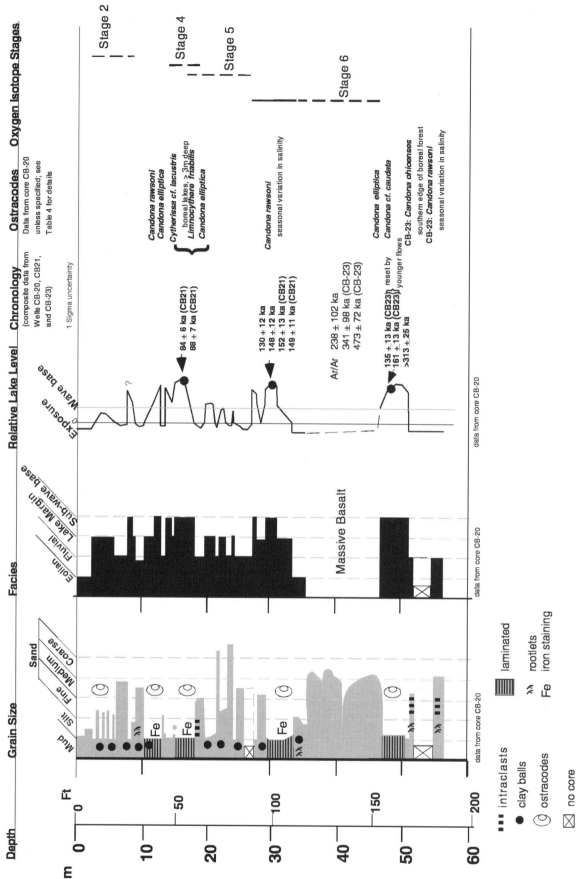

Figure 3. Summary of data for the Jefferson County Landfill (Circular Butte) wells, Mud Lake subbasin of Lake Terreton. Marine oxygen isotope stages (right column) are provided for reference, provisionally correlated by chronologic data to specific sections of core; dashed lines indicate uncertain correlation.

Chronologic data

The chronology of highstands recorded in the Circular Butte cores is defined by luminescence age estimates (TL and IRSL) (Table 3; Fig. 3). Amino acid racemization age estimates (Table 4) from core 2-2A in the Big Lost Trough permit limited correlation between basins. In addition, the age of basalt flows near the base of the cores is defined by TL and Ar/Ar age estimates, although the age estimates provided by these two techniques are not conclusive.

Luminescence data (Table 3) provide broad age information for the two major intervals of lacustrine sedimentation in the portion of the CB cores above the basalt flows (Fig. 3). Four TL and IRSL age estimates indicate that the lacustrine sediments at 28–31 m depth were deposited between ca. 120 and 160 ka (Table 3, lab numbers UIC-623 and UIC-624). The lacustrine sediment at 12–18 m depth yielded a TL age estimate of 88 ± 7 ka and an IRSL age estimate of 84 ± 6 ka (UIC-632). These dates place the two major lake highstands within marine oxygen isotope stage 6 and within stage 5 or 4, respectively. A stage 6 highstand is broadly correlative with the Little Valley lake cycle in the Bonneville basin (e.g., Scott et al., 1983). The stage 5 or 4 highstand does not have an apparent correlative in the Bonneville basin. The lacustrine sediment at 8 m depth lacks datable material. It may represent a latest Pleistocene highstand (oxygen isotope stage 2) documented in surficial sediments (Forman and Kaufman, 1997; see following) or an additional highstand not otherwise documented.

The TL and Ar/Ar age estimates define the age of the basalt near the base of the cores, and place minimum age controls on the underlying lacustrine sediments. The TL age estimates of 135 ± 13 ka (UIC-614), 161 ± 13 ka (UIC-612), in CB-23 and >313 ± 25 ka (UIC-613) in CB-20 indicate the timing of thermal resetting of the TL signal in lacustrine sediment by the overriding basalt flow. The oldest of the three TL dates may indicate different age flows in the two wells or incomplete zeroing of the TL signal. Because the TL ages date the timing of thermal resetting by the overlying lava, they provide only a minimum age on the underlying sediment. An Ar/Ar age of 238 ± 102 ka from the basalts in the CB cores (Hughes et al., this volume) overlaps with luminescence ages at 1σ and is thus not statistically different. Additional Ar/Ar dates of 341 ± 98 ka and 473 ± 36 ka were obtained on samples from the basalts in CB-20 and CB-23 (Hughes et al., this volume). It is not clear if this is a result of the Ar/Ar method yielding overestimates because of the small amount of evolved Ar in these K-poor basalts, or a result of other factors. Ostracodes in the lake deposits below the basalt were not suitable for age estimates via amino acid techniques due to the heating effect of the lava.

Correlation to the Big Lost Trough

Correlation of sediments between the Mud Lake and Big Lost Trough subbasins of Lake Terreton is limited by the pau-

city of cored sediment in the latter basin for much of the time interval represented by the Circular Butte cores. However, AAR data on the ostracodes *Limnocythere friabilis* and *Candona rawsoni*, found in silt and clay at 43 m depth in core 2-2A (Table 2), suggest a correlation of these sediments with the Little Valley alloformation in the Bonneville basin and, therefore, an age of ca. 150 ka. This age estimate indicates correlation with the lacustrine sediment in core CB-20 (28–31 m), dated with luminescence methods to ca. 120–160 ka (Table 4). The ostracode taxa and the fine-grained sediment that may represent either lacustrine or playa facies suggest a shallow, seasonally variable lake with the core site proximal to the shore.

Excavations in the North Lake embayment

Sediment data. In the North Lake study area, 2–4 m of sediment overlie a basal basalt from the Bovo Muerte eruptive centers northeast of the lake (Hughes et al., this volume). In most excavations on the now dry lake bottom, the stratigraphic succession overlying the basalt includes three distinct layers (Fig. 5): (1) basal sand, (2) locally fossiliferous mud, and (3) sequence-capping sand. The basal sand unit is a 0.3–0.4-m-thick, yellow-brown, nonfossiliferous, medium-grained, poorly sorted lithic arenite. Above a gradational contact, the middle layer consists of pale yellowish mud. Within this mud layer, discontinuous, matrix-supported, 0.10–0.15-m-thick layers contain basalt clasts ranging from 0.1 to 8 cm in diameter. In nine of the ten pits, this mud unit lacks mollusks and ostracodes. In contrast, the upper 10 cm of the mud in pit 4 contains a very pale orange bioturbated layer containing ostracodes (*Candona rawsoni*, *Limnocythere* sp., *Limnocythere staplini*, and *Cypridopsis vidua*) and gastropods (*Gyralus deflectus*, *Menetus* sp., *and Stagnicola* [*cf. Lymnaea*] sp.). The limited extent and reworked character of this deposit suggest that it postdates most of the presumed lacustrine mud. Mud cracks developed at the top of the layer penetrate both the bioturbated, fossiliferous horizon and the underlying nonfossiliferous mud to depths of 1.75 m. Mud-crack polygons have diameters of ~0.3 m. The mud cracks are filled with the overlying eolian sand and live grass rootlets from the surface.

We separated 75 ostracode valves from the pit 4 sediments (Table 5). Species abundances are dominated by the *Limnocythere* sp., followed by *Candona rawsoni*. Size variation within species is most apparent in *Limnocythere* sp. *Candona rawsoni* occurs in permanent ephemeral freshwater and saline lakes and ponds on the prairies and forest-prairie transition in the United States and Canada. Seasonal variabilities in chemistry and salinity are often large. *Limnocythere* lives mostly in saline lakes that exhibit seasonal variability in temperature and chemistry (Forester et al., 1987). *Limnocythere staplini* typically lives in carbonate-depleted saline waters. The two found in the North Lake sediments have ornate shells, which may imply a complex ecosystem within a geologically stable lake (Forester et al., 1987). *Cypridopsis vidua* lives in marshy, relatively warm areas

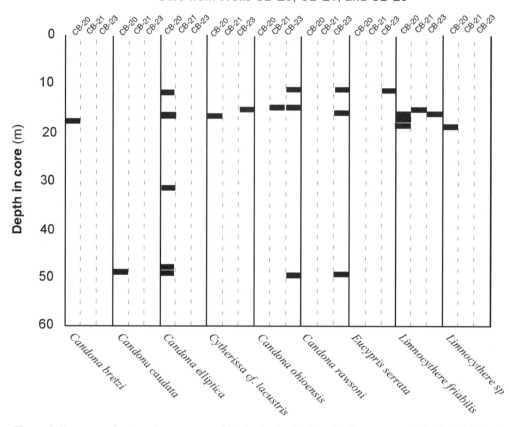

Figure 4. Presence of ostracode taxa at specific depths in the Circular Butte cores (CB-20, CB-21, and CB-23). Each bar denotes presence of taxon at that depth.

**TABLE 4. AMINO ACID RACEMIZATION AND ¹⁴C AMS DATA FOR WELL 2-2A
(BIG LOST TROUGH) AND PIT 4 AT NORTH LAKE**

Sample	Estimated age (ka)	Aspartic acid* (D/L)	Glutamic acid (D/L)	Genus	Isoleucine
UAL-2455†		0.43 ± 0.01 (4)	0.23 ± 0.01 (4)		
Bonneville	21	0.30 ± 0.01 (10)	0.08 ± 0.01 (10)		
Little Valley	150	0.42 ± 0.01 (7)	0.19 ± 0.01 (7)		
UAL-2591§				*Gyralus*	0.091 ± 0.003 (5)
UAL-2592§				*Stagnicola*	0.102 ± 0.011 (2)

Note: Numbers in parentheses are numbers of subsamples.
*Mean value ± 1 standard deviation of (n) subsamples.
†Results of amino acid racemization analyses on *Candona rawsoni* from Bonneville basin well 2-2A, 114.4′, Big Lost Trough subbasin of Lake Terreton compared to data from *Candona* spp. Samples from Little Valley, Utah; data from Oviatt and Miller (1997).
§Results of amino acid analyses on gastropods from North Lake pit 4 (Fig. 4), 0.6 m depth, Mud Lake subbasin of Lake Terreton, southeast Idaho.

containing fresh to saline water. Seasonal variability of both chemistry and temperature is often large (Forester, 1991). These tolerances suggest that the North Lake fauna lived in a shallow but rapidly fluctuating lake and/or marsh environment.

The stratigraphic sequence in the excavations is capped by a discontinuous, moderate yellow-brown, well-sorted, medium-grained lithic arenite sand that has a gradational contact with the underlying clay. This eolian sand forms longitudinal dunes that trend southwest to northeast.

Geochronology. Age estimates for the fossiliferous horizon were determined by AAR and AMS radiocarbon methods. Amino acid (isoleucine epimerization) ratios in snails (Table 4) indicate that the mud forming the reworked shelly unit in pit 4 is broadly correlative with the Bonneville alloformation in the Bonneville basin (Sherbondy and Gianniny, 1999). Because this sample was collected from only ~60 cm below the ground surface, however, it may have been affected by surface heating. Radiocarbon analysis on a gastropod (*Stagnicola*) shell from the same sample yielded an age of 11 600 ± 70 ^{14}C yr B.P. (NSRL-10574).

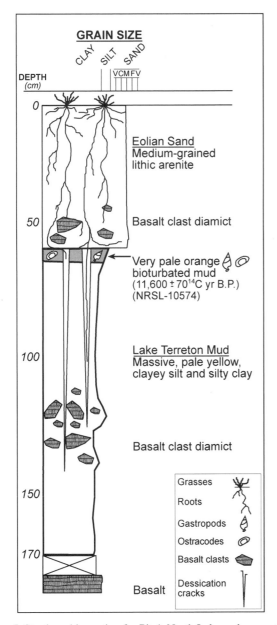

Figure 5. Stratigraphic section for Pit 4, North Lake embayment, Mud Lake subbasin of Lake Terreton. Chronologic data at ca. 60 cm depth were obtained via amino acid racemization (AAR) and radiocarbon methods, and are on material from the same stratigraphic level. See text for discussion of age discordance.

Interpretation of the North Lake sequence. The sediments, ostracode taxa, and chronologic data from pit 4 indicate that the North Lake sequence records a final phase of elevated lake levels in the Mud Lake basin. The floor of North Lake is 4–5 m above modern Mud Lake; thus, marshy conditions reflect a moderately elevated lake level ca. 11 600 ^{14}C yr B.P. The elevated lake level at that time reflects either a phase during regression from the oxygen isotope stage 2 highstand, or a rise in lake level following complete regression from the main stage 2 highstand. That is, we cannot determine from available data whether one or multiple transgressive-regressive events occurred during stage 2. The eolian sand overlying the ostracode-bearing mud indicates that the lake level subsequently retreated and effective precipitation diminished substantially.

DISCUSSION

Interpretation of the sediments and fossils

The dark brown, laminated mud with lacustrine ostracodes, contained in the Circular Butte cores, demonstrates the presence of relatively stable and fresh lakes at three distinct time intervals during the past 160 k.y., as well as during one earlier period. Two of those lacustrine intervals have been dated to ca. 120–160 ka and ca. 80–100 ka (Fig. 3). Pebble-rich quartz sand between depths of 20 and 21 m in the CB-20 core suggests the influx of fluvial sediment (and thus substantial regression) between those dated highstands. An additional, thin layer of lacustrine sediment (undated) is at 8 m depth. Ostracode-rich mud also is beneath the basalts (CB-20, 47–52 m, Fig. 3), but its depositional age has not been determined.

This study has employed alternative dating techniques to determine the ages of sediments that are too old for ^{14}C dating, and potentially too young for Ar/Ar dating in associated, low-potassium basalts. While we recognize the uncertainties associated with these techniques, we are satisfied that the new data presented here indicate Lake Terreton highstands ca. 120–160 ka and ca. 80–100 ka. It is not clear, however, whether Lake Terreton rose high enough during the oxygen isotope stage 6 highstand to connect the Mud Lake subbasin with the Big Lost Trough subbasin, or whether the lakes were isolated, but rose in response to the same climate forcing. We lack data to ascertain the presence or absence of the ca. 80–100 ka highstand in the Big Lost Trough.

The record of more recent Lake Terreton highstands, during oxygen isotope stages 2 and 1, was been outlined by Forman and Kaufman (1997), and is augmented by data presented herein. Forman and Kaufman's (1997) chronology is constrained by ^{14}C AMS dates, luminescence age estimates, and amino acid age estimates on sediments and fossils from Clay Butte on the northern margin of Mud Lake (Fig. 1). Those dates suggest that lake levels in the Mud Lake basin had begun to rise by ca. 22 ka. Here we have reported on a younger sequence of sediments in desiccated North Lake, which suggests that by

TABLE 5. TAXA AND ENVIRONMENTAL OCCURRENCES FOR OSTRACODES RECOVERED FROM NORTH LAKE PIT 4

Ostracode taxa (reference)	Number of samples	Percent	Occurrence
Candona rawsoni (Tressler, 1959)	24	32	*Candona rawsoni* occurs in permanent ephemeral freshwater and saline lakes and ponds on the prairies and forest-prairie transition in the United States and Canada. Seasonal variability in chemistry and salinity is often large (Forester et al., 1987).
Limnocythere sp.	41	55	*Limnocythere* lives mostly in saline lakes that exhibit seasonal variability in temperature and chemistry (Forester et al., 1987).
Limnocythere staplini	2	3	*Limnocythere staplini* typically lives in carbonate-depleted saline waters where high Cl- dominated salinity serves as a barrier. The two found had ornate tests, which may imply a complex ecosystem within a geologically stable lake (Forester et al., 1987).
Cypridopsis vidua	8	10	*Cypridopsis vidua* lives in marshy, relatively warm areas containing fresh to saline water. Seasonal variability of both chemistry and temperature is often large (Forester, 1991).

Note: Depth of pit 4 is 0.8 m below surface. Table is after Sherbondy and Giannininy (1999).

11.6 ka Mud Lake was significantly lower than during the stage 2 highstand documented by Forman and Kaufman (1997). Ostracodes and gastropods from the North Lake mud layer suggest a shallow lake or marsh environment at an elevation of 1460 m, ~4–5 m above the modern surface of Mud Lake. Forman and Kaufman (1997) also documented a subsequent Mud Lake highstand, occurring within the past 1 k.y.

The emerging chronology of Lake Terreton fluctuations correlates partially with lake-level records from the Bonneville basin (e.g., Scott et al., 1983; Oviatt and Miller, 1997). The ca. 120–160 ka and ca. 22–11 ka Lake Terreton high-water episodes are broadly correlative with the Little Valley and Bonneville lake cycles in the Bonneville basin. These correlations suggest that Lake Terreton responded to regional climatic controls at those times. Conversely, the ca. 80–100 ka highstand has no known counterpart in the Bonneville basin, and may reflect either regional differences in climatic patterns at that time between the eastern Snake River Plain and Bonneville basin, or unrecognized Lake Terreton basinal changes caused by lava flows or tectonic subsidence.

CONCLUSIONS

Ostracode species composition and sedimentary facies from the Mud Lake basin, from this study and that of Forman and Kaufman (1997), indicate at least four episodes of wetter climate and associated higher recharge during the past 160 k.y. High lake levels occurred ca. 120–160 ka, 80–100 ka, 22–11 ka, and <1 ka. The 120–160 ka highstand appears broadly correlative with the presence of a lake in the Big Lost Trough. An additional interval of lacustrine sedimentation, recorded below the basalts in the Circular Butte cores, has not been dated. These results suggest that extensive lake development on this portion of the eastern Snake River Plain is associated both with long-term glacial-interglacial cycles (10 k.y. to 100 k.y.) and with short-term climate changes (i.e., 100–1000 yr).

ACKNOWLEDGMENTS

This research was supported by Department of Energy contract KEK-078-98B. The Bucknell University Program for Undergraduate Research supported research by M. Sherbondy. Partial funding for the amino acid and ^{14}C analyses were provided by National Science Foundation grant EAR-9896251. We thank Jordon Bright for ostracode identifications and assistance in the Amino Acid Laboratory, Duane Champion and Bill McIntosh for Ar/Ar dates, and Tom Maeder and Mark Gamblin of Idaho Department of Fish and Game for permission and access to areas in and around Mud Lake. Richard Smith of the INEEL and Linda Davis of the U.S. Geological Survey provided access to the Jefferson County Landfill cores, and Holladay Engineering provided additional information on the cores. Reviews by Jack Oviatt, Paul Link, and an anonymous reader improved the manuscript substantially.

REFERENCES CITED

Aitken, M.J., and Xie, J., 1992, Optical dating using infrared diodes: Young samples: Quaternary Science Reviews, v. 11, p. 147–152.

Aitken, M.J., and Bowman, S.G.E., 1975, Thermoluminescent dating: Assessment of alpha particle contribution: Archaeometry, v. 17, p. 132–138.

Balescu, S., and Lamothe, M., 1992, The blue emissions of K-feldspar coarse grains and its potential for overcoming TL age underestimates: Quaternary Science Reviews, v. 11, p. 45–51.

Bartholomay, R.C., and Knobel, L.L., 1989, Mineralogy and grain size of surficial sediment from the Little Lost River and Birch Creek Drainages, Idaho National Engineering Laboratory, Idaho: U.S. Geological Survey Open-File Report 89-385 (DOE/ID-22082), 19 p.

Bartholomay, R.C., 1990, Mineralogical correlation of surficial sediment from area drainages with selected sedimentary interbeds at the Idaho National Engineering Laboratory, Idaho: U.S. Geological Survey Water-Resources Investigations Report 90-4147 (DOE/ID-22092), 18 p.

Benson, L.V., and Thompson R.S., 1987, The physical record of lakes in the Great Basin, *in* Ruddiman, W.F., and Wright, H.E., Jr., eds., North America and adjacent oceans during the last deglaciation: Boulder, Colorado,

Geological Society of America, The Geology of North America, v. K-3, p. 241–260.

Blair, J.J., and Link, P.K., 2000, Pliocene and Quaternary sedimentation and stratigraphy of the Big Lost trough from coreholes at the Idaho National Engineering and Environmental Laboratory, Idaho: Evidence for a regional Pliocene lake during the Olduvai normal polarity subchron, *in* Robinson, L., ed., Proceedings of the 35th Symposium on Engineering Geology and Geotechnical Engineering: Pocatello, Idaho, Idaho State University, p. 163–179.

Bright, R.C., and Davis, O.K., 1982, Quaternary paleoecology of the Idaho National Engineering Laboratory, Snake River Plain, Idaho: American Midland Naturalist, v. 108, no. 1, p. 21–33.

Delorme, L.D., 1970a, Freshwater ostracodes of Canada. 1. Subfamily Cypridinae: Canadian Journal of Zoology, v. 48, p. 153–168.

Delorme, L.D., 1970b, Freshwater ostracodes of Canada. 3. Family Candonidae: Canadian Journal of Zoology, v. 48, p. 1099–1127.

Delorme, L.D., 1970c, Freshwater ostracodes of Canada. 4. Families Ilyocyprididae, Notodromadidae, Darwinulidae, Cytherideidae, and Entocytheridae: Canadian Journal of Zoology, v. 48, p. 1251–1259.

Delorme, L.D., 1971, Freshwater ostracodes of Canada. 5. Families Limnocytheridae, Loxoconchidae: Canadian Journal of Zoology, v. 49, p. 43–64.

Folk, R.L. and Ward, W., 1957, Brazos river bar: A study in the significance of grain size parameters: Journal of Sedimentary Petrology, v. 27, p. 3–26.

Forester, R.M., 1991, Pliocene climate history of the western United States derived from lacustrine ostracodes: Quaternary Science Reviews, v. 10, p. 133–146.

Forester, R.M., Colman, S.M., Reynolds, R.L., and Keigwin, L.D., 1994. Late Quaternary limnological and climate history from ostracode, oxygen isotope, and magnetic susceptibility: Journal of Great Lakes Research, v. 20, p. 93–107.

Forester, R.M., Delorme, L.D., and Bradbury, J.P., 1987, Mid-Holocene climate in northern Minnesota: Quaternary Research, v. 28, p. 263–273.

Forman, S.L., 1999, Infrared and red stimulated luminescence dating of late Quaternary nearshore sediments from Spitsbergen, Svalbard: Arctic, Antarctic, and Alpine Research, v. 31, no. 1, p. 34–49.

Forman, S.L., and Kaufman, D.S., 1997, Late Quaternary oscillations of Lake Terreton, Eastern Snake River Plain, Idaho: Geological Society of America Abstracts with Programs, v. 29 no. 26, p. A253.

Forman, S.L., Pierson, J., Smith, R.P., Hackett, W.R., and Valentine, G., 1994a, Assessing the accuracy of thermoluminescence to date baked sediments beneath late Quaternary lava flows, Snake River Plain, Idaho: Journal of Geophysical Research, v. 99, no. B8, p. 15569–15576.

Forman, S.L., Lepper, K., and Pierson, J., 1994b, Limitations of infra-red stimulated luminescence in dating high arctic marine sediments: Quaternary Geochronology (Quaternary Science Reviews), v. 13, p. 545–550.

Forman, S.L., Pierson, J., and Lepper, K., 2000, Luminescence geochronology, *in* Sowers, J.M., Noller, J.S., and Lettis, W.R., eds., Quaternary geochronology: Methods and applications: Washington, D.C., American Geophysical Union Reference Shelf, p. 157–176.

Geslin, J.K., Link, P.K., and Fanning, C.M., 1999, High-precision provenance determination using detrital-zircon ages and petrography of Quaternary sands on the eastern Snake River Plain, Idaho: Geology, v. 27, no. 4, p. 295–298.

Gianniny, G.L., Geslin, J.K., Riesterer, J., Link, P.K., and Thackray, G.D., 1997, Quaternary and Holocene surficial sediments of the Test Area North (TAN) area, northeastern Snake River Plain: An actualistic guide to aquifer characterization, *in* Sharma, S., and Hardcastle, J.H., eds., Proceedings of the 32nd Symposium on Engineering Geology and Soils Engineering, Boise, Idaho: Moscow, Idaho, University of Idaho, Department of Civil Engineering, p. 29–44.

Godfrey-Smith, D.I., Huntley, D.J., and Chen, W.-H., 1988, Optical dating studies of quartz and feldspar sediment extracts: Quaternary Science Reviews, v. 7, p. 373–380.

Hackett, W.R., and Smith, R.P., 1992, Quaternary volcanism, tectonics, and sedimentation in the Idaho National Engineering Laboratory area, *in* Wilson, J.R., ed., Field guide to geologic excursions in Utah and adjacent areas of Nevada, Idaho, and Wyoming: Utah Geological Survey Miscellaneous Publication 92-3, p. 1–18.

Humphreys, E.D., Dueker, K., Schutt, D., and Saltzer, R., 1999, Lithosphere and asthenosphere structure and activity in Yellowstone's wake: Geological Society of America Abstracts with Programs, v. 31, no. 4, p. A17.

Huntley, D.J., Berger, G.W., and Bowman, S.G.E., 1988, Thermoluminescence responses to alpha and beta irradiations, and age determinations when the high dose response is non-linear: Nuclear Tracks and Radiation Measurements, v. 105, p. 279–284.

Hütt, G., Jaek, I., and Tchonka, J., 1988, Optical dating: K-feldspars optical response stimulation spectra: Quaternary Science Reviews, v. 7, p. 381–385.

Jerlov, N.G., 1976, Marine optics: New York City, Elsevier Scientific, 231 p.

Kaufman, D.S., 2000, Amino acid racemization in ostracodes, *in* Goodfriend, G., Collins, M., Fogel, M., Macko, S., and Wehmiller, J., eds., Perspectives in amino acid and protein geochemistry: New York, Oxford University Press, p. 145–160.

Kaufman, D.S., and Manley, W.F., 1998, A new procedure for determining enantiomeric (D/L) amino acid ratios in fossils using reverse phase liquid chromatography: Quaternary Science Reviews (Quaternary Geochronology), v. 17, p. 987–1000.

Lang, A., 1994, Infra-red stimulated luminescence dating of Holocene reworked silty sediments: Quaternary Science Reviews (Quaternary Geochronology), v. 13, p. 525–528.

Levish, D.R., Ostenaa, D.A., Klinger, R.E., and Simpson, D.T., 1999, Holocene lunettes on the Birch Creek Playa, Butte County, Idaho: Implications for a post-glacial Lake Terreton: Geological Society of America Abstracts with Programs, v. 31, no. 4, p. A21.

McQuarrie, N., and Rodgers, D.W., 1998, Subsidence of a volcanic basin by flexure and lower crustal flow: The eastern Snake River Plain, Idaho: Tectonics, v. 17, no. 2, p. 203–220.

Nace, R.L., Veogeli, P.T., Jones, J.R., and Morris, D., 1975, Generalized geologic framework of the National Reactor Testing Station, Idaho: U.S. Geological Survey Professional Paper 725-B, 49 p.

Oviatt, C.G., 1997, Lake Bonneville fluctuations and global climate change: Geology, v. 25, p. 155–158.

Oviatt, C.G., Currey, D.R., and Sack, D., 1992, Radiocarbon chronology of Lake Bonneville, eastern Great Basin, USA: Palaeogeography, Palaeoclimatology, Palaeoecology, v. 99, p. 225–241.

Oviatt, C.G., and Miller, D.M., 1997, New explorations along the northern shores of Lake Bonneville: Provo, Utah, Brigham Young University Geology Studies, v. 42, part 2, p. 345–371.

Prescott, J.R., and Hutton, J.T., 1988, Cosmic ray and gamma ray dosimetry for TL and ESR: Nuclear Tracks and Radiation Measurements, v. 14 (1–2), p. 223–227.

Scott, W.E., 1982, Surficial geologic map of the eastern Snake River Plain and adjacent areas, 111° to 115° W., Idaho and Wyoming: U.S. Geological Survey Miscellaneous Investigations Series Map I-1372, scale 1:250 000, 2 sheets.

Scott, W.E., McCoy, W.D., Shroba, R.R., and Rubin, M., 1983, Reinterpretation of the exposed record of the last two cycles of Lake Bonneville, western U.S.: Quaternary Research, v. 20, p. 261–285.

Sherbondy, M.J., and Gianniny, G.L., 1999, Late Pleistocene climate record from the North Lake embayment of Quaternary Lake Terreton, Idaho: Geological Society of America Abstracts with Programs, v. 31, no. 4, p. A55.

Smith, B.W., Aitken, M.J., Rhodes, E.J., Robinson, P.D., and Geldard, D.M., 1986, Optical dating: Methodical aspects: Radiation Protection Dosimetry, v. 17, p. 229–233.

Spinazola, J.M., 1994, Geohydrology and simulation of flow and water levels in the aquifer system in the Mud Lake area of the eastern Snake River Plain, eastern Idaho: U.S. Geological Survey Water-Resources Investigations Report 93-4227, 78 p.

Spooner, N.A., Aitken, M.J., Smith, B.W., Franks, M., and McElroy, C., 1990, Archaeological dating by infrared-stimulated luminescence using a diode array: Radiation Protection Dosimetry, v. 34, p. 83–86.

Stearns, H.T., Bryan, L.L., and Crandall, L., 1939, Geology and water resources of the Mud Lake region, Idaho: U.S. Geological Survey Water-Supply Paper 818, 125 p.

Tressler, W.L., 1959, Ostracoda, *in* Edmundson, W.I., ed., Fresh-water biology: London, John Wiley and Sons, p. 657–734.

Wetmore, P.H., 1998, An assessment of physical volcanology and tectonics of the central eastern Snake River Plain, based on the correlation of subsurface basalts at and near the Idaho National Engineering and Environmental Laboratory, Idaho [M.S. thesis]: Pocatello, Idaho, Idaho State University, 118 p.

MANUSCRIPT ACCEPTED BY THE SOCIETY NOVEMBER 2, 2000

Geological Society of America
Special Paper 353
2002

Holocene paleoflood hydrology of the Big Lost River, western Idaho National Engineering and Environmental Laboratory, Idaho

Dean A. Ostenaa
Daniel R.H. O'Connell
Seismotectonics and Geophysics Group, D-8330, U.S. Bureau of Reclamation,
Denver, Colorado 80225, USA
Roy A. Walters*
National Research Program, Water Resources Division, MS 413, U.S. Geological Survey,
Denver, Colorado 80225, USA
Robert J. Creed
U.S. Department of Energy, MS 1220, 850 Energy Drive, Idaho Falls, Idaho 83401, USA

ABSTRACT

Stratigraphic and geomorphic evidence along the Big Lost River at the Idaho National Engineering and Environmental Laboratory (INEEL) defines age and paleostage limits for a paleoflood ~400 yr ago with an estimated discharge of ~100 m³/s. The discharge for this paleoflood is ~40% larger than the flood of record from a gaging site near Arco where flow is regulated, but is smaller than 6 historical peak discharges from a gage in the unregulated upstream portion of the drainage basin. The paleoflood is the largest flood along the Big Lost River in the past ~400 yr and confirms that large downstream decreases in Big Lost River peak discharge predate historical stream diversion and regulation. Flow simulations indicate that discharges only slightly larger than ~110 m³/s will initiate extensive flow across the unmodified Pleistocene alluvial surfaces that flank the Big Lost River on the INEEL site. The geomorphology of these surfaces and two-dimensional flow simulations are the bases for establishing a paleohydrologic bound at a discharge of 150 m³/s for the past 10 k.y.

When the paleoflood and paleohydrologic bound data are included in peak-discharge-frequency analyses, they provide strong constraints on peak discharge for annual probabilities from $>10^{-2}$ to 5×10^{-5}. Sensitivity testing is used to assess the potential impacts of historical regulation of annual peak discharge and of alternative characterizations of the paleohydrologic information on discharge-frequency estimates. These tests demonstrate that for annual probabilities of 10^{-2} and 10^{-4}, the upper limits of peak discharge are unlikely to exceed ~110 m³/s and ~170 m³/s, respectively, as long as the long-duration paleohydrologic bounds are included in the analyses. In contrast, peak-discharge-frequency analyses using only annual peak-discharge discharge data result in estimates that range from ~105 m³/s to >170 m³/s for an annual probability of 10^{-2}. Adding paleohydrologic information to discharge-frequency analyses reduces the possible range of discharge estimates over a wide range of annual probabilities.

*Present address: National Institute of Water and Atmospheric Research Limited, P.O. Box 8602, Riccarton, Christchurch, New Zealand.

Ostenaa, D.A., O'Connell, D.R.H., Walters, R.A., and Creed, R.J., 2002, Holocene paleoflood hydrology of the Big Lost River, western Idaho National Engineering and Environmental Laboratory, Idaho, *in* Link, P.K., and Mink, L.L., eds., Geology, Hydrogeology, and Environmental Remediation: Idaho National Engineering and Environmental Laboratory, Eastern Snake River Plain, Idaho: Boulder, Colorado, Geological Society of America Special Paper 353, p. 91–110.

INTRODUCTION

Paleoflood hydrology includes the study of the geomorphic and stratigraphic record of past floods (e.g., Patton, 1987; Baker, 1989; Jarrett, 1991). Paleoflood data can help characterize historical floods, define limits on the magnitude and frequency of extreme floods, construct models of flood characteristics or frequency, and test models of extreme floods formulated from other data sets. At sites where paleoflood records are long, they can provide powerful data for evaluating hypotheses of flood frequency and magnitude (Stedinger and Cohn, 1986; Frances et al., 1994), especially for floods with long return periods, because the paleoflood record can be 10 to 100 times longer than typically available western United States stream-flow records. The paleoflood data provide a means of verifying return periods for floods that are many times longer than the length of annual peak-discharge or historical records (Costa, 1978) because the ages of stable geomorphic surfaces adjacent to streams and rivers are a direct indication of the potential risk of flooding in unregulated or unmodified channel systems.

This chapter describes results from a paleoflood study of the Big Lost River and the characterization of these data for use in applied flood hazard evaluations at the Idaho National Engineering and Environmental Laboratory (INEEL) (Ostenaa et al., 1999). The INEEL site includes a diverse array of facilities, and estimates of the annual probabilities of floods over a range from $>10^{-2}$ to 10^{-5} are required. In years with high Big Lost River stream flow, out-of-channel flow could affect facilities on the INEEL site. Long-term limits on Big Lost River peak discharge are one type of data that can be used to characterize the likelihood of stream flows exceeding the capacity of the Big Lost River at the INEEL over a wide range of annual probabilities. The paleoflood data are used in flood-frequency analyses to express flood hazard in terms of annual peak-discharge probability.

Setting

The INEEL site is located on the eastern Snake River Plain of Idaho (Fig. 1), a large area of Quaternary basaltic lava flows mantled with extensive, thin, wind-blown deposits and lesser areas of alluvium and lacustrine deposits (Kuntz et al., 1994). Middle to late Cenozoic extension in the Basin and Range Province mountains to the north and south is overprinted by the volcanic activity on the eastern Snake River Plain, presumably in response to passage of the Yellowstone hotspot (e.g., Pierce and Morgan, 1992).

The headwaters of the Big Lost River are in the glaciated mountains of the Idaho Basin and Range Province north of the Snake River Plain (Fig. 1). The upper basin includes peaks that exceed 3500 m elevation in the northeast-facing basins of the Pioneer Mountains and the southeastern portion of the steep

southwest-facing front of the Lost River Range. The river flows southeast for a distance of ~80 km through the Big Lost River Valley, a late Cenozoic structural basin filled with alluvium. Mackay Reservoir, ~30 km upstream of Arco, stores irrigation water for use downstream in the Big Lost River Valley. At the northern edge of the Snake River Plain near Arco, the drainage basin includes an area of ~3650 km^2 that is above an elevation of 1550 m. About 10 km downstream of Arco, the river flows onto the INEEL site where it turns northeast and flows another 35 km to its natural terminus in the Big Lost River Sinks and several playas at the northern edge of the INEEL site. The last portion of the river course parallels the axis of the Big Lost Trough, a late Cenozoic depositional center on the north side of the eastern Snake River Plain (Geslin et al., this volume). Subsidence along the Big Lost Trough has been more or less matched by the rate of volcanic and sedimentary infill (Geslin et al., this volume). Thus, on the Snake River Plain and the INEEL site, a sequence of late Pleistocene terraces along the Big Lost River records only a few meters of net incision in the past 95 k.y. (Ostenaa et al., 1999; Simpson et al., 1999).

Big Lost River historical stream flow

Average annual precipitation in the Big Lost River basin ranges from ~1250 mm/yr in the mountainous upper basin areas to ~200 mm/yr across much of the INEEL site on the Snake River Plain. This precipitation occurs mostly in the winter months and is largely derived from moisture from the northern Pacific Ocean (Kjelstrom, 1991). During the late spring and summer snowmelt period, the air flow from the Pacific generally consists of relatively dry, subtropical air that produces only sporadic thunderstorms across Idaho. Southeastern Idaho can be affected as well by summer monsoon flow from the south and southwest, which can cause increased precipitation (Kjelstrom, 1991). Meteorological conditions favorable for long-duration winter rainfall are uncommon (Kjelstrom, 1991), especially for large drainage basins.

The annual stream flow and the largest annual peak discharge in the Big Lost River are dominated by the spring and early summer snowmelt and runoff from the mountains in the upper drainage basin. Stream-flow records are available for the upper basin from 1904 and for the Arco area from 1947 (Table 1; Fig. 2). The timing of the snowmelt is regular, usually beginning in late May or early June of each year; significant flows extend into July. The magnitude of the annual peak discharge typically decreases in a downstream direction with increasing drainage area. Significant downstream decreases in peak discharge, even in the wettest years, indicate that the decrease is at least in part due to large amounts of natural channel infiltration and storage in the Big Lost River valley (Stearns et al., 1938). Additional decreases in peak discharge result from storage in Mackay Reservoir, ~65 km upstream of the INEEL site, and irrigation diversions upstream of Arco and the INEEL site.

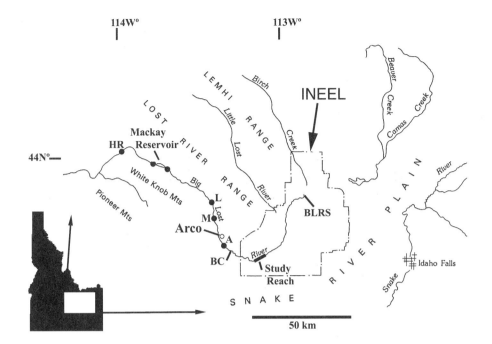

Figure 1. Regional location map of the Big Lost River and INEEL (Idaho National Engineering and Environmental Laboratory) area. The headwaters of the Big Lost River are in Pioneer Mountains. Flow terminates on Snake River Plain within INEEL site near Big Lost River sinks (BLRS). Principal stream-gaging stations on Big Lost River, shown with black dots, are located at Howell Ranch (HR), upstream and downstream of Mackay Reservoir, at Leslie (L), near Moore (M) and near Arco (A) (see Table 1). Study reach described in this chapter is on southwestern part of INEEL site, downstream of Box Canyon (BC).

TABLE 1. PRINCIPAL U.S. GEOLOGICAL SURVEY GAGING STATIONS ON THE BIG LOST RIVER, IDAHO

Gaging station		Period of record	Drainage Area	Discharge of Record	
Name	Number		(km²)	(m³/s)	Date
Big Lost River at Howell Ranch near Chilly, Idaho	13120500	1904–1998	1165	125	5/27/1967
Surface Inflow to Mackay Reservoir, near Mackay, Idaho	13125500	1919–1950	1984	78	6/12/1921
Big Lost River below Mackay Reservoir, near Mackay, Idaho	13127000	1903–1906; 1912–1915; 1919–1998	2106	85	6/10/1921
Big Lost River at Leslie, Idaho	13130500	1919–1921	2642	73	6/10/1921
Big Lost River near Moore, Idaho	13132000	1920–1926	3393	66	6/14/1921
Big Lost River near Arco, Idaho	13132500	1947–1961 1966–1980 1980 1982–1998	3652	71	6/29/1965

Note: Data from USGS WATSTORE (http://water.usgs.gov/nwis/sw). Station locations are in Figure 1.

Flood hazard studies

Guidance for the INEEL flood hazard studies was prescribed by the Department of Energy Natural Phenomena Hazards Characterization and Design Standards (U.S. Department of Energy [USDOE], 1996a, 1996b, 1996c, 1996d). These documents specify hazard characterization and design requirements for various DOE facilities using an approach where the design-flood return periods are increased as risk to the public, workers, and environment increases. For the most critical facilities, the standards require estimates of flood discharge for probabilities as low as 10^{-5}. The DOE standards include traditional approaches to flood hazard for the Big Lost River at the INEEL described in Bulletin 17B (Interagency Advisory Committee on Water Data, 1982) or FEMA 37 (Federal Emergency Management Agency, 1993) that rely on data from stream-gage records.

However, these approaches are complicated by the alteration of Big Lost River flow imposed by upstream irrigation diversions and Mackay Dam storage. For the evaluation of extreme floods, i.e., floods larger in magnitude than those of historical experience, alternatives to traditional approaches are required due to the effects of regulation and the limited length of the streamflow records.

Most prior flood hazard studies for the INEEL have focused on estimating floods with annual probabilities near 10^{-2} based on data from Big Lost River gaging stations. Kjelstrom and Berenbrock (1996) summarized the previous studies and applied a regional approach to estimating the 100 yr flows for the Big Lost River at the INEEL. Tullis and Koslow (1983) compared preliminary paleoflood data from sites upstream of Box Canyon to flood frequency estimated from the gaging records near Arco. Rathburn (1993) described evidence for a

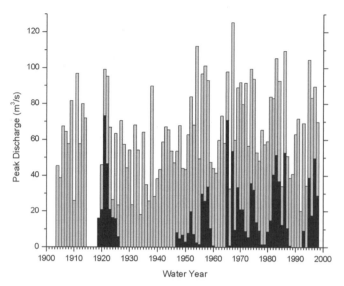

Figure 2. Annual peak discharge estimates for upstream (light shaded bars) and downstream gaging stations (dark shaded bars) on Big Lost River. Upstream estimates are from Big Lost River at Howell Ranch from 1904 to 1998. Downstream estimates are from Big Lost River near Arco from 1947 to 1998 except for period 1919 to 1926, which are peak discharge estimates from stations at Leslie and near Moore, early gaging stations located a short distance upstream of Arco. No peak discharge estimates are available from sites near Arco for periods 1905 to 1918, 1927 to 1945, and 1962 to 1964. Station data are described in Table 1. Station locations are shown in Figure 1.

Pleistocene glacial outburst flood on the Big Lost River, a flood mechanism not considered by the present study.

APPROACH AND RESULTS

The approach taken for this paleohydrologic analysis is similar to that used for recent Bureau of Reclamation Dam Safety flood hazard evaluations throughout the western United States (e.g., Ostenaa et al., 1996, 1997; Ostenaa and Levish, 1997). Flood-frequency analyses for these studies incorporate paleoflood estimates and paleohydrologic bounds (Levish et al., 1994, 1996, 1997; Ostenaa and Levish, 1996) into a Bayesian analysis that uses likelihood functions that incorporate both parameter and data (discharge and geologic age) measurement uncertainties (O'Connell et al., 1996, 1998).

A paleohydrologic bound is the time interval during which a given discharge has not been exceeded (Fig. 3). Paleohydrologic bounds are not actual floods, but instead are limits on paleostage over a measured time interval. These bounds represent stages and discharges that have not been exceeded since a geomorphic surface stabilized. Through hydraulic modeling, discharge for a paleohydrologic bound can be derived from stage, just as a discharge is derived from the paleostage indicators of past floods. Used appropriately, paleohydrologic bounds are powerful in flood-frequency analyses, even if the

number, timing, and magnitude of individual paleofloods are uncertain (Stedinger and Cohn, 1986).

In this context, the present analysis only assumes that for extreme floods, upstream regulatory structures and diversions do not increase flood magnitudes downstream compared to the unregulated natural flows, except for cases where upstream regulating structures might fail. Flood probabilities for such scenarios should be evaluated separately, and account for the overall failure probability of the structure under all conditions. The impacts of regulation and variations in smaller flows, such as those of historical experience, on frequency estimates of extreme floods are addressed through sensitivity analyses.

Geologic and geomorphic evidence of flooding

There are many different types of geologic and geomorphic information in fluvial systems that provide a direct indication of the magnitude and frequency of floods (e.g., Baker et al., 1988). Gravel bars and slackwater terraces indicate the minimum stage of past floods (Fig. 3). Likewise, evidence for past erosion such as channels on terrace surfaces and truncated soil profiles also indicates the minimum stage of past floods. The age and frequency of the floods that produced these features can be determined by the degree of soil development, the morphology and extent of weathering on surface features, and radiocarbon analysis of organic material within the deposits. For historical or more recent paleofloods, floated debris and subtle erosional scars are a shorter lived record of the maximum stage.

A complementary indication of the limits of past floods is the recognition of the amount of time during which floods have not modified geomorphic features or deposits. Soil development and the geomorphic evolution of deposits and surfaces are time-dependent processes (e.g., Birkeland, 1999). Thus, the age of stable geomorphic surfaces not modified by floods that are adjacent to streams is an indication of the minimum length of time since last flooding (Costa, 1978). Evidence of modification of these surfaces by floods includes the deposition of sediment resulting in burial of soils, erosion and truncation of soils, erosion of channels on the surfaces, or erosion of the deposits. Estimates of the stage required to modify these surfaces can come from empirical comparisons to data from historical floods or by comparison to hydraulic model results of observed flows. The minimum depth of flow required for the initiation of large-scale erosion or deposition on geomorphic surfaces can also be evaluated formally in terms of shear stress or stream power (e.g., Parker, 1978; Andrews, 1984; Baker and Costa, 1987).

Big Lost River–diversion dam study reach

South of Arco, the Big Lost River leaves the alluvium-filled Big Lost River valley (Fig. 1) and flows across middle to late Quaternary basalt on the Snake River Plain that is locally mantled with alluvium of varied thickness (Kuntz et al., 1994).

Near Box Canyon (Fig. 1), the river is incised from 5 to 30 m into the basalt, and only small areas of alluvium are preserved in the canyon. Downstream of Box Canyon, on the southwestern portion of the INEEL site, Kuntz et al. (1994) mapped extensive areas of alluvium along the Big Lost River. However, even in this reach, basalt exposures are common in the bed and banks of the river and as isolated outcrops on the alluvial surfaces near the channel, indicating that the alluvium overlying the basalt is relatively thin in this area.

The geologic and geomorphic descriptions and hydraulic modeling analyses focus on a 6-km-long study reach just downstream of the INEEL Diversion Dam (Fig. 4). Throughout the Diversion Dam reach, the Big Lost River is incised ~2–4 m into the relatively flat alluvial surfaces on either side. The areas of relief are associated with Pleistocene basalt outcrops above the alluvium. Previous mapping depicts most of this alluvium as Pleistocene on the basis of the degree of soil development and alluvial surface morphology (e.g., Rathburn, 1991; Kuntz et al., 1994). The Pleistocene alluvium is gravelly with a 0.5-1-m-thick cover of loess, mostly deposited before ca. 10 ka

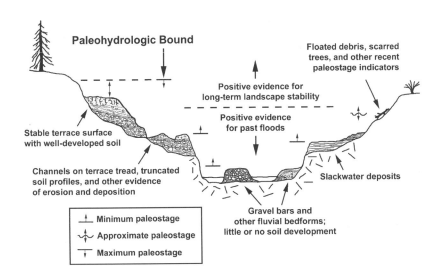

Figure 3. Schematic representation of different types of flood evidence. Most types of geologic and geomorphic data provide minimum paleostage limits. Paleohydrologic bounds represent maximum paleostage. To establish a paleohydrologic bound there must be sufficient depth of flow over terrace surface (double arrow) to result in either erosion of or deposition on terrace surface. Modified from Baker (1987).

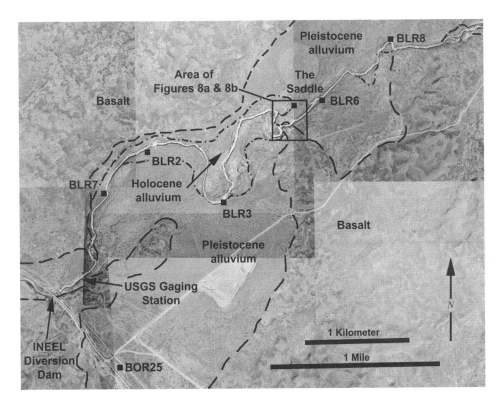

Figure 4. Photomosaic of Diversion Dam study reach with generalized surficial geology. In this area, Big Lost River channel is inset into thin deposit of Pleistocene alluvium that buries basalt. The generalized contact between the major areas of Pleistocene alluvium and basalt is shown with dashed line. Dot-dash line separates major areas of Holocene alluvium inset within Pleistocene deposits along Big Lost River. In several locations, channel flows directly on basalt (not shown at this scale), which provides hydraulic control for individual subreaches of river. Big Lost River flow is from left to right. Soil-stratigraphic study sites are shown with black squares and labeled. Flow simulations were conducted for reach extending from bend downstream of USGS gaging station to upper right edge of the Figure. Study Reach on Figure 1 shows extent of photomosaic along Big Lost River.

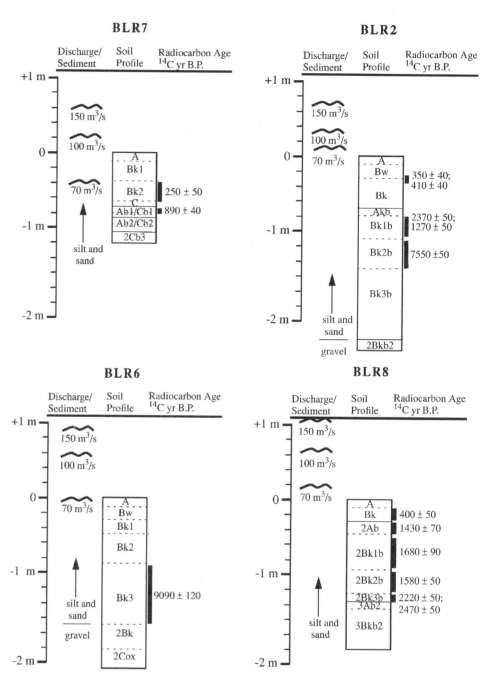

Figure 5. Soil and stratigraphic summaries of Big Lost River study sites BLR7, BLR2, BLR6, and BLR8. Dashed lines in each box are horizon boundaries for soil profile; solid lines are boundaries of buried soils. See Table 2 for calcium carbonate morphology and radiocarbon sample data. Thick vertical bars to right of each box depict interval sampled for datable material for adjacent radiocarbon age(s). Wavy lines to left of each box depict approximate stage of the indicated discharge relative to each soil description site. Scale shows approximate flow depth above ground surface and depth for each soil profile. Sediment sizes present in each soil profile are shown by labels and arrows that are adjacent to soil profile boxes, below discharge lines.

(Forman et al., 1993). Exposures in nearby gravel pits and stream banks show a moderately to well-developed soil with stage II or greater calcium carbonate accumulation in the gravel and loess.

Soils at sites BLR3, BLR6, and BOR25 all contain stage II or greater calcium carbonate accumulation (Fig. 5; Table 2). A radiocarbon age from site BLR6 indicates a minimum age for these deposits of at least 10 ka. The soil at site BOR25 has a very well developed calcic horizon (stage III), generally considered indicative of a late Pleistocene age (e.g., Scott, 1982; Birkeland, 1999). The Pleistocene surfaces adjacent to the Di-

version Dam study reach do not show evidence of having been overtopped by floods from the Big Lost River since the Pleistocene. Relic channels on these surfaces have a subdued morphology, consistent with a Pleistocene age. The surfaces also have a well-developed pattern of earth mounds, which follow and overprint the channels (Tullis, 1995). Soil development within and between the mounds indicates that the mounds developed during the late Pleistocene (Tullis, 1995) and many areas between the mounds have a weak to moderate gravel pavement (e.g., McFadden et al., 1998). The mounds consist largely of loess (Tullis, 1995) and would be highly erodible, as

TABLE 2. SOILS AND RADIOCARBON SUMMARY

Site Number	Soils data*			Radiocarbon sample and age data				
	Calcium carbonate stage	Maximum carbonate (wt%)	B horizon	Depth (cm)	Material dated	Laboratory number	Radiocarbon age (^{14}C yr B.P.) $\pm 1\,\sigma$	Calibrated age[†] (yr B.P.) $\pm 2\,\sigma$
BLR2								
(surface soil)	I−	5	Bw, Bk	25–35	*Artemisia* charcoal	Beta-121217	350 ± 40	490–300
				35	*Artemisia* charcoal	Beta-122946	410 ± 40	520–420; 400–310
(buried soil)	I++	12	Bk	80–90	Hardwood charcoal	Beta-121218	2370 ± 50	2750–2200
				80–111	Bark	Beta-122947	1270 ± 50	1290–1070
				111–150	Shell	Beta-122948	7550 ± 50	8370–8140
BLR3 (surface soil)	II+	na	Bw, Bk					
BLR6 (surface soil)	II+	21	Bw, Bk	100–160	*Artemisia* charcoal	Beta-124711	9090 ± 120	10360–9870
BLR7								
(surface soil)	I	9	Bk	40–60	*Artemisia* charcoal	Beta-124712	250 ± 50	470–250; 230–130; 30–0
(buried soil)				75–90	Salicaceae charcoal	Beta-124713	890 ± 40	970–700
BLR8								
(surface soil)	I−	8	Bk	11–29	*Artemisia* charcoal	Beta-124714	400 ± 50	520–310
(buried soil)	II−	7, 16	Bk	29–45	*Artemisia* charcoal	Beta-124715	1430 ± 70	1510–1180
				60–75	*Artemisia* charcoal	Beta-124716	1680 ± 90	1810–1390
				93–125	*Artemisia* charcoal	Beta-124717	1580 ± 50	1550–1340
				125–136	*Juniperus* charcoal	Beta-124718	2220 ± 50	2350–2110
				125–136	*Discus shineki* and *Stagnicola* shells	Beta-126509	2470 ± 50	2720–2360
BOR25 (surface soil)	II+ to III	34	Bw, Bk, K					

* Soils at all sites except BLR3 and BOR25 are developed in fine-grained fluvial deposits with <10% gravel. Soils at BLR3 and BOR25 are developed in loess and gravelly parent material. Soils nomenclature and calcium carbonate morphology follow Birkeland (1999).
[†] Radiocarbon ages were calibrated with OXCAL (Ramsey, 1995).

would the loess-covered Pleistocene alluvium, if there was significant surface flow around or over the mounds.

The overall channel configuration of the Big Lost River in this study reach (Fig. 4) is controlled by several locations where the river crosses outcrops of basalt. These outcrops form constrictions that locally create hydraulic controls on flow and indicate that the overall configuration of the channel has been stable since the river incised below the level of the Pleistocene surfaces. Well-developed soils on some of the inset fine-grained terraces, such as at BLR6, and a radiocarbon age from these deposits, indicate that this incision has an age of at least 10 ka (Fig. 5; Table 2).

In the central portion of the Diversion Dam study reach, the Big Lost River flows through a narrow basalt constriction (Fig. 6). This constriction is formed by a ridge of basalt that extends across the Big Lost River and protrudes above the level of the Pleistocene alluvial surface south of the river. Upstream of this constriction is a meandering reach of the river that is flanked by an extensive area of Holocene alluvium inset below the level of the Pleistocene alluvial surfaces on either side (Fig. 4). On the north side of the river, a low ridge apparently underlain by basalt forms the contact between the Holocene and Pleistocene alluvium (Fig. 6). The surface of the ridge is capped with gravel and eolian deposits. The low point along this ridge, informally known as the Saddle, is a location where high flows could spill over onto the Pleistocene surface downstream of the ridge due to elevated stage caused by hydraulic ponding at the basalt constriction. The morphology of the Pleistocene surface downstream of the Saddle is no different than elsewhere on the

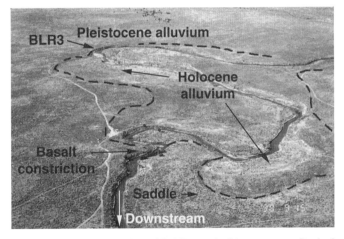

Figure 6. Aerial oblique view of Saddle area looking upstream. Dashed line shows approximate boundary between Pleistocene and Holocene alluvium upstream of the basalt constriction. Flow through Saddle would return to Big Lost River across Pleistocene surface in lower right center of the photo. BLR6 is just off bottom center of photo; scattered trees at top center are on basalt outcrops located northeast of USGS gaging station on Figure 1. Meander wavelength is about 1 km.

Pleistocene surfaces, indicating a limit for flow through the Saddle since the Pleistocene.

Holocene surfaces with distinctly different morphologies that are within and below the level of the Pleistocene surfaces are present throughout the Diversion Dam study reach. An extensive area of Holocene alluvium is just upstream of the Saddle

(Figs. 4 and 6), but elsewhere in the reach these deposits are of limited extent. Downstream of the Saddle, Holocene surfaces are limited to narrow terraces that mostly parallel the main channel, but that are somewhat wider at locations where high flows have cut across bends, such as near site BLR8 (Fig. 4). On the higher Holocene surfaces, ~2–3 m above the low-flow channel, the surface morphology is generally smooth, with only small, subdued channels evident. In contrast to the Pleistocene surfaces, earth mounds are absent. A 0.5–1-m-high terrace riser is often present at the back edge of these surfaces, defining the contact with the Pleistocene surfaces. Surfaces that are <2 m above the low-flow channel are distinctly channeled. Floated debris, such as milled timber, indicates that these lower surfaces and deposits have been flooded by recent flows.

Descriptions at three sites, BLR2, BLR7, and BLR8, in the Diversion Dam reach show that the Holocene terrace deposits of the Big Lost River are generally fine grained, consisting of sand and silt (Figs. 5 and 7). Gravel is generally present only as small bars in the channel, or underlying the fine-grained deposits in these terraces. Rathbun (1991) also noted that the Holocene deposits of the Big Lost River on the INEEL site are finer grained than the Pleistocene alluvium.

The uppermost unit at sites BLR2, BLR7, and BLR8 underlies Holocene terrace surfaces that are ~2 m above the low-flow channel (Fig. 5). This unit consists of sand and silt, interpreted as flood deposits, in which a weak calcic soil with up to stage I calcium carbonate is developed (Table 2). Radiocarbon ages from this upper deposit and soil calibrate to a range of ca. 400 ± 100 yr ago, a range that likely defines the minimum age for this deposit and soil. There is no stratigraphic evidence of younger floods modifying the surface soil or geomorphic evidence of paleostage above the surface of these deposits since that time. Hydraulic modeling results from the entire study reach (Ostenaa et al., 1999) indicate that for the discharge range of interest, flow at these sites is not affected by backwater from hydraulic constrictions such as at the Saddle. The upper unit

overlies buried soils in fine-grained fluvial deposits, which at some sites have stronger soil development than the surface soil. The youngest calibrated radiocarbon age from the buried deposits is ~800 yr ago (Fig. 5; Table 2). The youngest ages from the buried deposits are an upper limit for the maximum age of the overlying deposit and flood. The uppermost unit at sites BLR2, BLR7, and BLR8 is interpreted as the deposit of a paleoflood, ~400 ± 100 yr ago, which is the largest paleoflood since that time.

Discharge estimation

Modeling methods. Hydraulic models can be used to estimate discharge by comparing computed water-surface elevations and paleostage data. The precision of the discharge estimate can be assessed by bracketing paleostage indicators with water-surface elevations computed from a range of discharges. The specific model used depends on the actual problem to be solved and the desired results. O'Connor and Webb (1988) described the background and application of hydraulic modeling to paleoflood studies with a primary focus on the use of one-dimensional models, whereas Miller and Cluer (1998) included consideration of some two-dimensional models.

Although many factors influence the computed water-surface elevation, the primary factors are discharge and topography. For the Big Lost River, most flow is confined within banks of Pleistocene alluvium and basalt, but as discharge increases there are important areas of overbank flow. Adequate resolution of the flow characteristics for these overbank areas is an important factor in estimating the discharge for paleofloods and paleohydrologic bounds. In the Big Lost River study reach, there are cross sections where the flow at different points across the section is subcritical, transcritical, and supercritical. In addition, there are constrictions where the flow represents a balance between the surface pressure gradient and the local ac-

Figure 7. Big Lost River near site BLR2. BLR2 is description site for Holocene surface that is ~2-m above low-flow channel. Deposits of ~400 yr paleoflood form upper 59 cm of this bank. Note basalt outcrops across the channel downstream of site, in the upper left corner of photograph. Modeled stage for a discharge of 70 m³/s is near top of the bank (dashed line). Channel is about 20 m wide.

celeration (momentum advection) such that bottom friction is not important locally. For these problems, one-dimensional models are not appropriate because they cannot account for some of the important head losses such as at constrictions and meanders except by treating them empirically. Nor can one-dimensional models adequately treat overbank flow through meanders. Two-dimensional models can represent the physics of these flows without resorting to empirical relations. A three-dimensional model is not needed because a detailed knowledge of the flow and boundary-layer structure is unnecessary if water-surface elevation and discharge are the focus.

Thus, we have chosen specific two-dimensional models that are able to simulate transcritical flows in a stable and efficient manner. The two-dimensional simulations provide details of flow patterns and power distribution across the channel and in adjacent flooded areas. In addition to these details, the two-dimensional models can properly simulate the physics of subcritical, transitional, and critical flows with an adequate spatial resolution of the topography. With the use of relatively small grid cells, the model provides the spatial variation of power at a scale that can be related to geomorphic features of the channel and adjacent surfaces.

The two-dimensional numerical models used in the INEEL study are based on the shallow-water equations with a semi-implicit time integration scheme and a semi-Lagrangian approximation for advection (Casulli, 1990; Casulli and Cattani, 1994; Walters and Casulli, 1998). These models are accurate, robust, and very efficient, and hence can utilize large, high-resolution topographic data sets. Two separate implementations of this model were used: TRIMR2D (Transient Inundation Model for Rivers-2 Dimensional), which uses a finite difference spatial approximation and square grid cells (Casulli, 1990), and SISL2D (Semi-Implicit Semi-Lagrangian model-2 Dimensional), which uses a finite element spatial approximation with triangular elements (Walters and Casulli, 1998) and has a more flexible grid layout, but requires more computer memory. The two models were developed independently and were both used in the simulations to corroborate the results (Walters and Denlinger, 1999). The results for TRIMR2D are presented here because the computational efficiency allowed more extensive sensitivity testing. TRIMR2D has the same computational kernel as the model TRIM2D that has been used extensively and validated in simulations in estuaries and tidal rivers (e.g., Cheng et al., 1993; Lang et al., 1998). Both TRIMR2D and SISL2D have been tested with problems that have analytical solutions in order to assess the model accuracy and convergence rate (Walters and Casulli, 1998; Casulli and Walters, 2000). In addition, these models have been used for transient flood analysis on the Nisqually River, Washington (R. Walters, personal communication, 1999) where simulated stage was compared to high-water marks and a single gage, and for steady flows of various rivers where simulated stage was compared to high-water marks (R. Walters and R. Denlinger, personal communication, 1999).

Topographic data for the two-dimensional flow simulations in the Diversion Dam reach are based on a regular grid with 1 m grid spacing derived from a 0.6 m contour-interval map produced for the Department of Energy by Aerographics, Inc., from 1:10 000-scale aerial photographs taken in September 1993. There was no recorded flow in the river channel downstream of the INEEL Diversion Dam for more than two months prior to the time the photographs were taken. Thus, the map includes detailed channel geometry because the channel was dry. The simulation model subsamples the grid to obtain a computational grid with 2 m cell size. In this way, the velocity is defined at the midside of each cell face, and water-surface elevation is defined at the center of the cell (Casulli, 1990). For the plots shown in Figure 8, velocity was averaged so that all variables are reported at the cell centers and the contour plots were constructed from the grid made up of cell centers. The grid for the Diversion Dam reach contains 2800 columns and 1519 rows or ~4.2 × 10^6 cells. Of these, there are ~2.5 × 10^6 active cells in the calculations. These models include wetting and drying, so that an active cell is "wet" and is included in the calculations. Dry cells are inactive until they become flooded.

To initialize the model, a flow of ~10 m^3/s was specified at the upstream end and the simulation ended when the entire length of the river channel was wetted. Once the riverbed was wet, the waves traveled with the shallow-water wave speed rather than cell-by-cell such as during the initial wetting period. Hence, a larger model time step could be used once the channel was wetted. Steady flows of 70, 100, 150, 187, 263, and 400 m^3/s were simulated using the next smaller flow as the initial conditions. This procedure minimized the occurrence of a bore traveling downstream which tended to leave isolated patches of shallow water on the flood plain.

For each simulation, water-surface elevation, velocity, water depth, stream power (dissipation from bed friction), and Froude number (Fr = u/c, where u is water speed and c is the phase speed of water waves) were calculated at the computational cell centers. The model was run with a 3 s time step, which resulted in a Courant number of 4–6 in the main channel (Courant number is a measure of how many grid cells a water parcel will travel through during a time step). The only adjustable parameters in this model are the bottom friction coefficient (Manning's n), which was set at 0.038, and horizontal viscosity, which was set at 0.05 m^2/s^2, a value derived from turbulence theory. The results were locally insensitive to the former parameter and generally insensitive to the latter parameter. The sensitivity tests are described following discussion of the results. Because this model is robust and stable, it is not necessary to apply unrealistically large viscosity coefficients to stabilize the model.

Hydraulic modeling results. Simulation results indicate that discharges to 70 m^3/s do not result in significant overtopping of any of the study sites or of correlative geomorphic surfaces (Figs. 5 and 9A). At sites BLR2, BLR7, BLR6, and

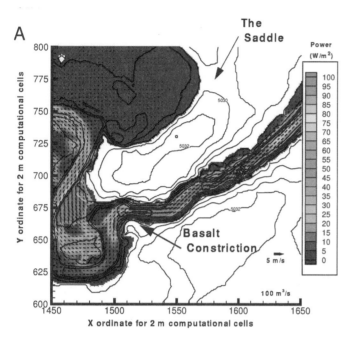

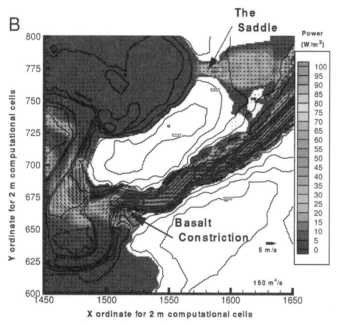

Figure 8. Big Lost River flow simulation results for portion of Diversion Dam study reach for discharges of 100 m³/s (A) and 150 m³/s (B). Color shading indicates extent of inundation and variations in unit stream power. Arrows indicate relative magnitude and primary direction of velocity. Simulation for 100 m³/s (A) shows high power values in main channel and on meander overflow channel upstream of basalt constriction. At higher discharges, power in these areas decreases due to backwater effects upstream of the constriction. For simulation in which flow overtops Saddle (B), high power values are associated with return flow across Pleistocene alluvium surface on left side of the channel. Base map is from U.S. Department of Energy 2 ft contour map of the Big Lost River area. Scale on each axis represents number of 2 m computational cells in hydraulic model; 100 units = 200 m. Flow direction is from left to right.

BLR8, this discharge is near the top of the streambanks or causes shallow, <15 cm, inundation. However, that discharge causes significant inundation outside the main channel in many other areas, such as the area of Holocene deposits that are upstream of the Saddle (Figs. 4 and 6). Most of these areas are characterized by many fresh channels and underlain by deposits with little soil development. Basalt outcrops adjacent to the channel that are inundated by this discharge are fresh, smoothed, and obviously rounded or fluted by stream flow. A discharge of 70 m³/s is approximately equivalent to the flood of record on the Big Lost River at the Arco gaging station 20 km upstream, although there is no estimate for the discharge of the 1965 flood in the Diversion Dam reach.

A discharge of 100 m³/s does not result in a stage sufficient to cause overflow at the Saddle (Figs. 8 and 9A), but does result in significant inundation, typically with depths of 25–90 cm, of areas that are underlain by deposits that appear to be ca. 400 yr old at sites BLR2, BLR7, and BLR8. This discharge is interpreted to be sufficient to result in either erosion or deposition of sediment on the Holocene terraces at these sites.

Simulations for a flow of 150 m³/s indicate significant overtopping of the Saddle and a prediction of high power values on the Pleistocene surface downstream of the Saddle where the local slope is relatively steep (Fig. 9B). These simulations imply that for flows of this size, the existing geomorphic surface of Pleistocene age would have been modified or eroded. Positive geomorphic indications of long-term surface stability on this surface, such as the well-developed soil and earth mounds, are strong evidence of no flood in that past 10 k.y. or more. The Pleistocene age of this surface is confirmed by the well-developed soil and radiocarbon age from the fine-grained deposits at BLR6 (Table 2), which are inset ~0.5 m below the flanking Pleistocene surfaces. The discharge of 150 m³/s inundates BLR6 by more than 0.5 m (Fig. 9A), and shallow flow begins to spread out onto the Pleistocene surfaces at many points throughout the study reach (Ostenaa et al., 1999) in addition to spilling over at the Saddle (Fig. 8B).

To assess whether geomorphic surfaces might be modified by various flows, power calculated in channel areas for historical flows in which transport of bedload occurs can be compared to power calculated from flow simulations that result in flow across geomorphic surfaces adjacent to the channel. Simulations for a discharge of 50 m³/s in another study reach upstream of Box Canyon (Fig. 1) resulted in calculated power of 50 to >100 W/m² (Ostenaa et al., 1999) in areas of channel erosion and gravel transport for recent flows of about this discharge. Bedload was not sampled in the modeled channel reaches, but fresh bars composed of 1–5 cm gravel are common. Andrews (1984) indicated that bedload transport is often initiated in channels at discharges near the bankfull discharge, a relatively frequent occurrence. The unregulated bankfull discharge for the Big Lost River near Arco and in the Diversion Dam study reach is difficult to assess from the available data (Fig. 2), but it is certainly much less than 50 m³/s, the smallest

A

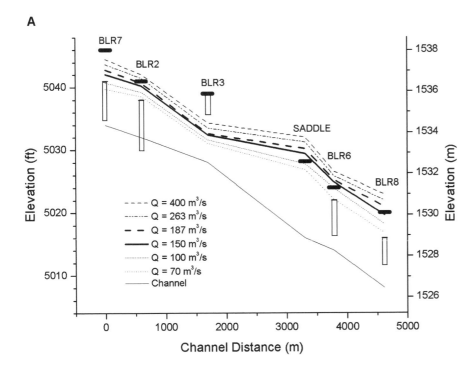

B

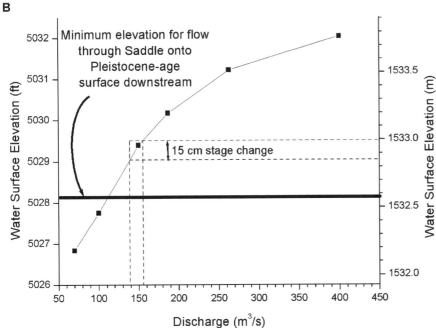

Figure 9. A: Summary profile of water-surface elevations at major study sites in Diversion Dam study reach. Continuous lines connect water surface elevations at each study site for simulated discharges. Vertical boxes show elevation extent of soil and stratigraphic descriptions at each site; thick horizontal bars show the projected elevation of the adjacent Pleistocene surface at each site. B: Simulated water-surface elevation versus discharge for Saddle area. Black squares connected by lines correspond to discharges shown on summary profile (A). Variations in model parameters that produce to 15 cm of change in computed water surface elevation correspond to discharge variations of <10 m³/s for discharges near 150 m³/s. Note that flow through the Saddle can begin at discharge as low as 110 m³/s.

discharge simulated by Ostenaa et al. (1999). For comparison, for discharges of 150 m³/s and larger, the calculated power in the Saddle and just downstream on the Pleistocene surface adjacent to the Big Lost River is 50 to >100 W/m² (Fig. 9B). Surficial deposits on the Pleistocene surfaces downstream of the Saddle consist mostly of sand, which would be subject to

erosion and transport at lower power than required to initiate mobility of gravel. Alternatively, shear-stress calculations based on the average reach slope (Paola and Mohrig, 1996) indicate potential mobility of 2 mm particles for flow depths of 5–18 cm. Local slopes on the downstream side of the Saddle and at locations where flow returns to the main channel are much steeper, resulting in more favorable conditions for erosion of surficial deposits in these areas in the event of such flows.

Hydraulic modeling uncertainty and sensitivity tests. Sensitivity tests were conducted by making changes in bottom friction, horizontal friction, and several model parameters such as grid spacing, time-step size, and implicit-weighting factor. The baseline for the tests is a simulated flow in the Diversion Dam reach with a discharge of 100 m³/s. This baseline run used these parameters: bottom friction coefficient $n = 0.038$, horizontal viscosity coefficient $A = 0.05$ m²/s², grid-spacing $\Delta x = 1$ m, time-step size $\Delta t = 3$ s, and implicit-weighting factor $\theta = 0.6$. The sensitivity tests with model parameters show small variations in water-surface elevation near the Saddle, indicating that the model has adequately converged for this range of values.

Refining the computational cell size from 2 m to 1 m resulted in a change of a few centimeters in water-surface elevation. This indicates that the model has fully resolved the input topography and any further improvements must come from maps that are more accurate. The time-step size was changed over the range of 1 s to 5 s. As the time step becomes larger, the semi-Lagrangian approximation becomes more accurate but the propagation of gravity waves becomes less accurate. The reverse is true as the time-step size decreases. Experience and theoretical considerations show that for these flow velocities and grid size, a time step of ~3 s should be optimal. Only a small change in water-surface elevation was noted for the range of time-step size examined here. The implicit-weighting factor defines where the numerical calculations are centered with respect to a time step. For transient calculations, large values ($\theta = 1.0$) tend to damp the flood wave, whereas centered values ($\theta = 0.5$) are the most accurate. The calculations presented here were run until steady state, so transient damping is not an issue.

Sensitivity tests with horizontal viscosity showed relatively little effect on water levels, although it should affect the velocity profiles across the flow. For this level of horizontal refinement viscosity is isotropic, so that the horizontal viscosity should roughly scale as u^*h, where $u^* = (\tau/\rho)^{1/2}$ is friction velocity, τ is bottom friction, ρ is water density, and h is water depth. For the flows considered here, this results in a horizontal viscosity of ~0.05 m²/s². Viscosity values from 0 to 0.25 m²/s² resulted in small changes in water levels near the Saddle. This was not pursued further because there are no velocity data available and the water levels are insensitive to this parameter.

The initial value for Manning's n, 0.038, was based on a typical value for similar rivers during large floods (e.g., Barnes, 1967; Costa, 1987). Our intention was to perform sensitivity tests about this value to determine if the results are sensitive to this parameter over a range of 0.030–0.060. They were not. The sensitivity tests with bottom friction coefficient led to two sets of results, one for the free-flowing channels where water-surface elevation depends on the value for bottom friction, and one for the backwater at the flow constrictions or control points where the dynamics are controlled by a balance between ac-

celeration and water-surface slope. At free-flowing sites upstream of the backwatered area at the Saddle (Figs. 4 and 9A), the water level increased over this range of n, as would be expected when bottom friction is important in the force balance. The other sites in the reach were much less sensitive to changes in Manning's n because they are located in places where advective effects rather than friction effects are important in the force balance. Near the Saddle, the water-surface elevation varied by ~10 cm over the range of Manning's n. Water levels at the Saddle are determined by the downstream constriction where the flow must accelerate from the upstream pond to the flow in the downstream channel. The overbank flow in the pond is insensitive to bottom friction coefficient because of the low velocity and low head gradient. For other sites, the overbank part of the flow is small relative to the total flow and results for stage are insensitive to choice of Manning's n. At site BLR7, the water-surface elevation is strongly affected by advective effects as the flow is accelerated around the bend. This site is not very sensitive to the value for n and varied by 7 cm over the range chosen. Overall, flow through the Diversion Dam reach is complex, with a central flooded area caused by a significant bedrock constriction near the near the Saddle (Figs. 4, 6, and 9A). Advective effects are the primary control on depth in this backwater.

The principal sources of uncertainty associated with the discharge estimates include flow model uncertainty and uncertainty in the characterization of the channel geometry and topography. The overall uncertainty in channel geometry is minimized because the channel is mostly confined by basalt bedrock and Pleistocene alluvium. Thus, for the periods of interest these elements fix the channel geometry at the critical hydraulic control locations. Characterization of the topography with a 0.6 m contour interval map provides elevation resolution to about one-half the contour interval, or ~0.3 m for the location of sites on the base maps and depiction of surface topography. Potential errors in channel capacity for grids derived from these maps are limited by the point-to-point relative accuracy of the maps and the size of the grids. The use of 1 m grid spacing in the hydraulic modeling samples each potential cross section with multiple cells; thus, uncertainties in the characterization of critical channel cross sectional reaches are considered to be unimportant.

Paleohydrologic summary for the Big Lost River—Diversion Dam study reach

The geologic observations, geochronology, and hydraulic modeling results can be combined into statements of time and discharge regarding paleofloods and paleohydrologic bounds. For this study, the available information defines the age and discharge for one paleoflood and two paleohydrologic bounds. Each component of the paleohydrologic information has its own independent characterization of uncertainty, which is used in the statistical analyses that follow.

Geomorphic and stratigraphic data define the age and paleostage for a large flood on the Big Lost River ~400 yr ago. Flow simulations suggest that the deposits and surfaces of this age have some shallow inundation at discharges of 70 m³/s, but are inundated by 25–90 cm at a discharge of 100 m³/s. The ~400 yr old deposits on the terraces are ~30–80 cm thick, which could indicate even larger maximum flow depths at these sites. Thus, a discharge of 100 m³/s and an uncertainty range of 90–110 m³/s are considered conservative representations of this paleoflood. Flow simulations indicate that discharges of ~110 m³/s reach stages where flow through the Saddle onto Pleistocene deposits might be initiated (Fig. 9B).

A paleohydrologic bound is also defined by the Holocene surfaces, ~2 m above the low–flow channel of the Big Lost River; these surfaces are capped by deposits of the ~400 yr old paleoflood. No recognizably younger fluvial deposits overlie the paleoflood deposits and the surface morphology of the ~2 m surface is distinct from channeled lower surfaces, on which there is floated historical debris, such as milled timber. The largest historical flood in the Diversion Dam reach is not known with certainty, but gaging information from Arco and other upstream sites suggests that floods in 1921 and 1965 were likely the largest (Table 1). No estimates of the discharge of these floods in the Diversion Dam study reach are available, but estimates for Arco and sites just upstream of 73 and 71 m³/s, respectively, are likely maximum values. Flow simulations in the reach for a discharge of 70 m³/s appear to inundate areas that are above the general level of the historical debris. This discharge also results in shallow inundation of some sites capped by the ~400 yr old paleoflood deposits. Discharges in the range of 90–110 m³/s result in substantial inundation of these same sites, as dicussed previously, and this range is used for the paleohydrologic bound discharge. The uncertainty range for the 2σ calibrated ages of the radiocarbon ages from the deposits of the paleoflood are typically <200 yr (Table 2). In addition, the youngest calibrated age from the underlying deposits is ~800 yr (Table 2 and Fig. 5). Thus, a range of 300–500 yr is the minimum amount of time before the present in which there was no flood similar in size to the paleoflood ~400 yr ago.

A second paleohydrologic bound is defined by the Pleistocene alluvial surfaces adjacent to the Big Lost River. Simulations for a discharge of 150 m³/s produce flow across large areas of the extensive Pleistocene surfaces flanking the Diversion Dam study reach, resulting in high power estimates at locations where flow is accelerated in areas of locally higher slope such as downstream of the Saddle (e.g., Fig. 8A). These results indicate that extensive modification of the Pleistocene surface would be likely if such a flow had happened in the past 10 k.y. The flow simulations indicate that the minimum discharge associated with flow through the Saddle is ~110 m³/s (Fig. 9B). For the statistical analyses, the uncertainty for the discharge associated with the Pleistocene paleohydrologic bound is assigned a range of 130–175 m³/s. This range represents ~30%

of the preferred discharge estimate of 150 m³/s. Simulations for 187 m³/s, as well as for larger discharges, indicate much greater inundation and flow areas, and higher power values, not only at the Saddle, but throughout the study reach (Ostenaa et al., 1999).

Floods that exceed the level of the Pleistocene surfaces, outside the channel of the Big Lost River, would have to flow across Pleistocene alluvial deposits and surfaces that range in age from older than 12 ka to as old as 95 ka (Ostenaa et al., 1999). Deposition of these deposits is generally associated with glacial periods of the Pleistocene (e.g., Pierce and Scott, 1982), a time when climate was substantially different than the present. For use in the flood-frequency analysis the time period for the Pleistocene paleohydrologic bound is limited to the Holocene or the past 10 k.y. The time for this bound is assigned a range of 9–12 k.y. to reflect uncertainty in the timing of the climate transition from the glacial to nonglacial period.

Peak-discharge-frequency analyses of the Big Lost River using paleoflood information

Methods and framework. Probabilistic estimates of flood-frequency are required for engineering risk assessments of critical structures. Frances et al. (1994) showed that paleohydrologic information can provide valuable flood-frequency bounds, particularly for small annual probabilities (the reciprocal of return period), which are of greatest concern for critical structures. Stedinger and Cohn (1986) and Stedinger et al. (1988) provided likelihood functions that combine annual-peak-discharge, historical, and paleohydrologic-bound data. They used numerical optimization techniques to estimate maximum likelihood, or best-fitting, flood-frequency models using parametric frequency functions. Observational measurement errors can seriously degrade the performance of maximum-likelihood estimation approaches (Kuczera, 1992). Standard deviations of annual-peak-discharge estimates increase significantly above a threshold discharge (Potter and Walker, 1981, 1985). Peak-discharge estimates from hydraulic models have 2σ uncertainties that are ~15%–20% for the larger historic floods (Cook, 1987) and range to 25% for paleostage measurements (O'Connor and Webb, 1988). Maximum-likelihood flood-frequency estimators that account for correlated measurement errors perform better than maximum-likelihood estimators that ignore rating errors (Kuczera, 1996). Thus, it is vital to rigorously incorporate discharge measurement uncertainties to develop realistic estimates of flood-frequency uncertainties using annual-peak-discharge, historical, and paleohydrologic-bound data.

A number of moment-based approaches have commonly been used to estimate flood frequency. The weighted-moments technique presented in Bulletin 17B (Interagency Advisory Committee on Water Data, 1982) is often used with annual-peak-discharge records, but incorporation of historical and paleohydrologic data is awkward. Alternatively, peak-discharge-frequency parameters can be estimated using sample

L-moments, as discussed in Hosking and Wallis (1997) and
Stedinger et al. (1993). The expected moments algorithm (Cohn
et al., 1997) is a moments-based estimation procedure developed to utilize historical and paleoflood information in a
censored-data framework. This approach explicitly acknowl-
edges the number of known and unknown peak discharges
above and below a threshold, similar to a maximum-likelihood
approach (Stedinger and Cohn, 1986; Jin and Stedinger, 1989).
However, these moment-based approaches do not account for
measurement errors and thus cannot provide probabilistic estimates of peak-discharge frequency.

Here, a Bayesian approach is used to incorporate measurement errors with annual-peak-discharge, historical, and paleohydrologic-bound data to calculate peak-discharge-frequency
probabilities. Common three-parameter frequency functions
(Hosking and Wallis, 1997) such as generalized extreme value
(GEV), generalized logistic (GLO), generalized normal (GNO),
generalized Pareto (GPO), Pearson type III (PE3), and log Pearson type III (LP3) are used. A Bayesian methodology (Tarantola, 1987) and likelihood functions modified from Stedinger
and Cohn (1986) are used to incorporate measurement errors
and parameter uncertainties. To remedy the problem of convergence to nonglobal optima (Kuczera, 1996), simulated annealing (Goffe et al., 1994; Goffe, 1997) and the downhill sim-
plex (Mathews, 1992) methods are used. Parameter and
peak-discharge-frequency likelihoods and probability intervals
are calculated directly by numerical integration to provide peak-
discharge-frequency probabilities suitable for risk assessment.
This approach is feasible using a modern personal computer; a
systematic, adaptive integration of a parameter space of four or
less can be completed without resorting to Monte Carlo methods of integration.

The annual-peak-discharge estimates from the gaging stations near Arco (Table 1; Fig. 2) were combined with the paleohydrologic information for the peak-discharge-frequency
analyses. The preferred characterization of the data includes
observational uncertainties for the annual peak-discharge estimates of ±10% for discharges to 40 m^3/s and ±25% for the
six largest discharges of 40–71 m^3/s. Characterization of uncertainty for the ~400 yr paleoflood and the two paleohydrologic bounds were described previously. Although the annual-
peak-discharge record from the Arco gaging station is highly
regulated, it is the only record close to the INEEL site with a
length >10 yr. Thus, the potential effects of regulation of the
historical peak-discharge estimates on the resultant frequency
estimates are investigated through sensitivity analyses in which
the input data were modified extensively.

Probabilistic   peak-discharge-frequency   estimates.
FLDFRQ3 (O'Connell, 1998), a public-domain peak-
discharge-frequency program incorporating these concepts, was
used to estimate individual maximum likelihood peak-
discharge-frequency models for each of six three-parameter frequency functions (GEV, GLO, GNO, GPO, PE3, and LP3). The
maximum-likelihood peak-discharge-frequency model was

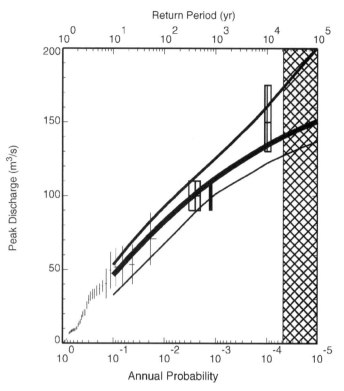

Figure 10. Peak discharge annual probability for Big Lost River at
INEEL based on the preferred characterization of the data. Thickest
curve is median model and the lower and upper solid curves are 2.5%
and 97.5% probability limits, respectively. Shaded region on right depicts annual probability ranges where insufficient observational information dictates caution in extrapolation based on sensitivity testing.
Observed peak discharges are thin crossed lines with the vertical line
showing ±2σ observational uncertainties and horizontal lines showing plotting position uncertainties. Short, thick vertical line shows discharge uncertainty and calculated plotting position for the 400 yr paleoflood. Boxes containing crossed lines are paleohydrologic bounds.
Intersecting lines within boxes depict uncertainty ranges of peak discharge and duration for each bound. Plotting position for paleohydrologic bounds is based on 1/n; where n is age of bound.

produced using the GPO frequency function. Only the LP3 and
PE3 frequency models produced likelihoods >10^{-7} of the GPO
maximum-likelihood peak-discharge-frequency model; the
GEV, GLO, and GNO frequency functions all produce best-
fitting peak-discharge-frequency models that had likelihoods
<10^{-7} of the GPO best-fitting model. Consequently, complete
parameter-grid integrations were performed using FLDFRQ3
with the GPO, LP3, and PE3 three-parameter frequency functions. The three three-parameter integration grids were combined into a single Bayesian analysis. Likelihoods and associated peak discharges from the GPO, LP3, and PE3 integration
grids were combined, sorted, and likelihoods integrated, to obtain probability regions for peak discharges as a function of
annual probability. Combination of multiple frequency functions provides a greater number of degrees of freedom in fitting
the observed data. Because no physical justification exists to

specify any particular frequency function as most representative of the physical system, it is prudent to incorporate the flexibility of several distinct frequency functions into a single unified analysis. 31 939 distinct three-parameter models were used in the Bayesian analysis.

The Bayesian peak-discharge-frequency analyses for the Big Lost River near the INEEL show that paleohydrologic data place strong limits on the range of peak discharges associated with a wide range of annual probabilities (Fig. 10). Median models fit the largest discharges estimated at the Arco gage as well as the longer duration paleoflood and paleohydrologic bounds (Fig. 10).

Sensitivity analysis-evaluation of alternative characterizations of data sets. Sensitivity analyses were conducted in order to quantify the influence of varying the characterizations of the annual peak-discharge and paleohydrologic data on peak-discharge-frequency estimates. Two basic issues are evaluated in the sensitivity tests: (1) effects of regulation, and (2) the addition of floods similar to the 400 yr paleoflood of 100 m^3/s. Diversions and regulation upstream of the INEEL clearly have the potential to reduce the peak discharge of floods at the INEEL. This effect is investigated by replacing the annual peak-discharge record with three scenarios for the number of large flows that might have been possible during the period of regulation. The effects of historical regulation on the peak-discharge record are not precisely known, and thus, no specific peak discharge can be associated with these floods. To explore potential regulation effects, a total of 0, 1, and 18 large peak discharges that exceed 70 m^3/s—a discharge comparable to the largest flood estimated at Arco since 1947, the 1965 peak discharge—are substituted for the annual peak-discharge record. This range for the number of exceedences is based on the number of large discharges estimated at the USGS gage at Howell Ranch in the upper part of the basin and assumes that there is no or little attenuation of flood peak discharge downstream of this gage.

An additional source of uncertainty is the possibility that the deposits interpreted as evidence of a single paleoflood ~400 yr ago are in fact the product of multiple floods of similar size. This scenario is investigated by varying the total number of paleofloods in the record to be 2, 5, and 10. The discharge assigned to these paleofloods was 100 m^3/s.

Full Bayesian peak-discharge-frequency analyses were used to evaluate each characterization of the input data. Combining the regulation and paleoflood scenarios produces nine different alternative descriptions of Big Lost River peak-discharge information. For each data input scenario, likelihoods and peak-discharges estimates from the GPO, LP3, and PE3 frequency functions were combined into a single Bayesian analysis. We considered 469 491 three-parameter models in conducting the 9 Bayesian peak-discharge-frequency analyses.

The results of the sensitivity analyses can be evaluated in terms of the change in the 97.5 percentile probability limits as compared to the probability limits developed from the preferred characterization of the data (Fig. 11). The sensitivity analyses

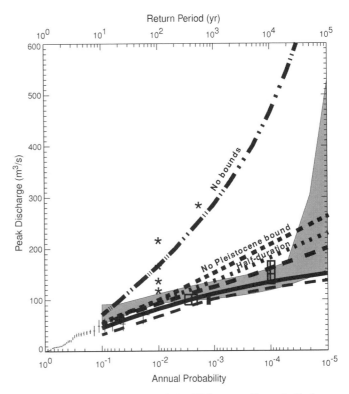

Figure 11. Sensitivity testing of the 97.5 percentile peak-discharge annual probability results for Big Lost River at INEEL. Shaded region shows range of 97.5 percentile peak discharge estimates obtained from 9 sensitivity-testing scenarios in which annual peak discharge data and number of paleofloods were varied. Thick solid curve is median model and lower and upper dashed curves are 2.5% and 97.5% probability limits, respectively, from Figure 10. Preferred data are plotted as in Figure 10, but discharge range is 3x larger. Three labeled, dashed lines are 97.5% probability limits for 3 additional sensitivity tests in which paleohydrologic bounds were varied in duration or deleted. Stars are previous estimates of 100 yr and 500 yr peak discharge from previous studies that did not include paleohydrologic information (from Kjelstrom and Berenbrock, 1996).

show that the 97.5 percentile probability limits are insensitive to variations in the characterization of the input data for annual probabilities in the 10^{-3} to 5×10^{-5} range relative to estimates from the preferred data characterization. In this probability range the paleohydrologic bounds place strict observational limits on peak discharge for return periods less than twice the length of the paleohydrologic record (~20 k.y.). In particular, the observational limit of nonexceedence of the Pleistocene surfaces for the past 10 k.y. is common to all scenarios. Conversely, there are no observational constraints on peak discharge for annual probabilities $<5 \times 10^{-5}$, and the predicted 97.5% percentile peak discharges for annual probabilities of 10^{-5} show a strong sensitivity to the characterization of the input data. Thus, significant cautions must be placed on extrapolations beyond an annual probability of 5×10^{-5}

All annual peak discharges, except the 1965 discharge of 71 m^3/s, were removed from the data sets constructed for the sensitivity analysis. Consequently, the largest dispersion in the sensitivity analyses results for annual probabilities $>10^{-3}$ is for annual probabilities near 10^{-1}, where the 97.5 percentile probability limits from the sensitivity analyses range to 40% larger than the preferred data characterization (Fig. 11). Peak discharges for annual probabilities of 10^{-2} are $<20\%$ larger than the preferred data characterization for the 97.5 percentile estimates, reflecting the constraints provided by the 400 yr, 100 m^3/s paleohydrologic bound. The absence of specific information on paleofloods between ~400 yr and 10 k.y. poses few constraints for estimates of annual probabilities between 10^{-3} and 10^{-4}. This allows many models to estimate 97.5 percentile probability limits that are smaller than peak-discharge estimates from the preferred data characterization in the same annual probability range (Fig. 11).

Sensitivity analysis–record length. Full Bayesian peak-discharge-frequency analyses were also conducted to assess the sensitivity of peak-discharge estimates for particular annual probabilities to changes in effective flood record length. For the purposes of this discussion, let nonstationarity be defined as the tendency of peak-discharge-frequency to vary with time. For example, some periods in the Holocene may have a greater number of floods than other periods (e.g., Knox, 1993; Ely et al., 1993). Although nonstationarity is a difficult issue that cannot be fully addressed in this paper, one simple test of the influence of this type of nonstationarity is to devalue the time extent of paleohydrologic-bound information by some factor and calculate peak-discharge changes for annual probabilities of interest. This approach might be considered analogous to saying that only during a portion of the Holocene has the frequency of extreme floods been comparable to the present. For this test, the durations of the two paleohydrologic bounds were halved. Thus, the duration of the 400 yr paleohydrologic bound was reduced to 200 yr, and the 10 k.y. paleohydrologic bound was cut to 5 k.y. Two additional sensitivity tests focused on assessing the statistical value of acquiring long-duration paleohydrologic information. In the first case, the longest duration paleohydrologic bound was deleted from the preferred data set, and in the second case, all paleohydrologic bounds were removed, leaving only the annual peak-discharge record.

When the duration of the paleohydrologic bounds are cut in half, the 97.5% probability limit increased by ~15% for an annual probability of 10^{-5} (Fig. 11), compared to the results from the preferred characterization of the data. For annual probabilities $>5 \times 10^{-5}$, much smaller changes were observed. Deleting the longest-duration paleohydrologic bound, the 10 k.y. Pleistocene bound, caused peak-discharge uncertainties to increase with the 97.5 percentile probability limits, increasing ~25% for annual probabilities between 10^{-3} to 10^{-4} compared to the preferred data characterization. In addition, the results of the previous section suggest that strong statistical confidence is confined to return periods of about twice the length of record. Thus, if only the 400 yr paleohydrologic bound was available, statistical confidence for annual probabilities $<10^{-3}$ would be

substantially degraded. For the case in which the statistical analyses use only the annual peak-discharge record available from Arco since 1947, peak-discharge uncertainties increase substantially. The estimated 97.5 percentile peak discharges increase by about a factor of 2 for an annual probability of 10^{-2}, consistent with the results of previous studies based only on the historic gaging information (Kjelstrom and Berenbrock, 1996). For this latter data set, the statistical basis for estimating peak discharges with annual probabilities $<10^{-2}$ is weak because there are no observational constraints for these small annual probabilities.

DISCUSSION AND CONCLUSIONS

Geologic data along the Big Lost River on the INEEL site provide key information for evaluating flood hazard. There is evidence of a paleoflood ~400 yr ago with a peak discharge of ~100 m^3/s. In addition, geomorphic surfaces along the Big Lost River define two paleohydrologic bounds. For the past ~400 yr, Holocene terraces capped by the deposits of the ~400 yr old paleoflood define a paleohydrologic bound with a discharge of ~100 m^3/s. Over longer times, the Pleistocene alluvial surfaces adjacent to the Big Lost River limit the paleostage of even larger floods. These Pleistocene surfaces define a paleohydrologic bound for the past 10 k.y. with a discharge of ~150 m^3/s.

The peak discharge of the ~400 yr old paleoflood, ~100 m^3/s, is ~40% larger than the largest peak discharge since at least 1947 near Arco, the nearest site with annual peak-discharge records upstream of the INEEL site. In the study reach on the INEEL site, evidence of this paleoflood is preserved as a geomorphic surface ~2 m above the low-flow channel of the Big Lost River and as a deposit underlying this surface in which a weak calcic soil is developed. Radiocarbon ages from this deposit have calibrated ages of ~400 ± 100 yr ago, a range that likely represents the minimum age for this deposit and the calcic soil. Calibrated ages from deposits with better developed soils that underlie the paleoflood deposit are no younger than ~800 yr. The paleoflood discharge of ~100 m^3/s appears to be the largest flood in the Diversion Dam study reach since at least ~400 ± 100 yr ago.

The Big Lost River channel, the deposits of the ~400 yr paleoflood, and other Holocene fluvial deposits are inset within a broad surface of late Pleistocene alluvium that in many areas is only ~3–4 m above the low-flow channel of the Big Lost River. This late Pleistocene alluvium buries the irregular topography on Pleistocene basalts of several ages, which locally create hydraulic constrictions to flow in the Big Lost River. This geologic setting provides a stable channel configuration for assessing Big Lost River peak discharge during the Holocene due to the low incision rate, and a channel confined by Pleistocene alluvium and basalt. The basalt constrictions provide hydraulic controls on flow that have been stable throughout the Holocene. At one of the constrictions, the Saddle, hydraulic ponding would result in flow spilling over a low ridge onto the Pleistocene alluvial surface for a discharge possibly as low as 110 m³/s. At a discharge of 150 m³/s, calculated power values on the Pleistocene surface are comparable to calculated power in flowing reaches of the main channel for flows that result in bedload transport. Under these conditions, it is highly likely that significant geomorphic modification of the loess-covered Pleistocene surface would result. However, geomorphic evidence at the Saddle indicates that the morphology of the Pleistocene surface is no different near the Saddle than at sites well removed from the Big Lost River, where such flows clearly have not occurred.

Concerns have been raised about the applicability of paleohydrologic information to flood hazard issues in light of past and future climate change (e.g., National Research Council, 1999). This is because a key assumption underlying all statistical hazard models is that of stationarity, the tendency of peak-discharge frequency to vary with time. For flood hazard decisions related to critical facilities, the key issue is the selection of data or records that provide robust predictive value for extreme floods. Proxy climate records for the past 10 k.y. demonstrate that strong decadal- to century-scale climate fluctuations are persistent (e.g., National Research Council, 1995). Likewise, the magnitude of climate fluctuations during the Holocene interpreted from these proxy records is small in comparison to the scale of pre-Holocene fluctuations (National Research Council, 1995). These data indicate the need for extreme caution in using predictions of extreme floods based on short records alone, because such short records may only record a small portion of a cycle of fluctuations in flood frequency. In contrast, long paleohydrologic records contain the runoff response to tens to hundreds of climate fluctuations and provide a direct record of the extreme floods that are of interest for hazard applications.

The paleohydrologic data from the Big Lost River provide a new perspective for use in flood hazard assessment of critical facilities at the INEEL, where estimates of peak discharge are needed for a wide range of probabilities, from $>10^{-2}$ to 10^{-5}. Because the annual peak-discharge records of the Big Lost River upstream of the INEEL are strongly affected by regulation and diversions, knowledge of the largest peak discharge in the past 400 yr provides a strong limitation for estimating flood hazards for annual probabilities at the higher end of this range. Previous studies recognized that natural infiltration and channel losses are significant issues for flow in the Big Lost River (e.g., Stearns et al., 1938; Kjelstrom and Berenbrock, 1996), but in the absence of prediversion and regulation annual peak-discharge data, it is difficult to define the relative importance of natural versus anthropogenic effects. With paleohydrologic observations spanning hundreds of years, the relative importance of these effects on estimates of floods with annual probabilities of 10^{-2} to 10^{-3} can be better assessed. The geologic limits on peak discharge from the Diversion Dam reach document the long-term persistence of large natural channel losses, even though six peak-discharge estimates in the gaging record

from the upper drainage basin (Fig. 2) exceed the preferred discharge estimate, 100 m³/s, of the ~400 yr ago paleoflood and paleohydrologic bound.

The preferred characterization of the input data produces nearly maximum estimates of peak discharges for annual probabilities in the 2×10^{-3} to 5×10^{-5} range compared to most scenarios considered in the sensitivity analyses (Fig. 11). Because the preferred characterization of the input data are most consistent with available observations of flood behavior on the Big Lost River near the INEEL, it is reasonable to use the probabilistic peak-discharge-frequency estimates based on the preferred characterization of the input data (Fig. 10) for applied hazard assessment. However, the sensitivity analyses demonstrate that little statistical confidence is present for discharge-frequency estimates beyond an annual probability of 5×10^{-5}. Estimates beyond this range are not constrained by the paleohydrologic data and this indicates the need for alternative approaches if discharge-frequency estimates for smaller annual probabilities are required. For more frequent probabilities, the upper limit of the 97.5 percentile from the sensitivity analysis in the 10^{-1} to 2×10^{-3} annual probability range in Figure 11 reflects multiple conservative assumptions, including 18 peak discharges exceeding 70 m³/s and 10 paleofloods exceeding 100 m³/s within the past 400 yr. These sensitivity analyses show that even when many large floods are artificially inserted into the statistical analyses, the constraints posed by the long-duration paleohydrologic information are essentially preserved. Compared to analyses in which paleohydrologic information is not included, the addition of paleohydrologic data substantially reduces the range of possible peak discharges associated with small annual probabilities. Analyses that include paleohydrologic information can rely on interpolation rather than extrapolation over a much greater range of flood discharge and annual probability. For critical facilities such as those at the INEEL that require discharge-frequency estimates over a wide range of annual probabilities, reduced uncertainty in these estimates is of great value in applied hazard assessment.

ACKNOWLEDGMENTS

The paleoflood study of the Big Lost River was completed as part of an interagency agreement between the Bureau of Reclamation and the Department of Energy, Idaho Falls, Idaho, to provide data for flood hazard assessment at the Idaho National Engineering and Environmental Laboratory (INEEL). V.R. Baker, University of Arizona, and U. Lall, Utah State University, provided many helpful comments on our studies for that assessment. The comments, cooperation and assistance of many U.S. Department of Energy and Lockheed Martin Idaho Technology Company (LMTCO) staff at the INEEL were also indispensable in the completion of that study. In particular, we thank R.P. Smith for sharing his knowledge of the INEEL site and previous studies and for assistance throughout the course of the study. D.R. Levish and R.E. Klinger of the Bureau of

Reclamation assisted with field studies and many aspects of the INEEL study. D.T. Simpson and T.E. Kolbe of URS Greiner Woodward-Clyde described soils at several sites. K. Puseman of PaleoResearch, Inc., processed soil samples for radiocarbon dating and provided macrofloral identifications of potentially datable material. E. Evanoff of the University of Colorado identified shell samples. Radiocarbon analyses were conducted by BETA Analytic. Soil carbonate and particle-size analyses were completed by the Soil, Water and Plant Testing Laboratory at Colorado State University. Review comments on drafts of this manuscript by J.E. Klawon, P.K. Link, R.D. Jarrett, and J.E. O'Connor were greatly appreciated and substantially improved the manuscript.

REFERENCES CITED

Andrews, E.D., 1984, Bed-material entrainment and hydraulic geometry of gravel-bed rivers in Colorado: Geological Society of America Bulletin, v. 95, p. 371–378.

Baker, V.R., 1987, Paleoflood hydrology and extraordinary flood events: Journal of Hydrology, v. 96, p. 79–99.

Baker, V.R., 1989, Magnitude and frequency of palaeofloods, in Beven, K., and Carling, P., eds., Floods: Hydrological, sedimentological, and geomorphological implications: New York, John Wiley and Sons, Limited, p. 171–183.

Baker, V.R., and Costa, J.E., 1987, Flood power, in Mayer, L., and Nash, D., eds., Catastrophic flooding: Boston, Massachusetts, Allen and Unwin, p. 1–21.

Baker, V.R., Kochel, R.C., and Patton, P.C., editors., 1988, Flood geomorphology: New York, John Wiley and Sons, 503 p.

Barnes, H.H., 1967, Roughness characteristics of natural channels: U.S. Geological Survey Water-Supply Paper 1849, 213 p.

Birkeland, P.W., 1999, Soils and geomorphology: New York, Oxford University Press, 430 p.

Casulli, V., 1990, Semi-implicit finite difference methods for the two-dimensional shallow water equations: Journal of Computational Physics, v. 86, p. 56–74.

Casulli, V., and Cattani, E., 1994, Stability, accuracy, and efficiency of a semi-implicit method for three-dimensional shallow water flow: Computers Mathematics Applications, v. 27, p. 99–112.

Casulli, V., and Walters, R.A., 2000, An unstructured grid, three-dimensional model based on the shallow water equations: International Journal for Numerical Methods in Fluids, v. 32, p. 331–348.

Cheng, R.T., Casulli, V., and Gartner, J.W., 1993, Tidal, residual, intertidal mudflat (TRIM) model and its applications to San Francisco Bay, California: Estuarine, Coastal and Shelf Science, v. 36, p. 235–280.

Cohn, T.A., Lane, W.L., and Baier, W.G., 1997, An algorithm for computing moments-based flood quantile estimates when historical information is available: Water Resources Research, v. 33, p. 2089–2096.

Cook, J.L., 1987, Quantifying peak discharges for historical floods: Journal of Hydrology, v. 96, p. 29–40.

Costa, J.E., 1978, Holocene stratigraphy in flood frequency analysis: Water Resources Research, v. 14, p. 626–632.

Costa, J.E., 1987, A comparison of the largest rainfall-runoff floods in the United States with those of the People's Republic of China and the world: Journal of Hydrology, v. 96, p. 101–115.

Ely, L., Enzel, Y., Baker, V.R., and Cayan, D.R., 1993, A 5000-year record of extreme floods and climate change in the southwestern United States: Science, v. 262, p. 410–412.

Federal Emergency Management Agency (FEMA), 1993, Flood insurance study guidelines and specifications for study contractors, FEMA 37: Washington, D.C., U.S. Government Printing Office, 61 p., 7 appendices.

Forman, S.L., Smith, R.P., Hackett, W.R., Tullis, J.A., and McDaniel, P.A., 1993, Timing of late Quaternary glaciations in the western United States based on the age of loess on the eastern Snake River Plain, Idaho: Quaternary Research, v. 40, p. 30–37.

Frances, F., Salas, J.D., and Boes, D.C., 1994, Flood frequency analysis with systematic and historical or paleoflood data based on the two-parameter general extreme values models: Water Resources Research, v. 30, p. 1653–1664.

Goffe, W.L., 1997, SIMANN: A global optimization algorithm using simulated annealing: Studies in Nonlinear Dynamics and Econometrics, v. 1, n. 3, p. 169–176.

Goffe, W., Ferrier, G.D., and Rogers, J., 1994, Global optimization of statistical functions with simulated annealing: Journal of Econometrics, v. 60, p. 65–99.

Hosking, J.R.M., and Wallis, J.R., 1997, Regional frequency analysis: An approach based on L-moments: Cambridge, Cambridge University Press, 224 p.

Interagency Advisory Committee on Water Data (IACWD), 1982, Guidelines for determining flood flow frequency, Bulletin #17B: Reston, Virginia, U.S. Department of the Interior, Geological Survey, Office of Water Data Coordination, 28 p., 14 appendices.

Jarrett, R.D., 1991, Paleohydrology and its value in analyzing floods and droughts, in Paulson, R.W., Chase, E.B., Roberts, R.S., and Moody, D.W., compilers, U.S. Geological Survey Water-Supply Paper 2375, p. 105–116.

Jin, M., and Stedinger, J.R., 1989, Flood-frequency analysis with regional and historical information: Water Resources Research, v. 25, p. 925–936.

Kjelstrom, L.C., 1991, Idaho, floods and droughts, in Paulson, R.W., Chase, E.B., Roberts, R.S., and Moody, D.W., eds., National Water Summary 1988–89: Hydrologic events and floods and droughts: U.S. Geological Survey Water-Supply Paper 2375, p. 255–262.

Kjelstrom, L.C., and Berenbock, C., 1996, Estimated 100-year peak flows and flow volumes in the Big Lost River and Birch Creek at the Idaho National Engineering and Environmental Laboratory, Idaho: U.S. Geological Survey Water-Resources Investigations Report 96-4163, 23 p.

Knox, J.C., 1993, Large increases in flood magnitude in response to modest changes in climate: Nature, v. 361, p. 430–432.

Kuczera, G., 1992, Uncorrelated measurement error in flood frequency inference: Water Resources Research, v. 28, p. 2119–2127.

Kuczera, G., 1996, Correlated rating curve error in flood frequency inference: Water Resources Research, v. 32, p. 183–188.

Kuntz, M.A., Skipp, B., Lanphere, M.A., Scott, W.E., Pierce, K.L., Dalrymple, G.B., Champion, D.E., Embree, G.F., Page, W.R., Morgan, L.A., Smith, R.P., Hackett, W.R., and Rodgers, D.W., 1994, Geologic map of the Idaho National Engineering Laboratory and adjoining area, eastern Idaho: U.S. Geological Survey Miscellaneous Investigation Map I-2330, scale 1:100 000.

Lang, G., Bergemann, F., Boehlich, M.J., Rudolph, E., Ruland, P., Seiss, G., and Winkel, N., 1998, HN-Verfahren TRIM-2D: Hamburg, Budesanstalt für Wasserbau, Validierungsdokument, Ver. 2.0.

Levish, D.R., Ostenaa, D.A., and O'Connell, D.R.H., 1994, A non-inundation approach to paleoflood hydrology for the event-based assessment of extreme flood hazards, in 1994 Annual Conference Proceedings, Association of State Dam Safety Officials: Lexington, Kentucky, p. 69–82.

Levish, D.R., Ostenaa, D.A., and O'Connell, D.R.H., 1996, Paleohydrologic bounds and the frequency of extreme floods, in Gruntfest, E., ed., Twenty years later what we have learned since the Big Thompson flood: Boulder, Colorado, University of Colorado, Natural Hazards Research and Applications Information Center, Special Publication No. 33, p. 171–182.

Levish, D.R., Ostenaa, D.A., and O'Connell, D.R.H., 1997, Paleoflood hydrology and dam safety: Waterpower '97, in Mahoney, D.J., ed., Proceedings of the International Conference on Hydropower: New York, American Society of Civil Engineers, p. 2205–2214.

McFadden, L.D., McDonald, E.V., Wells, S.G., Anderson, K., Quade, J., and Forman, S.L., 1998, The vesicular layer and carbonate collars of desert soils and pavements: Formation, age and relation to climate change: Geomorphology, v. 24, p. 101–145.

Mathews, J.H., 1992, Numerical methods for mathematics, science, and engineering (second edition): Englewood Cliffs, Prentice-Hall, Incorporated, 646 p.

Miller, A.J., and Cluer, B.L., 1998, Modeling considerations for simulation of flow in bedrock channels, *in* Tinkler, K.J., and Wohl, E.E., eds., Rivers over rock: Fluvial processes in bedrock channels: Washington, D.C., American Geophysical Union, Geophysical Monograph 107, p. 61–104.

National Research Council (NRC), 1995, Natural climate variability on decade-to-century time scales: Washington, D.C., National Academy Press, Climate Research Committee, 630 p.

National Research Council (NRC), 1999, Improving American river flood frequency analyses: Washington, D.C., National Academy Press, Committee on American River Flood Frequencies, 120 p.

O'Connell, D.R.H., 1998, FLDFRQ3: Three-parameter maximum likelihood flood-frequency estimation with optional probability regions using parameter grid integration: User's Guide (Release 1.0), ftp:// ftp.seismo. usbr.gov/pub/outgoing/geomagic/scr/fldfrq3/, June 2000.

O'Connell, D.R.H., Levish, D.R., and Ostenaa, D.A., 1996, Bayesian flood frequency analysis with paleohydrologic bounds for late Holocene paleofloods, Santa Ynez River, California, *in* Gruntfest, E., ed., Twenty years later what we have learned since the Big Thompson flood: Boulder, Colorado, University of Colorado, Natural Hazards Research and Applications Information Center, Special Publication No. 33, p. 183–196.

O'Connell, D.R.H., Levish, D.R., and Ostenaa, D.A., 1998, Risk-based hydrology: Bayesian flood-frequency analyses using paleoflood information and data uncertainties, *in* Proceedings of the First Federal Interagency Hydrologic Modeling Conference: Las Vegas, Nevada, v. 1, p. 4.101–4.108.

O'Connor, J.E., and Webb, R.H., 1988, Hydraulic modeling for paleoflood analysis, *in* Baker, V.R., Kochel, R.C., and Patton, P.C., eds., Flood geomorphology: New York, John Wiley and Sons, p. 393–402.

Ostenaa, D.A., and Levish, D.R., 1996, Event-based assessment of extreme flood hazards for dam safety *in* Proceedings, Association of State Dam Safety Officials Western Regional Conference: Lexington, Kentucky, Association of State Dam Safety Officials, p. 41–54.

Ostenaa, D.A., and Levish, D.R., 1997, Reconnaissance paleoflood study for Ochoco Dam, Crooked River Project, Oregon: Denver, Colorado, Bureau of Reclamation, Seismotectonic Report 96-2, 20 p., 1 folded plates, 2 appendices.

Ostenaa, D.A., Levish, D.R., and O'Connell, D.R.H., 1996, Paleoflood study for Bradbury Dam, Cachuma Project, California: Denver, Colorado, Bureau of Reclamation, Seismotectonic Report 96-3, 86 p., 1 folded plate, 4 appendices.

Ostenaa, D.A., Levish, D.R., O'Connell, D.R.H., and Cohen E.A., 1997, Paleoflood study for Causey and Pineview Dams, Weber Basin and Ogden River Projects, Utah: Denver, Colorado, Bureau of Reclamation, Seismotectonic Report 96-6, 69 p., 3 appendices.

Ostenaa, D.A., Levish, D.R., Klinger, R.E., and O'Connell, D.R.H., 1999, Phase 2 Paleohydrologic and geomorphic studies for the assessment of flood risk for the Idaho National Engineering and Environmental Laboratory, Idaho: Denver, Colorado, Bureau of Reclamation, Geophysics, Paleohydrology and Seismotectonics Group, Report 99-7, 112 p., 1 folded plates, 4 appendices.

Paola, C., and Mohrig, D., 1996, Palaeohydraulics revisited: Palaeoslope estimation in coarse-grained braided rivers: Basin Research, v. 8, p. 243–254.

Parker, G., 1978, Self-formed straight rivers with equilibrium banks and mobile bed. 2. The gravel river: Journal of Fluid Mechanics, v. 89, part 1, p. 127–146.

Patton, P.C., 1987, Measuring rivers of the past: A history of fluvial paleohydrology, *in* Landa, E.R., and Ince, S., eds., The history of hydrology: History of geophysics (volume 3): Washington, D.C., American Geophysical Union, p. 55–67.

Patton, P.C., Baker, V.R., and Kochel, R.C., 1979, Slack water deposits: A geomorphic technique for the interpretation of fluvial paleohydrology, *in* Rhodes, D.D., and Williams, G.P., eds., Adjustments of the fluvial system: Dubuque, Iowa, Kendal-Hunt, p. 225–253.

Pierce, K.L., and Morgan, L.A., 1992, The track of the Yellowstone hot spot: Volcanism, faulting, and uplift, *in* Link, P. K., Kuntz, M.A., and Platt, L.B., eds., Regional geology of eastern Idaho and western Wyoming: Geological Society of America Memoir 179, p. 1–53.

Pierce, K.L., and Scott, W.E., 1982, Pleistocene episodes of alluvial gravel deposition, Southeastern Idaho *in* Bonnichsen, B., and Breckenridge, R.M., eds., Cenozoic geology of Idaho: Idaho Bureau of Mines and Geology Bulletin 26, p. 685–702.

Potter, K.W., and Walker, J.F., 1981, A model of discontinuous measurement error and its effects on the probability distribution of flood discharge measurements: Water Resources Research, v. 17, p. 1505–1509.

Potter, K.W., and Walker, J.F., 1985, An empirical study of flood measurement error: Water Resources Research, v. 21, p. 403–406.

Ramsey, B.C., 1995, Radiocarbon calibration and analysis of stratigraphy: The OxCal Program: Radiocarbon, v. 37, no. 2, p. 425–430.

Rathbun, S.L., 1991, Quaternary channel changes and paleoflooding along the Big Lost River, Idaho National Engineering Laboratory: TriHydro Corporation report prepared for EG&G Idaho, Inc., subcontract no. C90-132903, 33 p.

Rathbun, S.L., 1993, Pleistocene cataclysmic flooding along the Big Lost River, east central Idaho: Geomorphology, v. 8., p. 305–319.

Scott, W.E., 1982, Surficial geologic map of the eastern Snake River Plain and adjacent areas, 111° to 115° W, Idaho and Wyoming: U.S. Geological Survey Miscellaneous Investigations Series Map I-1372, scale 1:250 000, 2 sheets.

Simpson, D.T., Kolbe, T.E., Ostenaa, D.A., Levish, D.R., and Klinger, R.E., 1999, Lower Big Lost River chronosequence: Implications for glacial outburst flooding: Geological Society of America Abstracts with Programs, v. 31, no. 4, p. A56.

Stearns, H.T., Crandall, L., and Steward, W.G., 1938, Geology and groundwater resources of the Snake River Plain in southeastern Idaho: U.S. Geological Survey Water-Supply Paper 774, 268 p.

Stedinger, J.R., and Cohn, T.A., 1986, Flood frequency analysis with historical and paleoflood information: Water Resources Research, v. 22, p. 785–793.

Stedinger, J.R., Surani, R., and Therival, R., 1988, The MAX users guide: Ithaca, New York, Cornell University, Department of Environmental Engineering (unpublished), 51 p.

Stedinger, J.R., Vogel, R.M., Foufoula-Georgiou, E., 1993, Chapter 18, Frequency analysis of extreme events, *in* Maidment, D.R., ed., Handbook of hydrology: New York, McGraw-Hill Incorporated, p. 18.1–18.66.

Tarantola, A., 1987, Inverse problems theory: Methods for data fitting and model parameter estimation: New York, Elsevier, 613 p.

Tullis, J.A., 1995, Characteristics and origin of earth-mounds on the eastern Snake River Plain, Idaho: Idaho Falls, Idaho, Lockheed Martin Idaho Technologies, Idaho National EngineeringLaboratory, INEL-95/0505, 90 p.

Tullis, J.A., and Koslow, K.N., 1983, Characterization of Big Lost River floods with recurrence intervals greater than 25 years: Idaho Falls, Idaho, EG&G Idaho, Inc., Internal Technical Report No. RE-P8-83-044, 19 p.

U.S. Department of Energy (USDOE), 1996a, Natural phenomena hazards design and evaluation criteria for Department of Energy facilities: Washington, D.C., U.S. Department of Energy, DOE-STD-1020-94, Change notice #1.

U.S. Department of Energy (USDOE), 1996b, Natural phenomena hazards performance categorization guidelines for structures, systems, and components: Washington, D.C., U.S. Department of Energy, DOE-STD-1021-93, Change notice #1.

U.S. Department of Energy (USDOE), 1996c, Natural phenomena hazards characterization criteria: Washington, D.C., U.S. Department of Energy, DOE-STD-1022-94, Change notice #1.

U.S. Department of Energy (USDOE), 1996d, Natural phenomena hazards assessment criteria: Washington, D.C., U.S. Department of Energy, DOE-STD-1023-95, Change notice #1.

Walters, R.A., and Casulli, V., 1998, A robust, finite element model for hydrostatic surface water flows: Communications in Numerical Methods in Engineering, v. 14, p. 931–940.

Walters, R.A., and Denlinger, R.D., 1999, Description of flood simulation models, Appendix C, *in* Ostenaa D.A., Levish, D.R., Klinger, R.E., and O'Connell, D.R.H., Phase 2 paleohydrologic and geomorphic studies for the assessment of flood risk for the Idaho National Engineering and Environmental Laboratory, Idaho: Denver, Colorado, Bureau of Reclamation, Geophysics, Paleohydrology and Seismotectonics Group, Report 99-7, 112 p., 1 folded plate, 4 appendices.

MANUSCRIPT ACCEPTED BY THE SOCIETY NOVEMBER 2, 2000

Geological Society of America
Special Paper 353
2002

Tension cracks, eruptive fissures, dikes, and faults related to late Pleistocene–Holocene basaltic volcanism and implications for the distribution of hydraulic conductivity in the eastern Snake River Plain, Idaho

Mel A. Kuntz*
U.S. Geological Survey, MS 913, Box 25046, Federal Center, Denver, Colorado 80225, USA
Steven R. Anderson
U.S. Geological Survey, Department of Geosciences, Idaho State University, P.O. Box 8072, Pocatello, Idaho 83209, USA
Duane E. Champion
Marvin A. Lanphere
U.S. Geological Survey, MS 910, 345 Middlefield Road, Menlo Park, California 94025, USA
Daniel J. Grunwald
U.S. Geological Survey, MS 913, Box 25046, Federal Center, Denver, Colorado 80225, USA

ABSTRACT

Tension crack–eruptive fissure systems are a key characteristic of most late Pleistocene–Holocene basaltic lava fields in the eastern Snake River Plain, Idaho. Models based on elastic displacements that accompany dike intrusion and the dimensions of tension cracks and eruptive fissures give new perspectives on the size and shapes of dike systems in the eastern Snake River Plain. Elastic-displacement models predict faults related to dike intrusion, but these are absent at the late Pleistocene–Holocene lava fields. Numerous faults in the Box Canyon area of the Arco–Big Southern Butte volcanic rift zone can be misinterpreted as being related to dike-emplacement processes. Our data strongly suggest that these faults are tectonic in origin and related to the Lost River range-front fault.

Data about size and shapes of dike systems, in conjunction with detailed mapping and regional paleomagnetic studies, are used to interpret the style of volcanism in a part of the Idaho National Engineering and Environmental Laboratory and for the entire eastern Snake River Plain. The mapping-paleomagnetic studies suggest that sections of dike systems as long as ~40 km can be active simultaneously or within periods of time as short as a few hundred years.

The characteristics and locations of dikes, eruptive fissure systems, and tension cracks have implications for the movement of groundwater and migration of radioactive and chemical wastes in the Snake River Plain aquifer at the Idaho National Engineering and Environmental Laboratory. Buried zones of northwest-trending dikes, eruptive fissures, and tension cracks, referred to as vent corridors, are perpendicular to the regional direction of groundwater flow and probably control some of the lowest and highest estimates of hydraulic conductivity in the aquifer.

*E-mail: mkuntz@usgs.gov

Kuntz, M.A., Anderson, S.R., Champion, D.E., Lanphere, M.A., and Grunwald, D.J., 2002, Tension cracks, eruptive fissures, dikes, and faults related to late Pleistocene–Holocene basaltic volcanism and implications for the distribution of hydraulic conductivity in the eastern Snake River Plain, Idaho, *in* Link, P.K., and Mink, L.L., eds., Geology, Hydrogeology, and Environmental Remediation: Idaho National Engineering and Environmental Laboratory, Eastern Snake River Plain, Idaho: Boulder, Colorado, Geological Society of America Special Paper 353, p. 111–133.

M.A. Kuntz et al.

INTRODUCTION

Like eruptions in Hawaii and Iceland, a characteristic aspect of the basaltic volcanism on the eastern Snake River Plain is that eruptive fissures, which represent the breaching of the Earth's surface by ascending dikes, dominate its early stages. Other structures associated with dike emplacement and fissure-dominated eruptions in the eastern Snake River Plain are tension cracks (Kuntz et al., 1992; Hackett and Smith, 1996; Smith et al., 1996) and faults (Hackett and Smith, 1992; Smith et al., 1996). Eruptive fissures and dike-induced tension cracks are clearly displayed at several late Pleistocene–Holocene lava flows in the eastern Snake River Plain, but our field investigations show that faults are absent at these localities. We have studied these late Pleistocene–Holocene lava fields in detail because they are remarkably preserved. The eruptive fissures and tension cracks at these lava fields are described in detail because they represent typical examples of the processes and products of basaltic volcanism of the eastern Snake River Plain and because the dimensions of the tension cracks and eruptive fissures can be used to interpret the geometry of dike systems at shallow crustal levels.

Faults are present in the Arco–Big Southern Butte and Spencer–High Point volcanic rift zones (Fig. 1) near the margins of the eastern Snake River Plain. At both localities, the faults are part of collinear extensions of major, range-front faults for basin-and-range mountains at the margins of the eastern Snake River Plain. Elastic-deformation models suggest that faults related to dike emplacement are aligned in the zone of tension cracks, but the absence of faults in the late Pleistocene–Holocene lava fields and the geometric association of faults with extensions of range-front faults suggest that faults at the two localities on the margins of the eastern Snake River Plain are of tectonic rather than volcanic origin. We discuss reasons for the near absence of faults and the dike-emplacement mechanisms that favor formation of tension cracks in place of faults.

Through detailed field mapping, paleomagnetic studies, and petrographic correlations, we have discovered that eruptions in one area of the eastern Snake River Plain are not typical of the style of most eastern Snake River Plain basaltic eruptions. The atypical eruption style is the result of an exceptionally long dike system, which produced several small shield volcanoes along a linear belt ~40 km long.

The style of volcanism and characteristics and locations of tension-crack and eruptive fissure systems also have implications for the movement of groundwater and migration of radioactive and chemical wastes in the Snake River Plain aquifer at the Idaho National Engineering and Environmental Laboratory (INEEL) (Fig. 1). These northwest-trending systems, referred to as vent corridors where they are inferred in the subsurface, are perpendicular to the regional direction of groundwater flow and probably control some of the lowest and highest estimates of hydraulic conductivity in the aquifer. Near-vent volcanic de-

posits, dikes, fissures, and cracks probably provide localized preferential pathways and barriers to groundwater flow.

GENERAL FEATURES OF TENSION CRACKS AND FAULTS ASSOCIATED WITH ERUPTIVE FISSURES

Previous studies

Recognition and physical description of the widespread and prominent tension cracks in the eastern Snake River Plain resulted from early analyses of aerial photographs and satellite imagery. Prinz (1970) first mentioned tension cracks in his discussion of the Idaho rift system, now termed the Great Rift volcanic rift zone (Kuntz et al., 1988, 1992). Prinz used the term Great Rift set for the eruptive fissures, tension cracks, and vents for lava flows of the composite Craters of the Moon lava field (Fig. 1). His term "Open Crack rift set" corresponds to two pairs of tension cracks that extend southeast from the southeast margin of the Craters of the Moon lava field (Figs. 1 and 4). Prinz's "Kings Bowl rift set" refers to an eruptive fissure and a flanking pair of tension cracks that are part of the Kings Bowl lava field, and his "Wapi rift set" refers to the inferred feeder dikes and elongated vent area for the Wapi lava field (Figs. 1 and 2). Greeley et al. (1977) and King (1977) discussed and showed in several oblique aerial photographs the eruptive fissure system and the flanking pair of tension cracks for the Kings Bowl lava field. A pair of tension cracks is exposed beyond the northwest margin of the Hell's Half Acre lava field (Figs. 1 and 3; Kuntz et al., 1992). Tension cracks have not been recognized at the Cerro Grande (Fig. 1) and Shoshone lava fields (Kuntz et al., 1992), but cracks at those localities may have been covered by their own lava flows or formed in rocks beyond the margins of the lava fields where they are not easily preserved or recognized.

Field aspects of tension cracks and faults

Our inspection of tension cracks in the eastern Snake River Plain shows that they exhibit evidence of purely extensional movement. The walls of the cracks have jagged surfaces that consist of blocks of lava that broke along preexisting fractures, mainly columnar joints. Projections of the fracture and/or joint surfaces on one side of the crack match perfectly with indentations on the other side of the cracks. Prinz (1970) described the tension cracks in his rift sets as having widths of 6–8 ft, and he stated that there is no evidence of vertical or rotational movements. We have found that crack widths are typically ~1 m but are as wide as 2 m; these widths reflect true amounts of dilation. Where crack widths are locally wider than 2 m, blocks of lava have fallen into the cracks, but this is not common.

Faults associated with tension cracks and eruptive fissures have not been recognized at the late Pleistocene–Holocene lava fields. At these localities, surfaces of lava on opposite sides of cracks are at the same elevation, except at localities where un-

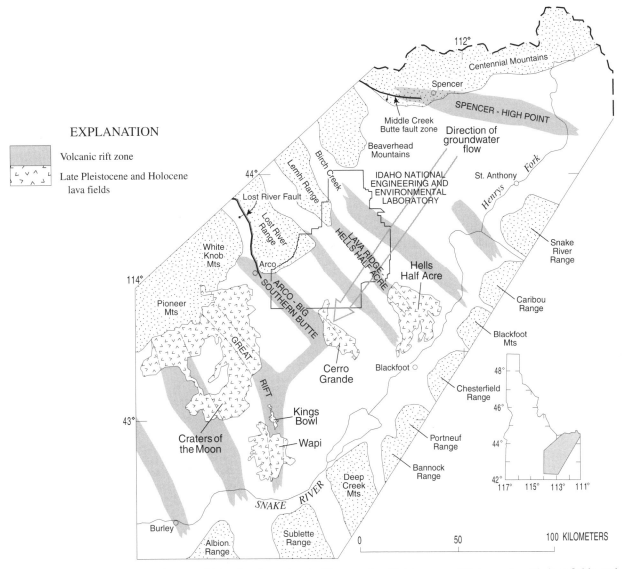

Figure 1. General map of part of eastern Idaho showing volcanic rift zones, late Pleistocene and Holocene basaltic lava fields, and other geologic and geographic features referred to in text. Only volcanic rift zones referred to in text are identified by name in this figure (modified from Kuntz et al., 1992).

supported blocks of lava have slumped into the crack. Faults having offsets of 0.5 m or more would be obvious from analysis of topography on 1:24 000 scale topographic maps having contour intervals of 10 ft, and faults have not been observed on aerial photographs. We suggest that dike-induced faults associated with the late Pleistocene–Holocene lava fields in the eastern Snake River Plain, if present, are extremely rare.

Laboratory methods of analysis of tension cracks and eruptive fissures

Tension cracks and eruptive fissures were identified on 1:30 000 and 1:40 000 scale black and white air photos and mapped on 1:24 000 scale map bases with the aid of a Kern PG-2 stereoplotter. Dimensions of these structures were measured from the 1:24 000 scale maps; those data are summarized in Table 1. The cracks and fissures were digitized and Figures 2, 3, and 4 herein were prepared using illustration software from the digitized data.

PRESERVATION AND RECOGNITION OF TENSION CRACKS

Several factors are key in identification and preservation of tension cracks. (1) The cracks must have formed in basalt flows that are relatively young and covered by little if any loess. (2) The cracks must have formed in basalt flows rather than in

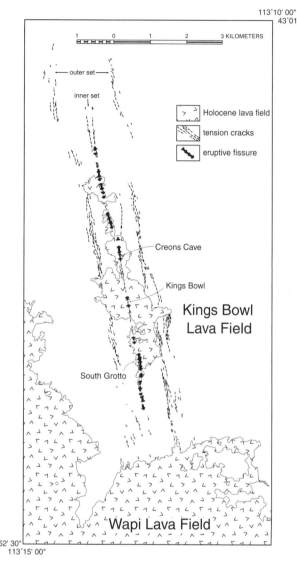

Figure 2. Map of tension cracks and eruptive fissures for Kings Bowl lava field, south-central Idaho. Figure prepared from Covington (1977) and from analysis of 1:40 000 scale black-and-white aerial photographs taken in 1973.

rocks that are friable or easily eroded. (3) At least some parts of the cracks were not resurfaced by younger lava flows.

The Hell's Half Acre volcanic field illustrates the first factor. Cracks northwest of the field (Fig. 3) formed in flows that are covered by 0.5–1 m of loess. Many of the cracks are wholly or partly covered by loess, but their size and extent can still be deciphered on aerial photos by recognizing openings in the cracks where loess has drained into them. Subtle linear belts of darker colors in the loess that overlies the cracks can also be used to recognize the Hell's Half Acre cracks. If the loess were thicker, it is possible that these cracks could not be recognized in the field or on air photos.

The second factor is illustrated by some cracks in the Craters of the Moon lava field. Eruptive fissures, tension cracks, and near-vent cinder and ash deposits are concentrated in a narrow belt in the northern part of the Craters of the Moon lava field. Tension cracks associated with the Trench Mortar Flat eruptive fissures formed along the eastern flank of Big Cinder Butte, where they are partly buried by cinders (Kuntz et al., 1989).

The third factor is illustrated by the cracks at the Kings Bowl, Hell's Half Acre, and the Open Crack rift set (Figs. 2, 3, and 4). At these three localities, some parts of the tension cracks are covered by younger flows, but the uncovered parts yield enough field data to make reasonable inferences about the nature of the original, complete set of tension cracks.

EXAMPLES OF ERUPTIVE FISSURES AND TENSION CRACKS IN THE EASTERN SNAKE RIVER PLAIN

We here discuss the character of eruptive fissures and tension cracks for three late Pleistocene–Holocene basaltic lava fields in the eastern Snake River Plain. We begin with the Kings Bowl lava field because the eruptive fissures and tension cracks there are easily identified and studied in the field and easily mapped on air photos. We then discuss eruptive fissures and tension cracks in the Hell's Half Acre lava field and the Open Crack rift set, where spatial relationships between the cracks and fissures are less obvious and where parts of the eruptive fissure–tension-crack systems are covered by young lava flows.

Kings Bowl lava field

The Kings Bowl lava field is the locus of source vents for the small, 2220 yr old Kings Bowl lava field that covers an area of 3.3 km² and has an estimated volume of 0.005 km³ (Kuntz et al., 1986, 1992; Fig. 2). Eruptive fissures and tension cracks extend over a distance of 11.3 km (Table 1). The eruptive fissure system, 6.1 km long, consists of 18 en echelon eruptive fissure segments that range from 220 m to 900 m long. The south end of the eruptive fissure system consists of five short (<100 m) eruptive segments that are surrounded by patches of lava that are each <200 m in longest dimension. During the initial phases of the Kings Bowl eruptions, some of the short segments may have been part of longer segments of eruptive fissures or they may represent en echelon fingers that merge at depth with a single dike having a length of ~6 km. Delaney and Gartner (1997) suggested that dikes in the San Rafael Swell, Utah, illustrate the latter case. In addition, areas between present-day eruptive fissure segments may have been covered by some of the youngest flows of the Kings Bowl lava field (Covington, 1977).

The eruptive fissure system is halfway between two sets of outer, parallel, tension cracks (Fig. 2). These cracks formed in

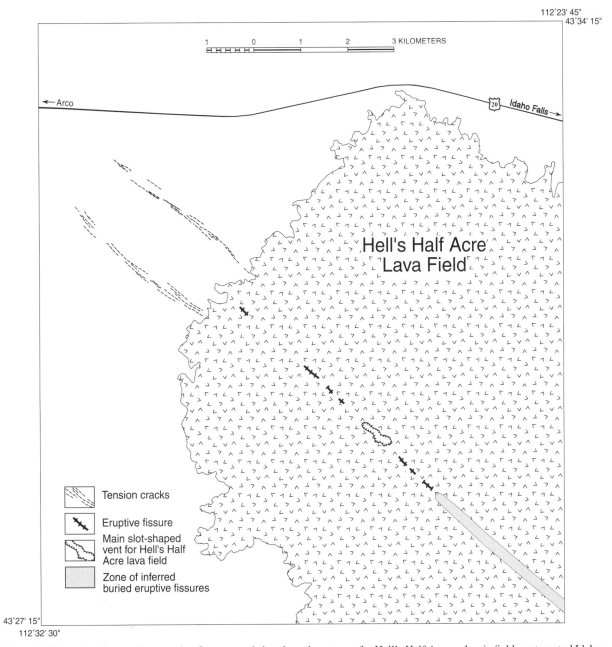

Figure 3. Map of tension cracks, eruptive fissures, and slot-shaped vent area for Hell's Half Acre volcanic field, east-central Idaho. Data from Kuntz et al. (1994) and from interpretation of 1:30 000 scale true-color aerial photographs taken in 1973.

relatively young lava flows that have not been covered by loess; thus they are easily identified in the field and on aerial photographs. The eastern set of cracks is nearly continuous along the length of the eruptive fissures except near Kings Bowl, where it is locally covered by Kings Bowl flows. The eastern set of cracks ranges in width from ~100 to 500 m and trends N10°W. There appears to be a 1-km-long gap in the eastern set of cracks ~2.5 km north-northeast of Creon's Cave. The western set of tension cracks is nearly continuous for a distance of ~9 km

north of South Grotto. South of that locality, the crack set is diffuse. The width of the western set of cracks ranges from 100 to 200 m north of South Grotto and is as wide as 800 m south of that locality. The western set of cracks trends N11°W. Along the trace of the eruptive fissures, the distance between the pair of tension cracks is ~1250 m. A second pair of inner tension cracks extends for a distance of ~1.5 km north of the northernmost eruptive fissures (Fig. 2). The distance between the inner pair of cracks ranges from ~50 m on the south to ~200 m on

116

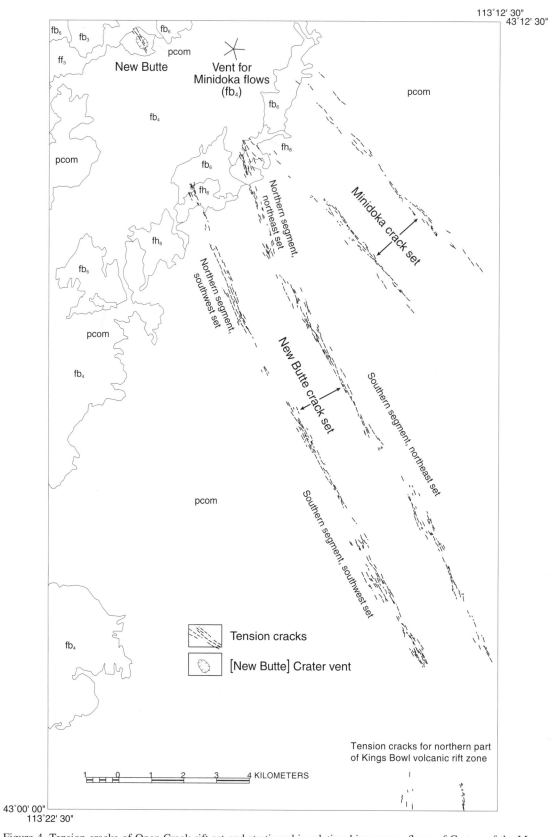

Figure 4. Tension cracks of Open Crack rift set and stratigraphic relationships among flows of Craters of the Moon lava field, south-central Idaho (from Kuntz et al., 1988). Flow designations as follows: fb3, Devils Cauldron; fb4, Minidoka; fb6, Rangefire; ff3, Bottleneck Lake; fh8, Brown; pcom, pre-Craters of Moon. See Kuntz et al. (1988) for details of stratigraphy and age designations for flows of Craters of the Moon lava field.

TABLE 1. DATA FOR TENSION-CRACK AND ERUPTIVE FISSURE SYSTEMS IN THE EASTERN SNAKE RIVER PLAIN, IDAHO

Crack systems	Measured length of cracks (km)	Average distance between crack pairs (m)	Length of eruptive fissures (km)	Inferred length of dikes (km)
Kings Bowl	11.3	1.250	6.1	15*
Hell's Half Acre	4.3	1450	5.5	24#
Open Crack rift set				
Minidoka set	8	2000	n/a	20*
New Butte set				
Northern segment	5	1800	n/a	16*
Southern segment	13	1810	n/a	13*
North and South segments	16	1800	n/a	40#

*Minimum estimate
#/Maximum estimate
n/a—not applicable

the north. A northward projection of the northernmost eruptive fissures bisects the inner pair of tension cracks.

Hell's Half Acre lava field

A set of parallel tension cracks extends ~4 km northwest of the 5200 yr old Hell's Half Acre lava field (Kuntz et al., 1986, 1992; Fig. 3). These tension cracks formed in loess-covered lava flows and thus are not as obvious in the field as are the tension cracks in the Kings Bowl area. Many of the Hell's Half Acre tension cracks are exposed only along short segments where loess has drained or been washed into the tension cracks. However, the extent and character of the crack system is clearly evident on 1:30 000 scale, true color aerial photographs taken in 1973. The tension cracks are denoted on the photos by short segments of open tension cracks, aligned pits in loess, and dark-toned lineaments that are parallel to or extensions of the exposed tension cracks.

Individual tension cracks average ~500 m long; the longest crack is ~1 km long and some cracks are as short as 100 m. Each of the crack sets consist of en echelon cracks; at any one locality, the crack set may consist of one to as many as five tension cracks. The width of the crack sets ranges between 60 m and 240 m. The average distance between the two crack sets is ~1450 m. The southwesterly set of tension cracks trends N56°W and the northeasterly set of tension cracks trends N52°W.

The main eruptive center for the Hell's Half Acre lava field is a slot-shaped vent, 800 m long and 100–200 m wide (Fig. 3). There is strong evidence of repeated episodes of piston-like draining and filling of the crater and overflow of the crater walls by thin, shelly, pahoehoe flows. Small spatter cones and eruptive fissures extend northwest and southeast of the main crater. Recent analysis of aerial photographs has identified a partly buried, 300-m-long, eruptive fissure segment just 400 m from the northwest margin of the lava field. Owing to a regional northwest to southeast slope direction, much lava from the main Hell's Half Acre vent area flowed to the southeast and probably covered parts of the southeastern end of the eruptive fissure

system. A lava-tube system that originates along the eruptive fissure system ~1000 m southeast of the main vent area has rootless vents that contributed lava that resurfaced at the southeast end of the eruptive fissure system (Kuntz et al., 1994). The exposed part of the entire eruptive fissure system for the Hell's Half Acre lava field is 5.5 km long. Assuming that the eruptive fissure–tension-crack system is symmetrical, i.e., that the main vent area for the Hell's Half Acre lava field is in the center of the system, then that system may be ~19 km long.

Open Crack rift set

The Open Crack rift set consists of two pairs of tension cracks that extend southeast from the southeast margin of the Craters of the Moon lava field (Fig. 4; Kuntz et al., 1988). We use the term Minidoka set for the northeast pair of cracks and the term New Butte set for the southwest pair of cracks. Compared to the Kings Bowl and Hell's Half Acre tension-crack sets discussed here, the Minidoka and New Butte crack pairs are unique because neither pair can unequivocally be related to an eruptive fissure system.

The Minidoka pair of cracks is the shorter and less well developed of the two pairs. In the Minidoka pair, the northeast set of cracks is 8 km long and trends N45°W. The southwest set of cracks is 6.5 km long and trends N41°W. The distance between the pair of cracks is 1920 m near the margin of the Craters of the Moon lava field and 2100 m near the southeast end. The central and southeast ends of each pair of cracks are easily identified in the field and on aerial photographs. The northwest ends of each pair of cracks are more diffuse, i.e., the cracks are fewer and less easily identified in the field and on aerial photographs because they formed in flows that have a significant cover of loess. Cracks of the Minidoka set cut Rangefire flows (fb_6 in Fig. 4) of the Craters of the Moon lava field, are believed to be about 4500 yr old. However, the cracks are covered by Minidoka flows (fb_4 in Fig. 4) of the Craters of the Moon lava field, also believed to be about 4500 yr old, but stratigraphically younger than Rangefire flows (Kuntz et al., 1986, 1988). These relations indicate that the Minidoka cracks formed during eruptive period B of the Craters of the Moon Lava field and are about 4500 yr old.

When cracks are projected northwest along strike, the vent area for the Minidoka flows bisects the Minidoka crack pair. This vent area is obscure, but its location is denoted as a topographically high, lava-filled depression from which lava-tube systems radiate in all directions. The distance from the Minidoka vent area to the southeast end of the tension-crack system is 10.2 km. Assuming the symmetry shown in Figure 2, the Minidoka eruptive fissure—tension-crack system has a presumed length of ~20 km.

The New Butte tension-crack set consists of two pair of tension cracks (Fig. 4). Each set of tension cracks consists of two segments; each segment has a slightly different trend. The northern segment of the southwest set of cracks extends 5 km from the southeast of the margin of the Craters of the Moon lava field and has a trend of N22°W. The average distance between the northern segments of the New Butte crack system is 1800 m. The southern segment of the southwest set of cracks has a length of 10 km and trends N27°W. The northern segment of the northeast pair of cracks is 4 km long and trends N16°W. The southern segment of the northeast set of cracks is 13 km long and trends N27°W. The average distance between the southern segments of the New Butte crack system is 1800 m. When projected northwest along strike, the New Butte set of cracks is on either side of the vent area for New Butte; thus New Butte is believed to be the site where the dike that formed the New Butte crack system breached the surface. Assuming the symmetry shown in Figure 2 and that the dike that breached the surface at New Butte is associated only with the northern segments of the New Butte set of cracks, the New Butte eruptive fissure—tension-crack system may have a length of ~16 km. If New Butte represents the center of an eruptive fissure—tension-crack system that includes both the northern and southern segments of the New Butte crack set, then the eruptive fissure–crack system may be as long as 40 km.

The differing trends of the two segments suggest that the New Butte crack system may consist of two sets representing two dike systems. In this case, the New Butte crack system may consist of the northwest set that has New Butte as the vent area, and the southeast set of cracks may represent a 13-km-long dike that failed to breach the surface.

FAULTS IN VOLCANIC RIFT ZONES OF THE EASTERN SNAKE RIVER PLAIN

Spencer–High Point volcanic rift zone

The Spencer–High Point volcanic rift zone extends ~70 km in a west-northwest–east-southeast direction on the southern flank of the Centennial Range in the northeastern part of the eastern Snake River Plain (Figs. 1 and 5; Kuntz et al., 1992). This zone has the highest concentration of vents of any volcanic rift zone in the eastern Snake River Plain. The vents at that locality are mainly of the eruptive fissure and lava-cone types, forming small-volume lava fields that are long and narrow because they moved down a relatively steep slope (1°–2°) to the south. The eruptive fissures and elongated vents in the zone have a remarkable degree of parallelism; they are aligned in a trend of N63° ± 2°W.

The western end of the volcanic rift zone consists of two small 4.1 Ma rhyolite domes (Indian Creek Buttes) and ~10 vents for small basaltic lava fields. There are also numerous, west-northwest–east-southeast–trending, down-to-the-south, normal faults that are part of the Middle Creek Butte fault zone (Fig. 5). The main part of the volcanic rift zone begins where the Middle Creek Butte fault zone enters the basalt-flow terrain of the eastern Snake River Plain near Spencer. The graben ~5 km south of Spencer and the graben ~3 km east of Experimental Crater are linear extensions of the Middle Creek Butte fault zone into the basalt-flow terrain. These grabens are very narrow (<100 m wide) when compared to the width of pairs of tension cracks and faults that are assumed to form in the zone of tension cracks (~1500–2500 m) in the late Pleistocene–Holocene lava fields in the eastern Snake River Plain (cf. Figs. 2, 3, 4, and 5). These relations suggest that these grabens cannot unequivocally be ascribed to emplacement of basaltic dikes, but rather they are likely the expression in overlying basalt flows of faults in bedrock beneath the flows.

A graben ~3 km north and a graben ~7 km west of Morgan's Crater (Fig. 5) may be related to dike emplacement. The northern graben is ~700 m wide and 1.4 km long. The northern boundary of this graben is very linear and has steep walls as high as 10 m. The southern wall is less steep and is as high as 5 m. Rubin and Pollard (1988) and Rubin (1990) showed that maximum displacement on faults related to single dike-injection events in Iceland, Hawaii, and Afar are typically <1 m. This relationship implies that numerous dikes were repeatedly injected beneath this graben to achieve the amount of fault offset observed. The graben is filled at both ends by younger lava flows, thus its length is a minimum value. It is collinear with Swan Butte, a major basaltic tephra cone 1.5 km to the east-southeast, and with an eruptive fissure just 300 m to the west. The eruptive fissure has several pit craters, suggesting that magma withdrawal may have led to formation of the pit craters and possibly the graben.

The graben west of Morgan's Crater is 2.5 km long and ranges from 200 m to 250 m wide. Its steep, linear southern wall is as high as 12 m; the northern wall is less steep and as high as 7 m. As for the graben cited previously, the amount of offset on the graben-bounding faults is an order of magnitude larger than the offset related to dike emplacement in Iceland, Hawaii, and Afar (Rubin and Pollard, 1988; Rubin, 1990). This graben is not collinear with any basaltic vents; it trends N76°W, which is significantly different than the N62°W trend of a 14-km-long eruptive fissure system located just 1 km north of the graben (Fig. 5). The divergence in trends of the dike-controlled vents and the large amount of total offset on the graben-

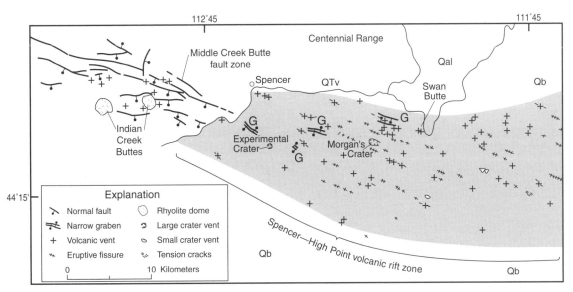

Figure 5. Generalized geologic map of Spencer-High Point volcanic rift zone, eastern Idaho. Abbreviations: QTv, Quaternary and Tertiary rhyolitic flows and ignimbrites of southern flank of Centennial Range; Qb, Quaternary basalt flows; Qal, Quaterary alluvium; G, grabens referred to in text. Modified from Kuntz et al. (1992).

bounding faults suggest that the graben west of Morgan's Crater may not be related to dike emplacement, but rather to tectonic faulting in bedrock beneath the flows.

Arco–Big Southern Butte volcanic rift zone

The Arco–Big Southern Butte volcanic rift zone extends 45 km southeast from the northwest margin of the eastern Snake River Plain at Arco (Fig. 6) to the long axis of the eastern Snake River Plain near Cedar Butte (Kuntz et al., 1994). This volcanic rift zone is the location of extensional faults, grabens, and fissure-controlled basaltic vents along the margins of the zone (see Kuntz et al., 1992, 1994).

Faults in this area, which are abundant and well developed along the Big Lost River ~8 km southeast of Arco, were discussed and/or mapped by Kuntz (1977, 1978a, 1978b), Kuntz et al. (1994), Smith et al. (1989, 1996), and Hackett and Smith (1992). Extensional faults having purely dip-slip movement mark the boundaries of a large graben in the Box Canyon area (Fig. 6). The faults have offsets of 1–8 m and form a broad graben that is ~10 km long and ranges from 2.7 to 3.5 km wide. The faults range in length from several tens of meters to as much as 4 km and form an en echelon pattern. Fault traces form curving, irregular, locally indistinct boundaries for the sides of the broad graben. At the end of some faults, the surface basalt is unbroken but shows substantial drag across linear extensions of surface ruptures, forming monoclines. Displacement on faults in the Box Canyon area generally decreases to the southeast, and ~15 km southeast of Arco, the faults pass into open cracks that have purely horizontal displacement.

There are no volcanic vents within the Box Canyon graben, but Coyote Butte, a small (<0.1 km³) fissure-dominated lava

field, and several other small lava cones are ~5–10 km southeast of the graben. Shield volcanoes such as Crater Butte and Sixmile Butte and fissure-dominated vents such as Teakettle Butte, Lavatoo Butte, and two unnamed vents ~7 southeast of Butte City are all found on the flanks of the graben. The lack of volcanic vents within the graben and the presence of many volcanic vents on the margins of the graben suggests that regional strain is accommodated by tectonic faulting and graben formation at the northwest end and by dike emplacement and formation of large lava fields along the margins and in the central and southeastern parts of the Arco–Big Southern Butte volcanic rift zone.

The northwest part of the Arco–Big Southern Butte volcanic rift zone contains faults that may qualify as dike-induced faults; we now discuss that locality. Along the southwestern side of the main graben is a smaller graben, known informally as Paula's graben, named after Paula LaPoint Iagmin, the first to study the area (Fig. 6). This graben is ~1.5 km long and 0.4 km wide. A down-to-the-east, curving, fault-monocline scarp trends north-south at its southern end and N31°W at its northern end, and forms the west wall of the graben. The eastern wall of the graben mirrors the west wall in trend, has offsets of 4–8 m, and may be considered as a paired antithetic fault to the main fault on the west side of the graben. Immediately south of Paula's graben is a north-south–trending monocline, 1.2 km long, which has ~0.5 m displacement, but is down to the west. There are also two north-south–trending cracks at this locality that are covered by sediment but observed on air photos.

The northeastern margin of the Box Canyon graben consists of two smaller grabens, which together form the Railroad graben (Fig. 6). The northwesternmost graben is 1.1 km long and 0.4 km wide. A curving fault scarp that trends roughly

120

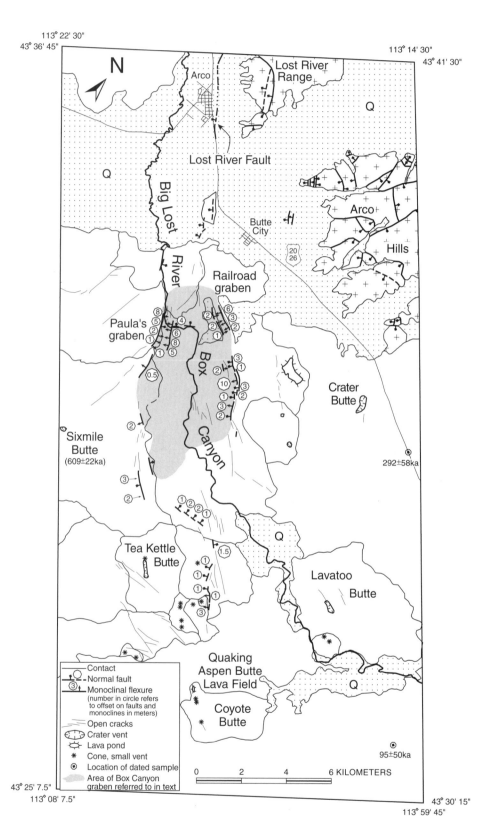

Figure 6. Generalized geologic map of northwestern part of Arco-Big Southern Butte volcanic rift zone, south-central Idaho. Surficial deposits are shown by dot pattern and labeled Q. Paleozoic and Tertiary rocks of Lost River Range and Arco Hills are shown by plus pattern. Basaltic lava flows have no pattern. See Kuntz et al. (1994) for more detailed geologic map of this area.

N70°W forms the northeast wall of the graben. Offsets along this scarp range from 2 to 6 m and the fault surface is vertical. The southwestern wall of the graben mirrors the northeastern wall in trend, has displacement of 1–2 m, and may be considered as a paired, antithetic fault to the main fault on the northeast side of the graben. The second graben, southeast of the previously described graben, is bounded on the northeast by a set of curving faults, 2.3 km in total length, which have displacements of 1–10 m. The southwest wall of this graben is short (0.8 km) and has displacement of 2 m. The general trend of the second graben is N42°W.

MODELS RELATING DIKES AND ASSOCIATED TENSION CRACKS AND FAULTS

Having described the field characteristics of eruptive fissures, tension cracks, and faults at various localities in the eastern Snake River Plain, we now examine models that describe the processes responsible for their formation and then focus on the description and interpretation of the field geometry between these structures. In view of the fact that tension cracks and eruptive fissures are such prominent structures and because faults are so rare in the late Pleistocene–Holocene lava fields we studied, the application of models offers explanations for these field relationships and possible explanations for the lack of dike-induced faults.

Elastic-displacement models for the genetic relationship between dikes, tension cracks and faults

Pollard et al. (1983) studied elastic displacements induced by dike emplacement and described the structures related to and formed by those displacements. Their models were derived to account for structures related to blade-like dikes (dikes that are tens of kilometers long and only a few kilometers high; see Fiske and Jackson, 1972) that were intruded into volcanic rift zones from a summit magma reservoir at Kilauea volcano, Hawaii. The dikes were approximated in the models by planar cracks in an elastic medium subjected to distributions of internal magma-driving pressure and remote stresses represented by gravitational loading.

The elastic-displacement model can be best summarized by the following (Pollard et al., 1983, p. 573–574):

Crack dilation perturbs this stress field especially over the top of the dike. Contours of the maximum principal stress outline a region of tensile stress that spreads outward and upward from the dike top (see Fig. 7B of this paper). Two maxima, one on each side of the dike plane, occur at the surface and are separated by a distance roughly equal to twice the depth to the dike top (see Fig. 7A). The tension decreases beyond the maxima, becoming a compressive stress at a distance somewhat greater than the depth to the crack center. Trajectories of potential tension cracks are vertical at the surface and dip steeply at depth (see Fig. 8). A surprising result of this analysis is the "bimodal" distribution of tensile stress at the free surface (see Fig.

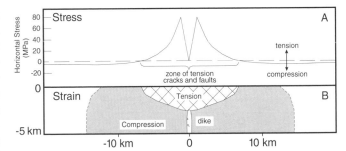

Figure 7. Diagrams illustrating relationship of horizontal stress (A) and zones of tensile and compressive strain (B) related to emplacement of dike into shallow crust, based on elastic displacement models of Pollard et al. (1983) and Rubin (1992). A: Example of theoretical horizontal stress based on January rifting event at Krafla volcano, Iceland (Pollard et al., 1983, Fig. 14), illustrating bimodal maxima of horizontal stress that predicts set of parallel tension cracks centered over dike. B: Zones of compressive and tensile stress in subsurface resulting from dike emplaced to depth of 2.5 km (Rubin, 1992, Fig. 6).

7A). The great tension at the crack tip induced by dilation is not transferred directly upward to the surface. Indeed the point immediately over the crack at the surface is stress free. This is a general result for nearly vertical pressurized cracks. It provides an explanation for the clustering of open cracks and normal faults into two parallel strips separated by relatively unbroken ground over the top of a dike (see Figs. 2, 3, 4 and 7A of this paper). It also enables predictions of the location and depth of a dike to be made from a map of surface structures.

Pollard et al. (1983) also showed that the elastic displacements accompanying emplacement of a single dike elevate the flanks of the area directly over the dike and may produce concomitant normal faulting (shown schematically in Fig. 8) in the zone of tension cracks. Rubin and Pollard (1988) and Rubin (1992) used elastic-displacement models to study faults and graben subsidence produced by the lateral propagation of basalt dikes that do not breach the surface over distances of tens of kilometers in volcanic rift zones in Hawaii, Iceland, and Afar. Their models indicate that lateral dike intrusion decreases compression beyond the ends of a laterally propagating dike, thus favoring faults at that locality. They also show that dike opening tends to compress and lock faults on either side of the dike, thus slip must occur on faults that intersect the dike near its top above the zone of dike-induced compression (zone of tension in Fig. 7B), or on faults that slip ahead of the dike as it propagates laterally (Rubin and Pollard, 1988). Pollard et al. (1983) documented vertical fault displacements of as much as 15 cm in the zone of tension cracks during fissuring events at Kilauea, and Rubin and Pollard (1988) and Rubin (1992) noted that slip of as much as 1 m occurred on graben-bounding faults during diking events in Iceland and Afar.

Because of the lack of erosion, observable dikes are nearly absent in the eastern Snake River Plain and thus structures such as offsets, steps, and ridges in the dikes cannot be used to determine the directions of dike propagation (Delaney and Gart-

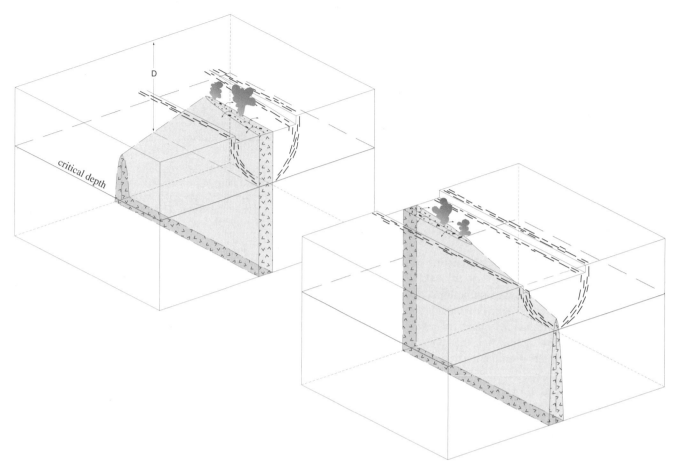

Figure 8. Geometric model showing relationships between tension cracks and faults formed by intrusion of vertical dike. D is critical depth (~750–1000 m in eastern Snake River Plain) at which tensional stresses caused by rising, vertical dike exceed strength of overlying rocks and create tension cracks and incipient faults. See text for discussion. Prepared from data of Pollard et al. (1983), Rubin (1992), and from field relationships shown in Figure 2.

ner, 1997; Delaney and Pollard, 1981). Several lines of evidence (admittedly weak) suggest that dikes in the eastern Snake River Plain ascended mostly vertically from great depth rather than having intruded almost exclusively by lateral motion from a high crustal level magma reservoir. First, caldera-like structures related to withdrawal of magma from shallow reservoirs are absent on the eastern Snake River Plain. Second, the vast majority of eastern Snake River Plain basalts are olivine basalts that contain olivine and plagioclase crystals that are typically <1 mm long; there is little evidence for the accumulation or growth of crystals in crustal reservoirs (Kuntz et al., 1992; Kuntz, 1992). Third, most eastern Snake River Plain basalts are highly uniform in textural, mineralogical, and chemical characteristics, and they show little if any evidence of differentiation or contamination. These features suggest that most eastern Snake River Plain basalt magmas represent near-primitive magmas tapped directly from a partially melted lithospheric mantle (Kuntz et al., 1992, and references therein) and that most magmas did not reside in magma reservoirs within the crust. We believe that the eastern Snake River Plain basaltic dikes were

emplaced in a manner similar to that of the mafic dikes of the San Rafael Swell, Utah, described by Delaney and Gartner (1997, p. 1185):

The horizontal propagation of the San Rafael dikes differed substantially from the lateral propagation of dikes along volcanic rift zones (Rubin and Pollard, 1987). Whereas rift-zone dikes propagate from high-level magma reservoirs, the San Rafael dikes ascended from great depth.

Geometrical and field relationships between eruptive fissures, tension cracks, and faults

The Kings Bowl lava field contains the best exposed and most complete set of eruptive fissures and tension cracks in the eastern Snake River Plain. The spatial association of the eruptive fissures and parallel set of tension cracks provide a classic example of the stress-displacement models of Pollard et al. (1983); i.e., the eruptive fissure bisects a set of parallel tension cracks and there is a nearly constant distance between the tension cracks (Fig. 2). These relationships suggest an ideal symmetry (Fig. 8) for an eruptive fissure-tension-crack system that may be applied to other of these systems in order to determine the dimensions of dikes in the eastern Snake River Plain. Based on the Kings Bowl model, we assume that other dike eruptive fissure-tension-crack systems in the eastern Snake River Plain had the same or similar geometry, i.e., that the ascending dike formed the tension cracks, and continued upward emplacement of the dike formed eruptive fissures or a central vent along the highest part (typically the midpoint) of the dike.

DIKE-EMPLACEMENT MECHANISMS AND THE FORM OF DIKES IN THE EASTERN SNAKE RIVER PLAIN

As the cupola of the eastern Snake River Plain dikes rose to a critical depth (depths of ~750–1000 m in Figs. 8 and 9), tension cracks formed as a result of the tensional stresses created above the dike. Pollard et al. (1983) showed that the critical depth is approximately half the average distance between the pair of tension cracks (Table 1). With continued rise, the shoulders of the dike progressively intersected the critical depth, causing the tension-crack system to migrate away from earlier formed cracks. Some dikes (e.g., Kings Bowl, Hell's Half Acre) continued upward migration and breached the surface, forming eruptive fissures (Fig. 9, A and B). Other dikes stagnated along most of their length in their upward migration after forming tension cracks, such as the dikes that formed the Minidoka and New Butte crack systems of the Open Crack rift set (Fig. 9, C–E).

The preceding discussion assumes that the dikes have a semicircular upper surface. This assumption is supported by the shapes of dike systems in the eastern Snake River Plain, which are shown at true scale in Figure 9. This figure shows longitudinal profiles of four dike systems based on the previously described field relationships between tension cracks and eruptive fissures or other single vent areas. We assume that the top of the dike slopes uniformly from the ends of the eruptive fissures or single vent areas to the critical depth below the end of the tension cracks. The nature of the ends of dikes remains equivocal in determining the form of the dikes. We do not know if the shoulders of the dike slope at similar dips for great distances beyond the position of the ends of the tension cracks or if the dike shoulders begin to dip steeply at or near that point. Perhaps future geophysical field work and modeling will reveal answers to this question.

DIKE-EMPLACEMENT MECHANISM THAT IMPEDES FORMATION OF FAULTS IN THE EASTERN SNAKE RIVER PLAIN

We believe that the absence of faults associated with tension cracks and eruptive fissures at the late Pleistocene-Holocene lava fields can be attributed to the fact that eastern Snake

River Plain dikes had a significant component of vertical ascent when compared to the mostly horizontally propagating intrusions of blade-like dikes in the Hawaiian and Icelandic examples cited herein. After formation of tension cracks and incipient faults, the continued rise of the dikes to shallower depths increased the compression adjacent to the dike walls and prevented the formation of faults or arrested the continued development of incipient faults. As can be visualized from Figure 7B, with continued vertical intrusion of the dike, the zone of compressive stress rises into the zone formerly occupied by the zone of tension (zone of incipient faulting) and inhibits continued or additional faulting.

VOLCANIC VERSUS TECTONIC ORIGIN OF FAULTS IN THE BOX CANYON AREA

Several factors suggest that the faults in the Box Canyon area were not formed as a result of dike injection. (1) The elastic-displacement models and data from the late Pleistocene-Holocene lava fields predict that faults should be extraordinarily straight and aligned with tension cracks in narrow zones, but in the Box Canyon area, faults are curvilinear and not associated with tension cracks in a narrow zone (cf. Figs. 2, 3, 4, 6, and 8). (2) The elastic displacement models predict that faults should show consistent dips facing inward toward the center of a graben, but in the Box Canyon area, faults show inconsistent senses of dip (cf. Figs. 6 and 8). (3) The scale of the faulted terrain in the Box Canyon area is not that expected from the examples from the late Pleistocene-Holocene lava fields; the distance between tension-crack pairs and supposed faults in the late Pleistocene–Holocene lava fields ranges from 1.4 to 2.2 km, but the distance between the boundaries of the major graben in the Box Canyon area is ~3 km (cf. Figs. 2, 3, 4, and 6). (4) It is significant that there are no eruptive fissures or pairs of tension cracks that would indicate that fault-forming dikes breached the surface or are at relatively shallow depth in the center of the Box Canyon area. (5) Fault offsets produced in dike-injection events in Iceland, Afar, and Hawaii are all ≤1 m (Rubin and Pollard, 1988; Rubin, 1992); the displacements on the various faults in the Box Canyon area (as much as 10 m) suggest that there must have been least 10 dike-injection events to achieve the total offsets observed in the Box Canyon area. This implies that the faults were produced by dike-forming intrusive events to the exclusion of extrusive events. We believe that at least some of the hypothesized fault-forming dike intrusions (Hackett and Smith, 1992) should have produced eruptive fissures and/or pairs of open tension cracks in the Box Canyon area, but these structures are not found there.

The Arco-Big Southern Butte volcanic rift zone is a southeastward extension onto the eastern Snake River Plain of the Lost River fault zone that forms the eastern margin of the Lost River Range (Kuntz et al., 1992; Figs. 1 and 6). We consider the Box Canyon area to be a southeasterly continuation of the Arco segment of the Lost River fault into the basalt-flow terrain

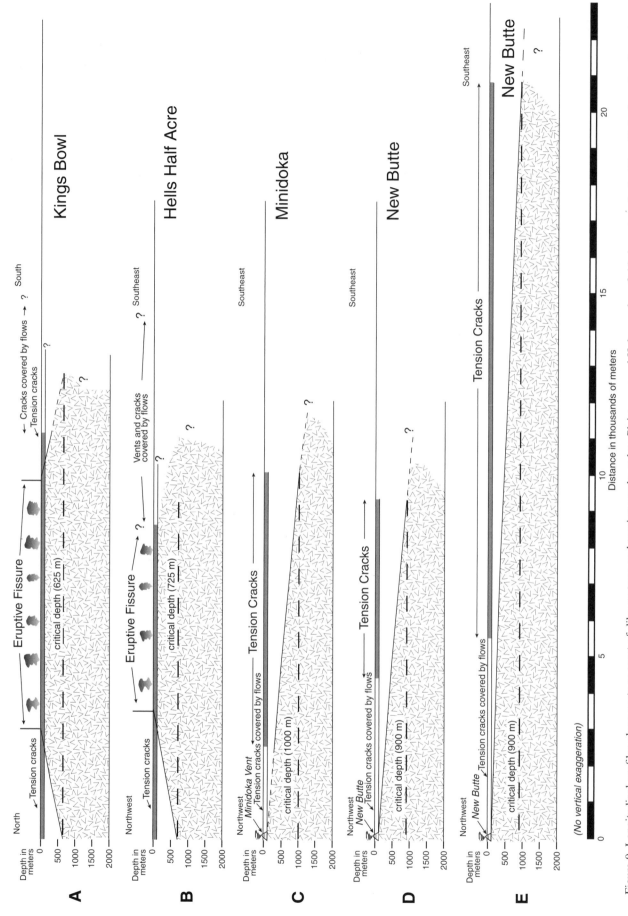

Figure 9. Longitudinal profiles along parts or most of dike systems and tension cracks at late Pleistocene and Holocene lava fields in eastern Snake River Plain. Lengths of eruptive fissures and tension cracks are based on field relationships between tension cracks, eruptive fissures, and single vents (Minidoka vent, New Butte) shown in Figures 2, 3, and 4. In all figures, dotted thick line at surface represents length of exposed tension cracks; open, thick line represents inferred tension cracks covered by basalt flows. Eruptive fissures show parts of cupola of dike that have breached surface. Critical depth is discussed in text. Shoulders of dike are assumed to slope smoothly from ends of eruptive fissures or single vent areas to critical depth beneath ends of tension cracks.

of the eastern Snake River Plain. The faults in the Box Canyon area pass southeastward into open cracks that have purely horizontal displacement. These geographical and structural relationships illustrate a fundamental difference in present tectonism between areas outside and within the eastern Snake River Plain: regional extension is manifested in nearly vertical strain that produces typical basin-and-range topography in mountains northwest and southeast of the eastern Snake River Plain, but only in horizontal extension near the axis of the eastern Snake River Plain, where extension is accommodated by dike emplacement (Parsons and Thompson, 1991). The Box Canyon area represents a transition between the areas of vertical and horizontal strain resulting from regional extension (Kuntz et al., 1992, 1994; Parsons et al., 1998).

Owing to the arguments presented here and because of the spatial relationship between the Lost River fault and the Box Canyon graben, we believe that the faults at the latter locality are mainly of tectonic origin related to the Lost River fault and have little if any genetic relationship to emplacement of dikes.

GEOMETRY OF DIKE SYSTEMS IN THE EASTERN SNAKE RIVER PLAIN

The elastic displacement model of Pollard et al. (1983), the longitudinal profiles of dikes in the eastern Snake River Plain shown in Figure 9, and the dimensions of eruptive fissures and tension cracks (Table 1) provide ways to interpret the geometry of basaltic dikes in the eastern Snake River Plain.

The length of the exposed part of the tension-crack system at Kings Bowl is ~12 km; the southern part of the crack system is covered by flows of the Wapi lava field (Figs. 2 and 9A). Because the Kings Bowl and Wapi lava flows have identical radiocarbon ages and paleomagnetic directions within analytical uncertainty (see Kuntz et al., 1986), flows from both lava fields probably erupted from a single, en echelon eruptive fissure system. Possible tension cracks and eruptive fissures associated with the Wapi lava field are covered by extensive Wapi flows, thus the length of the Kings Bowl–Wapi dike system cannot be determined. However, the Kings Bowl–Wapi dike system is at least 15 km long, when measured from the northern end of the Kings Bowl crack system to the main vent for the Wapi lava field at Pillar Butte (see Kuntz et al., 1988).

The length of the exposed part of the eruptive fissure system for the Hell's Half Acre lava field is 7 km, although part of the southeast extension of the eruptive fissure system may be covered by flows from the main vent area, as shown in Figures 3 and 9B. If it is assumed that the main slot-shaped vent is near the center of the eruptive fissures, the Hell's Half Acre eruptive fissure system may be as long as 17 km. The tension cracks extend ~4 km northwest of the northwesternmost section of eruptive fissures. Assuming a similar geometric relationship at the southeast end of the fissure-crack system, the length for the Hell's Half Acre dike system could be ~24 km.

The distance from the Minidoka vent to the southeast end

of the crack system is ~10 km (Figs. 4 and 9C), thus the Minidoka dike system may be ~20 km long. The maximum length for the Minidoka dike system would be ~40 km if it is assumed that the Minidoka vent is centered along the dike system.

The distance from the New Butte vent area to the southeast end of the northern segments of the pair of tension cracks is 10 km (Figs. 4 and 9D); thus the minimum length of the New Butte dike system may be ~20 km. If New Butte is at the center of a dike system that includes both the northwest and southeast pairs of tension cracks (Fig. 9E), the New Butte noneruptive fissure system may have a maximum length of ~40 km.

The maximum presumed lengths of the New Butte and Minidoka dike systems pose a problem for interpretation of dimensions of basalt dikes in the eastern Snake River Plain; i.e., is 40 km a reasonable length or upper length for basalt dikes in the eastern Snake River Plain and can dikes be as high as they are wide? We answer these questions by providing new field and paleomagnetic data about lava fields in the central and southern parts of the INEEL. In summary, dike systems in the eastern Snake River Plain are typically ~20 km long, but may be as long as ~40 km.

STYLE OF BASALTIC VOLCANISM IN THE TEST AREA NORTH REGION OF THE IDAHO NATIONAL ENGINEERING AND ENVIRONMENTAL LABORATORY

Detailed 1:50 000 scale geologic mapping, paleomagnetic, and petrographic studies have recently been completed on basaltic lava flows in the Test Area North (TAN) and central and southeastern parts of the INEEL. In the absence of detailed ^{40}Ar/^{39}Ar dating of all surface flows in this area, we have employed petrographic and paleomagnetic data (Table 2) as the main tools for regional correlation of flows and lava fields. These studies, in connection with the insight gained from analysis of tension cracks and dikes, give a new perspective on the character of basaltic eruptions in the TAN area and for the entire eastern Snake River Plain.

South and east of the TAN area, seven small shield volcanoes estimated to be 200–400 ka (Kuntz et al., 1994) were erupted from vents aligned northwest to southeast along the Lava Ridge–Hell's Half Acre volcanic rift zone (Figs. 1 and 10). Within this part of the volcanic rift zone, the seven lava fields can be assigned to as few as two, probably coeval, groups based on paleomagnetic data (Fig. 10; Table 2). A younger group (inclination, I = 62°, declination, D = 13°) consists of two vents (Qbc9 and Qbc8 in Fig. 10) that are 6 km apart. An older group (I = 41°–43°, D = 2°–6°) consists of five vents (Qbc5, Qbc6, Qbc4, Qbc2, Qbc1 in Fig. 10) in a linear trend ~40 km long. Because nearly identical inclination and declination values imply a very short span of time (~100–200 yr), it seems reasonable to assume that the five vents could have formed simultaneously. Thus, we assume that the dike system that fed the five coeval vents was ~40 km long.

M.A. Kuntz et al.

TABLE 2. PALEOMAGNETIC DATA FOR THE TAN AREA AND LAVA RIDGE-HELL'S HALF ACRE VOLCANIC RIFT ZONE

Unit Name	Site	Lat°N	Long°E	N/No	Exp.°	I°	D°	α95°	k	R	P lat.°	P long.°
Deuce Butte–Teat Butte area												
Qbc 10	357B8	43.718	247.360	12/12	Pl	62.3	13.2	2.0	975		80.5	333.5
Qbc 9	369B8	43.693	247.422	12/12	Pl	64.5	12.7	3.3	166		80.7	316.5
Deuce Butte Qbc 8	4B739	43.556	247.544	11/12	30+	64.5	13.0	1.7	746	10.98659		
Teat Butte Qbc 6*	4B895	43.763	247.257	10/12	20+	42.9	11.4	1.5	1104	9.99185	69.0	37.2
Qbc 5	345B8	43.788	247.317	10/12	Pl	44.3	5.9	2.1	737		71.6	50.3
Qbc 5*	4B907	43.761	247.264	11/12	30+	38.3	4.8	0.8	3318	10.99699	67.4	55.5
Qbc 5	A9619	43.783	247.244	12/12	20+	40.9	6.0	1.4	1041	11.98944	69.0	51.7
Qbc 4	465B8	43.664	247.376	11/12	Pl	41.0	2.8	5.4	65		69.7	60.0
Qbc 3	B7073	43.759	247.465	12/12	30+	42.5	8.4	1.3	1135	11.99031	69.7	45.0
Qbc 3	381B8	43.651	247.456	11/12	Pl	46.8	2.4	2.8	253		74.3	59.6
Qbc 2	453B8	43.617	247.428	11/12	30+	43.5	5.7	1.1	1693	10.99409	71.2	51.2
Qbc 1	477B8	43.552	247.506	11/12	30+	42.5	1.8	1.7	710	10.98592	71.0	62.6
Qbd 2	B7013	43.753	247.507	12/12	20+	64.3	30.1	1.8	595	11.98153	68.7	320.5
Qbd 1	B7085	43.724	247.462	9/12	40+	63.8	28.7	2.5	442	8.98189	69.6	322.5
Antelope Butte-Circular Butte area												
Circular Butte Qbe 9*	4B799	43.841	247.360	8/12	Pl	−67.6	189.8	2.8	1065		−80.6	108.9
Antelope Butte Qbd 2	A9631	43.885	247.387	12/12	30+	65.4	328.4	2.3	361	11.96951	67.8	177.9
Antelope Butte Qbd 2	B7109	43.819	247.398	12/12	Pl	64.4	334.9	3.0	236		72.2	173.9
Unnamed Qbd 3	A9643	43.878	247.344	10/12	Pl	58.8	328.1	1.9	630		66.0	158.1
Lava Ridge area												
Qbe 2	B7037	43.978	247.168	11/12	Pl	−53.0	157.8	1.5	1263		−69.9	313.5
Qbe 3*	4B787	43.948	247.241	12/12	40+	−47.2	158.5	2.1	434	11.97463	−66.8	302.2
Qbe 1	B7049	43.952	247.169	9/12	Pl	−77.7	199.0	2.5	481		−65.3	85.3
Qbe 4	B5500	43.924	247.211	11/12	30+	−56.0	201.0	2.4	360	10.9722	−72.5	174.9
Qbe 6*	4B883	43.886	247.201	12/12	30+	−50.6	203.1	1.7	676	11.98372	−67.9	184.0
Qbe 6	B7061	43.849	247.181	10/12	30+	−50.5	203.9	1.0	2305	9.99609	−67.4	183.0
Qbe 6	B7097	43.83	247.201	11/12	30+	−52.2	202.4	1.1	1657	10.99396	−69.4	182.0
Qbe 7	B7121	43.875	247.203	12/12	Pl	−71.9	161.1	2.2	568		−72.4	31.3
Richard Butte Qbe 5	B7025	43.963	247.241	12/12	30+	−65.8	194.0	2.1	426	11.97421	−79.4	129.3

Note: Unit name follows Kuntz (in prep.); *denotes data published in Kuntz et al. (1994); site is alphanumeric identifier; N/No is the number of cores used compared with the number originally taken at the site; Exp.° Is the strength of the peak cleaning field, P1 denotes planes solution for site; I° and D° are the remanent inclination and declination; α95 is the 95% confidence limit about the mean direction; k is the estimate of the Fisherian precision parameter; R is the length of the resultant vector; P lat.° and P long.° is the location in degrees north and degrees east of the virtual geomagnetic pole (VGP) calculated from the site mean direction.

Delaney and Gartner (1997, p. 1191) suggested that because of the primitive nature of the magmas, the dikes of the San Rafael Swell must have heights "that far exceed their <10 km lengths." They also suggested that in the absence of pressure gradients in magma density and in fracture toughness for host rocks, dikes ought to assume a circular shape. These relationships suggest that the 40-km-long dike system may represent the eruptive fissure that developed from an enormous dike having a height of ≥40 km. This height is close to the depth of the Moho in the eastern Snake River Plain (Kuntz, 1992), which suggests that dikes of this size may be connected directly to magma sources in the upper mantle.

Paleomagnetic data for three relatively small shield volcanoes (Circular Butte, Antelope Butte, and an unnamed shield just west of Antelope Butte) indicate that the shields belong to two groups (Fig. 10; Table 2). Antelope Butte and the unnamed shield (Qbd3) have nearly identical normal-polarity inclination-declination values and thus are considered to be coeval and to have formed simultaneously on the same eruptive fissure system. This conclusion contradicts our previously held view that long-lived, shield-producing eruptions in the eastern Snake River Plain emanated from only a single, slot-shaped vent

roughly centered along an initial eruptive fissure to produce a single shield volcano (Kuntz et al., 1992).

IMPLICATIONS OF FIELD AND PALEOMAGNETIC STUDIES IN THE TAN AREA FOR STYLE OF BASALTIC VOLCANISM IN THE EASTERN SNAKE RIVER PLAIN

In an earlier publication, we described what we believe to be the typical stages of basaltic eruptions, beginning with fissure eruptions that lasted a few hours or days, and ending with relatively long lived eruptions from a single, slot-shaped vent at a shield volcano that lasted perhaps several years (Kuntz et al., 1992). These ideas were based on study of the many monogenetic late Pleistocene and Holocene lava fields in the eastern Snake River Plain. The new field and paleomagnetic data from the Lava Ridge–Hell's Half Acre volcanic rift zone presented here depict a rather startling and unexpected supplement to the simple, monogenetic model. The data indicate that significant lengths of eruptive fissure systems, 40 km or more, were in eruption at the same time or within very short periods of geologic time and formed several shield volcano types of lava fields

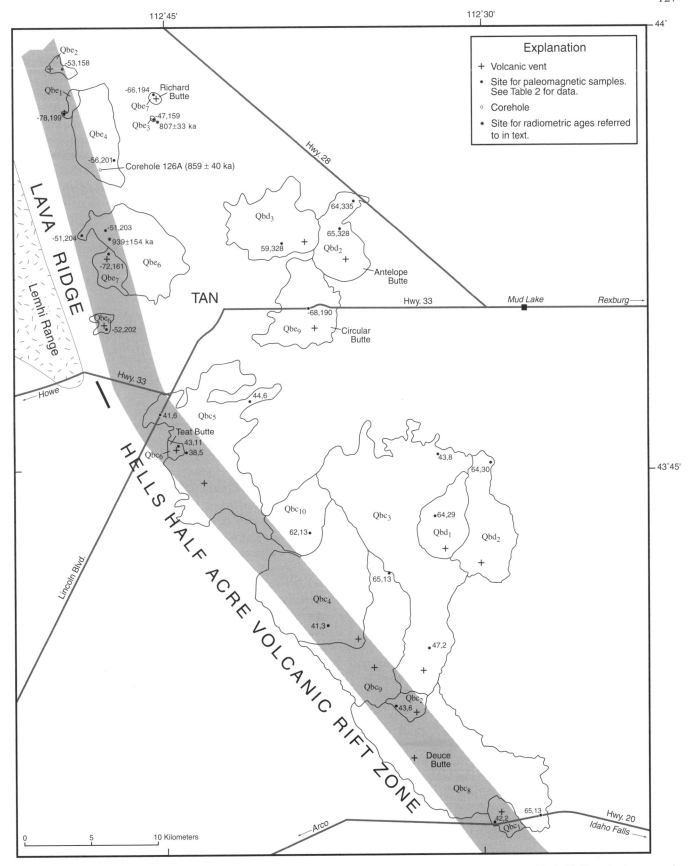

Figure 10. Generalized geologic map of Lava Ridge–Hell's Half Acre volcanic rift zone. Unit designations (e.g., Qbc8, Qbd3) and paleomagnetic inclination and declination values correspond to designations in Table 2.

at various sites along the eruptive fissure system. Other evidence tends to support such an idea. The presumed maximum lengths for inferred dike systems for the New Butte and Minidoka sets of tension cracks may be as long as 40 km (Fig. 4; Table 1). The northern and southern ends of the Great Rift volcanic rift (Fig. 1) zone erupted roughly simultaneously ~50 ka; a distance of ~60 km separates these two areas (Kuntz et al., 1988, 1992). In addition, studies of basalt-supergroups by Wetmore et al. (1997) suggest that considerable accumulations of basalt from several coeval source vents are present in the subsurface in the southern part of the INEEL.

As a result of these data, we now identify a new type of basalt eruption in the eastern Snake River Plain, which we term multiple-shield eruption style, which denotes several shield volcanoes that form simultaneously along an eruptive fissure system that may be 20–60 km long. Based on regional studies of basaltic volcanism in the eastern Snake River Plain (Kuntz et al., 1992), we believe that this style of eruption is a rare event and that the simple, short-length (10–20 km), fissure-dominated single-shield-type eruption is the typical eruption style for the eastern Snake River Plain.

Multiple-shield eruptions have important implications to be considered in evaluation of volcanic hazards for the eastern Snake River Plain. Multiple source vents and larger areas of inundation by basalt flows from a single eruption event increase volcanic hazards for certain areas of the eastern Snake River Plain. However, multiple-shield events also decrease the recurrence interval between eruptive events that were previously assumed to be widely spaced in time. The example from the Lava Ridge–Hell's Half Acre volcanic rift zone indicates that there is a strong need for applying the combined mapping-paleomagnetic-radiometric methods to clarify volcanic hazards elsewhere on the eastern Snake River Plain.

HYDROLOGIC IMPLICATIONS OF DIKES AND TENSION-CRACK AND ERUPTIVE FISSURE SYSTEMS

The characteristics and dimensions of dikes and tension-crack and eruptive fissure systems can be applied to hydrologic problems in the eastern Snake River Plain, specifically at the INEEL, where detailed hydrologic and volcanic-stratigraphic relations have been studied over the past 30 yr. The INEEL is operated by the U.S. Department of Energy, and covers ~2300 km^2 of the eastern Snake River Plain in eastern Idaho (Fig. 1). The INEEL is used in the development of nuclear and technological applications (Bartholomay et al., this volume). Liquid radionuclide and chemical wastes generated at the INEEL have been discharged to on-site infiltration ponds and disposal wells since 1952. Although the use of disposal wells was discontinued in 1984, past liquid-waste disposal has resulted in detectable concentrations of several waste constituents in water from the Snake River Plain aquifer underlying the INEEL (Bartholomay et al., 1997).

The migration of radioactive and chemical wastes is dependent on many factors, including those that control the hydraulic conductivity of basalt and interbedded sediment, the dominant rock units in the Snake River Plain aquifer. Primary controls of hydraulic conductivity include fracture networks and rubble zones of tube-fed pahoehoe lava flows, and grain size and sorting characteristics of sedimentary interbeds (Welhan et al., this volume, chapter 9; Mark and Thackray, this volume). In addition to these, Anderson et al. (1999) suggested controls related to dikes and tension-crack and eruptive fissure systems in the eastern Snake River Plain. These systems, which are most prevalent in volcanic rift zones (Fig. 1), are referred to as vent corridors where they are inferred in the subsurface. Deposits near volcanic vents are scoriaceous slab pahoehoe and shelly pahoehoe flows, spatter, and ash (Fig. 11) that compose zones of high permeability where they are present in the aquifer. Fissures and tension cracks are highly permeable vertical channels in layered sequences of basalt. In contrast, dikes are dense, vertical sheets of basalt that may impede the movement of groundwater. Most of these features are aligned along volcanic rift zones and vent corridors that are oriented generally northwestward, perpendicular to the regional direction of groundwater flow (Figs. 1 and 11). Many dikes, fissures, and tension cracks probably cut the entire thickness of the aquifer.

Where groundwater flows through volcanic rift zones and vent corridors at the INEEL, hydraulic conductivity probably is partly controlled by localized near-vent volcanic deposits, dikes, eruptive fissures, and tension cracks. These features are interpreted to underlie many parts of the INEEL and appear to coincide with estimates of hydraulic conductivity that vary by as much as six orders of magnitude. Many vents and most dikes, fissures, and tension cracks are concealed by surficial basalt flows and sediment, and cannot be verified by available cores and geophysical data. Because of this limitation, the relations between these features and hydraulic conductivity must be evaluated using indirect data. For example, estimates of hydraulic conductivity within vent corridors at the INEEL are similar to estimates from the islands of Hawaii, where dikes are known to control groundwater flow (Stearns, 1985; Meyer and Souza, 1995). We suggest that this simple comparison, coupled with available aquifer test data, stratigraphic interpretations, and inferred locations of individual tension-crack and eruptive fissure systems, can provide a useful first-order approximation of the geologic controls of hydraulic conductivity associated with volcanic rift zones and vent corridors at and near the INEEL.

Estimates of hydraulic conductivity used for a first-order approximation of geologic controls were derived from single-well aquifer tests in 114 wells (Ackerman, 1991; Bartholomay et al., 1997; Anderson et al., 1999). Wells are located throughout the INEEL and include 29 closely spaced wells at the Idaho Nuclear Technology and Engineering Center (INTEC) (Fig. 12). These wells were selected because their test data were assembled and analyzed in a uniform manner and could be compared with detailed stratigraphic interpretations (Anderson et

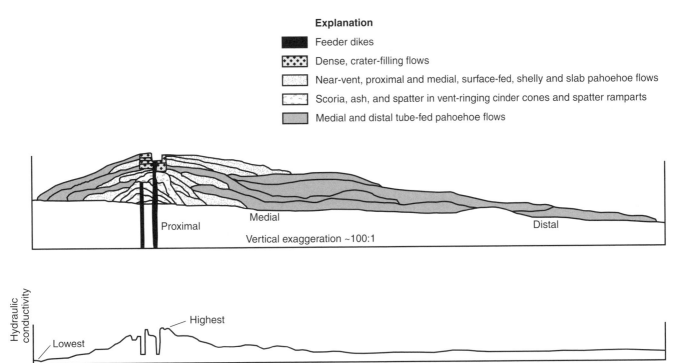

Explanation

■ Feeder dikes

▨ Dense, crater-filling flows

▢ Near-vent, proximal and medial, surface-fed, shelly and slab pahoehoe flows

▢ Scoria, ash, and spatter in vent-ringing cinder cones and spatter ramparts

▨ Medial and distal tube-fed pahoehoe flows

Figure 11. Basalt-flow morphology and inferred relative hydraulic conductivity through a typical shield volcano in Snake River Plain aquifer.

al., 1996). Estimates (available in Anderson et al., 1999, and summarized in Table 3) generally represent the bulk hydraulic conductivity of layered sequences of basalt. Although these estimates tend to average the down-hole variations of hydraulic conductivity observed in straddle-packer tests (Welhan et al., this volume, chapter 15), they are useful indicators of the overall permeability of any given rock sequence. This data-set samples effective open intervals that range in length from 5 to 145 m; the average length is 33 m. The intervals coincide with ~93% basalt and 7% sediment. Most wells are completed in thin pahoehoe flows, which are characterized by dense interiors separated by permeable interflow zones (Welhan et al., this volume, chapter 9).

Bulk hydraulic conductivity may vary to six orders of magnitude or more over small areas of the INEEL (Table 3). Estimates for the 114 wells range from 3.0×10^{-3} to 7.3×10^3 m/day. Two-thirds of these estimates are greater than 3.0×10^1 m/day and span two of the six observed orders of magnitude. Estimates of hydraulic conductivity for a subset of 29 wells at the INTEC range from 3.0×10^{-2} to 2.7×10^3 m/day. Two-thirds of these estimates also are greater than 3.0×10^1 m/day and span two of the five observed orders of magnitude.

Stratigraphic interpretations support the association of bulk hydraulic conductivity with different types of basalt flows and volcanic deposits. This is exemplified by a preponderance of thin (3–9 m) pahoehoe flows in 98 wells, thick (9–15 m), possibly ponded, pahoehoe flows in 6 wells, and near-vent volcanic deposits in 10 wells (Anderson et al., 1996, 1999). Estimates range from 2.4×10^{-2} to 7.3×10^3 m/day for thin flows,

from 3.0×10^{-3} to 4.9×10^0 m/day for thick flows, and from 4.9×10^1 to 1.5×10^3 m/day for near-vent volcanic deposits (Table 3). This overall range of estimates is similar to that reported by Welhan et al. (this volume, chapter 9, Fig. 3) for other estimates of field-scale hydraulic conductivity of basalt and sediment at the INEEL.

Estimates of hydraulic conductivity also are similar to estimates derived from aquifer tests and numerical simulations of groundwater flow in the islands of Hawaii (Meyer and Souza, 1995). Geohydrologic settings in Hawaii include dike complexes and dike-impounded aquifers in volcanic rift zones along the topographic crest of volcanoes. Basal aquifers, composed mainly of thin pahoehoe flows, occur along the flanks of volcanoes. Meyer and Souza (1995) suggested a plausible (measured and simulated) range of hydraulic conductivity of ~3.0 $\times 10^{-4}$ to 3.0×10^{-2} m/day for dike complexes, ~3.0 $\times 10^{-1}$ to 3.0×10^1 m/day for dike-impounded aquifers, and ~3.0 $\times 10^1$ to more than 3.0×10^3 m/day for basal aquifers. Although dikes in the Snake River Plain aquifer probably are more dispersed than those of volcanic rift zones in Hawaii, their hydraulic conductivity may be similar. If so, some estimates of $<3.0 \times 10^1$ m/day at the INEEL may indicate the presence of concealed dikes.

To test this hypothesis and to evaluate the range of hydraulic conductivity with respect to other generalized rock types and structures of volcanic rift zones, Anderson et al. (1999, p. 14) delineated 45 narrow, northwest-trending zones, referred to as vent corridors. They used the locations and trends of known and inferred volcanic vents to approximate possible locations

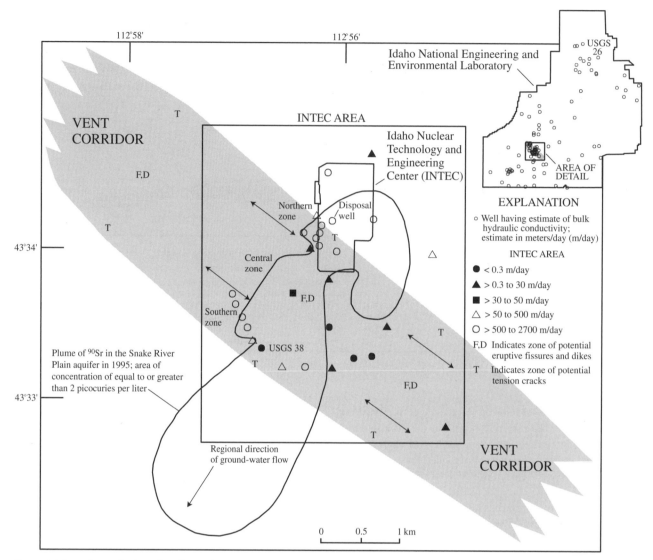

Figure 12. Locations of selected wells used to estimate bulk hydraulic conductivity in Snake River Plain (SRP) aquifer at Idaho Nuclear Technology and Engineering Center (INTEC) and Idaho National Engineering and Environmental Laboratory. Closed and open symbols at and near INTEC indicate ranges of estimates for 29 closely spaced wells open to similar stratigraphic intervals within and near a vent corridor crossed by plume of ^{90}Sr. Data are from Bartholomay et al. (1997) and Anderson et al. (1999).

of individual tension-crack and eruptive fissure systems in the subsurface. Although some zones are questionable because they rely heavily on inferred vent locations and interpretation, the dimensions of vent corridors are similar to the overall lengths and widths of exposed late Pleistocene-Holocene tension-crack and eruptive fissure systems (Table 1). Vent corridors average ~1–3 km in width and 8–24 km in length, dimensions that compare favorably with the distance between crack pairs, length of eruptive fissures, and inferred length of dikes of exposed late Pleistocene–Holocene lava fields. Although near-vent volcanic deposits and dikes likely occupy some parts of vent corridors, thin pahoehoe flows make up the largest part of their volume.

Estimates of hydraulic conductivity have a wider range

(from 3.0×10^{-3} to 7.3×10^{3} m/day for 99 wells) within vent corridors relative to outside of vent corridors (4.0×10^{0} to 1.7×10^{3} m/day for 15 wells). Differences in ranges might be the result of sampling bias; however, differences also might indicate the presence of near-vent volcanic deposits, dikes, fissures, and tension cracks within vent corridors. Near-vent volcanic deposits are estimated to cover only 10% of any given lava field, so their overall extent is small compared to that of pahoehoe flows. Estimates from 10 wells that penetrate locations near known and inferred vents suggest that near-vent volcanic deposits are as permeable as the most permeable pahoehoe flows. The same may be true of some tension cracks, based on their inferred locations with respect to some high estimates of hydraulic conductivity.

TABLE 3. ESTIMATES OF BULK HYDRAULIC CONDUCTIVITY (K) FOR SELECTED WELLS COMPLETED IN THE SNAKE RIVER PLAIN AQUIFER AT THE IDAHO NATIONAL ENGINEERING AND ENVIRONMENTAL LABORATORY (INEEL) AND IDAHO NUCLEAR ENGINEERING AND TECHNOLOGY CENTER (INTEC)

Location (area)	Number of estimates	Minimum K (m/d)	Maximum K (m/d)	Median K (m/d)	Remarks
INEEL (2,300 km²)	114	3.0×10^{-3}	7.3×10^{3}	1.6×10^{2}	K estimates mainly representative of thin (3–9 m) pahoehoe basalt flows. Sediment content averages 7%. Two-thirds of estimates are higher than 3.0×10^{1} m/d; many lower estimates probably controlled by concealed dikes. Some low and high estimates probably controlled by concealed fissures and tension cracks
	98*	2.4×10^{-2}	7.3×10^{3}	1.3×10^{2}	
	6†	3.0×10^{-3}	4.9×10^{0}	1.6×10^{-1}	
	10§	4.9×10^{1}	1.5×10^{3}	3.4×10^{2}	
INTEC AREA (13 km²)	29	3.0×10^{-2}	2.7×10^{3}	5.8×10^{2}	K estimates mainly representative of thin (3–9 m) pahoehoe basalt flows. Sediment content averages 4%. Two-thirds of estimates are higher than 3.0×10^{1} m/d; many lower estimates probably controlled by concealed dikes. Some low and high estimates probably controlled by concealed fissures and tension cracks

Note: Estimates for 29 wells at the INTEC are a subset of estimates for 114 wells at the INEEL. K, in meters per day (m/d). Estimates from Anderson et al. (1999). Effective open intervals exclude basalt flow group I (Morin et al., 1993) and other dense basalt flows below the effective base of the aquifer (Mann, 1986).
*Wells open mainly to thin (3–9 m) pahoehoe basalt flows and their associated interflow fracture networks and rubble zones.
†Wells open mainly to the dense interiors of thick (9–15 m), possibly ponded, pahoehoe basalt flows.
§Wells probably open to slab-pahoehoe and shelly-pahoehoe basalt flows and scoria, ash, and spatter near volcanic vents.

Dikes probably control some of the lowest estimates of hydraulic conductivity by limiting the flow of water to wells. Estimates of hydraulic conductivity for 35 wells open mainly to thin pahoehoe flows are $<3.0 \times 10^{1}$ m/day. Based on the comparison with Hawaii, some of these estimates may be controlled by concealed dikes. Open intervals in these wells range in length from 5 to 86 m. Open intervals are longer than 30 m in 23 (two-thirds) of the wells, where they potentially sample from 5 to 17 interflow zones, based on the average 5 m thickness of flow lobes measured in many INEEL cores (Welhan et al., this volume, chapter 9, Table 1). The large number of potential interflow zones seems inconsistent with the low estimates of hydraulic conductivity. Although other factors, such as inadequate well development or sediment-filled interflow zones, may contribute to some low estimates, we suggest that dikes are a more likely control.

This likelihood is based on four observations. (1) Of the 35 wells having low estimates of hydraulic conductivity, 34 are located within vent corridors where there is a high potential for dikes. (2) Low estimates tend to cluster in the central parts of vent corridors where dikes are most likely to be present. (3) High estimates tend to cluster along the edges of vent corridors where dikes are least likely to be present. (4) Clusters of low and high hydraulic conductivity seem relatively insensitive to the variable lengths and depths of open intervals in wells, suggesting that the overall distribution of estimates is not biased by factors related to well construction. Because our approach does not allow the discrimination of other possibilities, dikes also may include some open fissure segments and tension cracks that filled with lava and sediment during burial.

Geologic interpretations and estimates of hydraulic con-

ductivity suggest that concealed dikes are present within a vent corridor located immediately south of the INTEC (Fig. 12). This vent corridor was delineated on the basis of eight inferred vents in the unsaturated zone, is ~2.3 km wide and 8 km long, and is penetrated by 25 closely spaced wells having estimates of hydraulic conductivity that vary by five orders of magnitude. These wells are open to many thin pahoehoe flows in the uppermost part of the aquifer, and are located in zones of potential dikes, eruptive fissures, and tension cracks. Nine wells are clustered in a zone of potential eruptive fissures and dikes that is ~1 km wide and 3 km long in the center of the vent corridor. These wells have open intervals that average 37 m in length and a median hydraulic conductivity of 2.5×10^{0} m/day. Eight wells are clustered in a zone of potential tension cracks along the northern edge of the vent corridor. These wells have open intervals that average 38 m in length and a median hydraulic conductivity of 8.8×10^{2} m/day. An additional eight wells are clustered in a zone of potential tension cracks along the southern edge of the vent corridor. These wells have open intervals that average 30 m in length and a median hydraulic conductivity of 5.9×10^{2} m/day. Despite the similar average lengths of open intervals in wells, median values of hydraulic conductivity in the northern and southern zones of the vent corridor are two orders of magnitude larger than that of the central zone. Low estimates of hydraulic conductivity in the central zone are consistent with the presence of concealed dikes, and appear to control the distribution of a local plume of ^{90}Sr (Fig. 12), as delineated by Bartholomay et al. (1997).

The strength of this interpretation includes low estimates of hydraulic conductivity that are similar to those of dike complexes and dike-impounded aquifers in Hawaii, low estimates that cluster in the central zone of the vent corridor where dikes are most likely to be present, and large differences in the median values of hydraulic conductivity between the central zone and adjacent zones despite the similar average lengths of open intervals in wells. The weakness of this interpretation is the present lack of direct physical evidence for dikes within the INTEC vent corridor and an incomplete understanding of how dikes might be distributed in areas of focused volcanism in the eastern Snake River Plain.

Additional complexities are apparent in eruptive fissure systems, which typically have open segments where dikes either did not reach the surface or where magma drained back down into fissures after first reaching the surface. These fissure segments and associated tension cracks are similar to the openings of large tension fractures in basalt, and are capable of being filled with lava and sediment during burial (Welhan et al., this volume, chapter 9). Some fissures and cracks probably filled with enough lava and sediment to mimic the hydraulic effects of a dike. Other fissures and cracks probably were bridged by lava flows and rubble and retain much of their original void space. Locally, these features may affect some estimates of hydraulic conductivity.

Filled fissures and tension cracks may control some low estimates of hydraulic conductivity within the INTEC vent corridor (Fig. 12). Possible locations of this effect include the central zone of the vent corridor and near well USGS 38. This well is located in a zone of potential tension cracks and has one of the lowest estimates of hydraulic conductivity at the INEEL, 4.9×10^{-2} m/day.

Open fissures and tension cracks may control some high estimates of hydraulic conductivity, such as that in the INTEC disposal well (Fig. 12). This well also is located in a zone of potential tension cracks and has one of the highest estimates of hydraulic conductivity at the INEEL, 2.7×10^3 m/day. Well USGS 26, located within another vent corridor in the northern part of the INEEL (Fig. 12), has the highest estimate of hydraulic conductivity in this data set, 7.3×10^3 m/day. This estimate seems inconsistent with a short open interval of 10 m, and could be controlled by an open fissure or tension crack.

In conclusion, simple comparisons with Hawaii, coupled with available aquifer-test data, stratigraphic interpretations, and inferred locations of individual tension-crack and eruptive fissure systems, suggest that near-vent volcanic deposits, dikes, fissures, and tension cracks may locally control the range and distribution of hydraulic conductivity in the Snake River Plain aquifer at the INEEL. Near-vent volcanic deposits may be as permeable as the most permeable pahoehoe flows. Dikes probably control many estimates of low hydraulic conductivity of $<\sim3.0 \times 10^1$ m/day. Fissures and tension cracks may control some of the lowest and highest estimates of hydraulic conductivity. At the INEEL, these features may play an important role in the migration of radioactive and chemical wastes by providing localized preferential pathways and barriers to groundwater flow. Additional core drilling, detailed geophysical surveys, and other detailed methods will be required to refine these preliminary interpretations.

ACKNOWLEDGMENTS

We appreciate the careful and insightful reviews of this paper by Linda L. Davis, Bill Hackett, Margaret Hiza, Scott Hughes, Suzette Payne, David Sawyer, Dick Smith, Paul Wetmore, and an anonymous reader, and we thank them for their efforts. We have generally disregarded their advice to expand on certain implications of our studies for the sake of a reasonable-length paper.

REFERENCES CITED

Ackerman, D.J., 1991, Transmissivity of the Snake River Plain aquifer at the Idaho National Engineering Laboratory, Idaho: U.S. Geological Survey Water-Resources Investigations Report 91-4058 (DOE/ID-22097), 35 p.

Anderson, S.R., Ackerman, D.J., Liszewski, M.J., and Freiburger, R.M., 1996, Stratigraphic data for wells at and near the Idaho National Engineering Laboratory, Idaho: U.S. Geological Survey Open-File Report 96-248 (DOE/ID-22127), 27 p. and 1 diskette.

Anderson, S.R., Kuntz, M.A., and Davis, L.C., 1999, Geologic controls of hydraulic conductivity in the Snake River Plain aquifer at and near the Idaho National Engineering Laboratory, Idaho: U.S. Geological Survey Water-Resources Investigations Report 99-4033 (DOE/ID-22155), 38 p.

Bartholomay, R.C., Tucker, B.J., Ackerman, D.J., and Liszewski, M.J., 1997, Hydrologic conditions and distribution of selected radiochemical and chemical constituents in water, Snake River Plain aquifer, Idaho National Engineering Laboratory, Idaho, 1992 through 1995: U.S. Geological Survey Water-Resources Investigations Report 97-4086 (DOE/ID-22137), 57 p.

Covington, H.R., 1977, Preliminary geologic map of the Pillar Butte, Pillar Butte Northeast, Pillar Butte Southeast, and Rattlesnake Butte quadrangles, Bingham, Blaine and Power Counties, Idaho: U.S. Geological Survey Open-File Report 77-779, scale 1:48 000.

Delaney, P.T., and Gartner, A.E., 1997, Physical processes of shallow mafic dike emplacement near the San Rafael Swell, Utah: Geological Society of America Bulletin, v. 109, p. 1177–1192.

Delaney, P.T., and Pollard, D.D., 1981, Deformation of host rocks and flow of magma during growth of minnette dikes and breccia-filling intrusions near Ship Rock, New Mexico: U.S. Geological Survey Professional Paper 1202, 61 p.

Fiske, R.S., and Jackson, D.E., 1972, Orientation and growth of Hawaiian volcanic rifts; the effect of regional structure and gravitational stresses: Proceedings of the Royal Society of London, A-329, p. 299–326.

Greeley, R., Theilig, E., and King, J.S., 1977, Guide to the geology of the King's Bowl lava field, in Greeley, R., and King, J.S., eds., Volcanism of the eastern Snake River Plain, Idaho: A comparative planetary guidebook: Washington, D.C., National Aeronautics and Space Administration, p. 171–188.

Hackett, W.R., and Smith, R.P., 1992, Quaternary volcanism, tectonics, and sedimentation in the Idaho National Engineering Laboratory area, in Wilson, J.R., ed., Field guide to geologic excursions in Utah and adjacent areas of Nevada, Idaho, and Wyoming: Utah Geological Survey Miscellaneous Publications 92-3, p. 1–18.

King, J.S., 1977, Crystal Ice Cave and King's Bowl Crater, Snake River Plain, Idaho, in Greeley, R., and King, J.S., eds., Volcanism of the eastern Snake River Plain, Idaho: A comparative planetary guidebook: Washington, D.C., National Aeronautics and Space Administration, p. 153–164.

Kuntz, M.A., 1977, Extensional faulting and volcanism along the Arco rift zone, eastern Snake River Plain, Idaho: Geological Society of America Abstracts with Programs, v. 9, no. 6, p. 740–741.

Kuntz, M.A., 1978a, Geologic map of the Arco-Big Southern Butte area, Butte, Blaine, and Bingham counties, Idaho: U.S. Geological Survey Open-File Report 78-302, scale 1:62 500.

Kuntz, M.A., 1978b, Geology of the Arco-Big Southern Butte area, eastern Snake River Plain, and potential volcanic hazards to the Radioactive Waste Management Complex, and other waste-storage and reactor facilities at the Idaho National Engineering Laboratory, Idaho; with a section on statisitcal treatment of the age of lava flows by John O. Kork: U.S. Geological Survey Open-File Report 78-691, p.70.

Kuntz, M.A., 1992, A model-based perspective of basaltic volcanism, eastern Snake River Plain, Idaho, *in* Link, P.K., Kuntz, M.A., and Platt, L.P., eds., Regional geology of eastern Idaho and western Wyoming: Geological Society of America Memoir 179, p. 289–304.

Kuntz, M.A., Spiker, E.C., Rubin, M., Champion, D.E., and Lefebvre, R.H., 1986, Radiocarbon studies of latest Pleistocene and Holocene lava flows of the Snake River Plain, Idaho: Data, lessons, interpretations: Quarternary Research, v. 25, p. 163–176.

Kuntz, M.A., Champion, D.E., Lefebvre, R.H., and Covington, H.R., 1988, Geologic map of the Craters of the Moon, King's Bowl, and Wapi lava fields and the Great Rift volcanic rift zone, south-central Idaho: U.S. Geological Survey Miscellaneous Investigation Series Map I-1632, scale 1:100 000.

Kuntz, M.A., Levebvre, R.H., Champion, D.E., and Skipp, B.A., 1989, Geologic map of the Inferno Cone quadrangle, Butte County, Idaho: U.S. Geological Survey Geological Quadrangle Map GQ-16-32, scale 1:24 000.

Kuntz, M.A., Covington, H.R., and Schorr, L.J., 1992, An overview of basaltic volcanism of the eastern Snake River Plain, Idaho, *in* Link, P.K., Kuntz, M.A., and Platt, L.P., eds., Regional geology of eastern Idaho and western Wyoming: Geological Society of America Memoir 179, p. 227–267.

Kuntz, M.A., and twelve others, 1994, Geologic map of the Idaho National Engineering Laboratory and adjoining areas, eastern Idaho: U.S. Geological Survey Miscellaneous Investigations Series Map I-2330, scale 1:100 000.

Mann, L.J., 1986, Hydraulic properties of rock units and chemical quality of water for INEL-1: A 10,365- foot deep test hole drilled at the Idaho National Engineering Laboratory, Idaho: U.S. Geological Survey Water-Resources Investigations Report 86-4020 (IDO-22070), 23 p.

Meyer, W., and Souza, W.R., 1995, Factors that control the amount of water that can be diverted to wells in a high-level aquifer, *in* Hermann, R., Back, W., Sidle, R.C., and Johnson, A.I., eds., Water resources and environ-mental hazards: Emphasis on hydrologic and cultural insight in the Pacific Rim, Proceedings of the American Water Resources Association Annual Summer Symposium: Honolulu, Hawaii, American Water Resources Association, p. 207–216.

Morin, R.H., Barrash, W., Paillet, F.L., and Taylor, T.A., 1993, Geophysical logging studies in the Snake River Plain aquifer at the Idaho National Engineering Laboratory—wells 44, 45, and 46: U.S. Geological Survey Water-Resources Investigations Report 92-4184, 44 p.

Pollard, D.D., Delaney, P.T., Duffield, W.A., Endo, E.T., and Okamura, A.T., 1983, Surface deformation in volcanic rift zones, Tectonophysics, v. 94, p. 541–584.

Prinz, M., 1970, Idaho rift system, Snake River Plain, Idaho: Geological Society of America Bulletin, v. 81, p. 941–947.

Rubin, A. M., 1990, A comparison of rift-zone tectonics in Iceland and Hawaii: Bulletin of Volcanology, v. 52, p. 302–319.

Rubin, A.M., 1992, Dike-induced faulting and graben subsidence in volcanic rift zones: Journal of Geophysical Research, v. 97, p. 1839–1858.

Rubin, A.M., and Pollard, D.D., 1988, Dike-induced faulting in rift zones of Iceland and Afar: Geology, v. 16, p. 413–417.

Parsons, T., and Thompson, G.A., 1991, The role of magma overpressure in suppressing earthquakes and topography: Worldwide examples: Science, v. 253, p. 1399–1402.

Parsons, T., Thompson, G.A., and Smith, R.P., 1998, More than one way to stretch: A tectonic model for extension along the plume track of the Yellowstone hotspot and adjacent Basin and Range Province: Tectonics, v. 17, p. 231–234.

Smith, R.P., Hackett, W.R., and Rodgers, D. W., 1989, Geologic aspects of seismic-hazard assessment at the Idaho National Engineering Laboratory, southeastern Idaho *in* Proceedings, Second Department of Energy Natural Phenomena Hazards Mitigation Conference: Livermore, CA, Lawrence Livermore National Laboratory, p. 282–289.

Smith, R.P., Jackson, S.M., and Hackett, W.R., 1996, Paleoseismology and seismic hazard evaluations in extensional volcanic terrains: Journal of Geophysical Research, v. 101, p. 6277–6292.

Stearns, H.T., 1985, Geology of the state of Hawaii: Palo Alto, California, Pacific Books 335 p.

Wetmore, P.H., Hughes, S.S., and Anderson, S.R., 1997, Model morphologies of subsurface Quaternary basalts as evidence for a decrease in the magnitude of basaltic magmatism at and near the Idaho National Engineering and Environmental Laboratory, Idaho *in* Sharma, S. and Hardcastle, J.H., eds., Proceedings of the 32nd Symposium on Engineering Geology and Geotechnical Engineering: Boise, Idaho, p. 45–48.

MANUSCRIPT ACCEPTED BY THE SOCIETY NOVEMBER 2, 2000

Geological Society of America
Special Paper 353
2002

Morphology of inflated pahoehoe lavas and spatial architecture of their porous and permeable zones, eastern Snake River Plain, Idaho

John A. Welhan
Idaho Geological Survey, Department of Geology, Idaho State University, Pocatello, Idaho 83209, USA
Chad M. Johannesen*
Department of Geosciences, Idaho State University, Pocatello, Idaho 83209, USA
Kelly S. Reeves†
Department of Mathematics, Idaho State University, Pocatello, Idaho 83209, USA
Thomas M. Clemo
Center for Geophysical Investigations of the Shallow Subsurface, Boise State University, Boise, Idaho 83725, USA
John A. Glover§
Department of Geosciences, Idaho State University, Pocatello, Idaho 83209, USA
Kenneth W. Bosworth
Department of Mathematics, Idaho State University, Pocatello, Idaho 83209, USA

ABSTRACT

The internal vesicular structures of individual pahoehoe lava flow lobes observed in drill cores and in Holocene lava fields on the eastern Snake River Plain indicate that emplacement by inflation has operated throughout the Pleistocene and is dominant in Holocene monogenetic lava flow groups. This suggests that Holocene lava fields are representative analogs for understanding the morphology and internal elements of buried lava flows that determine spatial distribution of porosity and permeability in the eastern Snake River Plain aquifer.

Inflated pahoehoe lava flows are characterized by a hierarchical growth pattern of smaller lobes breaking out of larger lobes in a fractal pattern. Lobes <300 m long on the Hell's Half Acre and Wapi flow groups have a mean aspect ratio of ~3.2:1 (length:width) and mean length of the order of 85 m. The fractal dimension of lava flow units was analyzed with a modified box-counting method and found to be remarkably similar (1.3 to 1.6) in seven flow units from three different areas, and in flow units mapped at two very different spatial scales. These results suggest that information about lava morphology obtained from large-scale mapping can be extrapolated to understand inflationary lava flow features at smaller scales.

Our data suggest that in excess of 15% of interflow zone fracture porosity may survive burial by younger lava. Bridging by viscous lava may be the preservation mechanism, becoming effective at a mean fracture aperture threshold of ~10 cm. Because more than 50% of fractures measured on inflated lava flows have apertures of 10 cm or less, a significant amount of

*S.M. Stoller Corporation, 1780 First Street, Idaho Falls, Idaho 83401.
†Idaho National Engineering and Environmental Laboratory, Idaho Falls, Idaho, 83402.
§North Wind Environmental, Inc., 1843 W. Del Sol Lane, Yuma, Arizona 85364.

Note: Additional information for this chapter can be found in GSA Data Repository item 2002041. See footnote 1 on p. 136.

Welhan, J.A., Johannesen, C.M., Reeves, K.S., Clemo, T.M., Glover, J.A., and Bosworth, K.W., 2002, Morphology of inflated pahoehoe lavas and spatial architecture of their porous and permeable zones, eastern Snake River Plain, Idaho, *in* Link, P.K., and Mink, L.L., eds., Geology, Hydrogeology, and Environmental Remediation: Idaho National Engineering and Environmental Laboratory, Eastern Snake River Plain, Idaho: Boulder, Colorado, Geological Society of America Special Paper 353, p. 135–150.

fracture porosity may be retained after burial, particularly along the margins of inflated lava flow units where fractures are concentrated in long networks. If sufficiently interconnected, such fracture porosity could support permeable zones extending over distances of kilometers.

INTRODUCTION

The movement of ground water beneath the Idaho National Engineering and Environmental Laboratory (INEEL) in the hydrogeologically complex, layered basalt aquifer system of the eastern Snake River Plain is difficult to predict accurately at the scale of most groundwater remediation studies. The amount, type, and scale of subsurface permeability information are almost always insufficient to describe the actual spatial character of hydraulic heterogeneity that controls water and solute movement. In the eastern Snake River Plain basalt aquifer, such movement is controlled by the architecture of highly porous and permeable preferential flow zones (known as interflow zones), which occur along the contacts between individual pahoehoe lava flows. This study was conducted to develop quantitative constraints for modeling the architecture of pahoehoe lava and stochastically simulating the geologic heterogeneity of the lava-hosted aquifer system and the resultant spatial distribution of permeability.

Our approach focused on inflated pahoehoe lavas because most eastern Snake River Plain surface basalts are of this type (Kuntz et al., 1992), as are most subsurface basalts in the eastern Snake River Plain that have been cored (Welhan et al., 2002). Inflation occurs where lava is injected beneath a solidified or partially solidified crust (Self et al., 1998); growth occurs by breakouts at their margins (Hon et al., 1994) with a characteristic fractal pattern (Bruno et al., 1992; Welhan et al., 2002). In contrast, ponded lava that fills topographic depressions takes on the shape of its host basin and may also develop a different pattern of internal vesiculation. Self et al. (1998) proposed that the relative thickness of a lava flow lobe's upper vesicular zone may provide a guideline to distinguish an inflated versus a ponded origin. Using this criterion, it appears that more than 80% of these lava flows formed by inflation.

STATEMENT OF THE PROBLEM AND SCOPE

The intent of this research was (1) to quantitatively describe the geometry of inflated lava flow bodies; (2) to develop comparative data between inflated and ponded pahoehoe structure; and (3) to produce statistics and other quantitative measures of the morphology and geometric character of inflated pahoehoe lava flows and of the interflow zones they host. This information is required to describe and model the geometry of the architectural elements that host interflow zones in the east-

ern Snake River Plain aquifer, and will help to define the geostatistical structure of interflow zones in simulations of aquifer heterogeneity (see Welhan et al., this volume, Chapter 15).

This chapter builds on a conceptual framework outlined in Welhan et al. (2002). That work identified two different types of interflow zones: type-I, hosted in rubble at the predominantly horizontal contacts between lava flows; and type-II, hosted in long networks of tension fractures created by tensional stretching and rotation along the edges of inflating lava flows (Self et al., 1998). In this chapter we focus on (1) quantifying the size and shape of inflated lava flows; (2) establishing their fractal morphology; and (3) examining the spatial locations and apertures of tension fractures and their tendency to remain open after burial by lava. In addition, we present a refined conceptual model of inflated pahoehoe lava flow morphology and its impact on the spatial variability of permeability in the eastern Snake River Plain aquifer.

Three of the four study areas in which most of this work was carried out are from two of the largest Holocene monogenetic lava fields in the eastern Snake River Plain: Wapi and Hell's Half Acre (Fig. 1). Unlike the polygenetic Craters of the Moon Holocene lava field, these are thought to be representative of the great majority of low shield volcanoes composing the eastern Snake River Plain aquifer (Greeley, 1982; Kuntz et al., 1992). In addition, lava flows exposed in Box Canyon (Fig. 1) were examined for information on the characteristics of fracture filling by lava.

GEOLOGIC BACKGROUND

Most of the basalt that erupted into the eastern Snake River Plain structural trough during the past 1.4 m.y. is pahoehoe of olivine tholeiite composition (Kuntz et al., 1992). Almost all of this was derived from low shield volcanoes that characterize the plains-style volcanism of the eastern Snake River Plain (Greeley, 1982; Kuntz et al., 1992). Because of this long-lived basaltic volcanism, the aquifer is hosted predominantly in basalt, the composition and internal morphology of which is remarkably consistent, both areally and vertically. Figure 2A shows the manner in which inflated and ponded lavas are stacked; relatively thin sedimentary interbeds mark localized volcanic hiatuses.

Basalt lava flows examined in cores at the INEEL (Glover et al., GSA Data Repository[1]) have a predictable internal structure that is characteristic of inflated pahoehoe (Knutson et al.,

[1]GSA Data Repository item 2002041, Identification of basalt interflow zones with borehole geophysical and videologs at the Idaho National Engineering and Environmental Laboratory, Idaho, is available on the Web at http://www.geosociety.org/pubs/ft2002.htm. Requests may also be sent to editing@geosociety.org.

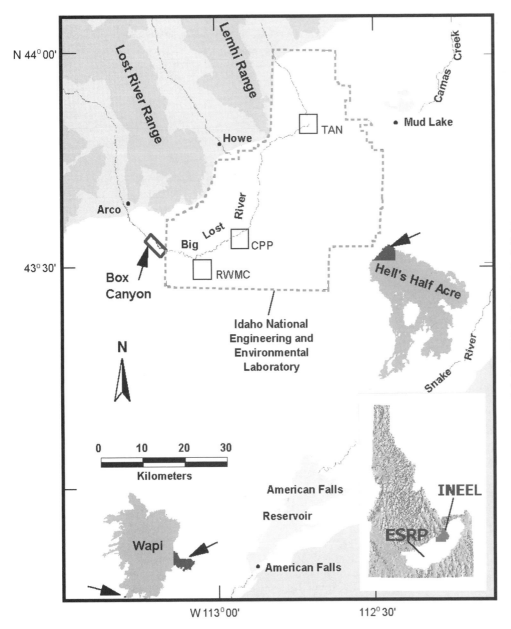

Figure 1. Location map of four study areas where mapping of lava flows was accomplished in Holocene lava fields (arrows) and in Box Canyon (rectangle). Labeled rectangles within Idaho National Engineering and Environmental Laboratory (INEEL) indicate areas within which detailed basalt core information has been examined. ESRP is eastern Snake River Plain; RWMC is Radioactive Waste Management Complex; CPP is Chemical Processing Plant; TAN is Test Area North.

1990; Self et al., 1997; Thordarson and Self, 1998). As shown in Figure 2B, inflated lava flows have a relatively massive flow interior of low permeability and vesicular zones at the upper and lower surfaces. The upper vesicular zone, the fractured upper entablature, and rubble and broken material at flow contacts, compose a type-I interflow zone (Welhan et al., 2002).

Self et al. (1998) postulated that the inflation mechanism of pahoehoe emplacement produces lava flows that have variable dimensions but display similar geometries at different spatial scales; e.g., thickness tends to increase with the lateral dimension of lava flows, and the areal geometry is similar at large and small scales. This is consistent with the findings of Bruno et al. (1992), who found that the geometry of the margins of

Hell's Half Acre lava field is fractal or self-similar, suggesting that the growth process of lava fields produces features with similar geometric characteristics over a variety of dimensional scales. This is evident in the similarity of mapped lava flows at large and small scales, shown in Figure 3.

Figure 3 also shows that tension fractures tend to develop along the edges of inflated pahoehoe lava flow units in a manner consistent with the inflation model summarized by Self et al. (1998). Tension fracturing occurs due to radial rotation of brittle crust on the surface of an inflating lava flow lobe. Rotation should be greatest in small inflating lobes and on the shoulders of larger flow units, and least on the relatively flat, plateau-like surfaces of large flow units.

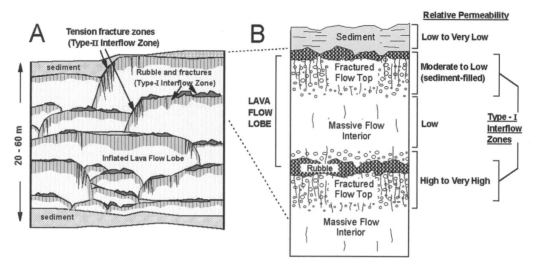

Figure 2. A: Representation of aquifer stratigraphy, showing predominantly inflated pahoehoe lava flows characterized by rubbly tops and intensely fractured upper surfaces, tension fractures, and sedimentary inter-beds. Modified from Welhan et al. (2002). B: Internal structure of inflated pahoehoe lava flows on eastern Snake River Plain; after Lindholm and Vacarro (1988) and Welhan et al. (1997).

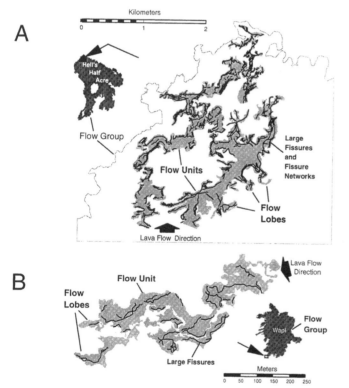

Figure 3. Comparison of subaerial shapes of inflated lava flow units mapped at two different scales. A: Large-scale lava flow units mapped from stereo aerial photo pairs, northern tip of Hell's Half Acre. B: Small-scale lava flow unit mapped by global positioning system on ground, on southwestern margin of Wapi flow group.

Terminology

Perhaps because of this scale-independent characteristic, some confusion in terminology has arisen among different workers describing similar features at different scales in inflated pahoehoe (e.g., the term lava flow applied to kilometer-scale features as well as to small-scale bodies, and the interchange-able use of terms such as lobes and toes for similar features). We use an internally consistent, scale-dependent terminology originally proposed by Welhan et al. (1997) and refined by Welhan et al. (2002) to describe lava bodies at the outcrop scale (flow lobes), the aerial photo scale (flow units), and the regional scale (flow groups and members).

A lava flow group is synonymous with a lava field, the low shield volcano that is characteristic of plains-style volcanism (e.g., Wapi or Hell's Half Acre in Fig. 1). Lava flow groups are kilometers to tens of kilometers in areal dimension and are the large-scale building blocks composing the thick and laterally extensive lava flow members (synonymous with the term su-pergroups used by Welhan et al., in press), which are vertical and lateral aggregates of multiple flow groups mapped in the subsurface (Anderson et al., 1996; Welhan et al., 1997). At a smaller scale, lava flow groups are built up from numerous, individual lava flow units derived from a single vent area; these are mappable at scales of hundreds of meters and greater. In-flated lava flow units, in turn, can be subdivided into a hierarchy of smaller inflated bodies, which we collectively call lava flow lobes. We do not use the term toes, but rather equate these to the smallest scale of flow lobes.

As shown in Figure 3, inflated lava flow lobes and their larger parent bodies, lava flow units, are identifiable at scales from tens to thousands of meters. The morphological terminology is consistent at both scales; however, the distinction between a lava flow unit at the smaller scale (Fig. 3B) and lava lobes at the larger scale (Fig. 3A) is arbitrary. For example, a lava flow body hosting an internal, insulated feeder conduit from a lava source would be distinguished as a flow unit, distinct from the flow lobes that were spawned from it. However, a flow unit may have been spawned from other, larger scale flow units. Intuitively, one can grasp how this geometry could be fractal.

A distinction between flow units and lobes has been retained, however, because it is useful for mapping exposed lavas at the surface. The entire two-dimensional morphology and the temporal hierarchy or sequence of lobe and flow unit growth can be deduced in surface exposures of lava flows. In cross-sectional outcrops and drill cores, however, the term flow unit is dropped because we lack information relating flow lobes in time or lateral space.

Ponded lava flows

In addition to forming inflated bodies, pahoehoe lava also ponds in topographic basins created between other lava flows or existing topography. Self et al. (1998) observed that the proportion of lava flow thickness composing the vesiculated upper zone compared to total lobe thickness is much smaller in ponded flows, and suggested that this ratio might be an indicator of the mode of emplacement. Applying a threshold proportion of 30% of lobe thickness to measurements of vesiculated zones observed in basalt cores of late Pleistocene age, ~80% of 396 lava flows logged in the INEEL subsurface appear to be of an inflated origin (Welhan et al., 2002). If it is assumed that basalt emplacement mechanisms in the Holocene were the same as in the Pleistocene, Holocene basalts also are expected to be dominated by an inflated origin.

This inference is consistent with the apparent paucity of ponded lava flows found in two Holocene flow groups in the eastern Snake River Plain (Johannesen, 2001). Because of their different modes of origin, it is hypothesized that ponded lava flows support substantially different degrees and styles of fracturing and different amounts and types of porous material on their surfaces compared to inflated lava flows. However, because ponded lavas appear to compose a relatively minor component of the aquifer, this communication focuses on inflated lava morphology. For more information on the few ponded flow units that have been mapped during this study, see Johannesen (2001).

Tension fracture networks

D. Champion (1997, personal commun.) posed a potentially significant question based on earlier work on eastern Snake River Plain basalt geology; i.e., whether the character-

istic tension fracturing observed on inflated flow could be hydraulically important in an aquifer co of fracture porosity by fine-grained sediment longed volcanic hiatus is certainly expected to reduce the permeability of fractures and fracture networks. However, most of the eastern Snake River Plain aquifer is composed of overlapping lava flows in which overlying lavas cover the fractures of older lava flows. The high spatial density of tension fractures along the edges of lava flow units suggests they could form important, large-scale conduits for groundwater flow if their primary porosity is retained during burial by younger lava (Welhan et al., 2002).

We are unaware of any previous work documenting whether fractures remain open following inundation by younger lavas. Knutson et al. (1990, 1992) alluded to the possibility that the high viscosity of lava causes it to bridge across large pore spaces in basalt rubble without destroying much of the primary rubble porosity. It seems reasonable to ask whether lava also may bridge across certain sizes of fractures, thereby preventing their filling and preserving fracture porosity after burial.

APPROACH AND RESULTS

Vertical lava flow geometry

A total of 396 individual lava flow lobes in almost 2700 m of basalt core were logged from two different areas on the INEEL (Glover et al., GSA Data Repository*).

This information was used to provide limitations on basalt lava flow thicknesses and internal flow structures. Table 1 summarizes these results and compares them with those of Knutson et al. (1990) from a third area of the INEEL. The internal structure of lava flow lobes in all areas sampled by coring can be grouped into four distinct elements that correspond to the structure shown in Figure 2B: (1) an upper vesicular zone and (2) a central massive zone, both composing relatively uniform average percentages of total lava flow lobe thickness (35%–50% and 44%–49%, respectively); (3) a lower vesicular zone, of more uniform absolute thickness (averaging <0.5 m, regardless of flow lobe thickness); and (4) a very highly porous zone of rubble and broken rock at the contact between lava flow lobes.

Table 1 shows that drill cores from different parts of the INEEL reveal remarkable consistency in the vertical geometry of these structural elements in lava flow lobes. The only appreciable differences between areas are the proportion of lava flow contacts having rubble and the average thickness of rubble. Although the numbers are very similar in the Test Area North (TAN) and Chemical Processing Plant (CPP) cores, the occurrence frequencies of rubble and mean rubble thickness in core from the Radioactive Waste Management Complex (RWMC) appear to differ substantially from those in the other two areas. This may be due to differences in core recovery between the

---

*See footnote 1, p. 136.

TABLE 1. DATA SUMMARY FOR CORE LOGGED FROM THE TEST AREA NORTH (TAN), CHEMICAL PROCESSING PLANT (CPP), AND RADIOACTIVE WASTE MANAGEMENT COMPLEX (RWMC)

	TAN* 7 wells	CPP* 6 wells	TAN + CPP* 13 wells	RWMC# 15 wells
Total core logged (m)	1030	1157	2187	884
Total number of flow lobes	194	202	396	114
Number of quantified lobes	167	188	355	114
Lobe thickness (m)	4.9	5.1	5.0	
Median lobe thickness (m)	3.3	3.1	3.2	3.5§
Thickness of upper vesicular zone (percent of lobe)	45.4	52.8	49.3	35.0
Thickness of massive basalt (percent of lobe)	48.4	41.8	44.9	49.0
Thickness of lower vesicular zone (percent of lobe)	6.3	5.4	5.8	11.0
Lower vesicular zone thickness (m)	0.25	0.20	0.22	0.45
Percentage of lobe contacts with rubble	43.2	42.3	42.7	11.4
Average rubble thickness (m)	0.46	0.69	0.58	
Median rubble thickness (m)	0.3	0.3	0.3	1.2
Total number of interbeds	14	26	40	20
Average thickness (m)	1.51†	4.55	5.38	n.a.

* Summarized from data of Glover et al. (Data Repository, this volume); includes well 2-2A outside the immediate TAN area.
# After Knutson et al. (1990).
† Within the immediate TAN area, only; 6.93 m including well 2-2A.
§ Recalculated from data in Table C-7, Appendix C, of Knutson et al. (1992) without normalization.

areas, although we have no information to substantiate this. Alternatively, differences in classification or nomenclature may be responsible. Knutson et al. (1990) variously referred to this structural element as a "collapse/rubble zone" and "complex vesicular zone"; this terminology implies that their classification may include fractured and broken rock within the uppermost part of a flow lobe as well as rubble and scoria at the flow contact. In our description of core features, we attempted to consistently classify broken material separately from rubble at lobe contacts, although in many cases the distinction was difficult to make in the core sample.

Overall, the results in Table 1 indicate that vertical lava flow geometry is remarkably consistent in the eastern Snake River Plain. Welhan et al. (1997) pointed out that comparisons of young and old lava flows represented in the 550-m-deep core hole C1A and in different INEEL boreholes intersecting different ages of lavas showed that the internal structure of lava flow lobes has remained unchanged over time. This implies that basalt lavas have been emplaced by similar mechanisms and under similar conditions over at least the past 1.4 m.y., the age of the oldest basalt dated from the eastern Snake River Plain aquifer (Anderson and Bowers, 1995).

Thickness and volumetric proportion of interflow zones

An important question concerning interflow zone geometry is their effective thickness. Interflow zones compose parts of several internal elements of a flow lobe, including rubble, jointed and fractured rock, and at least part of the vesicular zone (Fig. 2B). However, the thickness over which intense fracturing and jointing occur within the upper colonnade has not been quantified either in core descriptions or outcrop mapping be-

cause of the limitations imposed by sampling bias; only approximate estimates can be made. A minimum thickness is derived from the assumption that rubble composes the most important part of an interflow zone; this is a minimum estimate because core retrieval tends to be lowest in rubble zones and because other structural elements also compose the interflow zone. With this assumption, interflow zones have a minimum thickness of ~0.3 m (mean thickness of rubble observed in core).

An upper limit can be estimated from the thickness of the upper vesicular zone as one of the thickest elements contributing to the interflow zone. As Knutson et al. (1990) showed, and the data of Table 1 summarize, the thickness of the upper vesiculated zone varies with the thickness of the flow lobe, averaging ~45% of lobe thickness. It is similar to the proportion observed in other pahoehoe lavas, leading to the generalization that this is a characteristic of inflated pahoehoe lava flows (Self et al., 1998). Because the average flow lobe thickness is ~5 m (Table 1), the maximum average thickness of interflow zones, if they include all of the upper vesiculated zone, could be as much as 2.5 m (rubble and upper vesicular zone thicknesses combined).

However, based on our observations in basalt cores, the vesicular zone is usually not densely fractured except in the uppermost meter or less of the flow lobe. From outcrop observations, columnar jointing often extends throughout the upper vesiculated zone, but the most densely jointed and fractured rock usually appears in the uppermost ~1 m. Although we have no quantifiable measures of the thickness of the most densely fractured part of the upper vesiculated zone, a value of ~1 m appears to be realistic.

On this basis, it appears that the effective thickness of interflow zones is $> \sim 0.3$ m (rubble, only) and no greater than ~ 2.5 m overall (entire vesicular zone plus rubble); if interflow zones owe their permeability to rubble and the highly fractured upper crust, then it is < 2.5 m thick. For the sake of discussion, we adopt the approximate midpoint of these estimates, or ~ 1–1.5 m.

Lava flow lobe dimensions and aspect ratios

Data acquisition. The mapping approach and methods were outlined in Welhan et al. (2002) and Johannesen (2001). Field work focused on selected parts of Wapi and Hell's Half Acre (Fig. 1). Three areas in these lava fields were chosen as representative of the distal parts of flow groups, in which the morphology of flow units and flow lobes has not been obscured by later flows. Mapping of lava morphology was initially guided by a detailed geologic map of a 4.8 km^2 area of the Wapi eastern salient (Champion, 1973). These data were incorporated in a Geographic Information Systems (GIS) format and referenced to scanned aerial photographs to serve as a high-resolution training image. With this information, an aerial photo mapping approach was developed to identify and map lava flow units and large tension fractures from stereo photo pairs. This interpretation protocol was applied to selected areas of flow groups to map flow units and flow lobes, and checked with field traverses. Flow lobes and flow units were mapped as described in the following section. Using aerial photos, tension fractures $> \sim 1.5$ m in width and flow lobes $> \sim 30$ m in areal dimension were mappable.

A third area, on the southwestern tip of the Wapi flow group, was mapped by walking out the perimeter of a smaller scale lava flow unit with a Trimble GeoExplorer global positioning system (GPS) unit. The data were postprocessed with simultaneous base station GPS data obtained from an Idaho Falls location. The positional resolution of mapped features obtained with this approach is more than an order of magnitude improvement over that obtained with aerial photo mapping.

Visual identification of lava flow lobes. As Self et al. (1998) and others pointed out, identification of individual flow lobes and flow units can be very difficult in the field. Lava flow lobes in inflating pahoehoe form by a process of cracking of brittle lava crust and extrusion of molten lava with development of a cooling skin. The lobes generated, like the parent inflated lobe, also may grow in response to continued lava supply and spawn successive generations of lobes (Hon et al., 1994; Thordarson and Self, 1998; Self et al., 1998). The result is a spatial and temporal hierarchy of lava lobes with a characteristic geometry, as shown in the example taken from mapping work of Hon et al. (1994) during the growth of an inflating flow unit (Fig. 4).

The morphological expression of some lobes can be destroyed during inflation by coalescence of hot, plastic lobe margins, which leaves little or no trace of the original geometry of coalesced lobes (Hon et al., 1994; Self et al., 1998). For ex-

ample, in Figure 4 the lobes visible at 6 and 27 h eventually coalesced and were obliterated. At the risk of generalizing from a limited example, it appears that only the latest generation of individual lobes can be reliably identified by their characteristic lobate shapes. The areal dimensions of these lobes were estimated in several ways: (1) by tedious, visual approximations of length and average width (Fig. 4; Knutson et al., 1990; Welhan et al., 2002); (2) by a method of recursive calculation of an equivalent rectangular shape, the area and perimeter of which are equivalent to those of the lobe polygon; and (3) by least-squares fitting of an ellipse to the points defining the lobe polygon boundary.

Using method 1, Knutson et al. (1990) estimated the geometries of inflated lava flow lobes from outcrops in Box Canyon and Hell's Half Acre and found that individual lobes ranged in thickness from 1 to 21 m (in good agreement with the range of lobe thicknesses measured in drill cores); their median length: width: height aspect ratio was $> 8.7 : 4.6 : 1$.

An example of a small-scale Holocene flow unit mapped with GPS is shown in Figure 5. The identified lobes are highly subjective and are based primarily on an expectation that the inflation process created a hierarchy of lobes that is approximately defined by the current geometry. Comparison of Figures 4 and 5 suggests that many of the larger interior lobes shown are purely subjective guesses and perhaps should not even be classified as lobes as much as they are all part of a single lobe equivalent to the interior of the flow unit. Nevertheless, to remain consistent with the model of a temporally evolving hierarchy of inflated lobes, we have chosen to retain these larger interior lobes in order to better estimate the actual distribution of lobe sizes that results from the inflation process. The essential caveat is that all these lobe picks are approximations, because the actual hierarchy cannot be deduced accurately without knowledge of the temporal evolution of the flow unit.

For modeling purposes, estimating length:width (L:W) aspect ratios of lobes by method 1 is neither practical nor desirable. As shown in Figure 4, this is a tedious process of geometric approximations wherein the lengths and widths of individual lobes are estimated manually. An automated method for producing quantitative comparative measures between modeled and actual lava morphologies is preferable. Two alternative methods were developed based on geometric approximations. Although the geometric calculations are automated, both methods still rely on manual identification and specification of lobe boundaries and hence are only as reliable as the visually identified lobes.

Visually identified flow lobes were digitized and their areas and perimeters calculated by the Geographic Information System (GIS). Method 2 uses an iterative approximation to calculate an equivalent rectangle with the same area and area/perimeter ratio as the lobe, thereby defining an approximate equivalent length and L:W aspect ratio for each lobe.

Estimates derived by method 2 are summarized in Figure 6. As pointed out earlier, the dimensions of the largest lobes

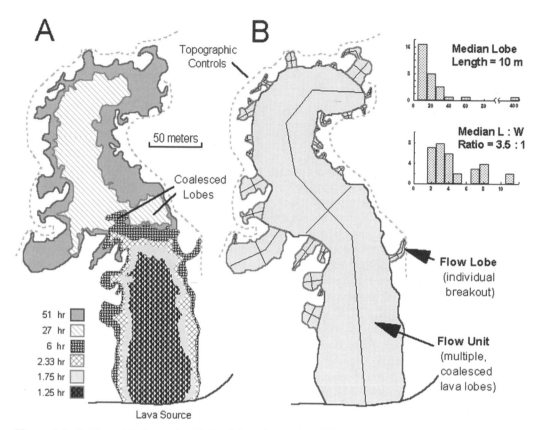

Figure 4. Left: Hierarchical growth of inflated lava flow unit on Kilauea, Hawaii, as mapped by Hon et al. (1994). On right, approximate visual identifications of flow lobes based on final shape of flow unit. Dimensions of flow lobes are shown estimated with a manual approximation method.

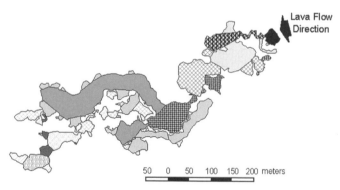

Figure 5. Example of visually picked lava flow lobes on lava flow unit shown in Figure 3B, showing apparent spatial and temporal hierarchy of lobe growth, as estimated from shape of flow unit. Different fill patterns represent temporal progression of lobes created during inflationary growth.

identified within the core of a large lava flow unit are probably overestimated due to coalescence of lobes during emplacement. On this basis, the results reported in Figure 6 and Table 2 are only for lobes <300 m long. Even with this correction applied, L:W aspect ratios in individual lobes vary widely, from 1:1 to >10:1. However, median L:W aspect ratios in four different study areas were remarkably similar (>2:1 to 4:1) and approximately normally distributed. The median of all available L:W estimates from all mapped areas of Wapi and Hell's Half Acre for lobes <300 m long is 3.2:1; the mean lobe length is 85 m.

The third method of estimating lobe dimensions uses an elliptical geometric approximation, by least-squares fitting of an ellipse with an area and perimeter equal to the GIS-derived values to the lobe boundary. A comparison of the elliptical L:W aspect ratio approximation versus the rectangular L:W approximation in Figure 7 reveals that the two methods produce estimates of aspect ratio that are statistically indistinguishable at the 95% confidence level (based on a nonparametric Kolmogorov-Smirnov test).

It is interesting that lobe aspect ratios estimated by either approximation are near-normally distributed, whereas lobe lengths are strongly skewed. Although this result may be real, it is also possible it is an artifact of the lobe censoring, as discussed previously. Without additional information, coalesced lobes could be mapped erroneously as a single large lobe, the apparent length of which would be unrepresentative of the length to which inflating pahoehoe lava extrudes before lobe branching occurs.

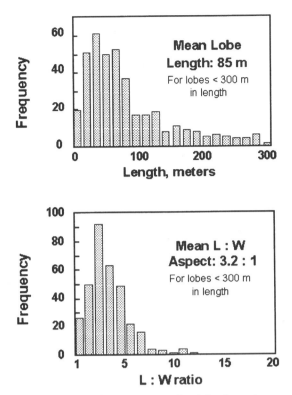

Figure 6. Summary of all Holocene lava flow lobe dimensions mapped on northern Hell's Half Acre and eastern Wapi lava fields, as estimated with rectangular approximation (method 2 in text) of visually identified lobes. Only lobes <300 m long are shown, to minimize effect of bias caused by censoring of coalesced lobes.

Automated lobe identification

The methods of quantifying lobe dimensions are not particularly useful for automated computations without the ability to identify lobes through objective criteria. In an attempt to eliminate the subjectivity of visual lobe identification, an algorithm was developed to automatically identify lobes on a mapped flow unit. With a polygonal representation of the flow unit, the algorithm uses an iterated, three-part process of identification, removal, and reattachment to partition the flow unit into lobes similar to those identified by visual criteria.

Given a polygonal representation of a flow unit boundary, the identification of flow lobes is automated in an iterative process of identifying and removing all protrusions from the flow unit boundary that satisfy particular criteria. During one iteration, all of the protrusions that can be identified are cut away and the remaining portion of the polygon is again subjected to the identification process. This continues until no further protrusions can be detected using the operational criteria. Once this is accomplished, some protrusions that had been cut into smaller pieces during multiple iterations are reconnected based on hierarchical comparisons, to provide an improved estimate of protrusions that represent lava flow lobes.

Figure 8 summarizes the lobe identification process. The cuts that remove tentative flow lobe candidates are identified by line segments (solid lines in Fig. 8, A, B, and C) joining two nodes of the flow unit polygon. There are four criteria used to identify a protrusion. (1) A protrusion may only be removed by cutting the polygon with a line segment joining two nodes that are part of separate reentrants in the polygon boundary. If a node can be used as an endpoint for more than one cutting line segment, the shortest line segment is used. (2) A protrusion must be at least 90% of its convex hull (smallest bounding convex polygon). This helps to avoid removing portions of the flow unit boundary that have more than one protrusion or a nested structure. (3) A protrusion must extend outward at least 80% of the length of the line segment used to separate it from the rest of the polygon. (4) If a protrusion is found to be a subset of a larger protrusion, it is not cut as a separate protrusion and the larger one is accepted.

The first generation of lobe candidates identified by these criteria is shown removed from the flow unit in Figure 8B; the process is repeated to allow identification of successive generations of lobes at successively larger scales (Fig. 8C). Once all of the protrusions have been identified in an iteration, they are checked against those from previous iterations.

Hierarchical criteria are used to rejoin partitions that may have been created as artifacts of local geometry. If a candidate from a later iteration shares a line segment with exactly one candidate from an earlier generation, the two candidates are joined. For example, the first generation of lobe candidates (Fig. 8D) is reattached to the second generation (Fig. 8, E and F); the third generation of lobe candidates (Fig. 8G) is not reattached because the latest defined lobe candidate shares line segments with more than one earlier lobe candidate. If it is found that a protrusion has a side common with no more than one other from a previous iteration, the two protrusions are rejoined.

The process was refined iteratively by comparison with visually identified flow lobes from the three Hell's Half Acre flow units. It was subsequently tested on the three Wapi flow units and performed reasonably well in comparison to subjective visual identification of lobes. Figure 9 shows a comparison of lobes determined by this algorithm and by visual analysis of one of the Hell's Half Acre flow units. Figure 10 summarizes the statistical comparison of lobe lengths and L:W ratios derived by the algorithm and by visual identification for all mapped flow units in Hell's Half Acre. Overall, many lobes identified by the two methods are similar. Compared to the visual estimation method the algorithm identified 15% more lobes, with a 22% lower mean length and 25% lower L:W aspect ratio. Although the histograms of the resulting lobe length distributions are statistically similar, those of the aspect ratio distributions are not. However, in light of the subjectivity of the visual lobe identification method, the degree of similarity is encouraging.

TABLE 2. SUMMARY OF FRACTURE-FILL DATA, BASED ON 28 MEASURED FRACTURES

	Total fracture depth (cm)	Depth of open fracture (cm)	Open depth (%)	Maximum surface aperture		Maximum bridging aperture in open fissures
				All fissures	Open fissures	
Median	221	27	17.1	24	24	12
Mean	226	59	26.8	37	32	10
Standard deviation	91	86	33.5	32	30	5

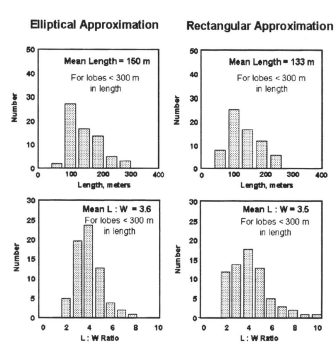

Figure 7. Comparison of lobe geometry estimates by rectangular (method 2 in text) and elliptical (method 3) geometric approximations for Hell's Half Acre lobes <300 m in long.

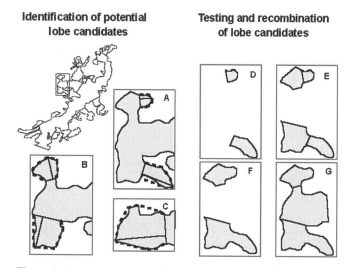

Figure 8. Graphical demonstration of multistep algorithm developed for automated identification and segregation of lava flow lobes from lava flow unit. Dashed boundaries represent smallest convex polygons containing lobe candidate, used to identify potential lobe candidates.

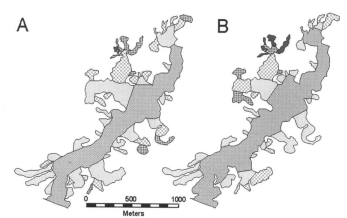

Figure 9. Comparison of visually picked (A) and algorithmically picked lava flow lobes (B) on one lava flow unit from Hell's Half Acre.

Analysis of fractal geometry of lava flow units

As is evident in Figure 3, the shape of flow units in plan view is similar at scales of tens, hundreds, or thousands of meters. Bruno et al. (1992) showed that a portion of the outer margin of the Hell's Half Acre flow group has a fractal dimension of 1.26. The fractal dimension is valued between 1.0 (for a simple nonfractal line) and 2.0 (a plane); values approaching 2.0 represent a curve so tortuous that it begins to fill the plane. The fractal dimension can be quantified by several methods, although that known as box-counting (Feder, 1988) is most easily adapted for analysis of digital map data. All inflated pahoehoe flow units that were mapped in this work have a fractal geometry that can be quantified by a fractal dimension.

Figure 11A summarizes the box-counting method. It involves overlaying a grid of square blocks of size r on the shape to be analyzed and finding the minimum number of grid blocks that contain a portion of the shape's edge. The process is repeated for different sizes, r, of grid blocks and the number of blocks, $N(r)$, is plotted against r in a double-log plot. The negative of the slope of a line through the plotted points is an approximation of the object's fractal dimension, D.

The fractal dimension should be invariant whether the set being measured is translated, rotated, or rescaled. However, for irregular objects such as lava flows such transformations can affect the fractal dimension slightly (Fig. 11, B and C), because the determination of the minimum number of grid blocks

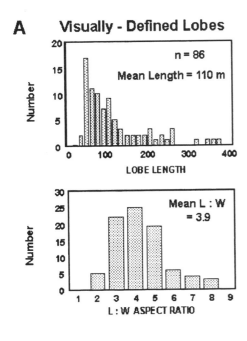

A   **Visually - Defined Lobes**

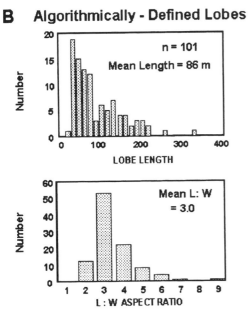

B   **Algorithmically - Defined Lobes**

Figure 10. Statistical comparison of dimensions of lava flow lobes identified visually (A) and by multistep algorithmic approach (B), for all Hell's Half Acre lava flow units (including those >300 m long). Elliptical geometric approximation (method 3 in text) was used to estimate lobe dimensions in both cases. Kolgomorov-Smirnov nonparametric test applied at 95% confidence level shows that distributions of lobe lengths are statistically indistinguishable, but that distributions of aspect ratio are significantly different.

containing a portion of the flow unit is not trivial. In order to better approximate the minimum box count, the standard box-counting technique was modified by plotting only the minimum $N(r)$ value at each box size to obtain the best approximation of D.

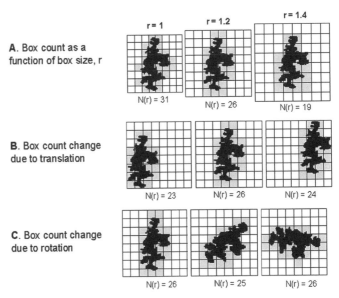

Figure 11. Implementation of box-counting method to determine fractal dimension of irregular lava flow unit shape. A: Minimum number of boxes required to cover boundary of shape is function of dimensions, r, of box. Modified box-counting method involves determining minimum number of boxes in each of several iterations using slightly different translations and rotations of grid (B and C), and plotting minimum $N(r)$ value found for each box size, r.

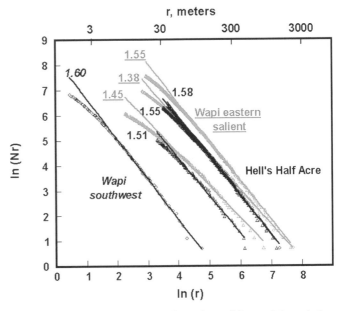

Figure 12. Comparison of fractal dimensions of flow unit boundaries, computed with modified box-counting method, for flow units in northern Hell's Half Acre, Wapi eastern salient, and Wapi southwestern sites.

The outlines of mapped flow units were analyzed with the modified box-counting algorithm. Figure 12 summarizes the fractal dimension results for all the large-scale inflated pahoehoe flow units mapped in Hell's Half Acre and the eastern salient of the Wapi lava field. The fractal dimension of the six kilometer-scale flow units in Hell's Half Acre and the eastern salient of the Wapi lava field varies between 1.3 and 1.6 with a mean of 1.48.

The relationship between ln(r) and ln[N(r)] can deviate from a straight line due to loss of resolution from mapping uncertainty and/or digitization error. This perceptibility limit (Barton and Scholz, 1995) varies from ~40 m to 120 m for the aerial photo mapped, large-scale flow units. The most detailed data on flow unit shape was obtained from the GPS survey of a 400-m-long flow unit in the southwest of the Wapi lava field (Fig. 3B), in which aerial photo map digitization error of ~30 m was replaced by GPS error of ~2 m and aerial photo interpretation error was eliminated by mapping the flow unit directly on the ground. The perceptibility limit of this data set is correspondingly smaller ~6 m (Fig. 12).

Johannesen (2001) found that the large-scale tension fractures mapped from aerial photos (with apertures >~1.5 m) also have a fractal dimension (with similar D values ranging from 1.36 to 1.59). This might be expected because these fractures tend to congregate along the edges of flow units (Fig. 3) that have fractal shapes.

Despite the loss of information due to digitizing and mapping errors, the good agreement between fractal dimensions estimated from large-scale mapping data and small-scale ground measurements indicates that inflated lava flow units are fractal over several orders of magnitude of spatial scale. This suggests that information derived from mapping of large-scale lava features from aerial photos can be used to extrapolate to smaller scale lava flow features, below the resolution of mapping.

Fracture-filling and fracture aperture distribution

To our knowledge, the hydrogeologic importance of large-scale tension fractures that develop on inflated pahoehoe has not been addressed by previous work on the eastern Snake River Plain. Welhan et al. (2002) proposed these may host high permeability (type-II interflow zones) if fracture porosity is preserved during burial by younger lava. In this chapter we address three specific questions. (1) Is fracture porosity in the upper surfaces of inflated lava preserved during burial? (2) Is there a characteristic fracture aperture that is preserved? (3) Are these apertures common on inflated lava flows? A fourth relevant question, i.e., whether the unfilled fracture porosity remains sufficiently interconnected to establish important conduits for preferential groundwater flow, was not addressed in this study.

At Box Canyon (Fig. 1), the Big Lost River canyon exposes at least two layers of lava flow lobes in vertical cross section. This section is essentially a surface analog of a lava flow member, because it contains one or more laterally contiguous flow groups overlying one or more older flow groups without intervening sedimentary interbeds. As shown in Figure 13, the canyon walls provide good exposures of fractures in the

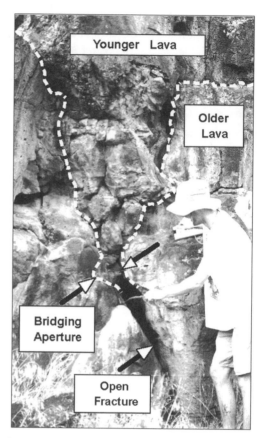

Figure 13. Example of partially filled, large-aperture fracture in buried lava flow lobe in Box Canyon, showing contact between younger and older lobes, massive lava filling fracture, and meter-long section of open fracture below open or bridging aperture.

lower flow lobes where they are covered by the upper flow lobes.

Fractures were located visually during traverses along ~5 km of canyon length. All fractures were photographed, and various measurements of fracture length, fracture aperture, fracture filling, and the nature of the contact between upper and lower lobes were recorded (Johannesen, 2001). Table 2 summarizes relevant information.

Of 28 fractures identified, 11 were completely filled by lava, 14 were partially filled, and 3 were completely open. Thus, a substantial proportion of fractures (61%) retained some or all of their original porosity after burial. Fracture apertures to 15 cm remained unfilled by lava. The fracture aperture below which lava filling did not occur (the bridging aperture) varied from 3 to 15 cm; it is approximately normally distributed with a mean of 10 cm and a standard deviation of 5 cm.

Two caveats should be noted. Because these fractures are exposed in a river canyon, we have no way to rule out the possibility that they were partly filled with sediment at the time of burial and that their sediment was subsequently eroded away. However, none of the lobe contacts (even those with rubble) showed evidence that sedimentation occurred in the time interval between the two layers of basalt. Selective plucking of fracture filling in the narrowest parts of fracture also could have biased the data. However, clear demarcations in texture, coloration, and pattern of microfractures between the host rock and lava infilling the fractures indicated that if fluvial erosion of basaltic fracture filling had occurred, rubbly material and not massive infill was probably removed. Additional evidence for open fractures has also been found in partially filled tension fractures exposed in road cuts at another location on the eastern Snake River Plain where erosion is not a factor (Johannesen, 2001).

The second cautionary note is that this sample set represents mostly larger fracture apertures because these were most readily identifiable in outcrop. The smallest fracture aperture identified at a lobe surface was 3 cm, and the largest was 106 cm. If the mean bridging aperture is 10 cm, then substantially more fractures with smaller apertures would be expected to remain open, and an even higher proportion of fracture porosity could be preserved than is represented in this small sample.

To determine the proportion of tension fractures with surface apertures greater and less than 10 cm, we carried out a systematic survey of aperture size distribution on 7 representative inflated lava flow units and lobes of different sizes in the Hell's Half Acre study area. Figure 14 shows the manner in which samples were collected along transects across the flow units and lobes.

Figure 15 shows the number of fractures of different aperture found on the edges and in the central portions of inflated lava flows. These results corroborate the distribution of large-aperture (>1.5 m) fractures mapped from aerial photos: a very large majority of tension fractures occur at the edges of inflated lava flows (e.g., Fig. 3).

Figure 16 shows a cumulative histogram of more than 1000 aperture measurements. More than half of these tension frac-

tures have apertures of 10 cm or less. Therefore, if the estimate of mean bridging aperture (10 cm) found in Box Canyon fractures is representative, most of the fracture porosity will be preserved during burial by younger lava. On the basis of these data sets, Johannesen (2001) estimated that at least 15% of total original fracture porosity in these Holocene lavas would be preserved if they were buried by lava.

Our data support Welhan et al.'s (2002) hypothesis that networks of tension fractures at the margins of inflated lava bodies impart a significant and heretofore unrecognized amount of porosity to the aquifer. If this fracture porosity is substantially interconnected, either through fracture to fracture com-

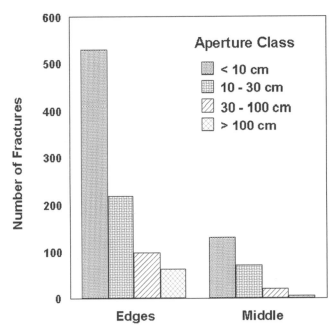

Figure 15. Comparison of number of fractures at edges and in middle of inflated lava flow units, for different fracture aperture classes.

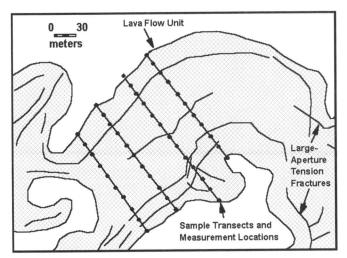

Figure 14. Arrangement of fracture sampling locations across inflated lava flow unit that was used to quantify spatial distribution and aperture sizes of tension fractures.

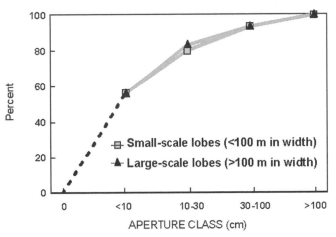

Figure 16. Cumulative frequency distribution of fracture aperture classes in lava flow units mapped on Hell's Half Acre.

munication or indirectly via rubble and type-I interflow zones, then highly permeable type-II interflow zones could extend hundreds of meters to kilometers along these fracture networks (e.g., Fig. 3).

Three-dimensional geologic model of eastern Snake River Plain stratigraphy

A preliminary version of a numerical simulator was developed to demonstrate how geologically realistic models of plains-style volcanism and subsurface geology could be created. Lava flows were modeled by simulating the growth of flow lobes, the creation of new lobes from existing lobes, and the inflationary mechanism. Superposition of multiple episodes of lava extrusion and inflationary growth created a spatial architecture, whose geometry and spatial characteristics can be constrained to match empirical data and mimic the temporal growth of the basalt pile.

The internal structure of the flow units (e.g., vesicular zones, rubble zones) was superimposed after the three-dimensional stratigraphy was defined. The accumulation of multiple lava flows from sequential eruptive sources, combined with a model of sedimentary interbed development during a volcanic hiatus, created a three-dimensional model of the resulting subsurface stratigraphy and volcanic structure (Fig. 17).

The simulator does not model the physics of the lava flow process. Instead, rules are used to cause the simulator to mimic the forms that tend to develop from inflationary pahoehoe flows. For example, a set of rules defines where and when vents

appear, and where new flow lobes may break out from an existing lobe. We cannot, at this point, select one set of rules that will result in simulated geometries that are similar to the Snake River Plain subsurface. Therefore a wide range of rules has been included in the simulator. The selection of the best rules and associated parameters that create morphologies similar to eastern Snake River Plain basalts is left to an optimization routine (see Clemo and Welhan, 2000, and Welhan et al., this volume, Chapter 15, for details).

While the rules defining evolution of a lava flow are intended to mimic the morphologies of lava flows and not the physics of the process, the physics and the rules must be related. As the results of the simulations are compared to the eastern Snake River Plain, refinements of the rules and equations defining the simulator will gradually take on the form of the physics of the processes. One example of this is ponded lava. The initial simulator did not have the capability of filling a depression until the lava overtopped a saddle and it became obvious that such a capability had to be included.

While the primary motivation for constructing such a simulator was to improve descriptions of hydraulic permeability structure in the eastern Snake River Plain aquifer, the simulator may help to identify gaps in our knowledge by drawing attention to geologic processes and structural features that are important in the evolution of the eastern Snake River Plain. For example, the simulator could be used to investigate the effect of constant versus variable eruptive rates on basalt stratigraphy (Kuntz et al., this volume); the impact of constructional-controlled versus subsidence-controlled growth on subsurface

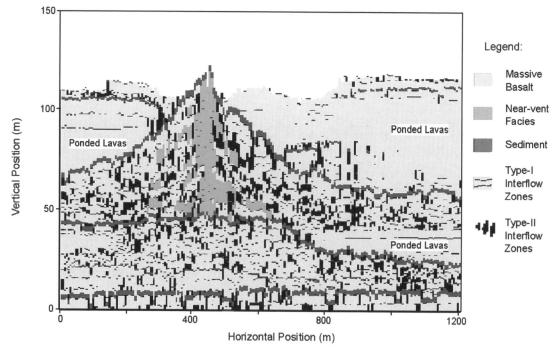

Figure 17. Example of structure-imitating model's simulated distribution of basalt and sediment, created after multiple eruptions over extended time period.

basalt stratigraphy; and volcanic controls on the development of sedimentary basins on the plain (Geslin et al., this volume).

SUMMARY AND CONCLUSIONS

Pahoehoe lava builds complex assemblages of inflated and ponded structures during the growth of a plains-style lava field (lava flow group). To describe the morphology of inflated pahoehoe, a scale-dependent terminology has been proposed for small-scale (lava flow lobes) and larger scale bodies (lava flow units). The emplacement of lava flow units and lobes during the creation of a flow group leads to a complex aggregate of morphologic entities that is difficult to map. We have developed methods for mapping, describing, and quantifying the morphology of inflated pahoehoe in order to help constrain models of the high-permeability elements in basalt that control aquifer heterogeneity.

Individual lava flow lobes are defined by their characteristic shapes and hierarchical spatial growth pattern, which reflects some of the hierarchical temporal growth of the lava flow unit that spawned them. However, morphology alone cannot distinguish lobes unambiguously due to coalescence of hot, plastic lobe surfaces during their emplacement. The median length:width aspect ratio of mapped flow lobes in the Wapi and Hell's Half Acre lava flow groups is 3.2:1, with a mean lobe length of 85 m. This aspect ratio may be characteristic of inflated pahoehoe growth and emplacement, because similar aspect ratios are seen in other eastern Snake River Plain lava fields (North and South Robbers; Welhan et al., 1997) and at Kilauea (Hon et al., 1994).

The hierarchical growth pattern of new lobes extruding from existing lobes produces lava flow units that have a fractal geometry. The fractal dimension of seven mapped flow units ranges from 1.4 to 1.6, and is remarkably similar in flow units having lengths from 0.5 to >3 km. This suggests that the inflation process produces a self-similar geometry, defined by the hierarchical growth pattern of lobes, from scales of <10 m to 10 km.

Figure 18 summarizes our conceptual model of an inflated lava flow lobe. Lobes have a characteristic internal structure in which a highly porous and permeable zone is supported within and on the upper crust of a lobe, with a relatively low-permeability, massive interior, and very porous fractured and rubbly margins. The combination of (1) highly porous and permeable upper crust and (2) rubble that is commonly observed on lava lobe surfaces and at the contacts between lobes in drill core is defined as a type-I interflow zone. The lateral dimensions and geometry of type-I interflow zones are assumed to be similar to the dimensions of the lava flow lobes on which they occur. Their vertical thickness is from 0.3 to 2.5 m and more likely less than 1 to 1.5 m.

Extensive networks of tension fractures are characteristic of the inflation process and tend to be particularly well developed along the margins of large-scale lava flow units. Tension fractures in inflated pahoehoe that have been buried by younger lavas in Box Canyon tend to remain open, possibly because high viscosity prevents lava ingress to fractures with smaller apertures. At Box Canyon, such a mechanism appears to be effective in preserving fracture porosity at apertures <10–15 cm. Because more than 50% of tension fractures on Holocene outcrops have apertures of 10 cm or less, we estimate that at least 15% of original total fracture porosity in the upper crust of inflated lavas is preserved during burial. If this fracture porosity is sufficiently interconnected, the networks of tension fractures in inflated pahoehoe could host highly permeable zones that extend over kilometer distances. These have been proposed as type-II interflow zones (Welhan et al., 2002), to distinguish them from type-I interflow zones.

The geometry of type-II interflow zones is determined by the geometry of tension fracture networks. Because tension fractures concentrate along the edges of flow units, they tend to be long, thin, and sinuous in plan view with a spatial architecture that is quite different from type-I interflow zones. Furthermore, because of the complex shape that characterizes the edge of inflated flow units, the geometry of type-II interflow zones is not described adequately by linear geostatistics. Type-II interflow zones may serve to interconnect type-I interflow zones over larger spatial scales than would be expected from the spatial geometry of type-I interflow zones alone.

A preliminary simulator of plains-style volcanism was developed to create stratigraphic models of the eastern Snake River Plain. The simulator is based on sets of rules and equations that undergo an optimization process to create models having internal characteristics that satisfy limitations derived from outcrop and core measurements. The models created by the simulator can be used to create stochastic maps of permeability structure in the aquifer, and may be particularly useful in modeling the geometry and hydrogeologic importance of type-II interflow zones. We suspect that a simulator of this type may eventually prove to be useful in understanding the overall

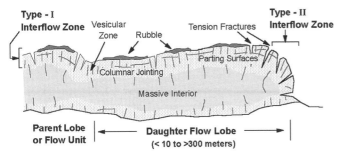

Figure 18. Longitudinal cross section of inflated lava flow lobe showing features that control porosity and permeability in the eastern Snake River Plain basalt aquifer. Type-I interflow zones develop along upper surface of lobe, whereas type-II interflow zones develop along fractured and broken margins of lobe and flow unit. Range of lobe lengths is approximate.

geologic evolution of the eastern Snake River Plain and improving visualization of the interplay between volcanism, structural evolution, and sedimentation.

ACKNOWLEDGMENTS

We are indebted to Duane Champion, U.S. Geological Survey (USGS), Menlo Park, for pointing out the potential importance of tension fractures in Holocene basalt studies, for the use of his detailed geologic maps of Wapi lava flows and aerial photograph collections, and for many stimulating discussions of basalt geology. We thank Dick Smith, Lockheed-Martin Idaho Technologies, Inc. (LMITCo), for much assistance in acquiring existing remote-sensed data and unpublished Idaho National Engineering and Environmental Laboratory (INEEL) reports and data. Cheryl Whittaker, LMITCo, and Linda Davis, USGS-INEEL, provided invaluable assistance in accessing digital data files and drill core, respectively. Steve Anderson and Mel Kuntz, USGS, Scott Hughes and Dennis Geist, University of Idaho, provided fruitful discussion, insights into basalt geology, and encouragement to pursue this research. We thank Stephen Self, University of Hawaii, and Craig White, Boise State University, for their helpful reviews and comments. This work was funded by the U.S. Department of Energy through grant DE-FG07-96ID13420 to the Idaho Water Resources Research Institute–Idaho Universities Consortium.

REFERENCES CITED

Anderson, S.R., and Bowers, B., 1995, Stratigraphy of the unsaturated zone and uppermost part of the Snake River Plain aquifer at Test Area North, Idaho National Engineering Laboratory, Idaho: U.S. Geological Survey Water-Resources Investigations Report 95–4130, 47 p.

Anderson, S.R., Ackerman, D.J., and Liszewski, M.J., 1996, Stratigraphic data for wells at and near the Idaho National Engineering Laboratory, Idaho: U.S. Geological Survey Open-File Report 96–248, 27 p. plus 1 diskette.

Barton, C.C., and Scholz, C.H., 1995, The fractal size and spatial distribution of hydrocarbon accumulations: Implications for resource assessment and exploration strategy, *in* Barton, C.C., and LaPointe, P.R., eds., Fractals in petroleum geology and Earth processes: New York, Plenum Press, p. 13–34.

Bruno, B.C., Taylor, G.J., Rowland, S.K., Lucey, P.G., and Self, S., 1992, Lava flows are fractals: Geophysical Research Letters, v. 19, p. 305–308.

Champion, D.E., 1973, The relationship of large scale surface morphology to lava flow direction, Wapi lava field, southeastern Idaho [M.S. thesis]: Buffalo, New York, State University of New York, 44 p.

Clemo, T., and Welhan, J., 2000, Simulating basalt lava flows using a structure imitation approach, *in* Bentley, L.R., et al., eds., Computational methods in water resources, Proceedings, 13th International Conference, Calgary: Rotterdam, A.A. Balkema, p. 841–848.

Feder, J., 1988, Fractals: New York, Plenum Press, 317 p.

Greeley, R., 1982, The style of basaltic volcanism in the eastern Snake River Plain, Idaho, *in* Bonnichsen, B., and Breckenridge, R.M., eds., Cenozoic geology of Idaho: Idaho Bureau of Mines and Geology Bulletin 26, p. 407–422.

Hon, K., Kauahikaua, J., Denlinger, R., and Mackay, K., 1994, Emplacement and inflation of pahoehoe sheet flows: Observations and measurements of active lava flows on Kilauea Volcano, Hawaii: Geological Society of America Bulletin, v. 106, p. 351–370.

Johannesen, C.M., 2001, Basalt surface morphology on the eastern Snake River Plain, Idaho: Implications for the emplacement of high-porosity elements in the Snake River Plain aquifer [M.S. thesis]: Pocatello, Idaho, Idaho State University, 140 p.

Johannesen, C.M., Funderberg, T., and Welhan, J.A., 1998, Basalt surface morphology as a constraint in stochastic simulations of high-conductivity zones in the Snake River Plain aquifer, Idaho: Eos (Transactions of the American Geophysical Union), v. 79, p. F243.

Johnson, N.M., and Dreiss, S.J., 1989, Hydrostratigraphic interpretation using indicator geostatistics: Water Resources Research, v. 25, p. 2501–2510.

Knutson, C.F., McCormick, K.A., Smith, R.P., Hackett, W.R., O'Brien, J.P., and Crocker, J.C., 1990, Radioactive Waste Management Complex vadose zone basalt characterization: Idaho Falls, Idaho, EG&G Idaho Report EGG-WM-8949, 126 p.

Knutson, C., McCormick, K.A., Crocker, J.C., Glenn, M.A., and Fishel, M.L., 1992, 3D Radioactive Waste Management Complex vadose zone modeling: Idaho Falls, Idaho, EG&G Idaho Report EGG-ERD-10246, 156 p.

Kuntz, M.A., Covington, H.R., and Schorr, L.J., 1992, An overview of basaltic volcanism of the eastern Snake River Plain, Idaho, *in* Link, P.K., Kuntz, M.A., and Platt, L.B., eds., Regional geology of eastern Idaho and western Montana: Geological Society of America Memoir, v. 179, p. 227–267.

Lindholm, G.F., and Vaccaro, J.J., 1988, Region 2, Columbia Lava Plateau, *in* Back, W., Rosenshein, J.S., and Seabar, P.R., eds., Hydrogeology: Boulder, Colorado, Geological Society of America, Geology of North America, v. O-2, Decade of North American Geology, p. 37–50.

Phillips, F.M., and Wilson, J.L., 1989, An approach to estimating hydraulic conductivity spatial correlation scales using geological characteristics: Water Resources Research, v. 25, p. 141–143.

Self, S., Keszthelyi, L. and Thordarson, T., 1998, The importance of pahoehoe: Annual Reviews of Earth and Planetary Sciences, v. 26, p. 81–110.

Self, S., Thordarson, T., and Keszthelyi, L., 1997, Emplacement of continental flood basalt lava flows, *in* Mahoney, J.J., and Coffin, M., eds., Large Igneous Provinces: American Geophysical Union Geophysical Monograph 100, p. 381–410.

Thordarson, T., and Self, S., 1998, The Roza Member, Columbia River basalt group: A gigantic pahoehoe lava flow field formed by endogenous processes?: Journal of Geophysical Research, v. 103, p. 27411–27445.

Vanmarcke, E., 1983, Random Fields: Analysis and synthesis: Cambridge, Massachusetts, Massachusetts Institute of Technology Press, 382 p.

Welhan, J.A., and Wylie, A.H., 1997, Stochastic modeling of hydraulic conductivity in the Snake River Plain aquifer. 2. Evaluation of lithologic controls at the core and borehole scales, *in* Sharma, S., and Hardcastle, J.H., eds., Proceedings of the 32nd Engineering Geology and Geotechnical Engineering Symposium: Boise, Idaho, p. 93–108.

Welhan, J.A., Funderberg, T.D., Smith, R.P. and Wylie, A., 1997, Stochastic modeling of hydraulic conductivity in the Snake River Plain aquifer. 1. Hydrogeologic constraints and conceptual approach, *in* Sharma, S., and Hardcastle, J.H., eds., Proceedings of the 32nd Engineering Geology and Geological Engineering Symposium: Boise, Idaho, p. 75–92.

Welhan, J.A., Johannesen, C.M., Glover, J.A., Davis, L.L., and Reeves, K.S., 2002, Overview and synthesis of lithologic controls on aquifer heterogeneity in the eastern Snake River Plain, Idaho, *in* Bonnichsen, W., White, C., and McCurry, M., eds., Tectonic and magmatic development of the Snake River Plain: Idaho Geological Survey Bulletin 30 (in press).

Manuscript Accepted by the Society November 2, 2000

Geological Society of America
Special Paper 353
2002

Geochemical correlations and implications for the magmatic evolution of basalt flow groups at the Idaho National Engineering and Environmental Laboratory

Scott S. Hughes*
Michael McCurry
Department of Geosciences, Idaho State University, Pocatello, Idaho 83209, USA
Dennis J. Geist
Department of Geological Sciences, University of Idaho, Moscow, Idaho 83843, USA

ABSTRACT

Whole-rock major element, trace element, and Sr-Nd isotopic analyses of samples taken from Idaho National Engineering and Environmental Laboratory (INEEL) core holes are used to correlate subsurface tholeiitic basalt lava flow groups in the Test Area North (TAN) region, and to evaluate time-space variations in their geochemical signatures. New $^{40}Ar/^{39}Ar$ ages are included to supplement existing dates in order to confirm major eruptive episodes. Incompatible trace elements range by factors of ~7–10; major oxides, especially MgO, range by factors of ~2–3; initial $^{87}Sr/^{86}Sr$ ratios range from ~0.706 to 0.708; and $^{143}Nd/^{144}Nd$ ratios range from ~0.5122 to 0.5125. These data suggest that each flow group represents a single shield-building monogenetic eruption derived from a separate magma batch.

The geochemical patterns of flow groups older than ca. 0.8 Ma at TAN are compared to younger ca. 0.2–0.6 Ma flow groups located in the southern INEEL (~30 km distance) to assess regional variations. No flow groups are known to correlate between these two regions although some TAN flow groups are found in core hole 2-2A located ~10 km south. Only the deepest flow group at TAN, with three measured radiometric ages ranging from 2.38 to 2.46 Ma, is correlated to core hole USGS 126-A located on Lava Ridge ~9 km northwest of TAN as well as core hole 2-2A. Recurrence of fairly primitive MgO-rich, La-poor flow groups in the northern and southern INEEL suggests that similar processes of magma genesis and shield growth are operative on eastern Snake River Plain volcanic rift zones regardless of age.

INTRODUCTION

Hundreds of small monogenetic tholeiitic basalt shields consisting of thousands of lava flow units dominate the Quaternary volcanic-sedimentary depositional sequence underlying the eastern Snake River Plain of Idaho (Kuntz et al., 1992). The province is an east-northeast–trending topographic depression, 400 km long and 100 km wide, extending from Twin Falls to Ashton, Idaho. Commingled lava fields and intercalated sediment compose the upper 1–2 km of the crust. Low-profile ba-

*E-mail: Hughscot@isn.edu

Hughes, S.S., McCurry, M., and Geist, D.J., 2002, Geochemical correlations and implications for the magmatic evolution of basalt flow groups at the Idaho National Engineering and Environmental Laboratory, *in* Link, P.K., and Mink, L.L., eds., Geology, Hydrogeology, and Environmental Remediation: Idaho National Engineering and Environmental Laboratory, Eastern Snake River Plain, Idaho: Boulder, Colorado, Geological Society of America Special Paper 353, p. 151–173.

saltic shields having subdued topography and shallow deposi-tional slopes overlap with each other and interfinger with sedimentary deposits to produce a complex, discontinuous stratigraphic sequence.

The correlation of basalts, either as individual lava flow units or as flow groups, which are packages of flow units de-rived from a single vent system, is essential to unravel the geo-metric attributes of the vadose zone and aquifer, to understand flow paths for subsurface contaminant migration to the ground-water, and to construct better models for the volcanic and tec-tonic evolution of the eastern Snake River Plain province. Vol-canic stratigraphy near the Idaho National Engineering and Environmental Laboratory (INEEL) remains the focus of re-search to better understand how volcanology influences the hy-drogeologic architecture of the eastern Snake River Plain, and to develop models for groundwater remediation (Anderson, 1991; Anderson and Bowers, 1995; Anderson et al., 1996; Hughes et al., 1997; Geist et al., this volume, chapter 4; Kuntz et al., this volume). Correlation of basalt flow units, and in some cases flow groups, has relied on various combinations of physi-cal and chemical properties such as natural gamma emissions, paleomagnetic polarity and inclination, radiometric age, pe-trography, and chemical composition (e.g., Champion et al., 1988, 1996; Anderson, 1991; Lanphere et al., 1993, 1994; An-derson and Bartholomay, 1995; Anderson and Bowers, 1995; Reed et al., 1997).

Analyses presented in this chapter, in conjunction with companion chapters in this volume (Bestland et al., Champion et al., and Geslin et al.), are intended to help characterize the volcanic and sedimentary depositional systems on the INEEL. Whole-rock major element, trace element, and isotopic analyses of samples taken from INEEL core holes (Fig. 1) are used to characterize and delineate major flow groups. During the sam-pling process, flow unit boundaries and thicknesses were mea-sured for primary stratigraphic controls. Surface samples from Circular Butte, a shield volcano east of Test Area North (TAN), and from other exposed units in the northern INEEL region (Casper, 1999) are also included to evaluate possible correla-tions between subsurface and exposed basalts (Fig. 2). Six new ^{40}Ar/^{39}Ar ages have been obtained to further define correlations and to supplement existing radiometric age information sum-marized by Lanphere et al. (1994), Anderson et al. (1997), and Champion et al. (this volume). These data are compiled along with literature data into a digital database to (1) correlate sub-surface basaltic lava flow groups that have been buried by younger volcanic or sedimentary deposits and (2) evaluate changes in geochemical compositions of flow groups over time and space to provide a regional assessment of compositional variations in the middle to late Pleistocene. A quantitative pet-rologic model is beyond the scope of this chapter and will be presented in a future contribution.

The focus of this study is on the northern part of the INEEL, where basalts exposed at the surface range in age from ca. 0.8 to 1.2 Ma, although we include samples from two re-

gions of the southern part of the INEEL, where basalts exposed at or near the surface range in age from ca. 0.2 to 0.6 Ma. Data from these regions, one surrounding the Idaho Nuclear Tech-nology and Engineering Center (INTEC; formerly called the Idaho Chemical Processing Plant, ICPP) and the other sur-rounding the Radioactive Waste Management Complex (RWMC), confirm many of the correlations by Anderson et al. (1991) and Anderson and Liszewski (1997). Chemical corre-lations are made only for lava flow groups at the INEEL and therefore represent only a fraction of the eastern Snake River Plain. We attempt to clarify some of the issues related to using whole-rock major and trace element geochemistry as a primary means of flow group correlation, the viability of which is equal to or greater than paleomagnetic, petrographic, radiometric age, and geophysical logging techniques. We also demonstrate the petrologic variability among eastern Snake River Plain tholei-itic basalts, previously thought to be rather homogeneous, that can be extended to other regions of the province.

This study builds upon extensive well and core-hole inter-pretations of natural gamma logs (Anderson et al., 1991, 1996; Anderson and Bowers, 1995); however, we correlate and clas-sify lava flow groups rather than the individual flow units that compose these packages. Each flow group in this classification is interpreted as a sequence of simple flow units erupted over a relatively short time span from a single shield volcano or a comagmatic eruptive center with several vents. A stratigraphic name is applied to each flow group based on its location, po-sition in the sequence, and time intervals supported by paleo-magnetic signatures and radiometric ages (Champion et al., 1988, 1996; this volume; Lanphere et al., 1993, 1994). The geochemical divisions of some groups are equivalent to the stratigraphic divisions by Anderson et al. (1991, 1996) and An-derson and Bowers (1995), although they generally indicated more subdivisions than we have been able to define. Thus, we do not attempt to revise their detailed alphanumeric nomencla-ture, but we recognize its importance for local correlations and possible subdivision of major sequences.

MAFIC VOLCANISM ON THE EASTERN SNAKE RIVER PLAIN

Pliocene to Quaternary basalt, eolian sand, loess, and al-luvial and lacustrine sediments were deposited on Miocene-Pliocene rhyolitic tuffs and lava flows that are exposed only along the margins of the eastern Snake River Plain. Basaltic lava flows exposed on the surface in and around the INEEL range in age from ca. 1.1 Ma to ca. 2.1 ka (Kuntz et al., 1992, 1994), and the radiometric (^{40}Ar/^{39}Ar and K-Ar) ages of basalts sampled in core holes are as old as 3.26 ± 0.06 Ma (Champion et al., this volume). Throughout the eastern Snake River Plain lava flows accumulated to form low-profile shields made of primitive olivine tholeiites (Kuntz et al., 1992; Hughes et al., 1999). A topographically high central axis of the eastern Snake River Plain, known as the Axial volcanic zone (Fig. 1), contains

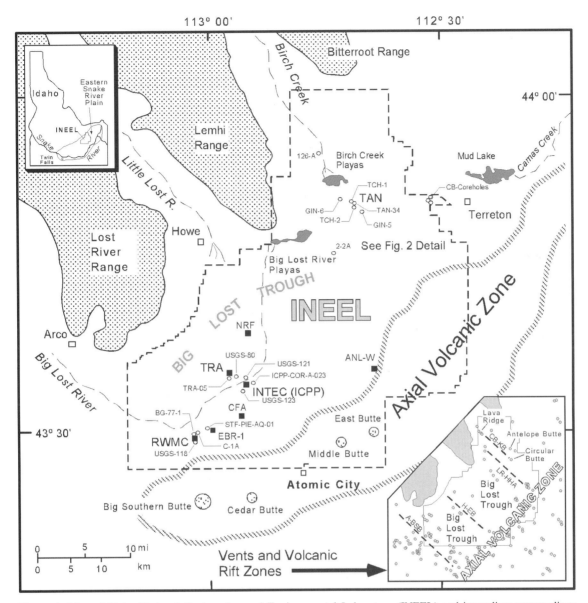

Figure 1. Map of Idaho National Engineering and Environmental Laboratory (INEEL) and immediate surroundings showing facilities and core holes from which samples of basalt flow groups were obtained. Inset in lower right illustrates exposed volcanic vents (after Hughes et al., 2000), approximate locations of central axes of volcanic rift zones (after Kuntz et al., 1992), and Big Lost Trough (Geslin et al., 1999) in relation to Axial volcanic zone. Abbreviations for volcanic rift zones are: CB-KB, Circular Butte–Kettle Butte; LR-HHA, Lava Ridge–Hell's Half Acre; H-EB, Howe–East Butte; and A-BSB, Arco–Big Southern Butte. Other abbreviations: TAN, Test Area North; INTEC (ICPP), Idaho Nuclear Technology and Engineering Center; RWMC, Radioactive Waste Management Complex; ANL-W, Argonne National Laboratory-West; CFA, Central Facilities Area; EBR-1, Experimental Breeder Reactor; TRA, Test Reactor Area; NRF, Naval Reactors Facility.

a compositionally diverse eruptive center (Cedar Butte), and silicic domes (Big Southern, Middle, and East Buttes), which have contributed to the volcanic sequence (Hackett and Smith, 1992; Kuntz et al., 1992). More in-depth discussions of eastern Snake River Plain physical volcanology and timing of volcanic and/or tectonic events were provided in Hackett et al. (2002) and Hughes et al. (2002).

Well-exposed basaltic eruptive vents and fissures occur mainly along northwest-trending volcanic rift zones such as the Great Rift and Arco–Big Southern Butte volcanic rift zones, which are roughly parallel to range-front faults of the adjacent Basin and Range province (Kuntz et al., 1992). The northwest orientation of vent clusters, eruptive fissures, and associated tension cracks are mainly due to lithospheric extension behind the Yellowstone hotspot track (Rodgers et al., 1990; Parsons et al., 1998; Kuntz et al., this volume). The margins of some vol-

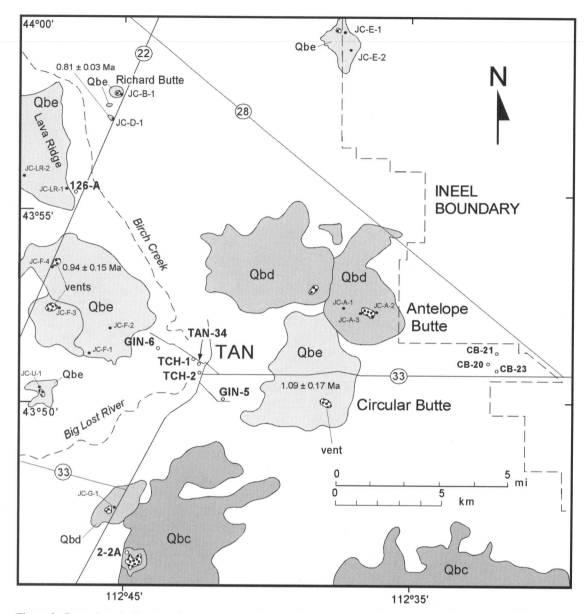

Figure 2. General geologic map of northern Idaho National Engineering and Environmental Laboratory (INEEL) Test Area North (TAN) showing core hole locations (small circles) and surface sample locations (JC numbers) from Casper (1999). Geology and radiometric K-Ar ages are from Kuntz et al. (1994) and Lanphere et al. (1994). Shaded and patterned areas are basaltic shields, including vent deposits and lava flows; unshaded areas are fluvial, lacustrine, and eolian surficial sediments. Estimated age ranges of lava fields and their associated vent deposits according to Kuntz et al. (1994) are: Qbe, older than 730 ka; Qbd, 400–730 ka; Qbc, 200–400 ka. Units younger than Qbc are not found in northern INEEL and TAN map region.

canic rift zones such as the Great Rift and the Arco–Big Southern Butte volcanic rift zones are clearly defined by a relatively high density of exposed vents and fissures compared to surrounding regions; however, several volcanic rift zones are less well defined and comprise more diffuse arrangements of subparallel fissures and exposed vents (Anderson et al., 1999; Hughes et al., 2002). This description befits the Lava Ridge–Hell's Half Acre and Circular Butte–Kettle Butte volcanic rift

zones somewhat, and the Howe–East Butte volcanic rift zone in particular (Fig. 1).

Low-profile coalescent shields and the eruptive mechanisms of eastern Snake River Plain basalts reflect lava plains–style volcanism (Greeley, 1982), in which low-volume monogenetic volcanoes represent numerous scattered vents. Each shield on the eastern Snake River Plain formed during a short time period of months or years (Kuntz et al., 1992), and each one probably formed from a series of small individual batches of magma (Hughes et al., 2002). The short duration of such monogenetic eruptions may be attributed to a rapid drop in magma pressure due to sluggish crustal response when the source is tapped (Kuntz et al., 1986; Kuntz, 1992).

Volcanic highlands along volcanic rift zones and the Axial volcanic zone control fluvial and lacustrine sediment deposition on the eastern Snake River Plain (Hughes et al., 1997), which occurs largely in playa-like sinks and intermittent streams (Kuntz et al., 1994; Gianniny et al., 1997; Geslin et al., 1999). One such system, the Big Lost Trough (Fig. 1) is a Pliocene–recent closed sedimentary basin bounded by the Arco–Big Southern Butte volcanic rift zone to the southwest, the Axial volcanic zone to the southeast, and the Lost River Range to the northwest (Wetmore, 1998; Geslin et al., 1999, this volume; Bestland et al., this volume; Blair and Link, 2000). The basin contains several thick sedimentary sequences in core hole 2-2A, and has received sedimentary deposits and lava flows throughout the Pleistocene. It is transected by the Howe–East Butte volcanic rift zone in the central INEEL and, to the northeast, it is separated from the Mud Lake basin by the volcanic highland that composes the Lava Ridge–Hell's Half Acre and Circular Butte–Kettle Butte volcanic rift zones.

Overall, the eastern Snake River Plain has been subsiding relative to the Basin and Range since mid-Miocene time (McQuarrie and Rodgers, 1998), yet the volcanic and sedimentary stratigraphy on the eastern Snake River Plain depicts local differences in the rate of subsidence, at least throughout the latest Pliocene–Quaternary (Wetmore, 1998). Evidence for differential vertical movement includes the development of basins such as the Big Lost Trough and the associated variations in sediment thickness found in core holes (e.g., Geslin et al., 1999), along with differences in surface ages of basalts (Kuntz et al., 1994) near TAN (older than 1 Ma) relative to those in the southern INEEL (ca. 200–400 ka), where older basalts are found at depth. Thus the rate of magmatism appears to be non-uniform and volcanic rift zones are active during different time periods. Thick volcanic sequences near a volcanic rift zone may have aggraded to keep up with the overall subsidence of the eastern Snake River Plain more so than less voluminous lava flows found in the sedimentary basins.

GEOCHEMICAL ANALYSES OF BASALTS

Rationale for geochemical correlations

Bulk rock major and trace element geochemistry, isotopic signatures, and petrographic characteristics supplemented by radiometric ages and paleomagnetic signatures have been effectively used for correlation in large flood basalt terranes such as the extensive Grande Ronde Formation of the Columbia River basalts (Reidel et al., 1989). Individual thick basalt flows are characterized on the basis of uniform element abundances or ratios, especially those that are not sensitive to posteruptive petrologic processes or secondary alteration. This approach is most effective if flow units are fairly homogeneous in chemical attributes, and if chemical signatures differ significantly from one major flow to the next.

Previous petrologic studies (Leeman and Vitaliano, 1976; Leeman, 1982a) showed that two types of olivine tholeiites exist on the eastern Snake River Plain, those that are either mildly enriched and fractionated, or those that are somewhat primitive with affinity to magmatic sources. Olivine tholeiites on the eastern Snake River Plain also were thought to have restricted chemical variability when compared to the wide range of chemical compositions associated with evolved latitic and silicic magmas at Craters of the Moon (Leeman et al., 1976; Leeman, 1982b; Kuntz et al., 1986) and Cedar Butte (McCurry et al., 1999), which have undergone various degrees of contamination, hybridization, or extensive fractionation. However, new data show that chemical variation among the eastern Snake River Plain olivine tholeiites is sufficient to distinguish lava flow groups, especially when both major and trace element data are considered (Hughes et al., 2002).

In some cases, the geochemical compositions of individual lava flow units can be recognized and used for correlation over short distances. Reed et al. (1997) correlated INEEL flow units on the basis of bulk major and trace element analyses across a distance of ~2 km beneath the INTEC (core holes USGS-121 and USGS-123, Fig. 1). Two flow units represented in both core holes had distinctive chemical compositions and were thus useful marker horizons. However, the remaining units displayed a gradient of compositions that required a more objective approach based on statistical K-cluster analysis of selected elemental (or oxide) abundances or ratios. While these correlations were adequate for the local area, they would be equivocal when considering the low probability of a single flow unit being sampled in core holes spaced over much greater distances. Regardless of this shortcoming, the Reed et al. (1997) study demonstrated the feasibility of geochemically correlating eastern Snake River Plain basalt flows with the caveat that other techniques, such as paleomagnetic signatures, petrography, natural gamma emission logs, and radiometric dating, should be combined when making detailed stratigraphic assessments.

Although many individual flow units may be homogeneous, recent geochemical evaluation of INEEL basalts (Reed et al., 1997; Casper, 1999; Hughes et al., 2002) demonstrates that conspicuous chemical variation is possible from one flow unit to the next within a given flow group. These studies further show that chemical variation in most flow groups follows a characteristic pattern, creating a regular trend such as increasing La or decreasing Cr with height in the sequence. Once the pattern is defined by numerous analyses it is possible to correlate a lava flow group between locations represented by any subset of flow units, but this requires the presence of several flow units in each section. Without knowing the geochemical pattern the correlation of major sequences based only on chemical uniformity is inappropriate, unless individual flow units can be uniquely and unequivocally identified in the sequence as marker horizons.

Correlation of lava flow groups in this study follows the premise that the geochemical pattern of each flow group can be identified and that overall variance can be established, although it is possible that any given pattern may recur in time and space. Subdivisions that coincide with sedimentary interbeds or those determined by previous stratigraphic interpretations (Anderson, 1991; Anderson and Bowers, 1995; Anderson and Liszewski, 1997) provide a first-order approximation of flow group boundaries from which the chemical range can be delimited. Generally, most flow groups are vertically continuous until broken by a thick sedimentary horizon, or by significant changes in paleomagnetic signature or radiometric age (Lanphere et al., 1993, 1994). When physical stratigraphic breaks are inconclusive, then subdivisions are made where sudden changes in the geochemical pattern occur with height in the section. In many cases, the shift in chemical pattern is evident in some elements (or ratios), but not in others. Nonetheless, the significance of a chemical shift depends on the overall variation, either due to analytical uncertainty or to real chemical differences, that would be expected in that particular flow group. We consider a chemical shift in a given element to represent a flow group division when the new value encountered in the sequence is higher or lower by more than a 2-sigma (two standard deviations) difference from the highest or lowest value in the group. Therefore, the chemical limits of a flow group can be determined only by high-precision analyses of most or all of its component flow units, and a reiterative assessment of the divisions between flow groups is necessary to clearly define the chemical pattern. The same kinds of limitations hold true for paleomagnetic signatures and radiometric ages (e.g., Lanphere et al., 1994; Kuntz et al., this volume), which have their own analytical uncertainties.

Analytical procedures

Basalts were sampled from cores listed in Table 1, which were made available at the INEEL Lithologic Core Storage Library (Davis et al., 1997). Cores were logged and visually inspected to avoid sampling from sections with secondary alteration, oxidation related to emplacement (especially near flow tops and bottoms), and clay-caliche contamination. Flow surfaces, lithologic descriptions, and sedimentary interbeds were recorded in a detailed stratigraphic log. Samples of ~0.5 kg were collected from the flow interiors and transferred to the Laboratory for Environmental Geochemistry (LEG) at Idaho State University for determination of elemental abundances by inductively coupled plasma-atomic emission spectroscopy (ICP-AES) and instrumental neutron activation analysis (INAA). Samples from TAN-34 were analyzed by X-ray fluorescence spectroscopy (XRF) in the geochemical laboratory at Washington State University for major elements and selected

trace elements in order to check for interlaboratory bias and to confirm analytical accuracy within the LEG. This was done primarily to produce a comprehensive database for TAN-34, and to evaluate in detail the chemical variation in a lava sequence (Geist et al., this volume, chapter 12).

Aliquots of selected samples from the northern INEEL were also sent for analyses of isotopic Sr and Nd ratios by mass spectrometry at either the Rice University laboratory, operated by J. Wright, or the University of California at Los Angeles laboratory, operated by J. Davidson. Isotopic analyses include 46 samples from core holes 2-2A, GIN-5, GIN-6, TAN-34, TCH-1, and TCH-1, plus two surface samples from Circular Butte volcano.

Radiometric $^{40}Ar/^{39}Ar$ ages were obtained for groundmass separates of two samples from core hole 2-2A, two from TAN-34, and one each from the proposed Circular Butte Landfill core holes, CB-20 and CB-23. The six samples were transferred to the New Mexico Geochronology Research Laboratory in Socorro, New Mexico (W.C. McIntosh, co-Director), where they were analyzed by the furnace incremental-heating age-spectrum method.

TABLE 1. SUMMARY OF NEW CHEMICAL ANALYSES OF MAJOR AND TRACE ELEMENTS IN CORE AND SURFACE BASALT SAMPLES FROM THE INEEL*

Location	Number of samples
TAN	
Core hole TCH-1	26
Core hole TCH-2	25
Core hole GIN-5	19
Core hole GIN-6	8
Core hole TCH-34[1]	16
Areas near TAN	
Core hole 2-2A	77
Core hole 126-A	35
Core holes west of Mud Lake (ML-1, CB-21, CB-21, CB-23)	15
Surface, Circular Butte, and TAN[2]	45
ICPP	
Core hole ICPP COR-A-023	31
Core hole USGS-80	9
Core hole TRA-5	11
RWMC	
Core hole USGS-118[3]	20
Core hole BG-77-1[3]	18
Core hole C-1A[3]	21
Core hole STF-PIE-AQ-01[3]	16
Other INEEL	
Core hole CH-1	23
Core hole Argonne-1[4]	10
Core hole WO-2[4]	5
Total	427

Note: Analyses were obtained by inductively coupled plasma-atomic emission spectrometry (ICP-AES) and instrumental neutron activation analysis (INAA), except major elements in TAN-34 were obtained by X-ray fluorescence spectrometry (XRF).
* INEEL—Idaho National Engineering and Environmental Laboratory.
References: [1]Geist et al. (this volume, chapter 12); [2]Casper (1999); [3]Wetmore (1998); and [4]Morse et al. (this volume).

Analytical results

Representative geochemical analyses of olivine tholeiites are reported in Table 2, and the comprehensive database of ~900 analyses of eastern Snake River Plain volcanic rocks is available in digital format from the authors. The database includes the 427 new analyses summarized in Table 1 (this study; Casper, 1999; Wetmore, 1998; Geist et al., this volume, chapter 12; Morse and McCurry, this volume) and analyses published elsewhere (e.g., Stout and Nicholls, 1977; Kuntz and Dalrymple, 1979; Leeman, 1982a, 1982b; Kuntz et al., 1985, 1992; Knobel et al., 1995; Shervais et al., 1994; Reed et al., 1997).

Measured ratios of $^{87}Sr/^{86}Sr$ and $^{143}Nd/^{144}Nd$, along with elemental Rb, Sr, Sm, and Nd values, are presented in Table 3. Initial age-corrected ratios, as well as Nd values, were calculated using values of the above elements determined by ICP-AES, INAA, or estimated values for Rb, and available radiometric ages or estimated ages based on their stratigraphic positions (e.g., Champion et al., this volume). Errors incurred during these calculations are negligible, as evident in the small differences between measured and initial isotopic ratios, due to the relatively young ages of these samples compared to the half-lives of parent Rb and Sm isotopes.

Results of $^{40}Ar/^{39}Ar$ age determinations are provided in Table 4 with 2-sigma analytical uncertainties, and where radiometric ages are cited from the literature their reported analytical errors are adjusted to the 2-sigma level if necessary. According to the internal analytical report (W.C. McIntosh, 2000, written commun.), each sample yielded a relatively flat age spectrum for at least 80% of the cumulative $^{39}Ar_K$ released, and, although the precision of individual analytical steps was generally poor with sigma values rarely better than ± 0.2 m.y., all of the age spectra met the plateau criteria of Fleck et al. (1977). The plateau ages of these six samples range from 0.03 ± 0.09 Ma to 1.20 ± 0.14 Ma, which are considered good estimates of the eruption ages.

TABLE 2. REPRESENTATIVE BULK ROCK GEOCHEMICAL ANALYSES OF INEEL BASALT COREHOLE SAMPLES

Flow group	(BC)	Mud Lake	(DE)	E	F upper	F lower	G	EFGH	IJ	Lava Ridge upper	Lava Ridge lower	Jaramillo	Circular Butte
Core hole	2-2A	CB 23	2-2A	BG 77-1	BG 77-1	BG 77-1	BG 77-1	COR A-023	COR A-023	USGS 126A	USGS 126A	2-2A	TAN TCH-1
Depth (m)	16.5	47.3	90.9	97.3	118.6	161.0	179.3	117.1	185.4	11.6	42.7	128.7	26.5
Major oxides (wt%-normalized to 100% anhydrous)													
SiO_2	47.1	47.9	45.6	47.6	46.6	47.5	47.9	47.6	47.1	47.1	47.8	47.1	47.2
TiO_2	2.70	2.70	2.92	1.18	2.94	2.81	1.30	1.15	1.86	3.24	1.61	3.18	3.10
Al_2O_3	15.0	14.8	14.4	16.1	14.5	14.9	15.7	16.5	15.5	14.7	15.4	14.2	14.1
FeO*	13.6	13.3	14.8	9.5	13.3	12.5	9.5	10.5	12.5	13.5	10.7	14.0	14.6
MnO	0.20	0.19	0.21	0.18	0.21	0.20	0.17	0.17	0.18	0.21	0.19	0.22	0.22
MgO	7.76	7.41	8.04	9.97	7.44	7.26	11.0	9.86	9.63	7.15	10.9	7.17	6.99
CaO	9.80	9.48	9.99	11.6	9.71	9.75	11.2	11.5	9.92	10.0	10.4	10.0	9.89
Na_2O	2.66	2.73	2.59	2.20	2.44	2.59	2.12	2.31	2.46	2.79	2.34	2.49	2.22
K_2O	0.63	0.88	0.65	0.26	0.76	0.74	0.34	0.22	0.46	0.59	0.42	0.69	0.69
P_2O_5	0.62	0.61	0.84	0.17	0.90	0.61	0.22	0.16	0.37	0.70	0.28	0.99	1.00
Trace elements (ppm)													
Sc	30.7	27.8	30.6	38.3	27.8	29.0	33.4	37.7	29.3	30.3	32.4	28.9	29.0
Cr	237	229	227	427	195	208	553	415	334	165	379	189	176
Co	49.4	47.5	53.6	54.1	51.8	49.8	53.0	53.2	56.4	48.6	56.5	48.1	50.5
Ni	56	90	61	112	115	98	170	123	102	nd	nd	85	nd
Zn	140	164	154	23	190	155	66	57	101	nd	nd	nd	nd
Rb	16	16	16	4	10	9	11	4	10	nd	1	4	11
Sr	305	296	278	172	348	351	209	177	250	318	211	319	313
Cs	0.22	0.25	0.29	0.02	0.26	0.25	0.14	0.07	0.21	0.18	0.14	0.10	0.62
Ba	372	448	393	125	522	452	185	118	250	441	215	451	599
La	25.3	28.5	30.4	6.97	38.2	26.1	9.71	7.3	15.7	26.0	11.7	42.1	43.2
Ce	56.7	59.8	69.8	16.8	82.1	58.4	26.7	17.8	36.7	57.2	29.4	88.6	94.6
Nd	31	30	38	9	54	39	13	10	22	25.9	14.4	47	56
Sm	7.11	7.30	8.00	2.67	9.42	7.27	3.13	2.74	4.61	7.96	4.28	9.78	10.8
Eu	2.50	2.45	2.81	1.10	3.34	2.63	1.19	1.08	1.68	2.93	1.50	3.28	3.57
Tb	1.30	1.16	1.32	0.79	1.37	1.26	0.68	0.64	0.81	1.31	0.77	1.60	1.72
Yb	3.39	3.01	3.96	2.63	3.88	3.60	2.33	2.25	2.40	3.48	2.34	4.01	4.02
Lu	0.48	0.45	0.56	0.36	0.52	0.48	0.36	0.35	0.39	0.50	0.37	0.55	0.61
Zr	258	243	293	92	383	301	114	96	160	306	135	357	346
Hf	5.87	5.61	6.45	2.09	7.40	6.17	2.39	2.32	3.74	6.17	2.99	7.16	7.03
Ta	1.42	1.52	1.76	0.39	1.85	1.35	0.57	0.40	0.90	1.37	0.53	2.05	2.05
Th	1.52	1.53	1.28	nd	0.82	0.75	nd	nd	1.07	0.74	0.61	1.18	1.59
U	1.1	1.8	nd	nd	0.4	0.5	0.5	0.3	1.0	1.1	1.3	1.2	0.3

(continued)

TABLE 2. REPRESENTATIVE BULK ROCK GEOCHEMICAL ANALYSES (continued)

Flow group	TAN Series-1	Q-R upper	Q-R lower	post-Olduvai	Series-2 upper	Series-2 lower	pre-Olduvai upper	pre-Olduvai lower	Kaena 541 m	(3.26 Ma) 581 m	2-2A 604 m	2-2A 625 m	2-2A 783 m
Core hole	TAN GIN-5	TAN TCH-2	2-2A	2-2A	TAN CH-2	TAN CH-2	TAN CH-2	USGS 126-A	2-2A	2-2A	2-2A	2-2A	2-2A
Depth (m)	59.1	141.8	237.5	277.1	157.9	193.9	275.3	131.7	541.8	582.6	610.7	702.7	795.4
Major oxides (wt%-normalized to 100% anhydrous)													
SiO_2	47.0	47.7	46.0	47.5	47.2	46.4	48.1	48.0	47.9	47.8	48.5	47.3	50.0
TiO_2	3.31	1.80	2.40	2.36	3.03	2.90	2.61	3.33	2.30	2.79	2.47	2.28	2.23
Al_2O_3	14.1	15.4	15.3	15.3	15.1	15.1	14.6	14.0	16.3	14.7	15.0	15.6	14.5
FeO^*	14.5	12.3	13.6	13.6	14.3	14.7	13.5	13.9	13.5	13.5	13.9	13.4	13.5
MnO	0.22	0.19	0.20	0.18	0.20	0.21	0.20	0.22	0.14	0.20	0.16	0.19	0.20
MgO	7.44	9.29	8.06	7.60	7.04	7.50	7.62	6.93	7.16	7.62	5.59	7.75	5.63
CaO	9.42	10.4	10.8	9.67	9.32	9.72	9.65	9.79	9.50	10.3	10.7	10.3	9.05
Na_2O	2.52	2.24	2.67	2.76	2.73	2.72	2.45	2.51	2.59	2.37	2.67	2.39	3.08
K_2O	0.68	0.39	0.42	0.60	0.64	0.41	0.73	0.64	0.27	0.32	0.47	0.34	1.40
P_2O_5	0.79	0.28	0.54	0.43	0.45	0.41	0.51	0.72	0.40	0.53	0.53	0.47	0.41
Trace elements (ppm)													
Sc	29.2	30.4	31.0	28.3	26.0	28.0	29.9	33.8	27.4	31.9	29.4	30.8	27.0
Cr	206	300	206	244	168	211	267	192	131	360	86	168	83
Co	52.3	56.5	55.2	55.3	53.9	53.6	53.9	49.3	53.5	58.3	48.7	52.1	41.8
Ni	86	nd	115	110	nd	nd	70	nd	nd	nd	50	100	nd
Zn	nd	nd	119	111	nd	nd	nd	nd	nd	nd	137	nd	159
Rb	9	7	2	3	4	5	18	6	5	7	7	nd	21
Sr	320	220	240	328	387	320	295	314	275	344	299	245	284
Cs	0.60	0.26	0.09	0.34	0.31	0.10	0.18	0.26	1.20	0.34	0.50	0.77	0.54
Ba	488	227	316	352	403	297	387	445	243	249	411	239	525
La	35.5	12.9	21.4	19.2	20.3	15.7	21.5	26.1	13.8	20.0	25.6	17.7	37.6
Ce	77.6	28.3	47.7	42.7	43.7	36.3	50.9	65.7	31.9	47.5	51.9	38.3	71.8
Nd	44	21	28	24	35	28	29	32.8	18	24	26	24	33
Sm	9.65	4.16	6.25	5.89	5.89	5.39	7.46	9.10	5.24	7.35	6.78	5.81	7.27
Eu	3.21	1.52	2.26	2.15	2.34	2.12	2.41	3.25	2.02	2.42	2.21	1.98	2.27
Tb	1.51	0.75	1.14	0.94	0.91	1.01	1.32	1.25	1.00	1.22	1.21	1.06	1.38
Yb	3.95	2.11	3.25	2.88	2.69	2.58	3.09	4.10	2.71	3.25	3.70	3.28	3.70
Lu	0.54	0.39	0.47	0.40	0.40	0.40	0.48	0.57	0.42	0.50	0.48	0.43	0.58
Zr	301	128	223	184	209	192	252	303	178	238	233	179	321
Hf	7.24	3.18	5.14	4.44	4.83	4.80	6.29	7.36	3.83	6.09	5.24	4.10	6.96
Ta	1.99	0.64	1.11	1.06	1.16	0.98	1.14	1.49	0.79	1.19	0.95	0.85	1.28
Th	1.23	0.62	0.25	0.54	0.81	nd	0.43	0.77	0.41	0.97	1.60	0.55	3.48
U	1.2	0.3	0.1	0.4	1.0	1.1	0.8	1.2	0.5	0.5	1.0	1.5	0.9

* Total Fe reported as FeO. Major elements and trace elements Sr, Ba, and Zr determined by inductively coupled plasma-atomic emission spectrometry (ICP-AES) (<2% uncertainty). All other trace elements determined by instrumental neutron activation analysis (INAA). Uncertainties: 1–2%, Sc, Co, La, Sm and Eu; 2–4% Cr, Ce, Yb, Hf and Ta; 5–10% Rb, Cs, Nd, Tb, Lu and Th; >10% Ni, Zn and U.

GEOCHEMICAL SEQUENCES

Classification of flow groups

Variations in TiO_2, Mg# (molar MgO/[MgO + FeO*]), P_2O_5, Cr, Ba, and La with depth (Figs. 3–5) exemplify their overall patterns, abrupt changes, and anomalous values in several example core holes near TAN. Flow group designations shown in the sequences represent a geochemical classification following the rationale outlined here, with limitations from core descriptions, paleomagnetic signatures, and radiometric ages. Covariant chemical diagrams of elements having different behaviors during igneous processes, such as La versus MgO (Fig. 6), are used to depict separation of chemical patterns.

Flow groups in the northern INEEL region are named according to the following priority: (1) known source vent (Circular Butte); (2) geographic location of thickest sequences

(TAN series 1 and series 2; (3) recognizable time period (Jaramillo, pre-Olduvai); and (4) adjacent sedimentary interbed names (Q-R). Although each division is regarded as a flow group, some ultimately may be recognized as formations. Upper and lower designations are applied to some flow groups where a definite geochemical change is recognized in the pattern. The term "member" may be applied to such subclassifications; however, that term also may be appropriate for some sequences that cannot be subdivided due to lack of data or inconclusive geochemical correlations (Bartholomay et al., this volume).

Flow groups represented in core holes from the southern INEEL are included for geochemical comparison to flow groups near TAN, so their correlations have not been subjected to a detailed assessment. Thus, we use the alphabetical nomenclature of Anderson and coworkers (Anderson, 1991; Anderson and Liszewski, 1997; Anderson et al., 1997) for these groups

TABLE 3. ISOTOPIC Sr AND Nd ANALYSES FOR INEEL* COREHOLE AND SURFACE BASALT SAMPLES

Sample	Depth (m)	Analysis Lab	$^{87}Sr/^{86}Sr$ measurement	$^{143}Nd/^{144}Nd$ measurement	Age[†] (ka)	Sm	Nd	Rb[†]	Sr	$^{87}Sr/^{86}Sr$ initial	$^{143}Nd/^{144}Nd$ initial	εNd
							(ppm)			initial	initial	
ICP121-125	38.1	Rice	0.706020 ± 10	0.512445 ± 5	(320)	11.0	44	34	286	0.706018	0.512445	−3.8
2-2A-54	16.5	Rice	0.705926 ± 10	0.512468 ± 5	(100)	7.1	31	16	305	0.705926	0.512468	−3.3
298	90.9	Rice	0.706673 ± 6	0.512385 ± 15	580	8.0	38	16	278	0.706672	0.512385	−4.9
393	119.8	Rice	0.707409 ± 9	0.512394 ± 8	(760)	6.1	27	(10)	266	0.707408	0.512393	−4.8
422	128.7	Rice	0.706934 ± 10	0.512399 ± 9	(850)	9.8	47	(10)	319	0.706933	0.512398	−4.7
482	147.0	Rice	0.706973 ± 9	0.512370 ± 5	(910)	8.4	40	12	328	0.706972	0.512369	−5.2
586	178.7	Rice	0.706860 ± 9	0.512357 ± 4	(1120)	7.3	35	(10)	345	0.706859	0.512356	−5.5
717	218.6	Rice	0.707235 ± 9	0.512350 ± 6	(1350)	4.4	19	(10)	223	0.707233	0.512349	−5.6
821	250.3	Rice	0.705994 ± 10	0.512472 ± 5	(1570)	6.5	26	(10)	247	0.705991	0.512470	−3.2
909	277.1	Rice	0.706520 ± 9	0.512415 ± 5	(1720)	5.9	24	(10)	328	0.706518	0.512413	−4.4
1269	386.9	Rice	0.707647 ± 8	0.512390 ± 6	2450	7.1	28	(10)	346	0.707644	0.512388	−4.8
1927	587.5	Rice	0.707362 ± 9	0.512407 ± 5	3260	6.5	20	(10)	316	0.707358	0.512403	−4.5
2023	616.8	Rice	0.710173 ± 8	0.512151 ± 6	(3350)	6.9	27	17	271	0.710164	0.512148	−9.5
2327	709.5	Rice	0.707356 ± 8	0.512402 ± 6	(3810)	5.6	24	(10)	274	0.707350	0.512398	−4.6
2349	716.2	Rice	0.707012 ± 8	0.512403 ± 5	(3870)	4.6	21	(10)	199	0.707004	0.512400	−4.6
GIN5-30	9.1	UCLA	0.707014 ± 10	0.512397 ± 11	(900)	12.5	51	(10)	313	0.707013	0.512396	−4.7
119	36.3	UCLA	0.706939 ± 11	0.512363 ± 12	1044	10.3	40	(10)	305	0.706938	0.512362	−5.4
241	73.5	UCLA	0.706669 ± 10	0.512462 ± 10	1248	7.9	32	(10)	321	0.706667	0.512461	−3.4
327	99.7	UCLA	0.706828 ± 10	0.512349 ± 11	(1314)	8.3	42	(10)	327	0.706826	0.512348	−5.6
403	122.9	UCLA	0.706725 ± 10	0.512379 ± 11	(1314)	6.5	29	(10)	277	0.706723	0.512378	−5.1
GIN6-88	26.8	UCLA	0.706905 ± 10	0.512408 ± 10	1044	10.8	55	(10)	303	0.706904	0.512407	−4.5
133	40.5	UCLA	0.706838 ± 10	0.512388 ± 11	(1215)	8.3	37	(10)	331	0.706836	0.512387	−4.9
190	57.9	UCLA	0.706836 ± 10	0.512349 ± 9	1248	7.7	39	(10)	335	0.706834	0.512348	−5.6
JC-C-01	0	UCLA	0.707027 ± 11	0.512264 ± 12	1200	12.6	64	(10)	326	0.707025	0.512263	−7.3
JC-C-02	0	UCLA	0.706953 ± 10	0.512288 ± 9	1200	10.5	50	(10)	327	0.706951	0.512287	−6.8
TAN34-215	65.5	UCLA	0.706868 ± 10	0.512362 ± 11	1010	9.2	42	(10)	327	0.706867	0.512361	−5.4
283	86.3	UCLA	0.706727 ± 10	0.512353 ± 10	1100	6.8	26	12	320	0.706725	0.512352	−5.6
334	101.8	UCLA	0.706584 ± 10	0.512381 ± 12	1100	6.5	27	11	300	0.706582	0.512380	−5.0
416	126.8	UCLA	0.706654 ± 10	0.512419 ± 10	1200	5.2	20	14	247	0.706651	0.512418	−4.3
TCH1-61	18.6	UCLA	0.707025 ± 11	0.512301 ± 11	1100	13.6	73	(10)	309	0.707024	0.512300	−6.6
124	37.8	UCLA	0.706932 ± 11	0.512405 ± 12	(1300)	9.9	53	(10)	339	0.706930	0.512404	−4.5
216	65.9	UCLA	0.706851 ± 11	0.512133 ± 13	1300	7.6	36	(10)	334	0.706849	0.512132	−9.9
250	76.2	UCLA	0.706836 ± 10	0.512261 ± 11	1400	7.7	35	(10)	332	0.706834	0.512260	−7.4
370	112.8	UCLA	0.706694 ± 10	0.512392 ± 9	(1500)	5.1	23	13	240	0.706691	0.512391	−4.8
596	181.7	UCLA	0.706712 ± 10	0.512451 ± 11	(2200)	5.8	29	12	322	0.706709	0.512449	−3.6
TCH2-58	17.7	Rice	0.707022 ± 7	0.512398 ± 7	(1100)	11.1	52	(10)	322	0.707021	0.512397	−4.7
58	17.7	UCLA	0.706984 ± 13	0.512415 ± 10	(1100)	11.1	52	(10)	322	0.706983	0.512414	−4.4
99	30.2	Rice	0.706930 ± 8	0.512384 ± 5	(1100)	10.0	57	(10)	306	0.706929	0.512383	−5.0
151	46.0	Rice	0.706913 ± 9	0.512384 ± 6	(1200)	10.3	50	(10)	305	0.706911	0.512383	−5.0
169	51.5	UCLA	0.706895 ± 10	0.512306 ± 11	(1200)	9.1	45	(10)	321	0.706893	0.512305	−6.5
197	60.1	Rice	0.706889 ± 9	0.512400 ± 4	(1200)	8.8	39	(10)	324	0.706887	0.512399	−4.6
252	76.8	Rice	0.706935 ± 7	0.512356 ± 4	1320	8.4	54	(10)	322	0.706933	0.512355	−5.5
252	76.8	UCLA	0.706890 ± 11	0.512384 ± 12	1320	8.4	54	(10)	322	0.706888	0.512383	−5.0
296	90.2	UCLA	0.706823 ± 10	0.512406 ± 9	(1340)	8.1	48	(10)	320	0.706821	0.512405	−4.5
366	111.6	UCLA	0.706569 ± 10	0.512383 ± 9	(1360)	6.4	28	(10)	292	0.706567	0.512382	−5.0
465	141.8	Rice	0.707171 ± 8	0.512326	1410	4.2	21	(10)	220	0.707168	0.512325	−6.1
636	193.9	UCLA	0.706608 ± 10	0.512410 ± 11	2110	5.4	28	(10)	320	0.706605	0.512408	−4.4
1113	339.3	UCLA	0.706401 ± 10	0.512421 ± 12	(2300)	6.3	30	18	420	0.706397	0.512419	−4.2

* INEEL—Idaho National Engineering and Environmental Laboratory.
† Estimated ages and Rb values in parentheses for the purpose of calculating initial isotopic ratios.

without attempting to reclassify them into formational sequences.

TAN flow groups

The deepest core hole near TAN, TCH-2 (Fig. 3), yields both gradual changes and sudden shifts in chemistry through a relatively long time period (ca. 3-1 Ma). For example, in TCH-2 La (~15–50 ppm) and Hf (~3–9 ppm) remain fairly low throughout the deeper regions from 335 to 150 m and gradually increase in abundance from 150 m to the top of the core. Cr (~100–400 ppm) gradually decreases upward from 335 to 150 m, increases sharply at 150 m, then gradually decreases from 128 to 15 m. The chemical break at ~146 m corresponds to a hiatus in basalt accumulation from ca. 2 to ca. 1.4 Ma (Champion et al., this volume).

Geochemical patterns observed in the upper part of core hole TCH-2 are also present in shallower core holes TCH-1 (Fig. 4), GIN-5, GIN-6, and TCH-34. They are essentially indistinguishable from those found in the TCH-2 sequence; how-

TABLE 4. SUMMARY OF ⁴⁰Ar/³⁹Ar AGES OF INEEL* BASALT GROUNDMASS CONCENTRATES DETERMINED BY THE NEW MEXICO GEOCHRONOLOGY RESEARCH LABORATORY,† SOCORRO, NEW MEXICO

Sample	Depth (m)	L#	Irradiation	Weight (mg)	K/Ca	n	plateau ages %³⁹Ar	Age (Ma)	±2 σ
2-2A-54	16.5	9411-01	NM-92	74.4	0.14	7	92.2	0.03	0.09
CB20-153	46.6	9410-02	NM-92	87.9	0.12	6	76.1	0.24	0.10
CB23-155	47.3	9409-02	NM-92	71.3	0.19	6	83.6	0.34	0.10
2-2A-298	90.9	9412-01	NM-92	75.7	0.11	6	93.0	0.58	0.15
TAN-34-1D	66.5	9408-02	NM-92	97.9	0.14	6	61.1	1.01	0.10
TAN-34-6E	126.8	9413-01	NM-92	70.5	0.09	7	95.7	1.20	0.14

Notes: L#—laboratory number, K/Ca—molar ratio calculated from reactor produced ³⁹Ar$_K$ and ³⁷Ar$_{Ca}$, n—number of heating steps, %³⁹Ar—percent of ³⁹Ar released included in plateau.
* INEEL—Idaho National Engineering and Environmental Laboratory.
† NMGRL: M.T. Heizler and W.C. McIntosh, co-directors; L. Peters and R.P. Esser, technicians.

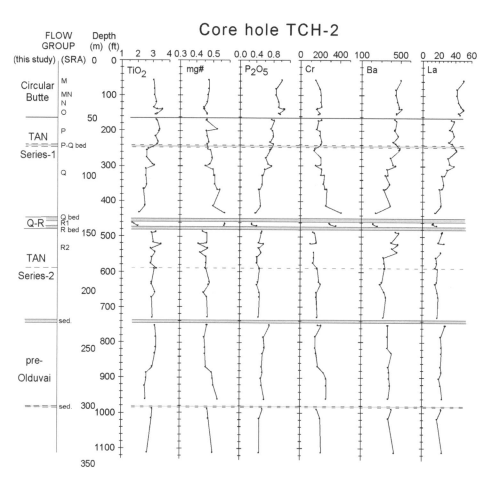

Figure 3. Geochemical log of selected elements (TiO$_2$ and P$_2$O$_5$ in weight percent; Cr, Ba, and La in ppm; Mg# is unitless) in basalt samples from core hole TCH-2 (Fig. 2) with radiometric ages and flow group designations. Sedimentary interbeds are shown as shaded horizons. Solid lines separate flow groups from each other or from sedimentary interbeds; dashed lines indicate chemical shifts or discontinuous sedimentary layers within flow groups. Chemical data include analyses from this study and from Knobel et al. (1995). Flow groups indicated by SRA (e.g., M, MN, N, O, P, etc.) are from Anderson and Bowers (1995) following nomenclature of Anderson (1991). References for radiometric ages (K-Ar or ⁴⁰Ar/³⁹Ar) are: 1, Lanphere et al. (1994); 2, M. Lanphere (1998, written commun.); 3, this study; values from Table 4. TAN is Test Area North.

ever, these chemical patterns are not observed in core hole 126-A (Fig. 4). Perhaps the most important observation is that individual lava flow groups are easily distinguished in covariation plots of La versus MgO (Fig. 6). The rationale for using these elements is their relative magmatic behaviors. Whereas La is highly incompatible in all major mineral phases in eastern Snake River Plain tholeiites and varies from ~7 to 70 ppm, Mg is a stoichiometric component of olivine and varies from about 4.5 to 11.5 wt% MgO.

The abundances of P$_2$O$_5$, La, Ba, and TiO$_2$ decrease with depth in the upper ~120 m of TAN core holes (Figs. 3 and 4). This geochemical trend has a punctuated shift at ~40 m, which implies two continuous episodes of magmatism. The sequence above the chemical shift is the Circular Butte flow group, based on geochemical comparisons with surface samples (Casper, 1999) and equivalent paleomagnetic measurements (Lanphere et al., 1994; D. Champion, 1998, written commun.). This correlation is also supported by radiometric K-Ar ages of 1.04 ±

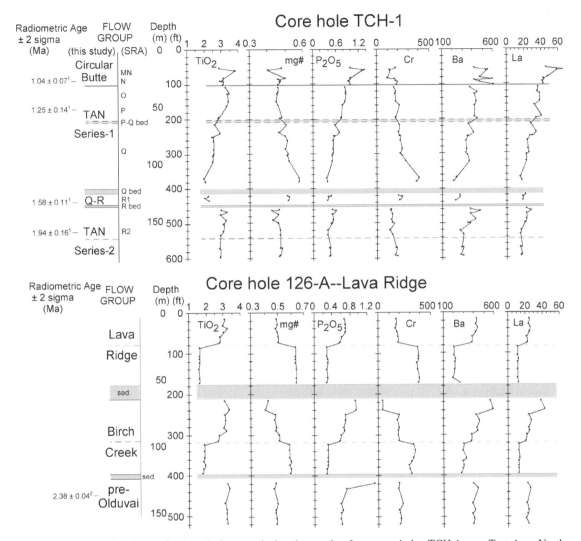

Figure 4. Geochemical logs of selected elements in basalt samples from core holes TCH-1 near Test Area North (TAN) and USGS 126-A located northwest of TAN (Fig. 2). Flow groups in 126-A are not recognized in other northern Idaho National Engineering and Environmental Laboratory core holes. Same parameters as Figure 3.

0.07 and 1.09 ± 0.17 Ma (uncertainty levels adjusted to 2-sigma) in TCH-1 and surface samples, respectively (Lanphere et al., 1994).

The lower ~80–90 m of the uppermost (120 m) sequence in the TAN core holes is designated the TAN series 1 flow group, and a thickness of ~50 m of this group (from ~145 to 195 m) is recognized in core hole 2-2A ~7 km south-southwest of TAN. Chemical gradients (Figs. 3 and 4) seem to be nearly continuous between the Circular Butte and TAN series 1 groups, but a slight chemical shift is apparent in La versus MgO plots (Fig. 6). Radiometric ages for the TAN series 1 flow group include the 1.20 ± 0.14 Ma ^{40}Ar/^{39}Ar age from TAN-34 (Table 4), a 1.25 ± 0.14 Ma K-Ar age from TCH-1 (Lanphere et al., 1994), and a 1.32 ± 0.04 ^{40}Ar/^{39}Ar age from TCH-2 (M. Lanphere, 1998, written commun.). An ^{40}Ar/^{39}Ar age of 1.01 ±

0.10 for a TAN-34 sample from 67 m depth (Table 4) is equivalent within analytical uncertainty to the 1.20 and 1.25 Ma ages of the chemically similar flows below (Geist et al., this volume, chapter 12).

TAN series 1 flow group correlation is somewhat complicated by the P-Q sedimentary interbed (Anderson and Bowers, 1995), which occurs in several core holes as much as 2 m thick in the middle of the lava flow sequence. The areal extent or depositional process of this interbed is not known; however, it is mainly fine sand, silt, and clay (Geslin et al., 1997). This may indicate local eolian deposition caused by dust storms, perhaps associated with wildfires, during an extended TAN series 1 eruptive episode. Thus, it may not be a single time-stratigraphic unit.

A major sedimentary layer, the Q-R interbed of Anderson

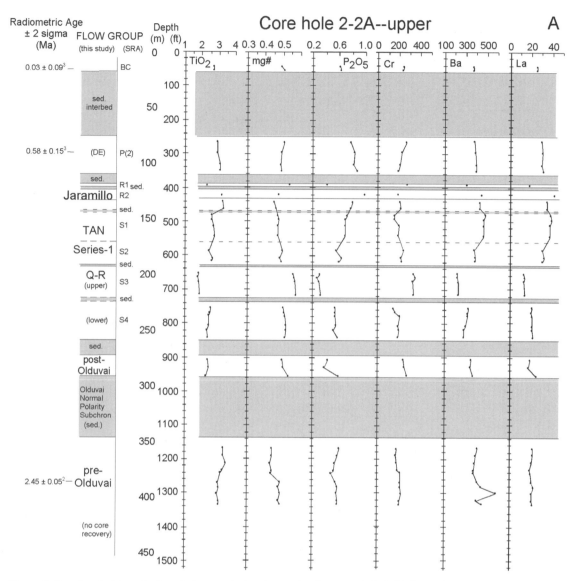

Figure 5. Geochemical log of selected elements in basalt samples from core hole 2-2A located south of Test Area North (TAN) (Fig. 2). Flow group designations are revised from previous studies. Sedimentary deposits include thick interbeds related to Big Lost Trough (Geslin et al., 1999; this volume; Bestland et al., this volume). Same parameters as Figure 3.

(Figure 5 continued on next page)

and Bowers (1995), underlies the TAN series 1 flow group. These sediments are penetrated in core holes TCH-1 and TCH-2 at depths of ~120–140 m, and an equivalent bed occurs in core hole 2-2A (Fig. 5) at a depth of ~190 m. This thick interbed is actually a compound unit representing at least two periods of nonvolcanic deposition intercalated by thin, discontinuous basalt lava flows. The intercalated lava flows constitute the Q-R flow group, which is a compound flow group chemically divisible into two sections (Fig. 6) and physically separated by a sedimentary layer in core hole 2-2A (Fig. 5). The upper flow of the Q-R flow group is observed in core hole TCH-2, and the lower flow is observed in TCH-1. Although the thickness of each flow group is ~30 m in core hole 2-2A, neither

part exceeds 10 m in the TCH core holes, which may indicate two separate flow groups. They have been dated by K-Ar as 1.41 ± 0.09 Ma and 1.58 ± 0.11 Ma (Lanphere et al., 1994).

Two thick lava flow sequences below the Q-R interbed portray fairly uniform chemistry, except for the uppermost 20–30 m. The upper sequence, named the TAN series 2 flow group, is ~80 m thick and is penetrated in the TCH-2 core (Fig. 3). It is also present in the lower 50 m of TCH-1 (Fig. 4). These two core holes represent the only known localities of this group.

The flow group below the TAN series 2 group, designated the pre-Olduvai flow group, is present in the lower 100 m of TCH-2 (Fig. 3), in the lower 25 m of 126-A (Fig. 4), and in the 345–400 m interval in 2-2A (Fig. 5). Some chemical overlap

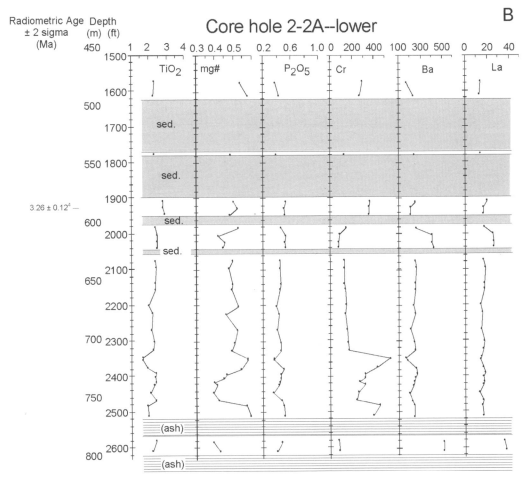

Figure 5. (*continued*)

occurs between these two groups (Fig. 6), but the TAN series 2 basalts form a tight linear trend that contrasts with a broad clustering of data for the pre-Olduvai flow group. Correlations of these two flow groups across the area are supported by radiometric ages. The TAN series 2 flow group has been dated as 2.05 ± 0.16 Ma in TCH-2 and as 1.94 ± 0.16 Ma in TCH-1 by Lanphere et al. (1994). K-Ar and $^{40}Ar/^{39}Ar$ ages confirm the pre-Olduvai flow group in each of the three sample locations: 2.56 ± 0.07 Ma and 2.46 ± 0.09 in TCH-2 (Lanphere et al., 1994); 2.38 ± 0.04 Ma in 126-A (Table 4); and 2.45 ± 0.05 Ma in 2-2A (Champion et al., this volume). This flow group is immediately beneath a thick sedimentary interbed in core hole 2-2A that Geslin et al. (this volume) describe as Olduvai Lake, representing the ca.1.77–1.95 Ma Olduvai normal polarity subchron within the Matuyama reversed zone. Thus, we designate this lava sequence the pre-Olduvai flow group primarily due to its regional occurrence, although other flow groups may be closer in age to the Olduvai subchron.

Several other distinctive flow groups are represented in the upper and lower parts of core hole 2-2A (Fig. 5), but are not sampled by the other available cores. The uppermost sequence

is <20 m of relatively young basalt dated as 0.03 ± 0.09 Ma (Table 4) and included in the BC flow group by Anderson et al. (1997). Below this group is ~50 m of sediment that overlies an ~30-m-thick basalt flow group dated as 0.58 ± 0.15 Ma (Table 4). Anderson and Liszewski (1997) originally designated this the P(2) flow group; however, due to age limitations and chemical similarity with the DE series of flows represented in wells near INTEC (Anderson, 1991), we tentatively reassign this sequence to the DE flow group. Below the DE layer is an ~20-m-thick sedimentary interbed that contains a thin intercalated basalt flow (R1; Anderson, 1991) of undetermined age. The Jaramillo flow group, a ~20 m basalt sequence intercalated between sedimentary layers beneath the R1 flow, is so designated by its normal paleomagnetic signature (D. Champion, 1998, written commun.) and is the only basalt representative of the Jaramillo normal polarity subchron within the Matuyama reversed zone.

Flow groups below the pre-Olduvai sequence in core hole 2-2A have not been named or correlated, although a thin flow found within a thick sedimentary sequence at ~545 m probably represents the Kaena subchron, and the lava sequence under-

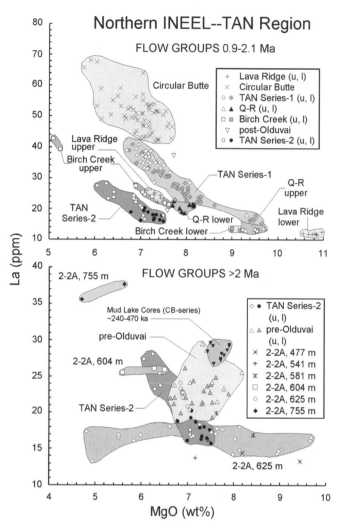

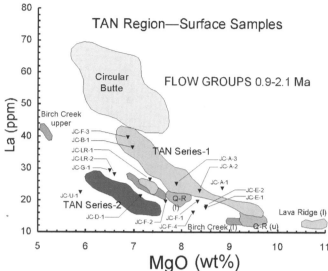

Figure 7. Covariation of La vs. MgO in surface basalt samples (Fig. 2) in northern Idaho National Engineering and Environmental Laboratory region compared to flow groups outlined in Figure 6. Samples collected north and west of Test Area North (TAN) are chemically more related to Lava Ridge and Birch Creek flow groups; samples from Antelope Butte have close affinity to TAN series 1 flow group.

Figure 6. Covariation of La vs. MgO in basalt flow groups from northern Idaho National Engineering and Environmental Laboratory (INEEL)-Test Area North (TAN) region. Surface exposures at least 0.9 Ma (Kuntz et al., 1994) and subsurface data, including radiometric ages and paleomagnetic signatures (Lanphere et al., 1994; Champion et al., this volume), were used to determine flow group fields. Upper and lower parts of some flow groups, designated u and l, indicate chemical or stratigraphic (interbed) breaks in otherwise continuous sequences. Basalt lava flow groups that are uncorrelated and/or unsubdivided are shown as core hole name followed by depth to top of series (e.g., 2-2A, 477 m). TAN series 2 basalts are shown in both diagrams due to age overlap; Mud Lake samples from CB-series core holes (Fig. 2) are shown for comparison and represent youngest stage of volcanism near TAN (ca. 240–270 ka).

core hole 126-A (Figs. 4 and 6) located at Lava Ridge northwest of the TAN facility (Fig. 2). Four distinct chemical shifts, two that are coincident with sedimentary interbeds, divide the more than 150 m of core into three recognized flow groups. The lowermost sequence is the pre-Olduvai flow group described herein. The upper two chemical patterns represent nearly identical flow groups designated the Lava Ridge and Birch Creek flow groups (Fig. 4). Each one directly overlies a sedimentary interbed and is composed of a lower and upper member. The two groups are observed only in core hole 126-A, and thus cannot be correlated to other flow groups in the northern INEEL. The lower part of each group is characterized by low TiO_2, P_2O_5, Ba, and La, and high Mg# and Cr, and is overlain without a break in volcanic deposition by a succession of more evolved basalts having higher and more variable TiO_2, P_2O_5, Ba, and La. Although each of the chemical patterns is similar overall to the succession from TAN series 1 to Circular Butte flows, the chemical transitions are dramatic, suggesting two distinct eruptive episodes within each flow group.

Regional correlations near TAN

Several flow groups in the TAN core holes can be correlated tentatively to exposed vents mapped by Kuntz et al. (1994) shown in Figure 2. Surface samples other than those from Circular Butte (Casper, 1999) are plotted in Figure 7 for comparison with possible flow groups recognized in core holes. The JC-F-3 sample obtained from a vent located ~7 km west of TAN is within the known chemical pattern of the TAN series

lying the sedimentary sequence at ~580 m has been dated as 3.26 ± 0.12 Ma (Champion et al., this volume).

Lava Ridge flow groups

In contrast to the smooth geochemical trends noted in the TAN lava flow groups (e.g., steadily increasingly P_2O_5 with height), geochemical patterns exhibit punctuated changes in

1 flow group. The exposed vent also has equivalent paleomagnetic inclinations (D. Champion, 1998, written commun.) and probably represents a source of TAN series 1 lava flows. Lava Ridge samples (JC-LR-1 and 2) plot along the well-defined trend shown for Lava Ridge and Birch Creek flow groups, and they probably are part of those groups. Antelope Butte samples plot with the TAN series 1 flow group, but there is an apparent relative age discrepancy. The vent was mapped (Kuntz et al., 1994) as being younger than the 1.1 Ma Circular Butte lava field, although the TAN series 1 flow group is stratigraphically lower. Thus, additional work is needed to test this hypothesis.

Regional correlations of basaltic flow groups in the northern part of the INEEL are summarized in Figure 8. The correlations are made on the basis of their chemical signatures in TAN core holes, along with data from sedimentary interbeds, radiometric ages, and paleomagnetic signatures (e.g., Lanphere et al., 1994).

Several flow groups in the northern INEEL region flowed southward due to the development of a topographic high along the Lava Ridge–Hell's Half Acre volcanic rift zone (Fig. 8). Some flows made it to the Big Lost Trough, represented in core hole 2-2A, but the basin apparently collected lava flows from other directions as well. TAN series 1 and the lower member of the pre-Olduvai flow groups are relatively thick and correlate between TAN and 2-2A. They probably originated from the Lava Ridge–Hell's Half Acre volcanic rift zone and flowed southwest toward the Big Lost Trough basin. The upper member of the pre-Olduvai flow group in TCH-2 and the TAN series 2 group do not occur in 2-2A and thus represent thick sequences near the eruptive center that were responsible for topographic construction. To the contrary, post-Olduvai basalts in core hole 2-2A do not correlate with TCH-2, and likely were derived from the Axial volcanic zone. The source of the Jaramillo flow found only in 2-2A is unknown, but younger (<1 Ma) flow groups sampled in 2-2A and in other core holes and wells located in the central INEEL (Anderson, 1991) were possibly derived from eruptive centers in southern INEEL.

Flow groups in southern INEEL regions

Chemical patterns in flow groups near INTEC are exemplified in core hole ICPP-COR-A-023 (Fig. 9) and in USGS 121 and 123 (Reed et al., 1997). Flow groups near the RWMC are exemplified in core holes BG-77-1, USGS-118, C-1A, and STF-PIE-AQ-01 (Fig. 10).

Previous stratigraphic assignments from the INTEC region (Anderson, 1991) are poorly defined by the chemical data (Fig. 9). These groups are younger than TAN flow groups (Kuntz et al., 1994; Champion et al., this volume) and portray less distinct chemical variation with depth. Subtle chemical shifts coincide with stratigraphic divisions of flow groups BC, DE, I, and J (Anderson, 1991; S.R. Anderson, 1999, written commun.), but flow groups E, FG, and H cannot be clearly distinguished from each other (Fig. 11).

Most of the flow groups in core holes near the INTEC might actually represent multiple sequences that cannot be subdivided without more detailed sampling. The groups are thinner than the flow groups in the northern INEEL and are thus not well characterized chemically. This is especially true for the DE flow group (Fig. 11), in which the chemical variation is large (e.g., see Reed et al., 1997) and apparently has numerous widely separated vents. Therefore, these flow groups may not correlate to any other sites at INTEC, a relation that was initially determined by Anderson and Liszewski (1997) and Wetmore (1998). Our interpretation is that fewer and smaller basaltic vents are near the INTEC along the Howe–East Butte volcanic rift zone (Fig. 1) and that many distal thin lava sequences accumulated in this region from different sources because of basin subsidence.

Conversely, flow group subdivisions in the RWMC region are easily recognized in chemical patterns (Figs. 10 and 11) and coincide well with the Anderson et al. (1996) correlations. The most obvious breaks occur above and below flow groups E and F. Flow group E is characterized by low TiO_2, P_2O_5, Ba, and La, with concomitantly higher Mg# and Cr. This pattern is similar to that observed in the lower members of the much older Lava Ridge and Birch Creek sequences found in core hole 126-A. Flow group F is chemically distinct from E in the RWMC core holes, and it has increasing TiO_2, P_2O_5, Ba, and La with height, along with relatively constant or slight decrease in Mg# and Cr.

IMPLICATIONS FOR MAGMATIC EVOLUTION

Geochemical data

Geochemical diversity, and the distinction of lava flow groups, is greater at TAN and the RWMC than at INTEC. This may reflect their proximity to the more well-defined volcanic rift zones, Arco-Big Southern Butte (RWMC) and Lava Ridge–Hell's Half Acre (TAN), compared to the Howe–East Butte volcanic rift zone, which is topographically and volcanically much less distinct. The similarities of Lava Ridge and the RWMC chemical patterns indicate the occurrence of similar, time-independent modes of magmatism on the eastern Snake River Plain. The melting processes and evolution of magmas in the crust must be regular and repetitive in space and time.

The wide range in incompatible elements at constant MgO (Figs. 6 and 11) clearly illustrates separate magma batches associated with each flow group, which is consistent with the idea that each flow group represents a separate shield-building magmatic episode. Separate magma batches are less evident on covariation plots of two incompatible elements, such as Ba and TiO_2 (Fig. 12), which show overlapping patterns for each group. For example, flow group F has two distinctly different La versus MgO patterns and a sharp break in Ba versus TiO_2 from the lower to upper parts. This represents either a change in chemical trend during the eruptive episode, or that the F

166

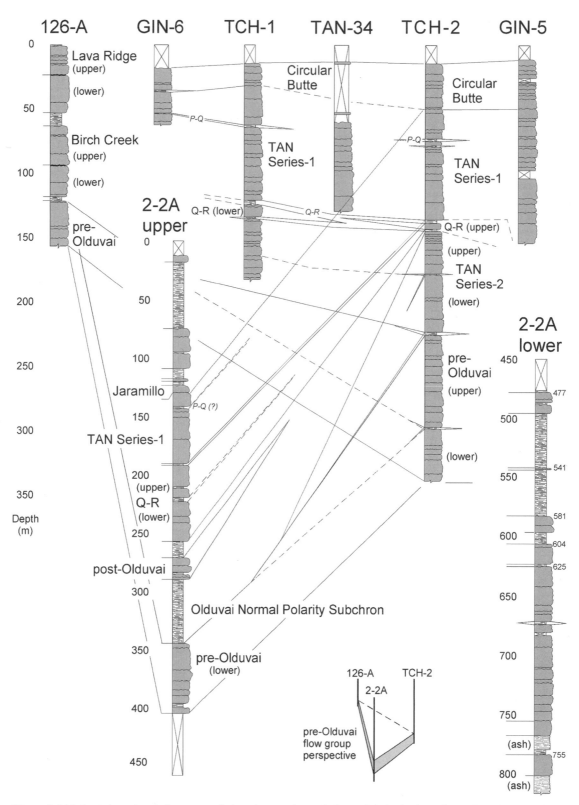

Figure 8. Lithologic logs, basalt flow group designations, and correlations of major horizons in northern Idaho National Engineering and Environmental Laboratory. Flow groups are shaded and sediment (or rhyolitic ash) interbeds are shown as light bedded layers. Several flow groups have upper and lower members distinguished by sudden changes in their geochemical pattern with stratigraphic height or sedimentary interbeds. Inset illustrates relative locations of three core holes shown in Figure 2 in which pre-Olduvai flow group was sampled. Compare to Figures 3–7.

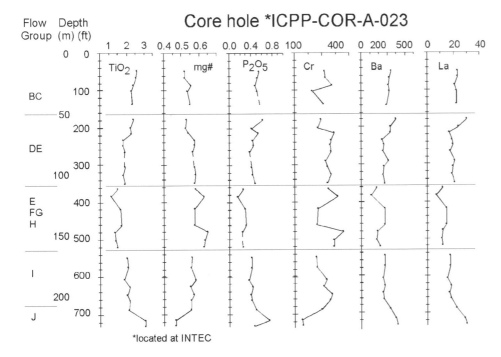

Figure 9. Geochemical logs of selected elements in basalt samples from core hole ICPP-COR-A-023 located near Idaho Nuclear Technology and Engineering Center (INTEC) (ICPP, Fig. 1). Flow group designations are from Anderson (1991). Flow groups DE and E-FG-H series represent distribution of many possibly unrelated flow units erupted from distant vents. They cannot be subdivided with current geochemical database. Same parameters as Figure 3.

group is actually two flow groups. The lower to upper F sequence is found in three of four core holes at the RWMC (Fig. 10). The sequence has been regarded in previous studies as a single unit because of its unique reversed polarity (Anderson and Liszewski, 1997), although Champion et al. (1988) subdivided the flow group into four parts based on paleomagnetic inclinations. Isopach contours by Wetmore (1998) based on geophysical well-log interpretations by Anderson (1991), suggest that the F flow group is associated with a single eruptive center, i.e., it represents a single buried shield volcano. The two separate chemical trends imply that this eruptive center, although constructed during a single episode of volcanism, may have been formed by two separate magma batches.

Flow groups with primitive chemical signatures seem to be recurrent in some places, despite widely different ages. As mentioned above, flow group E at the RWMC has strong chemical affinity with the lower members of the Lava Ridge and Birch Creek flow groups. The chemical transition from F to E is similar to that noted for the Lava Ridge groups and in the lower-to-upper Q-R group. The high-Mg and low-Ti samples of flow group E and the lower part of the Lava Ridge flow group are the most primitive basaltic compositions recorded on the eastern Snake River Plain, and they may represent parental magmas. High-Mg and low-Ti compositions from other flow groups may also represent parental magmas.

Notably, the upper and lower members of flow groups F and Q-R yield continuous patterns in Figure 12, which supports a single parental magma for these groups. Upper and lower members of the Lava Ridge and Birch Creek flow groups, on the other hand, plot as separate chemical clusters on the Ba versus TiO₂ diagram. Many of the groups in Figure 12 yield

scattered Ba abundances at higher TiO_2 values, especially Birch Creek and TAN series 2 flow groups. This attests, in part, to the open-system behavior (Geist et al., this volume, chapter 12) of the incompatible elements during magma genesis and ascent. It is also related to variability in the compositions of magmatic sources and the dynamics of partial melting associated with each batch of magma (Hughes et al., 2002).

Isotopic data

The overall $^{87}Sr/^{86}Sr$ and $^{143}Nd/^{144}Nd$ signatures of eastern Snake River Plain basalts (Fig. 13) support assimilation and variability in magmatic sources as causes for compositional diversity (Menzies et al., 1984), but the relative contribution of these processes remains unclear. For example, Circular Butte and TAN series 1 flow groups at TAN have essentially uniform $^{87}Sr/^{86}Sr$ ratios with more variable $^{143}Nd/^{144}Nd$ ratios (Fig. 13, detail). Variations in εNd and $^{87}Sr/^{86}Sr$ with Sm and Sr (Fig. 13, right side) reveal no systematic trend that could relate isotopic values with elemental abundances.

More important to this study, however, is the degree to which isotopic signatures help distinguish between basalt lava flow groups. The abundances of Sm within Circular Butte and TAN series 1 flow groups at TAN are clustered in Figure 13, as expected from La variations; yet they all have essentially the same range in εNd, except for one or two outliers. Similar clustering is apparent in $^{87}Sr/^{86}Sr$ ratios; each flow group has a unique Sr isotopic signature, although Sr abundances are less distinctive. The $^{87}Sr/^{86}Sr$ clusters are too close to each other and overlap outside of analytical uncertainty. It appears from these examples that isotopic signatures are decoupled from the

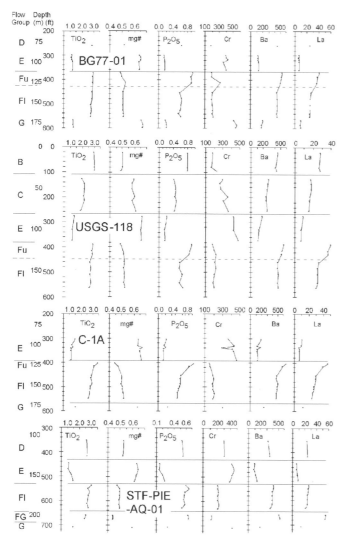

Figure 10. Geochemical logs of selected elements in basalt samples from core holes BG77-01, USGS-118, C-1A, and STF-PIE-AQ-01 located near Radioactive Waste Management Complex (RWMC) (Fig. 1). Flow groups from Anderson (1991) and Anderson and Liszewski (1997) are easily distinguished by geochemical patterns and significant shifts in element abundances. Flow group F has been subdivided into upper and lower member based on geochemistry. Same parameters as Figure 3.

TAN series 1 flow groups (Fig. 13, detail) could reflect either open-system differentiation subsequent to magma genesis (Geist et al., this volume, chapter 12) or different sources.

The observed variation in trace element abundance with little isotopic change, or no systematic trend, argues against simple upper crustal contamination. This is consistent with previous arguments that disregarded crustal contamination as a major contributor to compositional variability in eastern Snake River Plain olivine tholeiites (Leeman, 1982a, 1982b), although assimilation cannot be ruled out. Basalts represented by samples with anomalous isotopic signatures, such as TCH-1 at 65.7 m and 2-2A at 616.8 m, and the variations in the remaining samples that trend toward these samples (Fig. 13), could be produced by assimilation of Cretaceous granitoids (Clarke, 1990) or Archean lower crust (^{143}Nd/^{144}Nd ~0.5108–0.5114, ^{87}Sr/^{86}Sr ~0.705–0.727 and ^{143}Nd/^{144}Nd ~0.5104–0.5107, ^{87}Sr/^{86}Sr ~0.715–0.734) (Leeman et al., 1985). This would require, however, that rare earth element (REE) and Sr abundances are not sensitive to contamination, even though the isotopic values are more easily reequilibrated, which is unlikely.

Alternatively, the isotopic signatures could reflect compositional heterogeneity within mantle source regions (Hughes et al., 2002). Overall, the isotopic signatures imply an EM2-type enriched mantle, which is one of two major types of enriched mantle having higher ^{87}Sr/^{86}Sr and lower ^{143}Nd/^{144}Nd ratios compared to bulk earth (Zindler and Hart, 1986). The EM2 isotopic signature is characteristic of the circumcratonic domain in the western United States (Menzies, 1989). The weak grouping of flow groups shown in Figure 13 may be related to a source region that is heterogeneous on the scale of flow group melting regimes.

DISCUSSION

The wide range in elemental abundances and isotopic ratios in eastern Snake River Plain tholeiites is probably related to some combination of fractional crystallization, crustal assimilation, and source heterogeneity. Relatively minor isotopic variation in eastern Snake River Plain flow groups is apparently decoupled from trace element abundances (such as REE and Sr) as shown in Figure 13, and the entire range in incompatible elements in eastern Snake River Plain flow groups cannot be related to simple fractional crystallization. The La versus MgO values in Figures 6, 7, and 11 illustrate the inadequacy of simple crystal fractionation to account for all the variation in trace elements. Only a small amount of olivine or pyroxene fractionation will produce the range in MgO abundances, although the range in incompatible elements exemplified by La would require much greater amounts of fractional crystallization (e.g., Shervais et al., 1994). Nonetheless, fractional crystallization of olivine and plagioclase phenocrysts in these basalts (phases observed in eastern Snake River Plain basalts and stable at <10 kbar) has probably been important in eastern Snake River Plain petrogenesis (Leeman, 1982a). If parental magmas of eastern

trace element patterns and their distinction is only marginally sufficient for their use in flow group correlation. Thus, isotopic data alone cannot distinguish eastern Snake River Plain lava flow groups and probably are better suited for assessment of the characteristics of magma genesis and evolution.

Although the isotopic data for eastern Snake River Plain basalts are not nearly as complete as the geochemical database and a detailed discussion of petrogenesis is beyond the scope of this chapter, some consideration of how isotopic systematics relate to magmatic evolution is important to better understand eastern Snake River Plain basalt lava flow groups. The small spread of ^{143}Nd/^{144}Nd versus ^{87}Sr/^{86}Sr in Circular Butte and

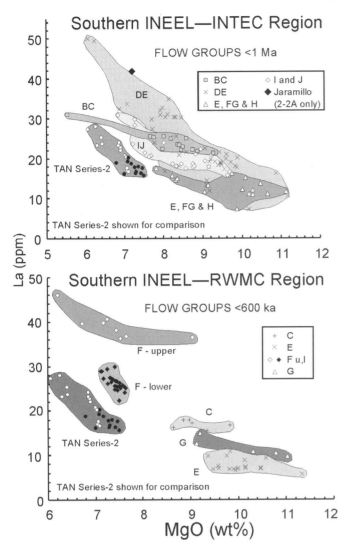

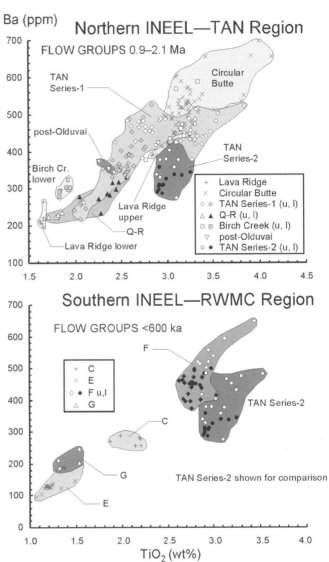

Figure 11. Covariation of La vs. MgO in southern Idaho National Engineering and Environmental Laboratory (INEEL) near Idaho Nuclear Technology and Engineering Center (INTEC [ICPP]) and Radioactive Waste Management Complex (RWMC). Flow group designations are from Anderson (1991). The Test Area North (TAN) series 2 flow group from northern INEEL core holes is shown for comparison. Flow groups near INTEC are not readily distinguished on their geochemical patterns, in contrast to those at RWMC, which are easily distinguished and represent separate magma batches. Compare to Figures 9 and 10.

Figure 12. Covariation of Ba vs. TiO$_2$ in Test Area North (TAN) and Radioactive Waste Management Complex (RWMC) flow groups illustrate continuous trends for all groups except Lava Ridge and Birch Creek groups. TAN series 2 flow group from northern Idaho National Engineering and Environmental Laboratory (INEEL) core holes is shown for comparison to southern INEEL flow groups. These covariation patterns illustrate minimum discrimination of magma batches by plotting two elements that are incompatible to phenocrysts and source mineral phases.

Snake River Plain tholeiitic basalts are compositionally similar after partial melting in the mantle, then derivative compositions must involve open-system assimilation in concert with fractional crystallization. In this case the open-system models proposed by Geist et al. (this volume, chapter 12) for a single series of TAN basalts (TAN series 1 flow group) would apply to the compositional variations within other eastern Snake River Plain flow groups.

An alternative and equally viable hypothesis is that isotopic and incompatible element signatures are generated in the source

region. Different incompatible element concentrations at high MgO values suggest a different parental magma for each flow group. Isotopic differences also would be expected if magma batches are generated from different sources in a heterogeneous subcontinental lithospheric mantle (e.g., Hart, 1985; Leeman and Vitaliano, 1976; Menzies et al., 1984; Hildreth et al., 1991; Reid, 1995; Hanan et al., 1997).

The recurrence of chemical types throughout time is apparently unrelated to the spatial distribution of eruptive centers

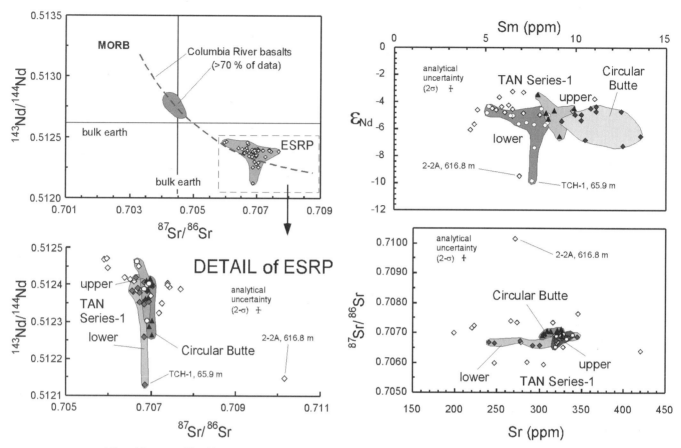

Figure 13. Isotopic $^{143}Nd/^{144}Nd$ and $^{87}Sr/^{86}Sr$ ratios for northern Idaho National Engineering and Environmental Laboratory (INEEL) basalt samples compared to mid-ocean ridge basalt (MORB) and Columbia River basalt fields (after Wilson, 1989). Eastern Snake River Plain (ESRP) samples essentially reiterate field defined by previous analyses of ESRP olivine tholeiites (Basaltic Volcanism Study Project, 1981) and plot in EM2-type mantle of Zindler and Hart (1986). Although some distinction is possible between Test Area North (TAN) series 1 and Circular Butte flow groups, there is no significant change in εNd or Sr isotopic signature with rare earth element or Sr ppm values. Small differences in isotopic signature between various flow groups are believed to be related to source heterogeneity. Two samples that plot away from general cluster, TCH-1 at 65.9 m and 2-2A at 616.8 m, represent probable crustal contamination.

(Hughes et al., 2002). This suggests that the petrogenetic processes responsible for each flow group operate repeatedly and consistently in space and time. Once a fertile source has undergone partial melting and magma has been extracted, further melting of the depleted residuum is unlikely without additional heat input. Hughes et al. (2002) propose that some of the chemical variability in a flow group may be related to variability in the amount of partial melting, such that interior regions of a partial melt zone would undergo greater degrees of melting than distal regions. Tapping of regions with successively lower degrees of melting would yield a wide range in incompatible trace elements and major oxides (O'Hara, 1985). Support for this model is found in chemical stratigraphy depicting overall increases in incompatible element abundances (e.g., La) with stratigraphic height in core samples, especially for older (>1 Ma) basalts in the northern part of the INEEL. Increases in incompatible elements with stratigraphic level also exist in individual flow groups, such as flow group F, Circular Butte, and TAN series 1. This suggests that some individual batches of

magma are derived from zoned melting regions and yield more than one chemical series in a single volcano.

The basalts from Lava Ridge core hole 126-A are compositionally similar to surface basalts in the region north and west of TAN (Fig. 6). This implies that this chemical type may be the product of a single source region beneath the northwestern part of the Lava Ridge–Hell's Half Acre volcanic rift zone. The same argument may apply to the Antelope Butte and Circular Butte areas east of TAN, which also display some chemical similarity, although they have a different age and magnetic polarity. These relations support the concept of isolated magmatic systems on the eastern Snake River Plain, even within a single volcanic rift zone.

This petrologic model of eastern Snake River Plain magmatism is consistent with geophysical data and structural interpretations of bedrock geology. Seismic measurements by Peng and Humphreys (1998) confirm the presence of an extensive 9-km-thick mid-crustal gabbroic sill proposed by Sparlin et al. (1982) based on gravity modeling. The sill, thought to be ca.

10 Ma, is responsible for much of eastern Snake River Plain subsidence (McQuarrie and Rodgers, 1998), which caused sediment-starved basins such as the Big Lost Trough to form between major volcanic rift zones. Seismic studies by Saltzer and Humphreys (1997) and Peng and Humphreys (1998) indicate a corridor of low-velocity, partially melted mantle extending from the lower crust to depths as great as 200 km. These indicate a great depth range where magmas are being actively generated. Saltzer and Humphreys (1997) attributed the low-velocity region to partially melted peridotite. Melting and melt extraction (depletion) are dominant processes influencing compositional variation within the upper mantle.

CONCLUSIONS

Geochemical and isotopic data suggest separate magma batches for individual shield volcanoes, each of which is probably derived by partial melting of enriched EM2-type subcontinental lithospheric mantle. Chemical variation over the entire eastern Snake River Plain system of tholeiites is substantial; there is seven- to ten-fold enrichment in incompatible trace elements and two- to three-fold variation in major element oxides, especially MgO. Systematic covariation of incompatible and compatible elements in single flow groups can be related to either different parental magmas (due to source heterogeneity and variable partial melting), or open-system assimilation and fractional crystallization during magma ascent.

Single flow units may not be correlated over great distances (cf. Geist et al., this volume, chapter 4), supporting the idea that single flows are spatially restricted to ~<10 km and very unlike flood basalts (e.g., Reidel et al., 1989). Major and trace element signatures can be used to distinguish and correlate basalt flow groups within their vertical and areal extents when core samples are available. Chemical correlations of different eastern Snake River Plain flow groups are, in some cases, equivocal because chemically equivalent magmas recur in time and space. Detailed correlations require companion studies using core logs, the presence or absence of sedimentary interbeds, radiometric ages, paleomagnetic signatures, geophysical logging, petrographic analyses, and surface mapping, to be viable.

Distinct chemical patterns enable correlation of all flow groups penetrated by core holes located near TAN except for those in core hole USGS 126-A, which are not recognized elsewhere. Flow groups are also compositionally distinguished in core holes at the RWMC and, in some cases, at the INTEC (ICPP); however, several previously assigned flow groups cannot be correlated in the INTEC core holes on the basis of geochemical data. These include the DE and E-FG-H sequences, which may represent piecemeal distribution of many flow groups erupted from distant vents or from small local eruptions. Thin narrow lobes flowed into the INTEC region due to subsidence of the Big Lost Trough.

Subsequent to the regional ca. 2.4 Ma pre-Olduvai flow groups recognized in core holes TCH-2, 126-A, and 2-2A, at least three separate magmatic episodes constructed the northern end of the Lava Ridge–Hell's Half Acre and Circular Butte–Kettle Butte volcanic rift zone. The TAN series 2 represents a ca. 2.0 Ma volcanic episode that produced one flow group. The TAN series 1, dated as ca. 1.3 Ma, and the ca. 1.1 Ma Circular Butte flow group (and possibly Antelope Butte lava field) represent a second episode that produced several magma batches characterized by a gradual increase in incompatible elements and decrease in compatible elements with stratigraphic height. The third volcanic episode occurred in the Lava Ridge–Birch Creek region, which is inferred from core hole 126-A and surface samples northwest of TAN. This episode represents a 0.8–0.9 Ma eruptive system that ranged from Richard Butte southward to areas southwest of TAN. Two lava flow groups, the Birch Creek and Lava Ridge groups, erupted in sequence near Lava Ridge. Each began with fairly primitive MgO-rich, La-poor flow units, then more chemically evolved lava flow units followed the chemical trend initiated by the more primitive types. Recurrence of this chemical pattern at the RWMC suggests that equivalent processes of magma genesis and shield growth are operative on eastern Snake River Plain volcanic rift zones regardless of age. Thus, the individual volcanic rift zones do not have unique geochemical characteristics.

ACKNOWLEDGMENTS

We are deeply indebted to Steven R. Anderson, U.S. Geological Survey, for his extensive work on basalt flow correlations, and for his untiring efforts to keep volcanic stratigraphy at the forefront of subsurface investigations at the Idaho National Engineering and Environmental Laboratory. This paper benefited greatly from numerous discussions with him, and with U.S. Geological Survey colleagues Mel Kuntz, Duane Champion, and Marvin Lanphere, who collectively helped mold the arguments presented in this paper. Insightful reviews by Bill Hackett, Bill Hart, Mel Kuntz, Steve Reidel, Dick Smith, and editor Paul Link significantly improved the quality of this manuscript, although the authors take full responsibility for any misrepresentations or lack of clarity. This research was supported by grant DE-FG07-96ID13420 from the U.S. Department of Energy to the Idaho Water Resources Research Institute, subcontract KEK066-97-B to Idaho State University. Support for neutron irradiation was provided by the Radiation Center at Oregon State University, the U.S. Department of Energy Office of New Production Reactors, and Assistant Secretary for Environmental Management under Department of Energy Idaho Operations Office contract DE-AC07-94ID13223.

REFERENCES CITED

Anderson, S.R., 1991, Stratigraphy of the unsaturated zone and uppermost part of the Snake River Plain aquifer at the Idaho Chemical Processing Plant and Test Reactors Area, Idaho National Engineering Laboratory, Idaho: U.S. Geological Survey Water-Resources Investigations Report 91-4058 (DOE/ID-22097), 35 p.

172   *S.S. Hughes, M. McCurry, and D.J. Geist*

Anderson, S.R., and Bartholomay, R.C., 1995, Use of natural-gamma logs and cores for determining stratigraphic relations of basalt and sediment at the Radioactive Waste Management complex, Idaho National Engineering Laboratory, Idaho: Journal of the Idaho Academy of Science, v. 31, p. 1–10.

Anderson, S.R., and Bowers, B., 1995, Stratigraphy of the unsaturated zone and uppermost part of the Snake River Plain aquifer at Test Area North, Idaho National Engineering Laboratory, Idaho: U.S. Geological Survey Water-Resources Investigations Report 95-4130 (DOE/ID-22122), 47 p.

Anderson, S.R., and Liszewski, M.J., 1997, Stratigraphy of the unsaturated zone and the Snake River Plain Aquifer at and near the Idaho National Engineering and Environmental Laboratory, Idaho: U.S. Geological Survey Water-Resources Investigations Report 97-4183, 65 p.

Anderson, S.R., Liszewski, M.J., and Cecil, L.D., 1997, Geologic ages and accumulation rates of basalt-flow groups and sedimentary interbeds in selected wells at the Idaho National Engineering Laboratory, Idaho: U.S. Geological Survey Water-Resources Investigations Report, 97-4010 (DOE/ID-22134), 39 p.

Anderson, S.R., Ackerman, D.J., Liszewski, M.J., and Feiburger, R.M., 1996, Stratigraphic data for wells at and near the Idaho National Engineering Laboratory, Idaho: U.S. Geological Survey Open-File Report 96-248 (DOE/ID-22127), 27 p.

Anderson, S.R., Kuntz, M.A., and Davis, L.C., 1999, Geologic controls of hydraulic conductivity in the Snake River Plain aquifer at and near the Idaho National Engineering and Environmental Laboratory, Idaho: U.S. Geological Survey Water-Resources Investigations Report 99-4033, 38 p.

Basaltic Volcanism Study Project, 1981, Basaltic volcanism on the terrestrial planets: New York, Pergamon Press, 1286 p.

Blair, J.J., and Link, P.K., 2000, Pliocene and Quaternary sedimentation and stratigraphy of the Big Lost Trough from coreholes at the Idaho National Engineering and Environmental Laboratory, Idaho: Evidence for a regional Pliocene lake during the Olduvai normal polarity subchron, in Robinson, L., ed., Proceedings of the 35th Symposium on Engineering Geology and Geotechnical Engineering: Pocatello, Idaho, Idaho State University, College of Engineering, p. 163–179.

Casper, J.L., 1999, The volcanic evolution of Circular Butte [M.S. thesis]: Pocatello, Idaho, Idaho State University, 113 p.

Champion, D.E., Lanphere, M.A., and Anderson, S.R., 1996, Further verification and $^{40}Ar/^{39}Ar$ dating of the Big Lost River reversed polarity subchron from drill core subsurface samples of the Idaho National Engineering Laboratory, Idaho [abs.]: Eos (Transactions, American Geophysical Union), v. 77, no. 46, p. 165.

Champion, D.E., Lanphere, M.A., and Kuntz, M.A., 1988, Evidence for a new geomagnetic reversal from lava flow in Idaho: Discussion of short polarity reversal in the Brunhes and late Matuyama polarity chrons: Journal of Geophysical Research, v. 93, p. 11667–11680.

Clarke, C.B., 1990, The geochemistry of the Atlanta Lobe of the Idaho Batholith in the Western United States Cordillera [Ph.D. thesis]: Milton Keynes, The Open University, 357 p.

Davis, L.C., Hannula, S.R., and Bowers, B., 1997, Procedures for use of, and drill cores and cuttings available for study at, the Lithologic Core Storage Library, Idaho National Engineering Laboratory, Idaho: U.S. Geological Survey Open-File Report 97-124, 31 p.

Fleck, R.J., Sutter, J.F., and Elliot, D.H., 1977. Interpretation of discordant $^{40}Ar/^{39}Ar$ age-spectra of Mesozoic tholeiites from Antarctica: Geochimica et Cosmochimica Acta, v. 41, p. 15–32.

Geslin, J.K., Link, P.K., and Fanning, C.M., 1999, High-precision provenance determination using detrital-zircon ages and petrography of Quaternary sands on the eastern Snake River Plain, Idaho: Geology, v. 27, no. 4, p. 295–298.

Gianniny, G.L., Geslin, J.K., Riesterer, J.W., Link, P.K., and Thackray, G.D., 1997, Quaternary surficial sediments near the Test Area North, northeastern Snake River Plain: An actualistic guide to aquifer characterization, in Sharma, S., and Hardcastle, J.H., Proceedings of the 32nd Symposium on Engineering Geology and Geotechnical Engineering: Boise, Idaho, p. 29–44.

Greeley, Ronald, 1982, The Snake River Plain, Idaho: Representative of a new category of volcanism: Journal of Geophysical Research, v. 87, p. 2705–2712.

Hackett, W.R., and Smith, R.P., 1992, Quaternary volcanism, tectonics and sedimentation in the Idaho National Engineering Laboratory area, in Wilson, J.R., ed., Field guide to geologic excursions in Utah and adjacent areas of Nevada, Idaho and Wyoming: Utah Geological Survey, Miscellaneous Publications 92-3, p. 1–18.

Hackett, W.R., Smith, R.P., and Khericha, S., 2002, Volcanic hazards of the Idaho National Engineering and Environmental Laboratory, southeast Idaho, in Bonnichsen, B., White, C., and McCurry, M., eds., Tectonic and magmatic evolution of the Snake River Plain volcanic province: Idaho Geological Survey Bulletin 30 (in press).

Hanan, B.B., Vetter, S.K., and Shervais, J.W., 1997, Basaltic volcanism in the eastern Snake River Plain: Lead, neodymium, strontium isotope constraints from the Idaho Idaho National Engineering Laboratory WO-2 core site basalts: Geological Society of America Abstracts with Programs, v. 29, p. A298.

Hart, W.K., 1985, Chemical and isotopic evidence for mixing between depleted and enriched mantle, northwestern U.S.A: Geochimica et Cosmochimica Acta, v. 49, p. 131–144.

Hildreth, W., Halliday, A.N., and Christiansen, R.L., 1991, Isotopic and chemical evidence concerning the genesis and contamination of basaltic and rhyolitic magma beneath the Yellowstone Plateau volcanic field: Journal of Petrology, v. 32, p. 63–138.

Hughes, S.S., Smith, R.P., Hackett, W.R., and Anderson, S.R., 1999, Mafic volcanism and environmental geology of the eastern Snake River Plain, in Hughes, S.S., and Thackray, G.D., eds., Guidebook to the Geology of Eastern Idaho: Pocatello, Idaho, Idaho Museum of Natural History, p. 143–168.

Hughes, S.S., Casper, J.L., Geist, D.J., 1997, Potential influence of volcanic constructs on hydrogeology beneath Test Area North, Idaho National Engineering and Environmental Laboratory, Idaho, in Sharma, S., and Hardcastle, J.H., eds., Proceedings of the 32nd Symposium on Engineering Geology and Geotechnical Engineering: Boise, Idaho, p. 59–74.

Hughes, S.S., Wetmore, P.H., and Casper, J.L., 2002, Evolution of Quaternary tholeiitic basalt eruptive centers on the eastern Snake River Plain, Idaho, in Bonnichsen, B., White, C., and McCurry, M., eds., Tectonic and magmatic evolution of the Snake River Plain volcanic province: Idaho Geological Survey Bulletin 30 (in press).

Knobel, L.L., Cecil, L.D., and Wood, T.R., 1995, Chemical composition of selected core samples, Idaho National EngineeringLaboratory, Idaho: U.S. Geological Survey Open-File Report 95-748, 59 p.

Kuntz, M.A., 1992, A model-based perspective of basaltic volcanism, eastern Snake River Plain, Idaho, in Link, P.K., Kuntz, M.A., and Platt, L.P., eds., Regional geology of eastern Idaho and western Wyoming: Geological Society of America Memoir 179, p. 289–304.

Kuntz, M.A., and Dalrymple, G.B., 1979, Geology, geochronology, and potential volcanic hazards in the Lava Ridge-Hell's Half Acre area, eastern Snake River Plain, Idaho: U.S. Geological Survey Open-File Report 79-1657, 70 p.

Kuntz, M.A., Covington, H.R., and Schorr, L.J., 1992, An overview of basaltic volcanism of theeastern Snake River Plain, Idaho, in Link, P.K., Kuntz, M.A., and Platt, L.P., eds., Regional geology of eastern Idaho and western Wyoming: Geological Society of America Memoir 179, p. 227–267.

Kuntz, M.A., Skipp, B., Lanphere, M.A., Scott, W.E., Pierce, K.L., Dalrymple, G.B., Champion, D.E., Embree, G.F., Page, W.R., Morgan, L.A., Smith, R.P., Hackett, W.R., and Rodgers, D.W., 1994, Geologic map of the Idaho National Engineering Laboratory and adjoining areas, eastern Idaho: U.S. Geological Survey Miscellaneous Investigations Series Map 1-2330, scale 1:100 000.

Kuntz, M.A., Champion, D.E., Spiker, E.C., and Lefebvre, R.H., 1986, Contrasting magma types and steady-state, volume-predictable volcanism along the Great Rift, Idaho: Geological Society of America Bulletin, v. 97, p. 579–594.

Kuntz, M.A., Dalrymple, G.B., Champion, D.E., and Doherty, D.J., 1980, Petrography, age, and paleomagnetism of volcanic rocks at the Radioactive Waste Management Complex, Idaho National Engineering Laboratory, Idaho, with an evaluation of volcanic hazards: U.S. Geological Survey Open-File Report 80-388, 63 p.

Kuntz, M.A., Elsheimer, N.H., Espos, L.F., and Klock, P.R., 1985, Major element analyses of latest Pleistocene-Holocene lava fields of the Snake River Plain, Idaho: U.S. Geological Survey Open-File Report 85-593, 64 p.

Lanphere, M.A., Champion, D.E., and Kuntz, M.A., 1993, Petrography, age and paleomagnetism of basalt lava flows in coreholes Well 80, NRF 89-04, NRF 89-05, and ICPP 123, Idaho National Engineering Laboratory: U.S. Geological Survey Open-File Report 93-327, 40 p.

Lanphere, M.A., Kuntz, M.A., and Champion, D.E., 1994, Petrography, age and paleomagnetism of basalt lava flows in coreholes at Test Area North, Idaho National Engineering Laboratory: U.S. Geological Survey Open-File Report 94-686, 49 p.

Leeman, W.P., 1982a, Olivine tholeiitic basalts of the Snake River Plain, Idaho, in Bonnichsen, B., and Breckenridge, R.M., eds., Cenozoic geology of Idaho: Idaho Bureau of Mines and Geology Bulletin 26, p. 181–192.

Leeman, W.P., 1982b, Evolved and hybrid lavas from the Snake River Plain, Idaho, in Bonnichsen, B., and Breckenridge, R.M., eds., Cenozoic geology of Idaho: Idaho Bureau of Mines and Geology Bulletin 26, p. 193–202.

Leeman, W.P., and Vitaliano, C.J., 1976, Petrology of the McKinney basalt, Snake River Plain, Idaho: Geological Society of American Bulletin, v. 87, p. 1777–1792.

Leeman, W.P., Vitaliano, C.J., and Prinz, M., 1976, Evolved lavas from the Snake River Plain, Craters of the Moon National Monument, Idaho: Contributions to Mineralogy and Petrology, v. 56, p. 35–60.

Leeman, W.P., Menzies, M.A., Matty, D.J., and Embree, G.F., 1985, Strontium, neodymium, and lead isotopic compositions of deep crustal xenoliths from the Snake River Plain: Evidence for Archean basement: Earth and Planetary Science Letters 75, 354–368.

McCurry, M., Hackett, W.R., and Hayden, K., 1999, Cedar Butte and cogenetic Quaternary rhyolite domes of the eastern Snake River Plain, in Hughes, S.S., and Thackray, G.D., eds., Guidebook to the Geology of Eastern Idaho: Pocatello, Idaho, Idaho Museum of Natural History, p. 169–179.

McQuarrie, N., and Rodgers, D.W., 1998, Subsidence of a volcanic basin by flexure and lower crustal flow: The eastern Snake River Plain, Idaho: Tectonics, v. 17, p. 203–220.

Menzies, M.A., 1989, Cratonic, circumcratonic and oceanic mantle domains beneath the western United States: Journal of Geophysical Research, v. 94, no. B6, p. 7899–7915.

Menzies, M.A., Leeman, W.P., and Hawkesworth, C.J., 1984, Geochemical and isotopic evidence for the origin of continental flood basalts with particular reference to the Snake River Plain, Idaho, U.S.A.: Philosophical Transactions of the Royal Society of London, v. 310, p. 643–660.

O'Hara, M.J., 1985, Importance of the "shape" of the melting regime during partial melting of the mantle: Nature, v. 314, p. 58–61.

Parsons, T., Thompson, G.A., and Smith, R.P., 1998, More than one way to stretch: A tectonic model for extension along the plume track of the Yellowstone hotspot and adjacent Basin and Range Province: Tectonics, v. 17, p. 221–234.

Peng, Xiaohua, and Humphreys, E.D., 1998, Crustal velocity structure across the eastern Snake River Plain and the Yellowstone swell: Journal of Geophysical Research, v. 103, no. B4, p. 7171–7186.

Reed, M.F., Bartholomay, R.C., and Hughes, S.S., 1997, Geochemistry and stratigraphic correlation of basalt lavas beneath the Idaho Chemical Processing Plant, Idaho National Engineering Laboratory: Environmental Geology, v. 30, p. 108–118.

Reid, M.R., 1995, Processes of mantle enrichment and magmatic differentiation in the eastern Snake River Plain: Th isotope evidence: Earth and Planetary Science Letters, v. 131, p. 239–254.

Reidel, S.P., Hooper, P.R., Beeson, M.H., Fecht, K.R., Bentley, R.D., and Anderson, J.L., 1989, The Grande Ronde Basalt, Columbia River Basalt Group; Stratigraphic descriptions and correlations in Washington, Oregon, and Idaho, in Reidel, S.P., and Hooper, P.R., eds., Volcanism and tectonism in the Columbia River flood-basalt province: Boulder, Colorado, Geological Society of America Special Paper 239, p. 21–53.

Rodgers, D.W., Hackett, W.R., and Ore, H.T., 1990, Extension of the Yellowstone Plateau, eastern Snake River Plain, and Owyhee Plateau: Geology, v. 18, p. 1138–1141.

Saltzer, R.L., and Humphreys, E.D., 1997, Upper mantle *P* wave velocity structure of the eastern Snake River Plain and its relationship to geodynamic models of the region: Journal of Geophysical Research, v. 102, p. 11829–11841.

Shervais, J., Vetter, S., and Hackett, W.R., 1994, Chemical stratigraphy of basalt in coreholes NPR-E and WO-2, Idaho National Engineering Laboratory, Idaho: Implications for plume dynamics in the Snake River Plain, in Proceedings of the 7th International Symposium on the Observation of the Continental Crust Through Drilling: Santa Fe, New Mexico, p. 93–96.

Sparlin, M.A., Braile, L.W., and Smith, R.B., 1982, Crustal structure of the eastern Snake River Plain from ray trace modeling of seismic refraction data: Journal of Geophysical Research, v. 87, p. 2619–2633.

Stout, M.Z., and Nicholls, J., 1977, Mineralogy and petrology of Quaternary lava from the Snake River Plain, Idaho: Canadian Journal of Earth Sciences, v. 14, p. 2140–2156.

Wetmore, P.H., 1998, An assessment of physical volcanology and tectonics of the central eastern Snake River Plain based on the correlation of subsurface basalts at and near the Idaho National Engineering and Environment Laboratory, Idaho [M.S. thesis]: Pocatello, Idaho, Idaho State University, 118 p.

Wetmore, P.H., Hughes, S.S., and Anderson, S.R., 1997, Model morphologies of subsurface Quaternary basalts as evidence for a decrease in the magnitude of basaltic magmatism at and near the Idaho National Engineering and Environmental Laboratory, Idaho, in Proceedings of the 32nd Symposium on Engineering Geology and Geotechnical Engineering: Boise, Idaho, p. 45–58.

Wilson, Marjorie, 1989, Igneous petrogenesis: London, Unwin Hyman, 466 p.

Zindler, A., and Hart, S., 1986, Chemical geodynamics: Annual Reviews of Earth and Planetary Science, v. 14, p. 493–571.

MANUSCRIPT ACCEPTED BY THE SOCIETY NOVEMBER 2, 2000

Geological Society of America
Special Paper 353
2002

Accumulation and subsidence of late Pleistocene basaltic lava flows of the eastern Snake River Plain, Idaho

Duane E. Champion*
Marvin A. Lanphere
U.S. Geological Survey, 345 Middlefield Road, Menlo Park, California 94025, USA
Steven R. Anderson
U.S. Geological Survey, Department of Geology, Idaho State University, Pocatello, Idaho 83209-8072, USA
Mel A. Kuntz
U.S. Geological Survey, Box 25046, Denver Federal Center, Denver, Colorado 80225, USA

ABSTRACT

Studies of cores from drill holes with detailed petrographic descriptions, paleomagnetic characterization and correlation, and conventional K-Ar and $^{40}Ar/^{39}Ar$ dating allow examination of the process of accumulation of basaltic lava flows in a part of the eastern Snake River Plain, Idaho. Core holes at various locations in the Idaho National Engineering and Environmental Laboratory (INEEL) demonstrate variable accumulation rates that can be fitted by linear regression lines with high correlation coefficients. Hiatuses of several hundred thousand years are represented in many of the core holes, but accumulation of flows resumed in most of the areas sampled by these core holes at rates nearly identical to previous rates. The studies show that an area of the eastern Snake River Plain north of its topographic axis, including the area of the INEEL, has undergone a hiatus in eruptive activity for the past ~200 k.y. The data also allow enhanced interpretations of the volcanic hazard to the INEEL with regard to lava flow inundation, prediction of lava flow thickness, and assessment of eruption recurrence-time intervals.

GEOLOGIC SETTING

The eastern Snake River Plain rises 1000 m in elevation from Twin Falls, Idaho, to the Yellowstone Plateau. The relatively flat surface has a northeast-southwest–trending central axis that connects those two areas. The plain is bounded on the northwest and southeast by Tertiary block-faulted mountains of the Basin and Range Province, underlain mainly by folded and thrust-faulted Precambrian, Paleozoic, and Mesozoic sedimentary rocks and Mesozoic and younger plutonic rocks. The plain terminates to the northeast in the Pliocene and Quaternary Yellowstone Plateau volcanic field. The Snake River flows along the southern boundary of the eastern Snake River Plain (Fig. 1) along a course defined in part by the extrusion of the lava flows;

there is no evidence that the river ever crossed the lava surface (Walker, 1964).

At the surface, the eastern Snake River Plain comprises basaltic lava flows, shield volcanoes, minor associated pyroclastic deposits, and more widespread sedimentary accumulations of variable thicknesses. The surface is locally covered by a thin discontinuous mantle of loess. Most of the surface flows have normal magnetic polarity associated with the Brunhes Normal Polarity Chron, and are thus younger than 780 ka. Pleistocene and Holocene lava fields cover large parts of the eastern Snake River Plain, suggesting that renewed lava shield formation must be considered a possibility anywhere along the length of the eastern Snake River Plain. Several andesitic to rhyolitic flows and domes crop out along the axial area of the eastern

*E-mail: dchamp@mojave.wr.usgs.gov

Champion, D.E., Lanphere, M.A., Anderson, S.R., and Kuntz, M.A., 2002, Accumulation and subsidence of late Pleistocene basaltic lava flows of the eastern Snake River Plain, Idaho, in Link, P.K., and Mink, L.L., eds., Geology, Hydrogeology, and Environmental Remediation: Idaho National Engineering and Environmental Laboratory, Eastern Snake River Plain, Idaho: Boulder, Colorado, Geological Society of America Special Paper 353, p. 175–192.

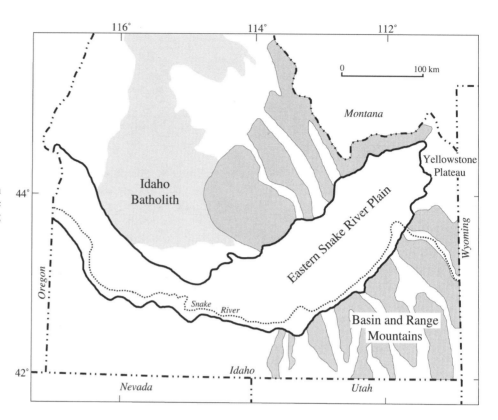

Figure 1. Generalized map of southern Idaho showing geologic and geographic features referred to in text, including eastern Snake River Plain.

Snake River Plain and have ages ranging from 1.4 to 0.3 Ma (Kuntz et al., 1994). Lava flows of evolved composition have been considered to be a very minor constituent of the eastern Snake River Plain, but recent work suggests that a larger amount of higher silica lava flows may be found at depth (Hughes et al., 1997). Outcrops of rhyolite pumice-fall deposits and welded tuff only occur at the margins of the plain; similar rocks occur at great depth (>730 m) in the deepest wells of the Idaho National Engineering and Environmental Laboratory (INEEL), but play no part in the surface or near-surface stratigraphy of the plain.

The distinctive low relief and low-slope morphology of basalt lava flows of the eastern Snake River Plain prompted Greeley (1982) to give the name "basaltic plains volcanism" to the eruptions of the eastern Snake River Plain to distinguish them from the more voluminous flood-basalt eruptions of plateau volcanism and from Hawaiian volcanism, which produces greater surface relief. Basaltic plains volcanism is dominated by low shield volcanoes of modest overall volume (5 ± 3 km^3) formed from fluid, gas-rich lavas. Eruptions may begin from a system of fissures and typically consolidate in long-lived eruptions to one or several vents along the original fissure system. The eruptions produce hundreds of lava flow units that extend tens of kilometers from the vent during a time no longer than a few decades. This type of eruption doesn't flood a surface 100 m thick with single flow units or pile up an equant mass of lava. Rather, flow directions follow subtle creases in the topography, moving great distances by endogenous flow when the effusion rate is appropriately maintained. The net effect is

to produce nearly planar surfaces puddled by successive effusions, and widespread overlapping of lava fields.

The subsurface record of the eastern Snake River Plain is generally similar to the lava and sediment accumulations at the surface environment (Walker, 1964; Anderson et al., 1996; Anderson and Liszewski, 1997). In many core holes in the southern half of the INEEL, fine-grained sediments separate flows of significantly different ages. Anderson and Liszewski (1997) suggested an average sediment content of 15% for wells drilled in or near the INEEL. Sands and gravels recovered in southern INEEL core holes suggest that these are fluvial deposits associated with buried courses of the Big Lost River. The course of that stream has been displaced frequently as the plain has subsided to positions peripheral to new lava fields. Thick accumulations of sediments are typical of wells drilled or cored in the sinks area of the Big Lost or Little Lost Rivers or in the Mud Lake area. Sediment contents range from 26% to 46% for the deepest of the wells from those areas (Lanphere et al., 1994). By contrast, the proportion of sediment in core holes and drill holes along the southern boundary of the INEEL near the topographic axis of the eastern Snake River Plain is generally low, typically <5%. For example, Core Hole #1, located on the axis of the plain between East and Middle Buttes, has <1% sediments over a depth range of 600 m.

Kuntz et al. (1992) suggested that as many as nine volcanic rift zones cross the eastern Snake River Plain following Basin and Range trends. Four of these, the Arco–Big Southern Butte, Howe–East Butte, Lava Ridge–Hell's Half Acre, and Circular Butte–Kettle Butte volcanic rift zones cross the INEEL. The

Lava Ridge–Hell's Half Acre volcanic rift zone can be documented through the Test Area North (TAN) area of the INEEL (Fig. 2); core holes show that the subsurface section is composed of sediment-poor, flow-on-flow sequences to depths >300 m. Topographic freeboard arising from frequent eruptions prevents accumulation of sediments, thus sediment-poor sequences indicate near-vent locations for nearly all flows. In comparison, core holes to the northeast and south of the TAN area contain more than 100 m of sediment over the same depth range, indicating that their locations are on the flanks of volcanic edifices located in the vent zone.

PURPOSE OF THE PRESENT STUDY

We studied the stratigraphy of basaltic lava flows and sediments in 20 core holes at the INEEL by conventional logging,

petrographic, paleomagnetic, [40]Ar/[39]Ar, and conventional K/Ar dating methods. These studies have yielded detailed information about the rates at which the lava flows and sediments have accumulated. The data indicate that the flows and sediments accumulate predictably with linear accumulation rates. The data for some core holes show that episodes of nearly constant accumulation rates were separated by hiatuses of several hundred thousand years. However, the hiatuses were followed by the resumed accumulation of lava flows and sediments, typically, but not universally, at a linear rate nearly identical to the prior rate. Accumulation rates also varied considerably in magnitude at various locations in the study area.

In this chapter we document the techniques used in this study, describe the accumulation rates found, and discuss the influences of these rates and their geographic distribution on the character of basaltic volcanism and on the potential volcanic hazards of the eastern Snake River Plain.

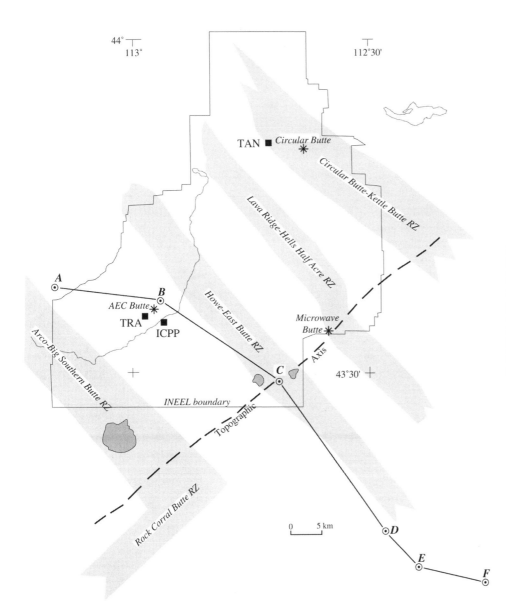

Figure 2. Map of part of eastern Snake River Plain at and near Idaho National Engineering and Environmental Laboratory (INEEL) showing geologic and geographic features referred to in text. Line A-F shows location of cross section in Figure 17. TAN, Test Area North; ICPP, Idaho Chemical Processing Plant; TRA, Test Reactor Area, RZ, rift zone.

ANALYTICAL TECHNIQUES

Each core hole was logged to establish tops and bottoms of individual lava flows, positions of sedimentary interbeds and unrecovered intervals, and generally to sample the cores for thin-section petrographic examination. Detailed petrographic descriptions were made of lava flows and flow units. Multiple paleomagnetic samples (typically seven for each flow or flow unit) were taken to establish mean remanent inclination values, and then compared one to another, assuming that lava flows of the same age share identical inclinations while lava flows of different ages typically have different mean inclinations or even different polarities. The inclination values were based on the assumption that each sampled core slug was vertical; this assumption can be tested using borehole deviation logs. The analytical techniques associated with the sampling, measurement, and demagnetization of basalt lava flow paleomagnetic specimens followed well-established protocols (Irving, 1964).

The K-Ar dating followed the established procedures of Dalrymple and Lanphere (1969). The application of those procedures to young basalts of very low potassium concentrations and the high success rate of our experiments merit some explanation. Careful thin-section evaluations were used to determine whether samples were suitable for dating (Mankinen and Dalrymple, 1972). The K_2O measurements were made in duplicate on each of two splits of powder by flame photometry (Ingamells, 1970). Argon mass analyses were done on a computerized multiple-collector mass spectrometer (described by Stacey et al., 1981). Weighted-mean ages for duplicate samples (some triplicate) were calculated by the method of Taylor (1982). These methods produced ages that stack well, generally in ascending age order with decreasing depth. In addition, ages from different core holes on flows correlated by petrography or remanent magnetization are the same, within analytical uncertainty.

Some basalt samples were also analyzed by the $^{40}Ar/^{39}Ar$ technique, which employs the radioactive decay of ^{40}K to ^{40}Ar as a chronometer in a different way (Dalrymple and Lanphere, 1971, 1974; Lanphere and Dalrymple, 1971). In the $^{40}Ar/^{39}Ar$ method, the sample is irradiated with fast neutrons, along with a monitor mineral of known age, to induce the reaction $^{39}K(n, p)^{39}Ar$. The age of the sample is calculated from the measured $^{40}Ar/^{39}Ar$ ratio after determining the fraction of ^{39}K converted to ^{39}Ar by analyzing the monitor mineral. The important difference between the two methods is that while quantitative measurements of the contents of radiogenic ^{40}Ar and ^{40}K are required by the conventional method, for the $^{40}Ar/^{39}Ar$ method only the ratios of Ar isotopes are measured. There is currently a widespread mistaken belief that $^{40}Ar/^{39}Ar$ dating has rendered conventional K-Ar dating obsolete. A direct comparison of the two dating techniques (Lanphere, 2000) finds consistency between ages determined by the two techniques. Whereas the mean K-Ar and $^{40}Ar/^{39}Ar$ ages are statistically identical, the precision of individual $^{40}Ar/^{39}Ar$ ages is generally better than the precision of K-Ar ages.

Accumulation rates were calculated from graphs of age of lava flows plotted against depth. A linear regression was fitted to the depth versus age data points, yielding the accumulation rate and regression coefficient, R.

CORE DESCRIPTIONS

Because most of the following cores have been described previously (Lanphere et al., 1993, 1994; Champion et al., 1988, 1981; Champion and Lanphere, 1997; Kuntz et al., 1980; Doherty, 1979), this section includes only information specific to the discussion of accumulation rates; i.e., the number of eruptive events, their K-Ar and $^{40}Ar/^{39}Ar$ ages, and thicknesses. In some cases, age information has improved since the original reports, either through the acquisition of new ages or by averaging multiple ages from what we have later learned to be the same flow in adjacent core holes. A diagram (Fig. 3) shows the locations of core holes examined for this study, and Tables 1–4 contain the detailed data pertinent to our study of lava flow and sediment accumulation through time. In addition, data on lava flow thicknesses and recurrence intervals are given in Tables 1–4 for use in a later discussion concerning volcanic hazards.

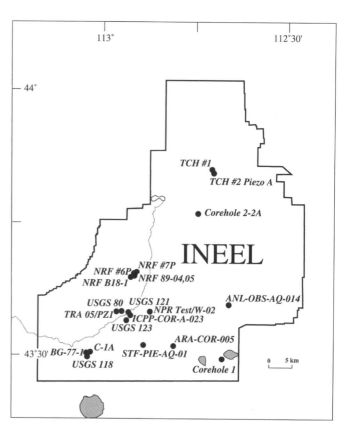

Figure 3. Map of Idaho National Engineering and Environmental Laboratory (INEEL) showing locations of core holes examined for this study (see text for designations).

TABLE 1. AGE, RECURRENCE INTERVAL, DEPTH AND THICKNESS DATA FOR
LAVA FLOWS SAMPLED BY DRILL CORES NEAR FACILITIES IN THE SOUTH-CENTRAL
PART OF THE INEEL*

Flow number	Age (ka)	Recurrence interval (1 k.y.)	Flow top depth (m/ft)	Thickness (m/ft)
USGS 80		*419+*		
1	419 ± 33	(−5)	13/*44'*	9/*29'*
2	[414 ± 8]	161	23/*76'*	18/*60'*
3	(575)	62	44/*144'*	5/*18'*
4	[637 ± 35]	—	54/*177'*	8/*27'* +
USGS 123		*229+*		
1	[229 ± 28]	73	10/*34'*	23/*74'*
2	(302)	4	34/*113'*	1/*4'*
3	(306)	44	36/*118'*	11/*36'*
4	[350 ± 40]	26	50/*163'*	13/*43'*
5	(376)	30	64/*209'*	12/*39'*
6	(406)	55	76/*248'*	9/*31'*
7	[461 ± 35]	30	87/*284'*	29/*98'*
8	[491 ± 80]	52	116/*382'*	12/*38'*
9	(543)	56	129/*424'*	22/*73'*
10	(599)	18	151/*497'*	4/*12'*
11	[617 ± 22]	20	155/*509'*	17/*55'*
12	[637 ± 35]	112	172/*564'*	37/*123'*
13	(749)	26	211/*691'*	10/*33'*
14	(775)	—	221/*724'*	5/*17'*
NPR Test/W-02		*229+*		
1	[229 ± 28]	82	2/*7'*	29/*96'*
2	(311)	39	34/*112'*	6/*21'*
3	350 ± 40	71	41/*136'*	33/*107'*
4	(421)	40	81/*267'*	9/*30'*
5	[461 ± 35]	30	92/*303'*	29/*96'*
6	491 ± 80	59	129/*424'*	10/*32'*
7	[550 ± 10]	67	139/*456'*	11/*35'*
8	[617 ± 22]	20	150/*492'*	20/*65'*
9	[637 ± 35]	—	182/*596'*	30/*98'* †
10	B/M-780 ± 5		216–234/*710–768'*	
ICPP-COR-A-023		*229+*		
1	[229 ± 28]	62	12/*41'*	16/*51'*
2	(291)	29	28/*92'*	3/*9'*
3	(320)	30	38/*126'*	5/*18'*
4	[350 ± 40]	22	52/*170'*	4/*12'*
5	(372)	89	57/*188'*	5/*18'*
6	[461 ± 35]	57	63/*206'*	46/*152'*
7	(518)	(−27)	110/*360'*	3/*10'*
8	[491 ± 80]	125	113/*371'*	6/*19'*
9	[616 ± 61]	(−6)	119/*390'*	24/*78'*
10	(610)	55	143/*468'*	19/*61'*
11	(665)	(−28)	162/*533'*	2/*7'*
12	[637 ± 35]	144	165/*540'*	40/*131'*
13	(781)	(−22)	205/*671'*	6/*21'*
14	[759 ± 12]	—	211/*692'*	14/*47'*
USGS 121		*77+*		
1	77 ± 39	—	10/*34'*	6/*19'*
2	—	—	16/*53'*	12/*39'*
3	—	—	28/*92'*	7/*22'*
4	—	—	35/*114'*	5/*16'*
5	376 ± 81	30	41/*134'*	16/*54'*

(continued on next page)

USGS 80

The stratigraphy, petrography, age, and paleomagnetic inclination stratigraphy of flows in core hole USGS 80 were originally reported by Lanphere et al. (1993), who found four lava flows of normal polarity to a depth of 62 m (Fig. 4; Table 1).

Pairs of K-Ar ages on three of the flows define an ∼220 k.y. span, from 637 to 419 ka. Anderson et al. (1997) fit the age versus depth data with a regression line of high correlation (R = 0.998), yielding an accumulation rate of ∼18 m/100 k.y. Additional age measurements have altered the accumulation-rate calculation. Flow #2 is paleomagnetically and petrograph-

**TABLE 1. AGE, RECURRENCE INTERVAL, DEPTH AND THICKNESS DATA FOR
LAVA FLOWS SAMPLED BY DRILL CORES NEAR FACILITIES IN THE SOUTH-CENTRAL
PART OF THE INEEL* (*continued*)**

Flow number	Age (ka)	Recurrence interval (1 k.y.)	Flow top depth (m/*ft*)	Thickness (m/*ft*)
USGS 121 (continued)		77+		
6	(406)	55	57/*188'*	17/*56'*
7	[461 ± 35]	82	76/*248'*	23/*76'*
8	543 ± 48	(−52)	99/*324'*	21/*68'*
9	[491 ± 80]	125	126/*412'*	2/*6'*
10	616 ± 61	(−12)	129/*422'*	14/*46'*
11	(604)	41	143/*468'*	16/*54'*
12	(645)	(−8)	160/*525'*	5/*15'*
13	[637 ± 35]	51	165/*540'*	14/*46'*
14	(688)	12	179/*586'*	5/*16'*
15	(700)	29	184/*603'*	12/*41'*
16	(729)	30	196/*644'*	18/*60'*
17	759 ± 12	—	215/*704'*	9/*30'*
TRA 05/PZ1		365+		
1	(365)	49	21/*69'*	5/*18'*
2	[414 ± 8]	120	27/*89'*	19/*61'*
3	(534)	59	47/*155'*	5/*15'*
4	(593)	44	56/*185'*	6/*21'*
5	[637 ± 35]	149	67/*220'*	16/*54'*
6	786 ± 23	—	84/*276'*	6/*21+*
ANL-OBS-AQ-014				
1			0/*1'*	15/*50'*
2			16/*51'*	3/*12'*
3	358 ± 46		19/*63'*	35/*115'*
4			55/*181'*	24/*80'*
5			80/*261'*	33/*110'*
6			113/*372'*	9/*29'*
7			122/*400'*	12/*40'*
8	565 ± 94		134/*440'*	37/*122'*
ARA-COR-005				
1	[229 ± 28]		3/*10'*	26/*85'*
2			30/*98'*	26/*86'*
3			56/*185'*	2/*6'*
4			59/*193'*	57/*187'*
5			116/*380'*	19/*61'*
6			135/*443'*	5/*15'*
7			140/*458'*	43/*142'*
8			183/*602'*	20/*64'*
9			202/*666'*	14/*46'*
10	[550 ± 10]		217/*712'*	16/*53'*
STF-PIE-AQ-01				
1			6/*20'*	16/*52'*
2	[229 ± 28]		26/*84'*	16/*54'*
3			43/*141'*	13/*43'*
4			56/*184'*	6/*21'*
5			63/*208'*	26/*86'*
6			90/*295'*	69/*228'*
7	[550 ± 10]		166/*543'*	30/*100'*

Note: Ages in square brackets from another core, or from averaging ages in multiple cores; ages in parentheses derived from accumulation rate line fits; negative recurrence intervals in parentheses due to errors in age determinations or in line fits to same data.
*INEEL—Idaho National Engineering and Environmental Laboratory.
†Thickness of flow 9, core hole NPR-Test derived from basal depth estimate in adjacent core W-02.

ically correlated with flow #2 in adjacent well TRA-05, and the pooled age estimate from the original K-Ar and new ^{40}Ar/^{39}Ar ages is 414 ± 8 ka, substantially younger than the original estimate of 461 ± 24 ka. Flow #4, which is correlated with flows from AEC Butte, also has a different age than originally presented; averaging two pairs of K-Ar ages from wells USGS 80 and USGS 123 and one pair of ages from a nearby surface outcrop yields a slightly younger age of 637 ± 35 ka instead of 643 ± 64 ka. These revised ages decrease the accumulation rate for this core hole to 16 m /100 k.y. and lower the regression slightly (Fig. 4), but do not significantly alter conclusions regarding accumulation rates.

USGS 123

The stratigraphy, petrography, age, and paleomagnetic inclination stratigraphy of core hole USGS 123 were reported by Lanphere et al. (1993), who found 14 lava flows of normal polarity to a depth of 226 m (Fig. 5; Table 1). Pairs of K-Ar ages on five of the flows defined an ~550 k.y. span, from 775 to 229 ka. Petrographic and paleomagnetic correlations of numerous lava flows in USGS 123 to flows in core hole NPR Test (Fig. 1) allow ages in NPR Test to be pooled with ages from USGS 123, yielding six points of accumulation-rate measure. The accumulation rate is 39 m/100 k.y.; the linear regression coefficient is 0.994.

NPR Test/W-02

The stratigraphy, age, and paleomagnetic inclination stratigraphy of core hole NPR Test (where NPR refers to New Production Reactor) were reported by Champion et al. (1988), who found eight lava flows of normal polarity, and one of reversed polarity (Big Lost Reversed Polarity Subchron of the Brunhes Normal Polarity Chron) to a depth of 186 m (Fig. 6; Table 1). A detailed description of the petrography of the core hole was provided by Morgan (1990), who divided the uppermost flow into three flow units, raising the overall flow count to 11. Four pairs of K-Ar ages, one set of four K-Ar ages, and one single K-Ar age for six flows, provide age control and define an ~500 k.y. span, from 637 to 229 ka. In addition, core hole W-02, located only 100 m from NPR Test, includes the Matuyama Reversed Polarity Chron–Brunhes Normal Polarity Chron boundary (hereafter referred to as the Brunhes-Matuyama boundary) between the top of reversed polarity flows at 234 m and the base of normal polarity flows at 216 m. Petrographic and paleomagnetic correlation of numerous lava flows in these core holes with flows in core hole USGS 123 allows age data in that well to be pooled with ages in NPR Test and W-02 to yield seven points of accumulation-rate measure. The accumulation rate is 43 m/100 k.y.; the high regression coefficient is 0.984.

ICPP-COR-A-023

The stratigraphy, hand-specimen petrography, and paleomagnetic inclination stratigraphy of flows in core hole ICPP-023 (where ICPP stands for Idaho Chemical Processing Plant) is reported on by Jobe and Champion (2002), who describe 14 lava flows of normal polarity to a depth of 225 m (Fig. 7; Table 1). Although no K-Ar ages have been measured on this core, correlations of flows in ICPP-023 with flows in adjacent core holes (USGS 121 and USGS 123) together with paleomagnetic and petrographic data allow ages to be assigned to seven flows. The ages define an ~530 k.y. span, from 759 to 229 ka. The accumulation rate is 36 m/100 k.y., and the linear regression coefficient is 0.964. The lower coefficient of regression sug-

gests errors in the dates, or possibly in correlations of flows from other wells.

USGS 121

The lithology and paleomagnetic inclination stratigraphy of core hole USGS 121 are reported by Jobe and Champion (2002), who describe 17 lava flows of normal polarity to a depth of 227 m (Fig. 8; Table 1). A single K-Ar age and four ^{40}Ar/^{39}Ar ages, as well as three K-Ar ages correlated from adjacent wells, establish a set of eight age and depth measures. The data define lava flow accumulation for the entire span of the Brunhes Normal Polarity Chron, ~780 k.y. The age versus depth graph displays a distinct "dog leg" in the upper 35 m of the well (not shown in Fig. 8), reflecting an interval of slower accumulation. An accumulation rate of 43 m/100 k.y., and a regression coefficient of 0.953, is derived for the majority of the core hole. This somewhat lower regression coefficient may reflect errors in the ages, or possibly in the correlations of flows from other wells.

TABLE 2. AGE, RECURRENCE INTERVAL, DEPTH AND THICKNESS DATA FOR LAVA FLOWS SAMPLED BY DRILL CORES NEAR THE NAVAL REACTOR FACILITY (NRF) AT THE IDAHO NATIONAL ENGINEERING AND ENVIRONMENTAL LABORATORY

Flow number	Age (ka)	Recurrence interval (1 k.y.)	Flow top depth (m/ft)	Thickness (m/ft)
NRF 89-04		303+		
1	[303 ± 30]	92	6/*21'*	26/*85'*
2	[395 ± 25]	126	36/*119'*	22/*71'*
3	521 ± 31	25	60/*196'*	7/*22'*
4	[546 ± 47]	—	67/*221'*	8/*27'* +
NRF 89-05		303+		
1	303 ± 30	92	6/*21'*	26/*86'*
2	[395 ± 25]	97	36/*118'*	23/*76'*
3	492 ± 56	29	60/*196'*	3/*12'*
4	521 ± 31	—	63/*208'*	10/*34'* +
NRF B18-1		303+		
1	[303 ± 30]	92	11/*35'*	16/*53'*
2	[395 ± 25]	97	27/*88'*	31/*102'*
3	[492 ± 56]	29	58/*190'*	10/*34'*
4	[521 ± 31]	—	68/*224'*	8/*27'* +
NRF #6P		303+		
1	[303 ± 30]	92	3/*11'*	24/*78'*
2	395 ± 25	126	27/*89'*	23/*76'*
3	[521 ± 31]	25	50/*165'*	14/*45'*
4	546 ± 47	181	64/*210'*	44/*143'*
5	727 ± 31	157	108/*354'*	29/*64'*
6	[884 ± 53]	—	127/*418'*	25/*83'* +
NRF #7P		303+		
1	[303 ± 30]	92	7/*25'*	20/*67'*
2	[395 ± 25]	126	32/*105'*	14/*46'*
3	[521 ± 31]	25	46/*152'*	6/*20'*
4	[546 ± 47]	181	52/*172'*	56/*184'*
5	[727 ± 31]	*(53)*	109/*358'*	13/*44'*
6	<780	*(104)*	123/*404'*	6/*21'*
7	884 ± 53	—	130/*428'*	22/*73'* +

Note: Ages in square brackets from another core, or from averaging ages in multiple cores; depths in parentheses from limiting age data in core hole NRF #7P.

TABLE 3. AGE, RECURRENCE INTERVAL, DEPTH AND THICKNESS DATA FOR LAVA FLOWS SAMPLED BY DRILL CORES NEAR THE RADIOACTIVE WASTE MANAGEMENT COMPLEX (RWMC) AT THE IDAHO NATIONAL ENGINEERING AND ENVIRONMENTAL LABORATORY

Flow number	Age (ka)	Recurrence interval (1 k.y.)	Flow top depth (m/*ft*)	Thickness (m/*ft*)
BG 77-1		95+		
1	95 ± 50	37	1/*4'*	12/*41'*
2	(132)	79	16/*51'*	16/*51'*
3	211 ± 16	19	32/*105'*	38/*126'*
4	230 ± 85	(285)	77/*252'*	10/*33'*
5	515 ± 85	35	88/*288'*	23/*76'*
6	550 ± 10	—	112/*367'*	57/*188'*
7	—	—	170/*557'*	13/*43'* +
C-1A		95+		
1	95 ± 50	25	2/*8'*	9/*28'*
2	(120)	91	11/*36'*	23/*75'*
3	211 ± 16	19	34/*111'*	36/*118'*
4	230 ± 85	(—)	73/*239'*	15/*50'*
5	—	(—)	89/*293'*	12/*39'*
6	515 ± 85	35	101/*333'*	20/*64'*
7	550 ± 10	133	121/*397'*	48/*158'*
8	(683)	37	173/*569'*	15/*50'*
9	(720)	19	189/*620'*	7/*25'*
10	(739)	41	197/*645'*	16/*53'*
11	B/M-780 ± 5	—	213/*700'*	—
USGS 118				
1	<180	—	4/*14'*	28/*92'*
2	211 ± 16	(—)	34/*112'*	34/*112'*
3	—	(—)	76/*250'*	4/*13'*
4	—	(—)	81/*265'*	10/*33'*
5	515 ± 85	35	91/*300'*	25/*81'*
6	550 ± 10	—	116/*381'*	57/*186'*

Note: Ages in parentheses derived from accumulation rate line fits; recurrence intervals in parentheses represent hiatuses in accumulation.

TRA 05/PZ1

In core hole TRA 05/PZ1, six lava flows of normal polarity occur to a depth of 91 m (Fig. 9; Table 1). The K-Ar ages on three lava flows, redating by ^{40}Ar/^{39}Ar on two of the flows, and correlation of lava flows to those in well USGS 80 define an ~400 k.y. span, from 786 to 365 ka. The three points of age control define an accumulation rate of 16 m/100 k.y., and a high regression coefficient of 0.994.

STF-PIE-AQ-01

There is no report on the stratigraphy, petrography, age, and paleomagnetism of a 218 m core hole drilled at the Safety Test Facility (STF). Paleomagnetic studies have designated nine lava flows; thicknesses and depths to tops of flows are listed (Table 1). A preliminary accumulation rate for the upper seven lava flows can be derived from age information correlated to this core hole for flows #2 and #7. Flow #2 can be paleomagnetically correlated with the top flow in core holes NPR Test/W-02 and USGS 123. The weighted mean K-Ar age of these two flows is 229 ± 28 ka. Flow #7 is part of the sequence of flow units of the Big Lost Reversed Polarity Subchron, which has an age of 550 ± 10 ka (Champion et al., 1996). The age and depth data suggest an average accumulation rate of 50 m/100 k.y. for this core hole.

ARA-COR-005

There is no report on the stratigraphy, petrography, age, and paleomagnetism of a 261 m core hole drilled at the Auxiliary Reactor Area (ARA). Paleomagnetic studies have designated 11 lava flows; thicknesses and depths to tops of flows are listed in Table 1. A preliminary accumulation rate for the upper 10 flows can be derived from age information correlated to this core hole for flows #1 and #10. Flow #1 can be correlated with the top flows in core holes NPR Test, W-02, and USGS 123 that have a weighted mean K-Ar age of 229 ± 28 ka. Flow #10 is one of the sequence of flow units of the Big Lost Reversed Polarity Subchron, which has an age of 550 ± 10 ka (Champion et al., 1996). The age versus depth data suggest an average accumulation rate of 67 m/100 k.y. for this location.

ANL-OBS-AQ-014

The age and paleomagnetic inclination stratigraphy of core hole ANL-OBS-AQ-014 were reported by Champion and Lanphere (1997), who described 27 lava flows of both normal and reversed polarity to a depth of 582 m. Four ^{40}Ar/^{39}Ar ages and one K-Ar age (Kuntz et al., 1994) suggest a history of alternating fast and slow accumulation at this location close to the central axis of the eastern Snake River Plain. Until more age information and petrographic details are available for this core hole, a detailed accumulation analysis is premature. A preliminary accumulation rate for the majority of the Brunhes Normal Polarity Chron lava flows can be derived from the ages of 358 ± 46 ka for flow #3 and 565 ± 94 ka for flow #8 (Table 1). The age and depth data suggest an average mid-Brunhes Normal Polarity Chron accumulation rate of 56 m/100 k.y.

NRF core holes

The stratigraphy, petrography, age and paleomagnetic inclination stratigraphy of two core holes at the Naval Reactor Facility (NRF) (Fig. 3) were reported by Lanphere et al. (1993). They found five lava flows of normal polarity in the NRF 89-04 and NRF 89-05 core holes to depths of 76 m and 74 m, respectively. Other core holes in and near the facility, NRF B18-1, NRF #6P, and NRF #7P, have now been studied by paleomagnetic inclination methods and ^{40}Ar/^{39}Ar ages have been obtained on five flows (Table 2). Ages have been measured on virtually all lava flows within the shared section in core holes NRF #6P and NRF #7P. The horizontal separation of the different core holes is only 1000 m, so they can be interpreted with a single accumulation rate for a single location. Core holes NRF #6P (Fig. 10A) and NRF #7P (Fig. 10B) were both cored

TABLE 4. AGE, RECURRENCE INTERVAL, DEPTH AND THICKNESS DATA FOR LAVA FLOWS SAMPLED BY DRILL CORES NEAR TEST AREA NORTH (TAN) AT THE IDAHO NATIONAL ENGINEERING AND ENVIRONMENTAL LABORATORY

Flow number	Age (Ma)	Recurrence interval (1 k.y.)	Flow top depth (m/ft)	Thickness (m/ft)
TCH #1		1.09+		
1	1.09 ± 0.09	0.06	13/44'	8/26'
2	(1.15)	0.10	21/70'	10/34'
3	1.25 ± 0.07	0.07	32/104'	37/121'
4	1.32 ± 0.02	0.26	69/226'	53/175'
5	1.58 ± 0.06	(0.42)	126/412'	9/31'
6	2.00 ± 0.06	0.12	137/448'	29/96'
7	2.12 ± 0.05	—	166/544'	17/56' +
TCH #2 Piezo A		1.09+		
1	1.09 ± 0.09	0.06	14/47'	9/28'
2	(1.15)	0.10	23/75'	18/59'
3	1.25 ± 0.07	0.07	41/134'	32/105'
4	1.32 ± 0.02	0.09	75/247'	61/201'
5	1.41 ± 0.05	(0.59)	138/452'	5/17'
6	2.00 ± 0.06	0.12	144/473'	34/111'
7	2.12 ± 0.05	0.40	179/586'	45/147'
8	2.52 ± 0.03	—	226/742'	107/351'
9			336/1102'	4/12' +
Corehole 2-2A				
1	{Brunhes Chronozone}		~3/10'	~17/55'
2	{top of Matuyama		76/249'	~16/54'
3	Chronozone}		~97/319'	~12/41'
4			120/393'	2/7'
5	{Jaramillo		124/407'	7/22'
6	Subchronozone}		131/431'	12/38'
7			144/471'	8/28'
8			152/499'	7/22'
9			159/521'	11/37'
10			170/559'	21/69'
11			194/636'	18/58'
12			212/697'	9/28'
13			226/742'	11/35'
14			237/778'	22/71'
15			273/895'	6/21'
16			280/918'	6/19'
17			287/942'	5/16'
	{Olduvai Subchronozone}			
18			350/1147'	22/73'
19			372/1220'	8/28'
20	2.45 ± 0.03		380/1248'	18/60'
21			399/1308'	4/13'
22			403/1322'	5/15'
23	{approx. top of Gauss		~477/1564'	~10/33'
24	Chronozone}		487/1598'	8/26'
25	{Kaena Subchronozone}		~541/1775'	~1/3'
26	3.26 ± 0.06		581/1906'	13/43'

Note: Ages in parentheses derived from accumulation rate line fits; recurrence intervals in parentheses represent hiatuses in accumulation; approximate depths and thicknesses in 2-2A from unrecovered cored intervals.

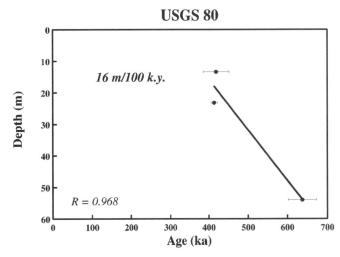

Figure 4. Graph of ages of lava flows plotted against depths to flow tops in core hole USGS 80 from Idaho National Engineering and Environmental Laboratory. One-sigma errors are plotted for age determinations, but small uncertainties in depth to tops of lava flows not shown. Accumulation rate in m/100 k.y. is calculated by straight line fit through data points; regression coefficient is also shown.

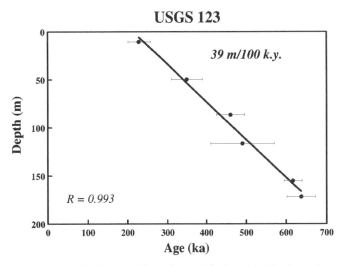

Figure 5. Graph of ages of lava flows plotted against depths to flow tops in core hole USGS 123 from Idaho National Engineering and Environmental Laboratory. Data are as in Figure 4.

Radioactive Waste Management Complex core holes

to 153 m, deeper than the other NRF core holes, and have six and seven identified lava flows, respectively. The two core holes bottom in a reversed polarity lava flow, and thus have the most complete record of accumulation rate. All five core holes share similar accumulation rates, between 22 m and 27 m/100 k.y.; the shallowest and most southerly core holes record the higher rates.

The stratigraphy, petrography, age, and paleomagnetic inclination stratigraphy of core hole BG-77-1 were reported by Kuntz et al. (1980) and Champion et al. (1981), who found seven lava flows, six having normal polarity and one having reversed polarity, to a depth of 183 m (Fig. 11; Table 3). The single reversed polarity flow, situated within the normal polarity lava flows, was initially miscorrelated with the Emperor Reversed Polarity Subchron of the Brunhes Normal Polarity Chron, but later identified as a new polarity event, the Big Lost

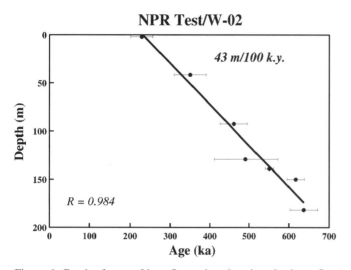

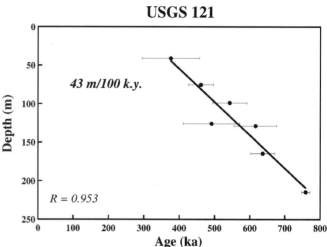

Figure 6. Graph of ages of lava flows plotted against depths to flow tops in core holes NPR Test and W-02 from Idaho National Engineering and Environmental Laboratory. Data are as in Figure 4.

Figure 8. Graph of ages of lava flows plotted against depths to flow tops in core hole USGS 121 from Idaho National Engineering and Environmental Laboratory INEEL. Data are as in Figure 4. Data point at 77 ka is not plotted nor used in linear regression.

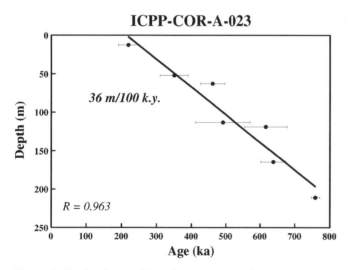

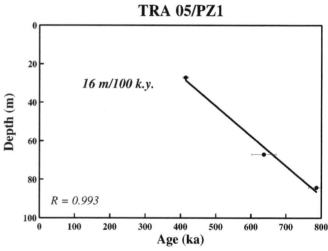

Figure 7. Graph of ages of lava flows plotted against depths to flow tops in core hole ICPP-COR-A-023 from Idaho National Engineering and Environmental Laboratory. Data are as in Figure 4.

Figure 9. Graph of ages of lava flows plotted against depths to flow tops in core hole TRA 05/PZ1 from Idaho National Engineering and Environmental Laboratory. Data are as in Figure 4.

Reversed Polarity Subchron (Champion et al., 1988). The event was assigned an age of 565 ± 14 ka on the basis of K-Ar dating, but subsequent $^{40}Ar/^{39}Ar$ dating improved the age assignment to 550 ± 10 ka (Champion et al., 1996). Core hole BG-77-1 is near the Radioactive Waste Management Complex (RWMC) of the INEEL (Fig. 3). Like the Naval Reactor Facility area, dozens of wells have been drilled and cored to various depths in the vicinity of that facility. We have analyzed results from three of the deeper core holes, BG-77-1, C-1A, and USGS 118, that form a short north-south transect across the RWMC. These three core holes sample the same stratigraphic interval with minor variations. The Matuyama-Brunhes boundary in core hole C-1A occurs at a depth of about 213 m at the

boundary between reversed and normal polarity flows (Fig. 12; Table 3). As many as six age versus depth pairs can be assigned to flows in these core holes, and accumulation rates can be calculated. Accumulation rates range between 20 m and 28 m/ 100 k.y. and have relatively low regression coefficients. The low (<0.95) coefficients of regression suggest these are apparent accumulation rates, to use a term coined by Anderson et al. (1997) to reflect hiatuses related to vent construction, periods of decreased volcanism, and differential subsidence and uplift. In particular, a hiatus was recognized at the ~90 m level in many RWMC wells by Anderson and Lewis (1989), and they assigned a duration of 285 k.y. to the hiatus based on flow ages in Champion et al. (1981). Consequently, we fit the ages to

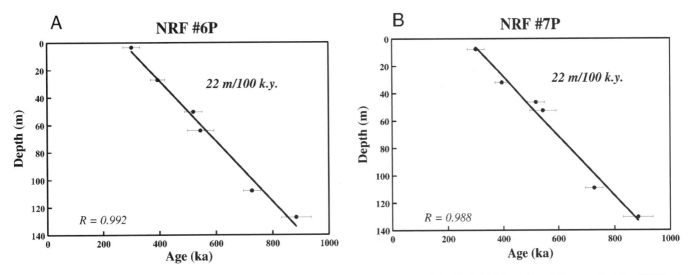

Figure 10. A: Graph of ages of lava flows plotted against depths to flow tops in core hole NRF #6P from Naval Reactor Facility (NRF) of Idaho National Engineering and Environmental Laboratory. B: Graph of ages of lava flows plotted against depths to flow tops in core hole NRF #7P. Data are as in Figure 4.

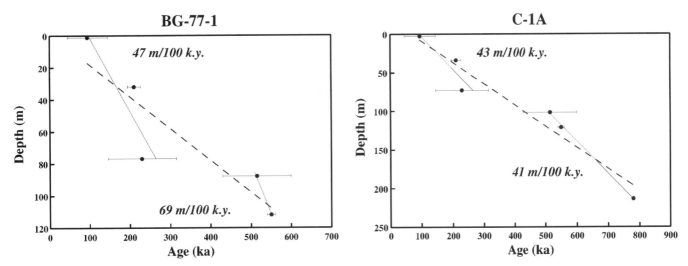

Figure 11. Graph of ages of lava flows plotted against depths to flow tops in core hole BG-77-1 at the Radioactive Waste Management Complex. Accumulation rate calculated without consideration of a hiatus between flows 4 and 5 shown as dashed line, accumulation rates calculated with a hiatus between those flows shown as two solid lines. Data shown as in Figure 4.

Figure 12. Graphs of ages of lava flows plotted against depths to flow tops in core hole C-1A at Radioactive Waste Management Complex. Accumulation rate calculated without consideration of hiatus between flows 4 and 6 is shown as dashed line; accumulation rates calculated with hiatus between those flows is shown as two solid lines. Data are as in Figure 4.

separate regression lines above and below the hiatus. Data from core hole C-1A define accumulation rates of 41 m/100 k.y. and 43 m/100 k.y. below and above the hiatus interval, respectively (Fig. 12).

TCH #1

The stratigraphy, petrography, age, and paleomagnetic inclination stratigraphy of core hole TCH #1 were reported by Lanphere et al. (1994), who described 14 lava flows of reversed polarity to a depth of 183 m (Fig. 13; Table 4). Similar mean-inclination values and the absence of intercalated sediments indicate that the number of flows in the core hole is fewer than earlier reported, based on the presumption that a feature observed in the cores represents multiple flow units within each of several eruptive events. An accumulation rate of 14 m/100 k.y. is derived from the six dated flows in the core hole, but the regression coefficient of 0.95 suggests that this should be considered an apparent accumulation rate. Thus, rates of 24 m/100 k.y. calculated for the periods before and after a hiatus of 420

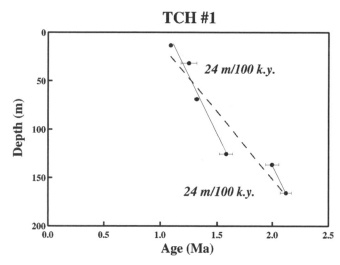

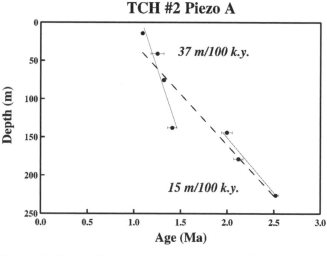

Figure 13. Graphs of age of lava flows plotted against depths to flow tops in core hole TCH #1 at Test Area North. Accumulation rate calculated without consideration of hiatus between flows 5 and 6 is shown as dashed line; accumulation rates calculated with hiatus between those flows are shown as two solid lines. Data are as in Figure 4.

Figure 14. Graphs of ages plotted against depths to flow tops in core hole TCH #2 Piezo A at Test Area North. Accumulation rate calculated without consideration of hiatus between flows 5 and 6 is shown as dashed line; accumulation rates calculated with hiatus between those flows are shown as two solid lines. Data are as in Figure 4.

k.y. at a depth of ~137 m might be more realistic (Fig. 13). No flows of Brunhes age, younger than 780 ka, are known in the immediate TAN area. Normal polarity flows located just north of "TCH #1" and on top of the 1.09 Ma reversed polarity flows of Circular Butte are likely to have been erupted during the Jaramillo Normal Polarity Subchron of the Matuyama Reversed Polarity Chron, 1.07–1.0 Ma. It is unclear that the accumulation rates of the northern (TAN) INEEL area can be directly compared to the younger sections to the south.

TCH #2 Piezo A

The stratigraphy, petrography, age, and paleomagnetic inclination stratigraphy of core hole TCH #2 Piezo A were reported by Lanphere et al. (1994), who described 20 lava flows of reversed polarity to a depth of 340 m (Fig. 14; Table 4). Again, owing to similar mean-inclination values and the absence of intercalated sediments, fewer flows in the core have been interpreted on the presumption that some multiple flow units formed within each of several eruption events. A regression coefficient of 0.93 suggests that an accumulation rate of 13 m/100 k.y. derived from seven dated flows in the core hole (Fig. 14) is an apparent rate. Thus, a rate of 37 m/100 k.y., after a hiatus of 590 k.y. at a depth of ~137 m, preceded by a rate of 15 m/100 k.y. deeper in the core, are probably more realistic values (Fig. 14).

Core hole 2-2A

There is no detailed report on the stratigraphy, petrography, age, and paleomagnetism of the 915 m core hole 2-2A in the north-central part of the INEEL (Fig. 3). A general description of the stratigraphy of the core hole was produced by Doherty (1979). Core hole 2-2A is unique among the longer core holes at the INEEL in that more than one-third of the section consists of fine-grained lake and playa sediments that can span 100 m. The large amount of sediment in this core hole complicates the lava accumulation rate for the core hole. Preliminary paleomagnetic studies have designated 26 lava flows within the upper 604 m of the core. Paleomagnetic studies below this depth have been more cursory. Preliminary determination of flow thicknesses and depths to tops of flows are listed (Table 4). An accumulation rate for the top two-thirds of core hole 2-2A has been derived from a combination of four K-Ar ages on four lava flows within the core, the ages of flows correlated to the core hole, and the depths of magnetic polarity boundaries between lava flows (Fig. 15). These 11 points of age control yield an accumulation rate of 20 m/100 k.y., and a correlation coefficient of 0.998. This calculation of accumulation rate is different from that of the other data sets of this report in that the reported depths may have considerable error, whereas the ages may constitute proportionately a lower source of error. This results from the age being assigned to the top of a lava flow, so that for a thick flow the assigned depth may have a considerable uncertainty; instead of knowing that accumulation proceeded to a certain depth, we know that a polarity change of the magnetic field occurred at a particular time (±0.01 m.y.), but that no lava accumulation occurred between flows or while sediments accumulated. The accumulation rate for this core must be considered an apparent rate, averaging accumulation bursts and hiatuses into a longer term rate. It is remarkable, however, that the accumulation of lavas at this location progressed rather smoothly for nearly 5 m.y.

Core hole 2-2a (polarity)

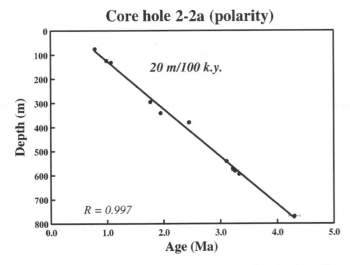

Figure 15. Graph of ages of lava flows plotted against depths to flow tops in core hole 2-2A in north-central part of Idaho National Engineering and Environmental Laboratory. See text for discussion of error sources.

RESULTS

Accumulation rate

The accumulation rates found in drill cores vary by a factor of 4, from 16 m/100 k.y. to 67 m/100 k.y. Regression coefficients of age versus depth plots are typically greater than 0.98 (Table 5). The graphical representations of accumulation rates are compelling and suggest a regular and controlled process of lava flow accumulation for the eastern Snake River Plain. This is well shown by those core holes for which complete or nearly complete K-Ar and ^{40}Ar/^{39}Ar age data are available.

Although the rates of accumulation varied from core hole to core hole, a contour map of their values suggests a simple overall pattern (Fig. 16). Only rates of accumulation of lava flows erupted during the Brunhes Normal Polarity Chron are included in the map; the rates from older sections in wells farther north in the INEEL may vary broadly through time. The data suggest a northwest-southeast gradient; contour trends are broadly parallel to the axis of the eastern Snake River Plain, and accumulation rates are greater at sites closer to the axis of the plain. The accumulation rates (Fig. 16) would require that lava accumulates in the vicinity of the axis of the eastern Snake River Plain, unless some other process is involved. This process must be subsidence, because the eastern Snake River Plain generally is rather flat. The higher accumulation rates toward the axis of the plain are compatible with subsidence that compensates for the higher accumulation rates. Proximity to lava flow source vents along the axis of the eastern Snake River Plain could explain the faster accumulation and/or subsidence of these areas, with waning lava flow accumulation to the northwest (Fig. 17).

Within the overall gradient of accumulation rates, there is an abrupt increase to higher values, centered in the TRA and ICPP area. The anomaly near TRA and ICPP suggests a zone of relative structural movement, such as a monocline or a fault. A cross section drawn through core locations TRA 05/PZ1, USGS 80, USGS 121, and NPR Test/W-02 (Fig. 18) shows the structural effect on the otherwise nearly horizontal flat-bedded age horizons of the lava section of the INEEL. Although this cross section has a vertical exaggeration of 10 for clarity, the parallelism of surfaces between core locations TRA 05/PZ1 and USGS 80 and between USGS 121 and NPR Test/W-02 is striking, compared to the tilted west to east zone between USGS 80 and USGS 121, where successively older surfaces are dropped proportionately to greater depths. The only lava flow present in all four of the core holes TRA 05/PZ1, USGS 80, USGS 121, and NPR Test/W-02 is the 640 ka flow from a vent exposed at AEC Butte (Fig. 2). This upper surface of this flow pitches downward on the southeast side >100 m between USGS 80 and USGS 121. The differences in depth of this upper surface of the flow could be explained by a great thickness of AEC Butte flow units in the near-vent locations of core holes TRA 05/PZ1 and USGS 80, the next deeper flow having a more nearly horizontal surface. Anderson (1991) suggested that AEC Butte flows are possibly 75 m thick beneath core hole USGS 80, which is 1.3 km from the vent at AEC Butte. The AEC Butte flow units are only 16 m thick in TRA 05/PZ1 and 14 m thick in USGS 121, core holes that are 1.9 km and 2.0 km, respectively, from the vent. If the AEC Butte flow units represent a thick accumulation of basalt forming a shield near these core holes, it is necessary to explain how thin flows in examples between 400 and 500 ka overlapped the edge of the shield to an aggregate depth of 100 m so soon after eruption of AEC Butte lavas.

Thickness

Mean thicknesses of basalt lava flows at INEEL vary from 10 to 39 m per eruptive event (Table 5), each eruptive event being defined by a combination of paleomagnetic and petrographic criteria. A given event may be a single lava flow or several flow units that together constitute a single lava accumulation. Changes in thicknesses define a north-northwest–trending decreasing gradient (Fig. 19); thus lava flows are thicker close to the vent locations than farther away. The thickness gradient decreases toward the areas of ICPP and TRA (Fig. 2), where mean thickness values are <12 m. The gradient of decreasing flow thicknesses (Fig. 19) is similar to the pattern of accumulation rates (Fig. 16); the core holes having thinner flows generally yield lower accumulation rates (USGS 80, TRA 05/PZ1), whereas most core holes having thicker flows yield high accumulation rates (USGS 121, ICPP-COR-A-023, USGS 123). In the vicinity of AEC Butte, the average flow thickness is <12 m; Anderson et al. (1997) speculated that this is an area of possible uplift. Whether by uplift or by a slower rate of

TABLE 5. MEAN RECURRENCE INTERVALS, MEAN THICKNESSES, ACCUMULATION RATES, AND LINEAR REGRESSION COEFFICIENTS FOR CORED WELLS OF THE IDAHO NATIONAL ENGINEERING AND ENVIRONMENTAL LABORATORY

Well name	Recurrence interval (1 k.y.)	Mean thickness (m)	Accumulation rate (m/100 k.y.)	Regression coefficient
USGS 80	73 ± 84	11 ± 7	16	0.968
USGS 123	42 ± 28	15 ± 10	39	0.994
NPR Test/W-02	51 ± 22	20 ± 11	43	0.984
ICPP-COR-A-023	41 ± 56	14 ± 14	36	0.964
USGS 121	32 ± 46	12 ± 6	43	0.953
TRA 05/PZ1	84 ± 47	10 ± 7	16	0.994
STF-PIE-AQ-01	(54)	25 ± 21	50	(1.0)
ARA-COR-005	(36)	23 ± 17	67	(1.0)
ANL-OBS-AQ-014	(41)	21 ± 13	56	(1.0)
NRF 89-04	81 ± 51	18 ± 10	24	0.992
NRF 89-05	73 ± 38	18 ± 12	26	0.994
NRF B18-1	73 ± 38	19 ± 11	27	0.990
NRF #6P	116 ± 61	25 ± 11	22	0.993
NRF #7P	97 ± 55	19 ± 19	22	0.989
BG-77-1	43 ± 26	26 ± 18	47/69	0.878/(1.0)
C-1A	50 ± 41	20 ± 13	43/41	0.898/0.999
USGS 118	—	26 ± 19	—	—
TCH #1	122 ± 81	25 ± 18	24/24	0.981/(1.0)
TCH #2 Piezo A	140 ± 129	39 ± 33	37/15	0.938/0.979
Core hole 2-2A	—	(11 ± 6)*	20	0.998

Note: Recurrence interval in parentheses are from incompletely studied core holes and preliminary; mean thickness for core hole 2-2A is preliminary due to insufficient data and an absolute minimum value; accumulation rates and regression coefficients separated by slash are for core holes with significant hiatuses in their sequence and represent before and after figures; regression coefficients in parentheses are on two-point lines and are unity by definition.

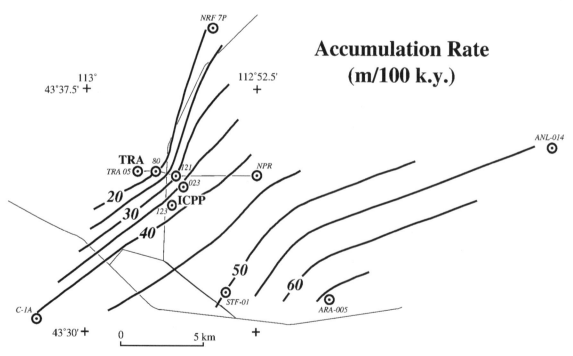

Figure 16. Contour map of basalt lava accumulation rates early to midway in Brunhes Normal Polarity Chron from core hole data in southern part of Idaho National Engineering and Environmental Laboratory. Values are in m/100 k.y., and contour interval is 5 m. Abbreviations: TRA, Test Reactor Area; ICCP, Idaho Chemical Processing Plant.

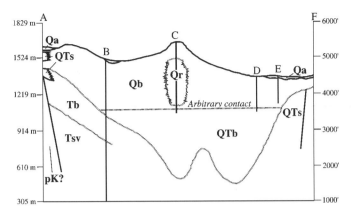

Figure 17. Northwest-southeast cross section showing subsidence of lava flows parallel to topographic axis of eastern Snake River Plain. Geologic units: Qa, Quaternary alluvium; Qr, Quaternary rhyolite; Qb, Quaternary basalt; QTs, Quaternary-Tertiary sediments; QTb, Quaternary-Tertiary basalts; Tb, Tertiary basalts; Tsv, Tertiary silicic volcanic rocks; pK, preCretaceous basement rocks. See Figure 2 for location of sections; after Whitehead (1992).

subsidence, the AEC Butte area seems to have served as a shoal for successive lava flow inundations.

Recurrence interval

The recurrence rate for eruptions can be estimated if we assume a linear age versus depth relationship in combination with a given number of flows within each core hole (Table 5). The recurrence intervals vary by a factor of four, from an eruption every 32 k.y. at USGS 121, to an eruption every 140 k.y. at TCH #2 Piezo A. Uncertainties in the ages of flows in USGS 118, and of age and flow-unit groupings in core hole 2-2A, prevent the calculation of mean recurrence intervals in those core holes. Estimates of mean recurrence intervals in core holes ANL-OBS-AQ-014, ARA-COR-005, and STF-PIE-AQ-01 were made by dividing the time interval between dated lava flows by the number of independent lava flows.

The ages used to calculate recurrence intervals come from two assessment processes. If the lava flow was directly dated or correlated to a flow in an adjacent core hole that was directly dated, then that age was used to calculate the recurrence interval between successive flows. For a lava flow for which no ages have been measured, an age estimate (parentheses in Tables 1–4) was generated from the linear age versus depth relationship in that core hole. In well-studied core holes such as NPR Test/ W-02, TCH #2 Piezo A, and USGS 123, the recurrence intervals vary. For incompletely studied core holes, such as USGS 121 and ICPP-COR-A-023, in which much of the age information is provided by correlation with other core holes, negative apparent recurrence intervals are paired with long positive recurrence intervals that immediately precede or follow the negative calculated intervals. Only the mean recurrence value

is used in geologic interpretations; we believe that the errors of age assessment only increase the standard deviation of the mean recurrence interval and not the mean value itself.

The contour pattern of the mean recurrence intervals is broadly smooth and roughly parallels the axis of the eastern Snake River Plain (Fig. 20). One would expect the mean recurrence interval to be shorter close to vents in fast accumulation zones and longer at locations farther from vents in zones of lower accumulation rate. This pattern is approximated (Fig. 20) with the important additional feature that a sharp gradient in recurrence interval mimics the strong gradient in accumulation rate (Fig. 16). The higher accumulation rates of USGS 121, ICPP-COR-A-023, and USGS 123 arise in part due to short recurrence intervals compared to core holes USGS 80 and TRA 05/PZ1. The recurrence intervals to the east and southeast of the sharp gradient suggest an area of relatively uniform recurrence intervals between 40 and 50 k.y. Mean recurrence intervals vary significantly, typically by 50%–100% of the mean values, and clearly indicate that recurrence of lava flow inundation at a given locality is nonuniform. These recurrence intervals pertain to a time from early to midway in the Brunhes Normal Polarity Chron. It is important to note that most of the eastern Snake River Plain at or near the INEEL underwent a hiatus in lava flow accumulation for the past 200 ka. However, it is important to note that the limited data from core holes that capture hiatuses suggest that these same recurrence intervals would again exist when eruptions resume.

APPLICATION OF ACCUMULATION-RATE DATA TO VOLCANIC HAZARDS AT INEEL

The data on accumulation rates, mean thicknesses, and mean-recurrence intervals generated from the INEEL core holes present an opportunity to further assess the volcanic hazard to the INEEL. Volcanic hazards were discussed by Hackett and Smith (1994) and Hackett et al. (2000), mostly on the basis of geologic data derived from the geologic map of the INEL (Kuntz et al., 1994). Hackett and Smith (1994) estimated recurrence intervals using vent density, lava flow area, and lava flow age. They estimated recurrence intervals for core holes NPR Test and BG-77-1 of 45 k.y., in close agreement with the respective values of 51 ± 22 k.y. and 43 ± 26 k.y. calculated in this study. They assigned single values for recurrence interval to individual volcanic rift zones as defined by Kuntz et al. (1992) and expressed the belief that recurrence intervals increase northward in the northwest-trending volcanic rift zones with increasing distance from the axis of the eastern Snake River Plain and decreasing vent density.

From our study, recurrence intervals do not seem to be related to the northwest-trending volcanic rift zones, possibly due to the distribution of core holes we studied. However, recurrence intervals are relatively short near the axis of the eastern Snake River Plain and relatively long in the north part of the

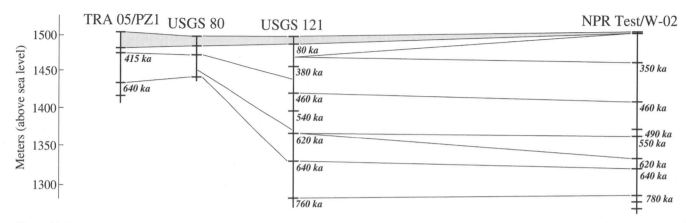

Figure 18. East-west vertical cross section between core holes TRA 05 and NPR Test/W-02 showing absolute depth position (in meters above sea level) of lava flows from AEC Butte (640 ka) and other dated flows correlated between these core holes. Gray halftone denotes surface sediment layer; numbers show age of horizons rounded to nearest 10 ka.

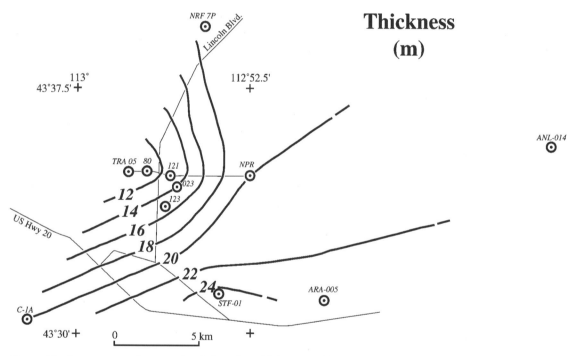

Figure 19. Contour map of average basalt lava flow thicknesses of lavas erupted early to midway in Brunhes Normal Polarity Chron, from core hole data in southern part of Idaho National Engineering and Environmental Laboratory. Values are in meters; contour interval is 2 m.

study area. Specific values computed from this study correspond to the 125 k.y. value estimated by Hackett and Smith (1994) only in the northern part of the Howe–East Butte, Lava Ridge–Hell's Half Acre, and Circular Butte–Kettle Butte volcanic rift zones, and generally for lava flows older than 780 ka. Appraisal of lava flow inundation hazards to sites on the INEEL must be viewed in light of the context that volcanic recurrence intervals are nonuniform, and that a hiatus of eruptions dominates the INEEL at present.

CONCLUSIONS

Our studies show that the accumulation of basaltic lava flows at 20 studied core holes in the eastern Snake River Plain is uniform at a given locality over very long periods of time. We also find that adjacent core holes yield remarkably different accumulation rates for strata of the same age (Fig. 16). The accumulation rates are highest near the axis of the eastern Snake River Plain and near volcanic rift zones, and lowest near the

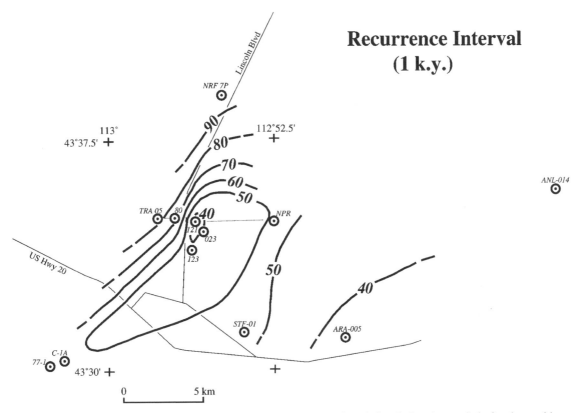

Figure 20. Contour map of average recurrence interval of basalt lava inundation during time period of early to midway through Brunhes Normal Polarity Chron from core hole data in southern part of Idaho National Engineering and Environmental Laboratory. Values are in k.y.; contour interval is 10 k.y.

margins of the eastern Snake River Plain and away from volcanic rift zones. In some core holes, the relatively steady accumulation history has been interrupted by hiatuses lasting several hundreds of thousands of years. When accumulation of flows began again at sites that underwent such a hiatus, the accumulation typically resumed at rates nearly identical to those before the hiatus.

These data provide strong evidence that basaltic volcanism in the eastern Snake River Plain is temporally and spatially predicable when viewed over hundreds of thousands or a few million years. Volcanism has occurred repeatedly along long-lived volcanic rift zones; areas away from the topographic axis of the eastern Snake River Plain and between volcanic rift zones have undergone fewer inundations by basaltic lava flows. The data can be used for long-term predictions about where and when future eruptions will occur and thus form a basis for volcanic-hazard evaluations for the INEEL area of the eastern Snake River Plain.

REFERENCES CITED

Anderson, S.R., 1991, Stratigraphy of the unsaturated zone and uppermost part of the Snake River Plain aquifer at the Idaho Chemical Processing Plant and Test Reactors Areas, Idaho National Engineering Laboratory: U.S. Geological Survey Water-Resources Investigations Report 91-4010 (DOE/ID-22095), 71 p.

Anderson, S.R., and Lewis B.D., 1989, Stratigraphy of the unsaturated zone at the Radioactive Waste Management Complex, Idaho National Engineering Laboratory, Idaho: Geological Survey Water-Resources Investigations Report 89-4065 (DOE/ID-22080), 54 p.

Anderson, S.R., Ackerman, D.J., Liszewski, M.J., and Freiburger, R.M., 1996, Stratigraphic data for wells at and near the Idaho National Engineering Laboratory, Idaho: U.S. Geological Survey Open-File Report 96-248 (DOE/ID-22095), 27 p. and 1 diskette.

Anderson, S.R., and Liszewski, M.J., 1997, Stratigraphy of the unsaturated zone and the Snake River Plain aquifer at and near the Idaho National Engineering Laboratory, Idaho: U.S. Geological Survey Water-Resources Investigations Report 97-4183 (DOE/ID-22095), 65 p.

Anderson, S.R., Liszewski, M.J., and Cecil, L.D., 1997, Geologic ages and accumulation rates of basalt-flow groups and sedimentary interbeds in selected wells at the Idaho National Engineering Laboratory, Idaho: U.S. Geological Survey Water-Resources Investigations Report 97-4010 (DOE/ID-22095), 39 p.

Champion, D.E., and Lanphere, M.A., 1997, Age and paleomagnetism of basaltic lava flows in corehole ANL-OBS-AQ-014 at Argonne National Laboratory-West, Idaho National Engineering and Environmental Laboratory: U.S. Geological Survey Open-File Report 97-700, 34 p.

Champion, D.E., Dalrymple, G.B., and Kuntz, M.A., 1981, Radiometric and paleomagnetic evidence for the Emperor Reversed Polarity Event at 0.46 ± 0.05 M.Y. in basalt lava flows from the eastern Snake River Plain, Idaho: Geophysical Research Letters, v. 8, p. 1055–1058.

Champion, D.E., Lanphere, M.A., and Kuntz, M.A., 1988, Evidence for a new geomagnetic reversal from lava flows in Idaho: Discussion of short polarity reversals in the Brunhes and late Matuyama polarity chrons: Journal of Geophysical Research, v. 93, p. 11667–11680.

Champion, D.E., Lanphere, M.A., and Anderson, S.R., 1996, Further verification and ^{40}Ar/^{39}Ar dating of the Big Lost reversed polarity subchron from drill core subsurface samples of the Idaho National Engineering Laboratory, Idaho: Eos (Transactions of the American Geophysical Union), v. 77, no. 46, p. F165.

Dalrymple, G.B., and Lanphere, M.A., 1969, Potassium-argon dating: New York, W.H. Freeman, 258 p.

Dalrymple, G.B., and Lanphere, M.A., 1971, ^{40}Ar/^{39}Ar technique of K-Ar dating: A comparison with the conventional technique: Earth and Planetary Science Letters, v. 12, p. 300–308.

Dalrymple, G.B., and Lanphere, M.A., 1974, ^{40}Ar/^{39}Ar age spectra of some undisturbed terrestrial samples: Geochemica et Cosmochemica Acta, v. 38, p. 715–738.

Doherty, D.J., 1979, Drilling data from exploration well 2-2A, NW1/4, sec. 15, T. 5 N., R. 31 E., Idaho National Engineering Laboratory, Butte County, Idaho: U.S. Geological Survey Open-File Report 79-851, 1 oversize plate.

Greeley, R., 1982, The Snake River Plain, Idaho: Representative of a new category of volcanism: Journal of Geophysical Research, v. 87, p. 2705–2712.

Hackett, W.R., and Smith, R.P., 1994, Volcanic hazards of the Idaho National Engineering Laboratory and adjacent areas: INEL-94/0276, 31 p.

Hackett, W.R., Smith, R.P., and Khericha, Soli, 2002, Volcanic hazards of the Idaho National Engineering and Environmental Laboratory, Southeast Idaho, *in* Bonnichsen, B., White, C.M., and McCurry, M., eds., Tectonic and magmatic evolution of the Snake River Plain Volcanic Province: Idaho Geological Survey Bulletin 30 (in press).

Hughes, S.S., Smith, R.P., Hackett, W.R., McCurry, M., Anderson, S.R., and Ferdock, G.C., 1997, Bimodal magmatism, basaltic volcanic styles, tectonics, and geomorphic processes of the eastern Snake River Plain, Idaho: Salt Lake City, Utah, National Geological Society of America Meeting, Fieldtrip Guidebook, p. 423–458.

Ingamells, C.O., 1970, Lithium metaborate flux in silicate analysis: Annals Chemica Acta, v. 52, p. 323–334.

Irving, E., 1964, Paleomagnetism and its application to geological and geophysical problems: New York, John Wiley & Sons, 399 p.

Jobe, S.A., and Champion, D.E., 2002, Petrography and paleomagnetism of core ICPP-COR-A-023 and correlation of a basalt lava flow sequence, Idaho National Engineering and Environmental Laboratory, Idaho: U.S. Geological Survey Open-File Report, 27 p. (in press).

Kuntz, M.A., Dalrymple, G.B., Champion, D.E., and Doherty, D.J., 1980, Petrography, age and paleomagnetism of volcanic rocks at the Radioactive Waste Management Complex, Idaho National Engineering Laboratory, Idaho, with an evaluation of volcanic hazards: U.S. Geological Survey Open-File Report 80-388, 63 p.

Kuntz, M.A., Covington, H.R., and Schorr, L.J., 1992, An overview of basaltic volcanism of the eastern Snake River Plain, Idaho, *in* Link, P.K., Kuntz, M.A., and Platt, L.B., eds., Regional geology of eastern Idaho and western Wyoming: Geological Society of America Memoir 179, p. 227–267.

Kuntz, M.A., Skipp, B., Lanphere, M.A., Scott, W.E., Pierce, K.L., Dalrymple, G.B., Champion, D.E., Embree, G.F., Page, W.R., Morgan, L.A., Smith, R.P., Hackett, W.R., and Rodgers, D.W., 1994, Geologic map of the Idaho National Engineering Laboratory and adjoining areas, eastern Idaho: U.S. Geological Survey Miscellaneous Investigations Map I-2330, scale 1:100 000.

Lanphere, M.A., 2000, Comparison of conventional K-Ar and ^{40}Ar/^{39}Ar dating of young mafic volcanic rocks: Quaternary Research, v. 53, p. 294–301.

Lanphere, M.A., and Dalrymple, G.B., 1971, A test of the ^{40}Ar/^{39}Ar age spectrum technique on some terrestrial materials: Earth and Planetary Science Letters, v. 12, p. 359–372.

Lanphere, M.A., Champion, D.E., and Kuntz, M.A., 1993, Petrography, age and paleomagnetism of basalt lava flows in coreholes Well 80, NRF 89-04, NRF 89-05, and ICPP-123, Idaho National Engineering Laboratory: U.S. Geological Survey Open-File Report 93-327, 40 p.

Lanphere, M.A., Kuntz, M.A., and Champion, D.E., 1994, Petrography, age and paleomagnetism of basaltic lava flows in coreholes at Test Area North (TAN), Idaho National Engineering Laboratory: U.S. Geological Survey Open-File Report 94-686, 49 p.

Mankinen, E.A., and Dalrymple, G.B., 1972, Electron microprobe evaluation of terrestrial basalts for whole-rock K-Ar dating: Earth and Planetary Science Letters, v. 17, p. 89–94.

Morgan, L.A., 1990, Lithologic description of the "Site E Corehole," Idaho National Engineering Laboratory, Butte County, Idaho: U.S. Geological Survey Open-File Report 90-487, 7 p.

Stacey, J.S., Sherrill, N.D., Dalrymple, G.B., Lanphere, M.A., and Carpenter, N.V., 1981, A five-collector system for the simultaneous measurement of argon isotope ratios in a static mass spectrometer: International Journal of Mass Spectrometry and Ion Physics, v. 39, p. 167–180.

Taylor, J.R., 1982, An introduction to error analysis: Mill Valley, California, University Science Books, 270 p.

Walker, E.H., 1964, Subsurface geology of the National Reactor Testing Station, Idaho: U.S. Geological Survey Bulletin 1133-E, 22 p.

Whitehead, R.L., 1992, Geohydrologic framework of the Snake River Plain Regional aquifer system, Idaho and eastern Oregon: U.S. Geological Survey Professional Paper 1408-B, 32 p.

MANUSCRIPT ACCEPTED BY THE SOCIETY NOVEMBER 2, 2000

Geological Society of America
Special Paper 353
2002

Open-system evolution of a single episode of Snake River Plain magmatism

Dennis J. Geist*
Elisa N. Sims
Department of Geological Sciences, University of Idaho, Moscow, Idaho 83844, USA
Scott S. Hughes
Michael McCurry
Department of Geosciences, Idaho State University, Pocatello, Idaho 83209, USA

ABSTRACT

The Test Area North (TAN) is located on the Idaho National Engineering and Environmental Laboratory near the northern margin of the Snake River Plain. The shallow bedrock at the site comprises a sequence of basaltic lavas that erupted in quick succession and covered a small area with ~100 m of lava in <1 k.y. approximately 1.1 Ma. The youngest lavas have magnesium numbers of ~45 to the oldest, ~57, representing cooling from ~1207 to ~1130 °C with time. Although much of the compositional variation can be attributed to ~20% crystallization of olivine and plagioclase, open-system processes must have also occurred. Phosphorus is enriched more than the other incompatible trace elements, and all incompatible elements are enriched more than can be produced by fractional crystallization alone. If the evolved magmas are related to the primitive ones by assimilation coupled with fractional crystallization, the assimilant had to have been rich in phosphorus and iron and poor in silica, potassium, barium, and scandium. This rules out lower crustal granulites and mid-crustal granites. The sedimentary Phosphoria Formation has many of these characteristics, but it is also exceedingly rich in vanadium and chromium, and the TAN lavas have no excess of either of these elements. Another rock type with these compositional characteristics is ferrogabbro. A number of workers have called on a mid-crustal mafic sill in the Snake River Plain, and this intrusion may have differentiated strongly to form phosphorus-rich ferrogabbro. Another possibility is that the evolving magma chamber was periodically replenished by primitive phosphorus-rich magma that was produced by low degrees of melting. The TAN basalts have isotopic and trace element concentrations that are similar to the EMII ocean-island basalt source. Because it has been suggested that the EMII source results from ancient subduction of continent-derived sediments, it is difficult to test for a plume component in these lavas.

INTRODUCTION

The Snake River Plain is a broad, flat, arcuate physiographic province that covers much of southern Idaho. The prevailing interpretation for the formation of the eastern Snake River Plain is that it is the trace of the Yellowstone hotspot, which may result from a mantle plume (e.g., Anders et al., 1989; Pierce and Morgan, 1992; Geist and Richards, 1993; Smith and Braile, 1994). The principal evidence supporting a hotspot origin is the time-transgressive siliceous volcanism that

*E-mail: dgeist@uidaho.edu

Geist, D.J., Sims, E.N., Hughes, S.S., and McCurry, M., 2002, Open-system evolution of a single episode of Snake River Plain magmatism, *in* Link, P.K., and Mink, L.L., eds., Geology, Hydrogeology, and Environmental Remediation: Idaho National Engineering and Environmental Laboratory, Eastern Snake River Plain, Idaho: Boulder, Colorado, Geological Society of America Special Paper 353, p. 193–204.

parallels the absolute motion of the North American plate (Armstrong et al., 1975). Furthermore, upper mantle seismic tomography reveals a low-velocity region composed of a partially molten peridotite surrounded by a high-velocity region composed of depleted residuum (Saltzer and Humphreys, 1997). Elevated $^3He/^4He$ ratios in the Yellowstone hydrothermal system and Snake River Plain basalts also support a mantle plume origin (Craig, 1993).

Numerous studies have been directed to understanding the regional geochemistry, age relations, tectonic environment, and petrogenetic processes responsible for creating the volcanic rocks of the Snake River Plain (Kuntz et al., 1992; Lum et al., 1989; Menzies et al., 1984; Leeman, 1982a, 1982b, 1982c; Thompson, 1975; Hildreth, et al., 1991). However, these studies have dealt with deciphering the geology of the Snake River Plain on a regional scale. The most important studies of a single Snake River Plain volcanic center have been those of the Craters of the Moon, but this center is atypical in the unusual volumes of evolved rocks that have erupted and unequivocal evidence for assimilation (Leeman, 1982c; Stout et al., 1994). In addition, Craters of the Moon may be a longer lived center than is typical for the Snake River Plain (Kuntz et al., 1986).

This study is a detailed petrologic investigation of a single, short-lived basaltic eruptive episode from a young volcanic center in the eastern Snake River Plain in the vicinity of the Test Area North (TAN). The samples were taken from a series of core holes that were drilled within several kilometers of one another to characterize a contaminant plume (Fig. 1).

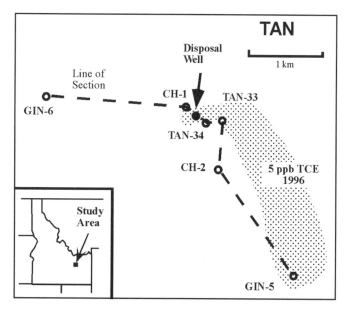

Figure 1. Location of study area and drill holes from which samples were taken. Study area is located in northern part of eastern Snake River Plain. Map shows locations of six core holes that were sampled and line along which cross section of Figure 2 was made. Stippled area shows limits of contaminant plume (from Geist et al., chapter 4, this volume). TAN is Test Area North.

Local geologic setting

The TAN site is located on the northern margin of the eastern Snake River Plain on the Idaho National Engineering and Environmental Laboratory (Fig. 1). Details of the volcanic history of the sequence of lavas dealt with herein are presented in Geist et al. (this volume, chapter 4). This study is restricted to a short, well-defined stratigraphic interval below the lavas erupted from the younger, nearby Circular Butte (Casper, 1999) and above the prominent Q-R sedimentary interbed (Anderson and Bowers, 1995). The age of the lavas is ca. 1.1 Ma, and the time span of this eruptive period is defined by paleomagnetic analyses and $^{39}Ar/^{40}Ar$ dating to be short, probably <1 k.y. (Lanphere et al., 1994; Geist et al., this volume, chapter 4). Examination of a single eruptive episode from a monogenetic center removes regional-scale differences in mantle processes and crustal character and variability associated with long-term processes that often complicate the petrogenetic histories proposed by Leeman (1982a, 1982b), Kuntz et al. (1992), and Hildreth et al. (1991). Ideally, after several centers have been studied in such detail, the petrogenetic relationships can be applied on a regional scale to better understand the igneous and tectonic processes that may have influenced the formation of the entire eastern Snake River Plain.

Detailed core logging, paleomagnetic inclinations, and geochemical fingerprinting enabled the construction of a cross section (Fig. 2) that shows that only two lava flows extend across the area. Estimations of flow direction suggest that the lavas might have erupted from two vents, one to the west and the other to the north-northeast of the TAN site. The continuity of the compositional trends discussed in this work indicates that if there were two vents, then they must have been connected to the same evolving supply of magma.

Analytical techniques

Major and trace element abundances were measured by X-ray fluorescence (XRF) spectroscopy at Washington State University using the technique of Johnson et al. (1999; see Geist et al., this volume, chapter 4). Other trace elements were measured by neutron activation analysis and are reported in Hughes et al. (this volume). Sr and Nd isotopic ratios are also reported by Hughes et al. (this volume). Analytical uncertainties are estimated to be <2% for the elements analyzed by XRF, <5% for the instrumental neutron activation analysis data used here, and ±0.00003 for the isotopic data.

PETROGRAPHIC CHARACTERISTICS OF TAN BASALTS

Detailed petrographic descriptions of lavas from the entire stratigraphic sequence in holes TAN CH-1 and TAN CH-2 were reported by Lanphere et al. (1994). The basaltic lavas in this study are predominantly microcrystalline, although a few flows

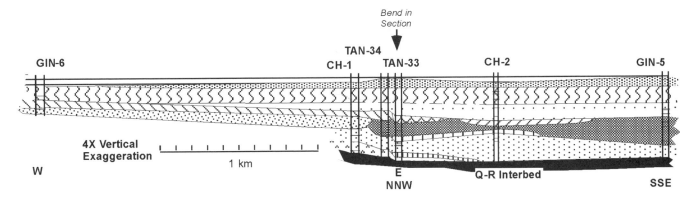

Figure 2. Cross section through Test Area North (from Geist et al., this volume, chapter 4). Section illustrates local distribution of cooling units (indicated as horizontal lines through drill holes). Flow units are correlated on basis of compositions, age determinations, paleomagnetic inclination, and petrographic inspection.

are sparsely microporphyritic to porphyritic. In this respect, they are readily distinguishable from the plagioclase-phyric lavas of Circular Butte. Of the nine flows identified in this study, only the lowermost flow of the interval occurring in holes TAN-33, TAN-34, and TAN CH-1 is notably porphyritic. The phenocrysts in this flow are 0.5–2 mm subhedral to euhedral vitreous olivine and compose ~10% of the rock. The other flows are microporphyritic; microphenocrysts constitute 5%–15% of the rock. The phenocryst assemblage consists of 1–4.4 mm subhedral to euhedral, prismatic to tabular, plagioclase (An_{25} to An_{79}) crystals and 5%–25%, 0.6–2 mm, subhedral to euhedral olivine crystals. Much of the plagioclase is glomeroporphyritic. Some of the plagioclase microphenocrysts contain oscillatory zoning patterns. All of the lavas contain a groundmass composed of plagioclase + olivine + augite + opaques + apatite ± glass. The textures of the groundmass range from intersertal to intergranular to diktytaxitic. Incipient alteration of olivine to iddingsite and clinopyroxene to chlorite has occurred along sparse cracks.

Several important conclusions about these lavas can be made based upon their textural and mineralogical characteristics. Perhaps the most striking of these features is the paucity of sizable phenocrysts, which suggests that the magmas did not reside and cool a significant amount in a shallow chamber, or if they did, the crystals segregated very efficiently and were not included in any of the magmas that erupted in this episode of volcanism. The microphenocryst assemblage indicates that the magmas were saturated with olivine and plagioclase at low pressures, but not with clinopyroxene.

COMPOSITIONAL CHARACTERISTICS OF THE TAN BASALTS

The lavas of this study are all tholeiitic basalts with normative hypersthene. Magnesium numbers (Mg# = molecular MgO/[MgO + FeO*], in %) range from ~43 to 59. Major element variation diagrams show a consistent increase of FeO*, TiO_2, P_2O_5, and K_2O with decreasing Mg# (Fig. 3). SiO_2 and Al_2O_3 correlate with Mg#. There is no clear trend between CaO and Na_2O with Mg#. With the exception of the most primitive lavas intercepted by three of the core holes, CaO/Al_2O_3 increases with decreasing Mg# (Fig. 4). K_2O/P_2O_5 appears to correlate with Mg# (Fig. 5) and shows an unusually wide variation, more than two fold. The unusually large variation in this ratio is important, because it indicates that the magmas could not be related simply by fractional crystallization.

In each of the core holes, the lavas display a general decrease in Mg# with time (Fig. 6), although there are smaller excursions to higher values upward through the package.

Ni and Cr correlate well with Mg#, which probably indicates control by olivine and chromite crystallization (Fig. 7). Although the range of Sc concentrations is beyond analytical uncertainty, the variation is not simply related to Mg#. Incompatible trace elements (e.g., Ba, Nb, Zr, Y, rare earth elements) display negative correlations with Mg# (Fig. 8). Incompatible trace elements also correlate very well with each other (Fig. 8). Sc/Y, which is diagnostic of augite control, correlates with Mg#, as does Sr/Y, which indicates that plagioclase also crystallized (Fig. 4). An abundance diagram normalized to mid-ocean ridge basalt of a primitive TAN lava shows that it is similar to a typical ocean-island basalt (Sun and McDonough, 1989), although the TAN basalts are rich in Ba and P and poor in Sr (Fig. 9). Such Ba enrichment is common among recent lavas in the intermountain west and has been attributed to contributions from the continental lithosphere (Lum et al., 1989; Hooper and Hawkesworth, 1993).

The $^{87}Sr/^{86}Sr$ isotope ratios range from 0.70652 to 0.70693. The $^{143}Nd/^{144}Nd$ values range from 0.51213 to 0.51246 (Fig. 10). The isotopic ratios do not display the negative correlation that is typical of oceanic basalts. With the exception of a single primitive sample with unusually radiogenic Sr, there is an excellent relationship between increasing $^{87}Sr/^{86}Sr$ and decreasing Mg# (Fig. 11). Although the relationship between Mg# and $^{143}Nd/^{144}Nd$ is less clear, the most primitive samples have higher neodymium isotopic ratios.

D.J. Geist et al.

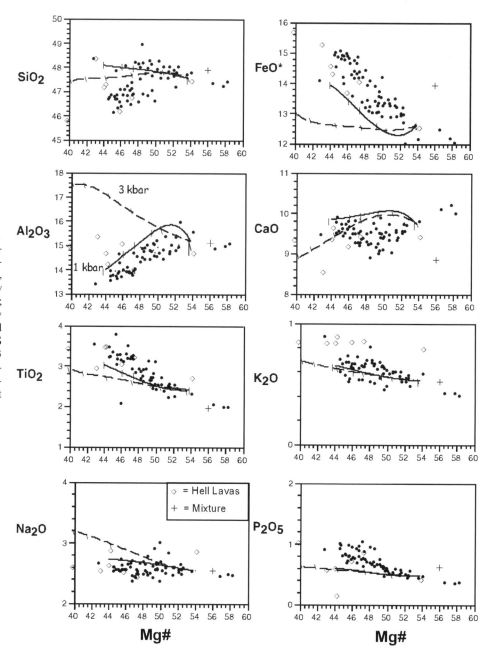

Figure 3. Major element variation diagrams of Test Area North lavas, as indicated by solid circles. For comparison, major element variation of Quaternary monogenetic center (Hell's Half Acre; Kuntz et al., 1992) is also shown by open diamonds. Solid line shows liquid line of descent predicted by MELTS program at 1 kbar, and dashed line for 3 kbar. Tick marks indicate 5% crystallization intervals. Cross shows composition of hypothetical parental magma that possibly mixed with evolving lava.

PETROGENETIC RELATIONSHIPS

Fractional crystallization

Previous workers have suggested that the dominant process controlling the evolution of Snake River Plain olivine tholeiites is fractional crystallization (e.g., Leeman, 1982b). This process is also consistent with most of the major element trends of the TAN lavas (Fig. 3): Al_2O_3 correlates with Mg# owing to the simultaneous fractionation of olivine and plagioclase, which are the observed phenocryst phases. Trace element ratios that are diagnostic of the crystallization of plagioclase (e.g., Sr/Y; Fig.

4) decrease with Mg#, further indicating that crystallization of olivine and plagioclase controls most of the observed compositional variation of the TAN suite. In addition, the concentrations of some of the most incompatible trace elements (e.g., La vs. Nb; Fig. 8) are well correlated, which is consistent with a fractional crystallization model (e.g., Hanson, 1989).

Evidence for crystallization of augite along with plagioclase and olivine is equivocal. Augite does not occur as a microphenocryst in the lavas, but thermodynamic calculations (Ghiorso and Sack, 1995) indicate that augite is stable in melts of these compositions at pressures greater than about 3 kbar. Thus, augite may have crystallized when the magmas cooled at

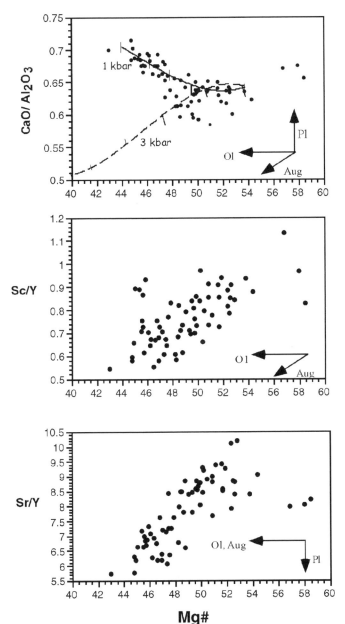

Figure 4. Diagnostic elemental ratios versus Mg# (= 100*MgO/ MgO + FeO*). Vectors indicate trends produced by fractionating each of individual minerals indicated. Curves on top diagram indicate liquid line of descent predicted by MELTS program.

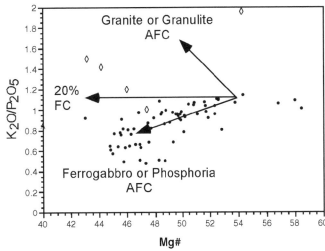

Figure 5. Variation of K_2O/P_2O_5 ratio of Test Area North and Hell's Half Acre basalts. Symbols as in Figure 3. Arrows show trends produced by 20% crystallization of olivine + plagioclase (±augite), assimilation-fractional crystallization (AFC) of granite or granulite with K_2O/P_2O_5 of 6 (typical of Idaho batholith and granulites; Clarke, 1990; Rudnick and Fountain, 1995), and assimilation-fractional crystallization of ferrogabbro or phosphatic sediment with K_2O/P_2O_5 of 0.19.

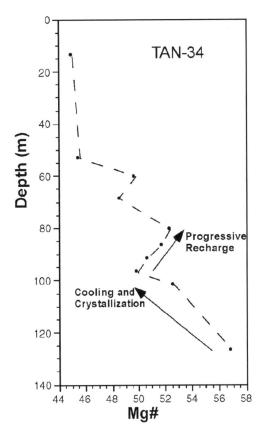

Figure 6. Variation of Mg# with depth in core hole TAN-34. Overall decrease in Mg# with short excursions to more primitive compositions is characteristic of each core hole.

depth but became unstable at shallow levels. Sc/Y ratios decrease with decreasing Mg#, which is usually indicative of augite crystallization (Albarede et al., 1997; Geist et al., 1998). In contrast, the CaO/Al_2O_3 ratios preclude significant crystallization of augite (Fig. 4). It therefore appears that Sc concentrations are controlled by some other process.

Several features indicate that fractional crystallization of plagioclase and olivine could not be the only process controlling the compositional evolution of these magmas. The com-

D.J. Geist et al.

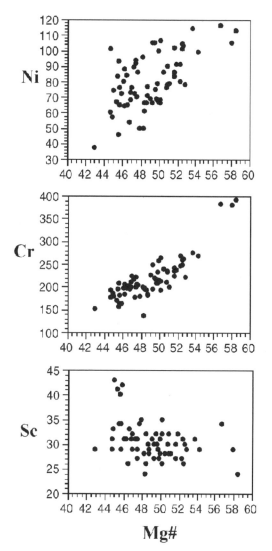

Figure 7. Variation of compatible elements with Mg#.

positional variation to be expected from fractional crystallization was established using the MELTS program, which is a
free-energy minimization algorithm (Ghiorso and Sack, 1995).
The parental composition chosen is from 120 m depth in TAN-
33, which is one of the most magnesian lavas in this suite, with
the exception of a cluster of three samples with anomalously
high magnesium concentrations that tend to cluster in all of the
major element variation diagrams. The shallow-level simulation
was run at 1 kbar; initial oxygen fugacities were set at the QFM
buffer, then the system was closed to oxygen as cooling began.
The total range of Mg# is accounted for by ~20% crystallization and cooling from 1207 to 1131 °C. This simulation fits
most elements well, with the exception of SiO_2 and P_2O_5 (Fig.
3). Phosphorus shows greater extents of enrichment with decreasing Mg# than is predicted by the model. MELTS predicts
slight enrichment of silica with decreasing Mg#, whereas there

is a notable decrease in silica concentrations in the lavas. One
possibility is that silica depletion is caused by higher pressure
crystallization, which enhances the stability of augite (Albarede
et al., 1997; Geist et al., 1998), and a model run at 5 kbar better
simulates the observed trend of silica depletion (Fig. 3). However, the higher pressures also suppress plagioclase crystallization, which results in alumina enrichment and increasing
CaO/Al_2O_3 (Fig. 4); both of these trends are opposite that observed in the natural samples.

Even more convincing evidence against the closed-system
crystallization hypothesis is the isotopic and trace element evidence that precludes fractional crystallization as being the sole
evolutionary process. The variations of Sr and Nd isotopic data
range well beyond analytical uncertainty (Figs. 10 and 11), and
are best explained by some crustal assimilation. The 20% crystallization predicted by the crystallization model would produce
only a 25% increase in the incompatible element concentrations; the observed more than two-fold variation would require
>50% crystallization, far more than what the major elements
indicate (Fig. 8). Paradoxically, the K_2O/P_2O_5 decreases
strongly (more than two fold) with decreasing Mg# (Fig. 5),
and this ratio cannot be strongly altered by crystallization of
the phases observed in the lavas, or any other phase that might
reasonably crystallize from a basaltic magma, except apatite,
which would cause the opposite trend. Although there is a general correlation between some of the incompatible trace elements, there is also much dispersion in, for example, Ba versus
Nb (Fig. 8). Fractionation of the incompatible trace elements is
also reflected in wide ranges of their ratios, such as Nb/Zr,
which greatly exceeds that possibly generated by fractional
crystallization (Fig. 8).

The three samples with highest Mg# are not on the same
trends as the bulk of the suite, but rather are on extensions of
the 3 kbar fractionation path on most of the diagrams (e.g.,
CaO/Al_2O_3, Sc/Y, and Sr/Y; Fig. 4), where augite crystallization
is enhanced and plagioclase suppressed. These samples also
have incompatible trace element ratios similar to those of the
lavas with Mg# of 50–54 and lack anomalous K_2O/P_2O_5 ratios
(Fig. 5). Thus, they may be compositionally similar to the parental magma of the entire suite and related to the main group
of lavas by essentially closed-system fractional crystallization
in the mid-crust. However the high-Mg# flow may be related
to the other lavas, it is clear that the most of the suite has
evolved by other processes.

Assimilation coupled with fractional crystallization

Much of the complexity described herein might be accounted for if crystallization were accompanied by assimilation
of crustal rocks (e.g., DePaolo, 1981). We review and dismiss
several logical possibilities before proposing some outlandish
ideas.

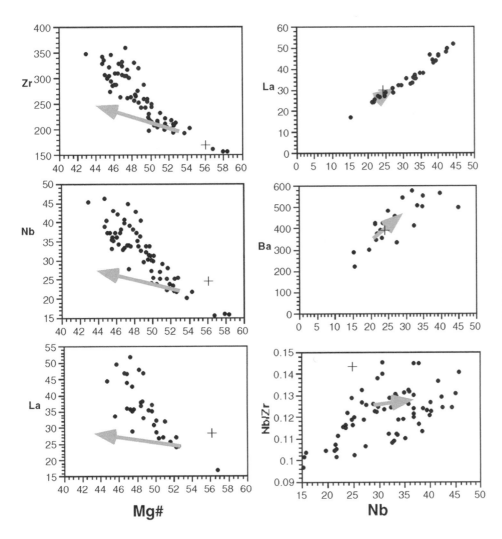

Figure 8. Variation of incompatible elements with Mg#. Arrow shows effect of crystallizing 20% olivine + plagioclase from primitive magma. Distribution coefficients are from Nielsen (1990). Cross shows composition of hypothetical parental magma that possibly mixed with evolving lava.

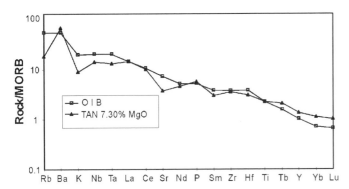

Figure 9. Typical Test Area North (TAN) basalt normalized to composition of normal mid-ocean ridge basalt (MORB) (Sun and McDonough, 1989). Also shown is composition of typical ocean-island basalt (OIB).

The most obvious possibility is that the basaltic magma assimilated rocks similar to those of the Idaho batholith as it was crystallizing. Alternatively, the assimilant may have been the Precambrian crustal rocks that melted to form the Idaho batholith (Mueller et al., 1995; Clarke, 1990), which would be compositionally and isotopically similar to the granites. Granitoids from the southern part of the Idaho batholith have a wide range of $^{87}Sr/^{86}Sr$ and $^{143}Nd/^{144}Nd$, but these ratios are generally higher and lower than those of the TAN basalts (Fig. 10). Therefore, assimilation of these rocks coupled with fractional crystallization could explain the isotopic variation of TAN basalts. However, overwhelming major and trace element evidence argue against this hypothesis. First, once a magma is saturated with both olivine and plagioclase, isenthalpic assimilation of granite by a crystallizing tholeiitic magma results in enrichment in silica and suppression of iron enrichment (e.g., Reiners et al., 1995), and the opposite trend is observed in the TAN suite (Fig. 3). The Idaho batholith, like virtually every

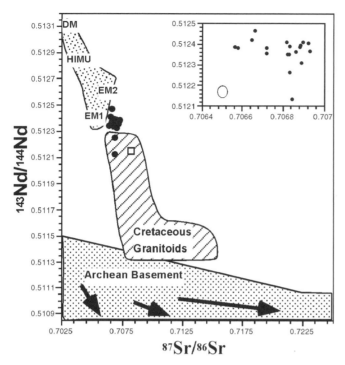

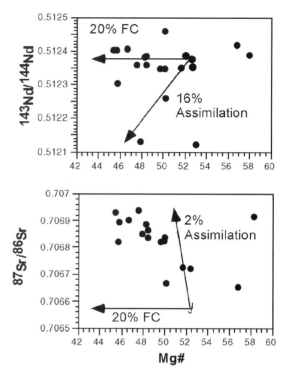

Figure 10. Sr and Nd isotopic variation of Test Area North (TAN) basalts; data are reported in Hughes et al. (this volume). Stippled field indicates isotopic reservoirs of ocean-island basalts (from Hofmann, 1997). Idaho batholith field and median granitoid (shown by open square) are from Clarke (1990). Field for Archean basement is from Leeman et al. (1985). Inset is closeup of TAN field with error ellipse.

Figure 11. Sr and Nd isotopic variation as function of Mg#. Also shown are model curves for 20% fractional crystallization (FC) of olivine + plagioclase and assimilation of typical granulite xenolith of Leeman et al., 1985.

other granite known, has very high K_2O/P_2O_5. Assimilation of even a small amount of these rocks concurrent with fractional crystallization should strongly increase the K_2O/P_2O_5 ratio, the opposite of the observed trend (Fig. 5). Moreover, like most continental granites, the Idaho batholith is characterized by very high Ba concentrations and high Ba/Nb ratios (Fig. 12; Clarke 1990). Assimilation of these granites would cause an increase in the Ba/Nb ratios of the daughter magmas and larger increases in the Ba concentration than those caused by fractional crystallization alone. Both of these predictions are opposite what is observed.

A second possibility is that the compositional variation is due to assimilation and fractional crystallization in the lower crust. Leeman et al. (1985) studied xenoliths of granulites in this region and found that they have a very large range in $^{143}Nd/$$^{144}Nd$ and $^{87}Sr/^{86}Sr$. The median value of $^{87}Sr/^{86}Sr$ of the granulites is 0.725 and $^{143}Nd/^{144}Nd$ is 0.5108. Thus, assimilation of ~2%–16% of typical granulite could explain the entire isotopic variation in the TAN basalts (Fig. 11).

Assimilation of granulite during fractional crystallization does not explain the dramatic decrease in K_2O/P_2O_5 or the lack of any increase in Ba/Nb. Although data for these elements were not reported by Leeman et al. (1985), a compilation of the compositions of granulites from throughout the world (Rudnick and Fountain, 1995) indicate that granulites also have high Ba/

Nb and K_2O/P_2O_5, although not quite as high as those reported for the Idaho batholith (Clarke, 1990). Thus, assimilation of granulites as a basaltic magma crystallizes should cause a steady increase in K_2O/P_2O_5 and Ba/Nb.

In summary, although assimilation of the most logical crustal rocks, such as the granites of the Idaho batholith or lower crustal granulites, can explain the changes in $^{87}Sr/^{86}Sr$ and $^{143}Nd/^{144}Nd$ with Mg#, they cannot replicate several of the most characteristic features of the TAN basalts. Especially perplexing is the enrichment in P_2O_5 beyond that possible by fractional crystallization and the lack of crustal signatures such as elevated Ba/Nb ratios. Another paradox that may be related to these trends is the lack of enrichment in Sc with decreasing Mg#, as reflected in steadily decreasing Sc/Y ratios (Fig. 4), despite the evidence against augite crystallization. Thus, if assimilation of crustal materials during crystallization controls the compositional variation, the assimilant must first, be rich in Fe and P and poor in Si; second, have mantle-like Ba/Nb ratios; and third, be rich in Sc.

The bewildering question is what reasonable rocks have these compositional characteristics?

Assimilation of ferrogabbros. Many tholeiitic intrusions differentiate to form iron- and phosphorus-rich rocks. The best known example of this is Upper Zone b of the Skaergaard intrusion, which averages 1.88% P_2O_5 and 26.6% FeO* (Wager and Brown, 1967; McBirney, 1989). These rocks have $K_2O/$

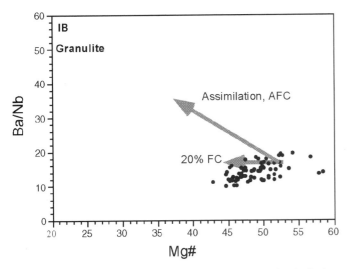

Figure 12. Ba/Nb vs. Mg#. Ba/Nb is sensitive indicator of assimilation; this ratio is very high in crustal rocks of Idaho batholith (Clarke, 1990) and granulites (Rudnick and Fountain, 1995). Arrows show trends expected by 20% crystallization of olivine + plagioclase, assimilation, and assimilation coupled with fractional crystallization (DePaolo, 1981).

P_2O_5 ratios of ~0.19. The Skaergaard is not a unique case; iron- and apatite-rich rocks are also associated with many anorthosites and other gabbroic intrusions (e.g., Wager and Brown, 1967; Mitchell et al., 1996). Although the Skaergaard ferrogabbros are not particularly rich in other incompatible trace elements (e.g., Ba = 81 ppm; La = 15 ppm), other iron-rich differentiates are very rich in incompatible trace elements (e.g., Mitchell et al., 1996).

Seismic, gravity, and topographic data indicate the presence of a 5-km-thick sill in the mid-crust of the Snake River Plain (Mabey, 1976; Smith and Braile, 1994; Peng and Humphreys, 1998). These dense, reflective rocks are thought to be gabbros that resulted from the crystallization of early-stage Snake River Plain olivine tholeiitic magma that crystallized in the middle crust. It is possible that this giant sill-like body has differentiated, creating a thick sequence of apatite-rich ferrogabbros. The implication is that the magmas that eventually erupted at TAN cooled and crystallized within the mid-crustal sill, assimilating ferrogabbros as they did so, resulting in anomalous enrichment in P, Fe, and the incompatible trace elements. Assimilation of ferrogabbro could not account for the small but regular variation of $^{143}Nd/^{144}Nd$ and $^{87}Sr/^{86}Sr$, which would require minimal assimilation (~2%) of older felsic crustal rocks in addition to the young ferrogabbro.

The evidence for assimilation of ferrogabbro is extremely circumstantial; there is little reason to suspect that other Snake River Plain magmas should undergo crystallization at the exact depth of the sill. That is, magmas that underwent crystallization at shallower or deeper depths would have encountered different rocks (e.g., granulites, granite, or normal gabbros) and not have such unusual compositions, but the enrichments in phosphorus

are not unique to the TAN suite and occur in many other Snake River plain basalts (Hughes et al., this volume). If contamination by ferrogabbros is the mechanism of phosphorus enrichment in TAN basalts, then this process must be regionally important.

Could the Phosphoria Formation be involved? The Phosphoria Formation is exposed in the mountains surrounding the Snake River Plain. As its name suggests, the Phosphoria Formation is very rich in phosphorus, and is mined in the region for phosphate. Therefore, and although circumstantial, it is possible that the TAN basalt assimilated Phosphoria rocks as it was crystallizing in the upper crust. Nd isotopic analyses of two of the phosphate-rich units have $^{143}Nd/^{144}Nd$ isotopic ratios of 0.51227 and 0.51214 (Manning et al., 1991). This is about the same range as the TAN basalts; therefore assimilation of these rocks could not account for all of the isotopic variation.

A number of compositional features of the Phosphoria go a long way toward solving the problems addressed here (Herring et al., 1999). Assimilation of these phosphatic ore deposits would cause substantial increase in P_2O_5, and they have very low K_2O/P_2O_5 ratios (average = 0.16). Furthermore, they have low Sc (mean = 6.5 ppm) and high Y (mean = 880 ppm) concentrations and low Sc/Y (mean = 0.007), which are also characteristics of the more evolved TAN basalts.

Two lines of evidence, both based on extreme compositions of these deposits, suggest to us that assimilation of Phosphoria Formation is not a reasonable hypothesis. First, they are very rich in V (mean = 643 ppm) and poor in Nb (<4 ppm) (Herring et al., 1999). Fractional crystallization of basaltic magma will slightly decrease this ratio, because V is slightly less incompatible than Nb. Assimilation of the vanadium-rich sediments should cause unusually large enrichments in V relative to other incompatible trace elements: in fact, the ratio decreases. Another extreme compositional characteristic of the Phosphoria is its very high chromium concentration, ~1000 ppm. Assimilation of even 10% of these rocks would cause strong excesses in Cr, which are not observed (Fig. 7).

Replenished magma chamber? Another possibility is that a cooling and crystallizing magma chamber was periodically replenished with a second parental magma. Slight excursions to higher Mg# with time (Fig. 6) indicate that the evolving magma chamber beneath TAN may have been episodically replenished with a primitive magma. For example, between 130 and 95 m depth in TAN-34, the Mg# decreases from 57 to 50 (Fig. 6). This can be accounted for by ~10% crystallization caused by cooling ~30° (as calculated by MELTS; Ghiorso and Sack, 1995). Upward from this interval, the Mg# increases progressively through a sequence of four flows to a Mg# of 52.5. This could be accounted for by intrusion of a more primitive magma into the magma that had evolved to a Mg# of 50. If the magma replenishing the chamber had a Mg# of 57, it would have to account for ~36% of the volume of the original magma. In addition to incompatible elements being enriched beyond that possible by fractional crystallization alone (O'Hara, 1977), evidence for magma mixing during crystallization is usually zoning or other disequilibrium textures in phenocrysts; owing to the paucity of phenocrysts in these lavas, such a test cannot be applied to the TAN sequence. Thus, in order to test this hypothesis, we are limited to mass-balance calculations.

Our procedure is to calculate the composition of a liquid produced by 10% crystallization of a TAN basalt with a Mg# of 54, using the MELTS program. We then calculate the composition of a melt that would have to be added to this hypothetical residual liquid to produce the TAN basalt with a Mg# of 52.5, with the limitation that 36% of the more primitive magma was replenished. This composition can then be compared to natural Snake River Plain magmas. This is not a very strong test of the replenishment hypothesis, but one of the few viable ones.

The results of this model calculation are plotted in the variation diagrams (Figs. 3 and 8). As might be predicted from the preceding discussions about the discrepancies between the predicted fractional crystallization model and the actual trends, such a parental magma would have to be characterized by very low Sc/Y (0.63) and K_2O/P_2O_5 (0.84) and high Nb/Zr (0.14). We reemphasize that the model is exceedingly circumstantial; it simply forces the hypothetical replenishing magma to account for the discrepancies between the trends produced by crystallization and the actual data. Nevertheless, some Quaternary lavas of the Snake River Plain have the requisite compositional characteristics: sample 84KS2 from the North Robbers field has 9.27% MgO, $K_2O/P_2O_5 = 0.71$, and Nb/Zr = 0.07 (Kuntz et al., 1992). This indicates that petrogenetic processes involved in the production of primitive Snake River Plain basalts, source composition differences, and different extents of partial melting, are capable of producing a magma that upon intrusion into an evolving magma chamber would generate the observed compositional variation of the TAN basalts.

IS THERE EVIDENCE OF THE YELLOWSTONE PLUME AT TAN?

Although it is generally accepted that the eastern Snake River Plain resulted from the passage of the North American plate over the Yellowstone hotspot (e.g., Smith and Braille, 1994; Pierce and Morgan, 1992; Anders et al., 1989; Geist and Richards, 1993), there has been very little evidence in the compositions of the magmas to support this contention (cf. Reid, 1995). As Hildreth et al. (1991) pointed out, because basalts have very low abundances of Pb, Nd, and Sr, contamination with even small amounts of crust or enriched mantle lithosphere can overwhelm the isotopic composition of the magmas.

We note, however, the similarity of the trace element concentrations in the TAN basalts to those of a typical ocean-island basalt (Fig. 9; island basalt is that of Sun and McDonough, 1989). The close correspondence between the trace element composition of Snake River Plain basalt and ocean-island basalt has been noted (Menzies et al., 1984; Reid, 1995) and was used

to provide evidence that contamination by continental crust has not been significant. The principal discrepancy is in the enrichment in Ba, which is probably a regional characteristic of the western United States subcontinental mantle, because most basalts of various ages and from a wide variety of tectonic provinces have this characteristic (e.g., Lum et al., 1989; Hooper and Hawkesworth, 1993; Hart et al., 1997). In addition, the Sr and Nd isotopic compositions of the TAN basalts are similar to those of the group of ocean island basalts rich in what is known as the "EMII" isotopic component (Fig. 10). Menzies (1989) showed that this isotopic character is common to young basalts from the circumcratonic environment of the western United States. Pb isotopic data (Leeman, 1982a; Hildreth et al., 1991) are consistent with either the EMI (Reid, 1995) or EMII sources. Melt inclusions in olivine in Snake River Plain basalts have high $^3He/^4He$ ratios (Craig, 1993), which are extremely rare in continental basalts but a common characteristic of most oceanic hotspots; this almost certainly indicates a sublithospheric contribution.

Other than the helium isotopes, positive identification of an ocean-island basalt or plume component in Snake River Plain basalts may be intractable. The compositions of EMII (and EMI) have been attributed to ancient subduction of oceanic lithosphere, including a small (<5%) volume of pelagic or terrigenous sediment (White, 1985; Zindler and Hart, 1986; Weaver, 1991; Hofmann, 1997). It would be impossible to tell this type of plume source from simple minor contamination by ancient crustal rocks or circumcratonic lithospheric mantle during ascent of the magma. A further complication is that Snake River Plain basalts have very radiogenic $^{187}Os/^{188}Os$ ratios, interpreted as being controlled by melting of mafic material in the mantle lithosphere (Hart et al., 1997).

CONCLUSIONS

A single eruptive sequence of basalt from beneath the TAN site reveals the petrogenesis of a single batch of Snake River Plain magma, which effectively removes influences involving regional crustal differences or the temporal evolution of the tectonic setting. The major process controlling the compositional variation is fractional crystallization, although open-system processes must also be involved. The crystallization occurred steadily through time and involved cooling from ~1207 to 1130°C and ~20% crystallization of olivine and plagioclase.

The exact cause of the open system behavior is difficult to pinpoint, but involves material with low K_2O/P_2O_5, low Sc, modest Ba/Nb, and high concentrations of incompatible trace elements. The obvious crustal contaminants, such as Idaho batholith or Archean granulites, are ruled out. Assimilation of the compositionally bizarre Phosphoria Formation is also ruled out. Instead, it is suggested that the magmas assimilated ferrogabbros of a large, differentiated mid-crustal sill as they crystallized. Alternatively, the cooling and evolving magma was periodically mixed with a magma having these compositional attributes, one that either came from a different source than the initial batch or segregated from a similar source that had been melted less. It is difficult to test either of these hypotheses until other volcanic centers are studied in such detail. The regional attributes of Snake River Plain magmatism were established by the pioneering works of Leeman (1982a, 1982b, 1982c), Kuntz et al. (1992), and Hildreth et al. (1991); it is clear that that detailed study of single centers will lend the most insight into the petrogenesis of these magmas.

ACKNOWLEDGMENTS

This work was funded by U.S. Department of Energy grant DE-FG07-96ID13420 through the Idaho Water Resources Water Research Institute. We thank Linda Davis and Travis McCling for help with access to core curated by the U.S. Geological Survey and INEEL. Insightful reviews by John Shervais and Craig White improved the presentation, and White is especially thanked for his tolerance of a history of hypocrisy on the part of the first author.

REFERENCES CITED

Albarede, F., Luais, B., Fitton, G., Semet, M., Kaminski, E., Upton, B.J.G., Bachelery, P., and Cheminee, J.-L., 1997, The geochemical regimes of Piton de la Fournaise volcano (Réunion) during the last 530 000 years: Journal of Petrology, v. 38, p. 177–210.

Anders, M.H., Geissman, J.W., Piety, L.A., and Sullivan, J.T., 1989, Parabolic distribution of circumeastern Snake River Plain seismicity and latest Quaternary faulting: Migratory pattern and association with the Yellowstone hotspot: Journal of Geophysical Research, v. 94, p. 1589–1621.

Anderson, S.R., and Bowers, B., 1995, Stratigraphy of the unsaturated zone and uppermost part of the Snake River Plain aquifer at Test Area North, Idaho National Engineering Laboratory, Idaho: U.S. Geological Survey Water-Resources Investigation Report 95-4130, 47 p.

Armstrong, R.L., Leeman, W.P., and Malde, H.C., 1975, K-Ar dating, Quaternary and Neogene volcanic rocks of the Snake River Plain, Idaho: American Journal of Science, v. 275, p. 225–251.

Casper, J.L., 1999, The volcanic evolution of Circular Butte [M.S. thesis]: Pocatello, Idaho, Idaho State University, 113 p.

Christiansen, R.L., and McKee, E.H., 1978, Late Cenozoic volcanic and tectonic evolution of the Great Basin and Columbia Intermontane regions, *in* Smith, R.B., and Eaton, G.P., eds., Cenozoic tectonics and regional geophysics in the Western Cordillera: Geological Society of America Memoir 152, p. 283–312.

Clarke, C.B., 1990, The geochemistry of the Atlanta Lobe of the Idaho Batholith in the Western United States Cordillera [Ph.D. thesis]: Milton-Keynes, The Open University, 357 p.

Craig, H., 1993, Yellowstone hotspot: A continental mantle plume: Eos (Transactions, American Geophysical Union), v. 74, p. 602.

DePaolo, D.J., 1981, Trace element and isotopic effects of combined wallrock assimilation and fractional crystallization: Earth and Planetary Science Letters, v. 53, p. 189–202.

Geist, D., and Richards, M., 1993, Origin of the Columbia plateau and Snake River plain: Deflection of the Yellowstone plume: Geology, v. 21, p. 789–792.

Geist, D., Naumann, T., and Larson, P.B., 1998, Evolution of Galápagos magmas: Mantle and crustal level fractionation without assimilation: Journal of Petrology, v. 39, p. 953–971.

Ghiorso, M.S., and Sack, R.O., 1995, Chemical mass transfer in magmatic processes. 4. A revised and internally consistent thermodynamic model for the interpolation and extrapolation of liquid-solid equilibria in magmatic systems at elevated temperatures and pressures: Contributions to Mineralogy and Petrology, v. 199, p. 197–222.

Hanson, G.N., 1989, An approach to trace element modeling using a simple igneous system as an example, *in* Lipin, B.R., and McKay, G.A., eds., Geochemistry and mineralogy of rare earth elements: Mineralogical Society of America, Reviews in Mineralogy, v. 21, p. 79–97.

Hart, W.K., Carlson, R.W., and Shirey, S.B., 1997, Radiogenic Os in primitive basalts from the northwestern U.S.A.: Implications for petrogenesis: Earth and Planetary Science Letters, v. 150, p. 103–116.

Herring, J.R., Desborough, F.A., Wilson, S.A., Tysdal, R.G., Grauch, R.I., and Gunter, M.E., 1999, Chemical composition of weathered and unweathered strata of the Meade Peak phosphatic shale member of the Permian Phosphoria formation: A. Measured sections A and B, central part of Rasmussen Ridge, Caribou County, Idaho: U.S. Geological Survey Open-File Report 99-147-A, 24 p.

Hildreth, W., Halliday, A.N., and Christiansen, R.L., 1991, Isotopic and chemical evidence concerning the genesis and contamination of basaltic and rhyolitic magma beneath the Yellowstone plateau volcanic field: Journal of Petrology, v. 32, p. 63–138.

Hofmann, A.W., 1997, Mantle geochemistry: The message from oceanic volcanism: Nature, v. 385, p. 219–229.

Hooper, P.R., and Hawkesworth, C.J., 1993, Isotopic and geochemical constraints on the origin and evolution of the Columbia River basalt: Journal of Petrology, v. 34, p. 1203–1246.

Johnson, D.M., Hooper, P.R., and Conrey, R.M., 1999, XRF analysis of rocks and minerals for major and trace elements on a single low dilution Litetraborate fused bead: Advances in X-ray Analysis, v. 41, P. 843–867.

Kuntz, M.A., Covington, H.R., and Schorr, L.J., 1992, An overview of the basaltic volcanism of the eastern Snake River Plain, Idaho, *in* Link, P.K., Kuntz, M.A., and Platt, L.B., eds., Regional geology of eastern Idaho and western Wyoming: Geological Society of America Memoir 179, p. 227–267.

Kuntz, M.A., Spiker, E.C., Rubin, M., Champion, D.E., and Lefebvre, R.H., 1986, Radiocarbon studies of latest Pleistocene and Holocene lava flows of the Snake River Plain, Idaho: Data, lessons, interpretations: Quaternatry Research, v. 25, 163–176.

Lanphere, M.A., Kuntz, M.A., Champion, D.E., 1994, Petrography, age and paleomagnetism of basaltic lava flows in coreholes at Test Area North (TAN), Idaho National Engineering Laboratory: U.S. Geological Survey-Open File Report 94-686, 47 p.

Leeman, W.P., 1982a, Olivine tholeiitic basalts of the Snake River Plain, Idaho, *in* Bonnichsen, B., and Breckenridge, R.M., eds., Cenozoic geology of Idaho: Idaho Bureau of Mines and Geology Bulletin 26, p. 181–191.

Leeman, W.P., 1982b, Development of the Snake River Plain-Yellowstone Plateau Province, Idaho and Wyoming: An overview and petrologic model, *in* Bonnichsen, B., and Breckenridge, R.M., eds., Cenozoic geology of Idaho: Idaho Bureau of Mines and Geology Bulletin 26, p. 155–177.

Leeman, W.P., 1982c, Evolved and hybrid lavas from the Snake River Plain, Idaho, *in* Bonnichsen, B., and Breckenridge, R.M., eds., Cenozoic geology of Idaho: Idaho Bureau of Mines and Geology Bulletin 26, p. 193–202.

Leeman, W.P., Menzies, M.A., Matty, D.J., and Embree, G.F., 1985, Strontium, neodymium, and lead isotopic compositions of deep crustal xenoliths from the Snake River Plain: Evidence for Archean basement: Earth and Planetary Science Letters, v. 75, p. 354–368.

Lum, C.C.L., Leeman, W.P., Foland, K.A., Kargel, J.A., and Fitton, J.G., 1989, Isotopic variations in continental basaltic lavas as indicators of mantle heterogeneity: Examples from the western U.S. Cordillera: Journal of Geophysical Research, v. 94, no. B6, p. 7871–7884.

Mabey, D.R., 1976, Interpretation of a gravity profile across the western Snake River Plain, Idaho: Geology, v. 4, p. 53–55.

Manning, L.K., Frost, C.D., and Branthaver, J., 1991, A neodymium isotopic study of crude oils and source rocks: Potential applications for petroleum exploration: Chemical Geology, v. 91, p. 125–138.

McBirney, A.R., 1989, The Skaergaard layered series. 1. Structure and average compositions: Journal of Petrology, v. 30, p. 363–397.

Menzies, M.A., Leeman, W.P., and Hawkesworth, C.J., 1984, Geochemical and isotopic evidence for the origin of continental flood basalts with particular reference to the Snake River Plain, Idaho, U.S.A.: Royal Society of London Philosophical Transactions, v. 310, p. 643–660.

Menzies, M.A., 1989, Cratonic, circumcratonic and oceanic mantle domains beneath the western United States: Journal of Geophysical Research, v. 94, no. B6, p. 7899–7915.

Mitchell, J.N., Scoates, J.S., Frost, C.D., and Kolker, A., 1996, The geochemical evolution of anorthosite residual magmas in the Laramie Anorthosite complex, Wyoming: Journal of Petrology, v. 37, p. 637–660.

Mueller, P.A., Shuster, R.D., D'Arcy, K.A., Heatherington, A.L., Nutman, A.P., and Williams, I.S., 1995, Source of the northeastern Idaho batholith: Isotopic evidence for a Paleoproterozoic terrane in the northwestern U.S.: Journal of Geology, v. 103, p. 63–72.

Nielson, R.L., 1990, Simulation of igneous differentiation processes, *in* Nicholls, J., and Russell, J.K., eds., Modern methods of igneous petrology: Understanding magmatic processes: Mineralogical Society of America, Reviews in Mineralogy, v. 24, p. 63–105.

O'Hara, M.J., 1977, Geochemical evolution during fractional crystallization of a periodically refilled magma chamber: Nature, v. 266, p. 503–507.

Peng, X., and Humphreys, E.D., 1998, Crustal velocity structure across the eastern Snake River Plain and the Yellowstone swell: Journal of Geophysical Research, v. 103, no. B4, p. 7171–7186.

Pierce, K.L., and Morgan, L.A., 1992, The track of the Yellowstone hot spot: Volcanism, faulting and uplift, *in* Link, P.K., Kuntz, M.A., and Platt, L.B., eds., Regional geology of eastern Idaho and western Wyoming: Geological Society of America Memoir 179, p. 1–53.

Reid, M.R., 1995, Processes of mantle enrichment and magmatic differentiation in the eastern Snake River Plain: Th isotope evidence: Earth and Planetary Science Letters, v. 131, p. 239–254.

Reiners, L.P., Nelson, B.K., and Ghiorso, M.S., 1995, Assimilation of felsic crust by basaltic magma: Thermal limits and extents of crustal contamination of mantle-derived magmas: Geology, v. 23, p. 563–566.

Rudnick, R.L., and Fountain, D.M., 1995, Nature and composition of the continental crust: A lower crustal perspective: Reviews of Geophysics, v. 33, p. 267–309.

Saltzer, R.L., and Humphreys, E.D., 1997, Upper mantle *P* wave velocity structure of the eastern Snake River plain and its relationship to geodynamic models of the region: Journal of Geophysical Research, v. 102, p. 11829–11841.

Smith, R.B., and Braile, L.W., 1994, The Yellowstone hotspot: Journal of Volcanology and Geothermal Research, v. 61, p. 121–187.

Stout, M.Z., Nicholls, J., and Kuntz, M.A., 1994, Petrological and mineralogical variations in 2500-2000 year B.P. lava flows, Craters of the Moon lava field, Idaho. Journal of Petrology, v. 35, p. 1681–1715.

Sun, S.-S., and McDonough, W.F., 1989, Chemical and isotopic systematics of oceanic basalts: Implications for mantle compositions and processes, *in* Saunders, A.D., and Norry, M.J., eds., Magmatism in the ocean basins: The Geological Society [London] Special Publication 42, p. 313–345.

Thompson, R.N., 1975, Primary basalts and magma genesis. 2. Snake River Plain, Idaho, U.S.A.: Contributions to Mineralogy and Petrology, v. 52, p. 213–232.

Wager, L.R., and Brown, G.M., 1967, Layered igneous rocks: San Francisco, Freeman, 588 p.

Weaver, B.L., 1991, The origin of ocean island basalt end-member compositions: trace element and isotopic constraints: Earth and Planetary Science Letters, v. 104, p. 381–397.

White, W.M., 1985, Sources of oceanic basalts: Radiogenic isotope evidence: Geology, v. 13, p. 115–118.

Zindler, A., and Hart, S.R., 1986, Chemical geodynamics: Annual Reviews of Earth and Planetary Sciences, v. 14, p. 493–571.

MANUSCRIPT ACCEPTED BY THE SOCIETY NOVEMBER 2, 2000

Geological Society of America
Special Paper 353
2002

Chemical characteristics of thermal water beneath the eastern Snake River Plain

Travis L. McLing*
Robert W. Smith
Idaho National Engineering and Environmental Laboratory, P.O. Box 1625,
Idaho Falls, Idaho 83415, USA
Thomas M. Johnson
University of Illinois at Urbana-Champaign, 245 Natural History Building, MC-102,
Urbana, Illinois 61801, USA

ABSTRACT

The eastern Snake River Plain aquifer is among the largest and most productive aquifers in the United States. Protection of this resource requires an understanding of the dominant mechanisms that control groundwater chemistry in the eastern Snake River Plain aquifer. To assess the chemistry of the deeper waters of the aquifer, two deep thermal wells and numerous thermal springs were evaluated. The results of this study indicate that the eastern Snake River Plain aquifer is composed of two systems, the upper aquifer, from which most water is produced, and a deep thermal aquifer. The chemistry of the upper aquifer is dominated by $Ca-Mg-HCO_3$, typical of groundwater in the arid west. The deep thermal system consists of $Na-K-HCO_3$ water. The difference in water chemistry between the upper and lower aquifers is the result of longer residence times and more water-rock interaction within the deep system. Differences in composition of the deep thermal waters may reflect the variety of types of aquifer host rock.

INTRODUCTION

The eastern Snake River Plain aquifer in eastern Idaho is among the largest and most productive aquifers in the United Sates. The aquifer is used as a groundwater resource and also provides significant recharge to the Snake River. Past and current activities at the Idaho National Engineering and Environmental Laboratory (INEEL), including reactor research, nuclear fuel reprocessing, nuclear waste storage, and other nuclear research, represent real or perceived risks to the eastern Snake River Plain aquifer. Quantification of these risks requires improved understanding of local (e.g., waste disposal practices) and regional (e.g., groundwater recharge and mixing) processes that influence the quality of groundwater in the aquifer. En-

hanced characterization of large-scale rapid groundwater flow paths and the improved quantification of the role of water mixing and geochemical reactions in the evolution of water quality are of particular interest.

The eastern Snake River Plain is an arcuate structural depression 50–100 km wide by 300 km long and encompasses ~12 700 km^2 in southeastern Idaho (Fig. 1). The eastern Snake River Plain aquifer is composed of a sequence of basalt flows interlayered with fluvial sediments and loess. Individual basalt flows of relatively small volume were extruded primarily from northwest-trending fracture systems or from numerous small shields. At depths greater than ~1 km, welded rhyolite tuff and tuffaceous sediments dominate the eastern Snake River Plain (Pierce and Morgan, 1992). Recent volcanic activity on the

*E-mail: TM1@inel.gov

McLing, T.L., Smith, R.W., and Johnson, T.M., 2002, Chemical characteristics of thermal water beneath the eastern Snake River Plain, *in* Link, P.K., and Mink, L.L., eds., Geology, Hydrogeology, and Environmental Remediation: Idaho National Engineering and Environmental Laboratory, Eastern Snake River Plain, Idaho: Boulder, Colorado, Geological Society of America Special Paper 353, p. 205–211.

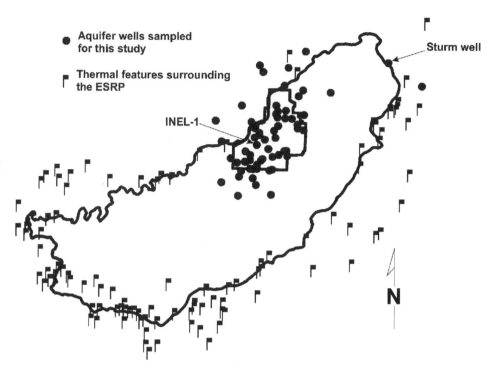

Figure 1. Location map showing eastern Snake River Plain (ESRP) and location of aquifer wells sampled for this study and thermal features located proximal to boundary of plain.

eastern Snake River Plain has resulted in a very high geothermal gradient of 40–50 °C/km (Blackwell, 1992), manifested by numerous thermal springs around the periphery of the eastern Snake River Plain.

Recharge to the aquifer is primarily from the drainage of highlands north of the plain (McLing, 1994). Groundwater is discharged to the Snake River through a series of springs near Hagerman, Idaho, ~260 km southwest of the INEEL (Wood and Low, 1988).

The aquifer is composed of two systems. The shallow, or effective, portion of the aquifer occurs from the water table (60–200 m below land surface [bls]), to a depth of 300–500 m bls. Fast-moving, (1.52–10.51 m/day), cold (9–15 °C), Ca- and Mg-rich waters characterize this part of the aquifer (Mann, 1986). The deeper portion of the aquifer is characterized by slower moving (0.006–0.091 m/day), warm (>30 °C) water (Mann, 1986). Although a sharp contact between these two systems is not always observed, changes in geothermal gradients can be used to delineate the two systems, as shown in Figure 2. The composition and extent of the lower system and the extent of its interactions with the upper system are the subjects of this chapter.

Thermal water derived from beneath the effective portion of the eastern Snake River Plain aquifer (the upper 100–400 m) has long been suspected of being a source of dissolved groundwater constituents (Robertson et al., 1974). Early investigators focused on numerous thermal springs that occur along the margins of the eastern Snake River Plain and the lack of such features on the plain to infer deep groundwater composition (Robertson et al., 1974; McLing, 1994). As part of INEEL efforts

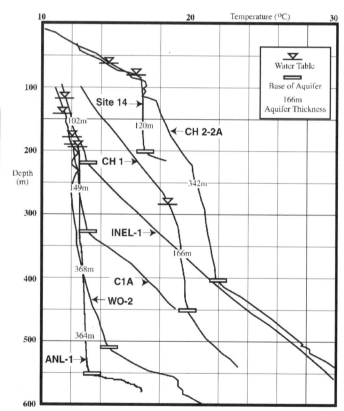

Figure 2. Temperature logs of wells and exploration drill holes that penetrate base of aquifer (from Blackwell, 1992).

in the early 1980s to assess the geothermal potential of the eastern Snake River Plain, the 3400 m INEL-1 well (Fig. 1) was drilled, and groundwater samples were collected at several depths (Mann, 1986). These groundwater samples show a progression from deep, warm, Na-HCO$_3$ waters to cooler Ca-Mg-HCO$_3$ waters at shallower depths (Table 1). Furthermore, the range in composition with depth in INEL-1 is similar to the range of compositions seen in thermal springs (Fig. 3) (Mariner et al., 1995). Although Mann's work (1986) provides significant insight into the nature of deep thermal water at a single location, samples of deep groundwater from additional wells and springs are needed to access spatial variability of deep groundwater composition. One such well and a subject of this paper is the 1000-m-deep Sturm well located near Ashton, Idaho (Fig. 1). Water samples collected from this well were analyzed for major and minor chemical species and for strontium isotopes. In addition, archived water samples from INEL-1 were analyzed for strontium isotopes. The results reported here are part of an ongoing study to delineate regional aquifer flow paths through the use of natural chemical tracers and isotopic ratios (Smith and McLing, 1998; Lou et al., 2000).

METHODS

Filtered (0.45 μm) groundwater samples were collected from the wells listed in Table 1, after purging three well-bore volumes. Because the Sturm well continually produces water under natural flow, purging was not required. Field measurements of pH, temperature, conductivity, and alkalinity were conducted by standard methods. Water samples were preserved using ultrapure HNO$_3$ and were shipped on ice for laboratory analyses of strontium isotopic ratios, cations, trace metals, and anions.

Selected thermal springs were also sampled using a peristaltic pump and 0.45 μm filter. The pH, temperature, and alkalinity were measured in the field, and samples were preserved and handled as described here. In addition, archived groundwater samples from INEL-1, collected by the U.S. Geological Survey in 1986, were used for strontium isotope and trace metal analyses.

Determination of 14 cations and trace metals (Li, B, Na, Mg, Al, Si, P, K, Ca, Mn, Fe, Sr, Ba, and U) was conducted by inductively coupled plasma mass spectroscopy (ICP-MS) with standard operating procedures using Ge and In as internal standards. Nominal uncertainty for the ICP-MS measurements is 5%, although duplicate measurements suggest that the reproducibility is generally much better than this. Anions were determined by ion chromatography. Strontium for isotope ratio analysis was concentrated from the water samples using Sr-specific cation exchange resin and determined by thermal ionization mass spectroscopy.

RESULTS

The strontium isotopic ratios, cation, and anion concentrations of selected aquifer and thermal water collected in this study are presented in Table 1, as are anion concentrations for INEL-1 reported in Mann (1986). These results, along with the results of Mariner et al. (1995) for thermal springs, are presented in Figure 3.

DISCUSSION

Although there are few thermal springs on the Snake River Plain, the high heat flow (Blackwell, 1992) and recent volcanic activity suggest the presence of geothermal potential beneath the eastern Snake River Plain. However, it is difficult to quantify the chemical contribution of geothermal waters to the upper portion of the aquifer, because there are a limited number of wells that sample the lower portion of the aquifer. These wells include INEL-1 and the rhyolite-hosted Sturm well. In addition, the chemistry of deep water may be inferred from the thermal springs on the margins of the eastern Snake River Plain. The new chemical data for thermal springs that are reported here complement previously reported compositions for thermal springs reported by Mariner et al. (1995).

Figure 4 is a Piper diagram showing the composition of eastern Snake River Plain groundwater collected from wells (including INEL-1 and Sturm). Most of these wells are completed in the upper portion of the aquifer, and specific sampling

TABLE 1. CHEMISTRY OF SELECTED AQUIFER WATERS AND THERMAL WATERS NEAR THE INEEL MG/L

Location	Ca	Mg	K	Na	SO$_4$	Cl	HCO$_3$	^{87}Sr/^{86}Sr
Condie HS	61.3	11.5	18	58.1	26.6	13.6	361.5	0.71440
Liddy HS	88	16	15	27	200	6.7	174	0.71082
Warm Spring	60.2	23.1	3.66	11.11	175	8.9	158.6	0.71434
Sturm	3.8	0.02	0.84	32.13	4.56	3.0	74	0.70871
INEL-1 > 1460 m (Anion data from Mann, 1986)	7.0	0.5	7.3	385	99	12	740	0.70980
INEL-1 1066–1460 m (Anion data from Mann, 1986)	8.9	1.1	8.1	370	97	13	670	Not analyzed
INEL-1 460–670 ft (Anion data from Mann, 1986)	8.2	2.0	10	92	32	17	210	0.70935
Big Lost River	37.5	9.9	1.4	5.80	18	4.8	200	0.71056
Little Lost River	31.1	12.6	1.2	6.51	16	8.8	177	0.71256
Birch Creek	41.2	14.3	1.5	6.07	4.5	25	164	0.71198
Yellowstone Plateau	10.1	6.2	9.06	4.52	3.5	3.7	63	0.70930
Site 17	54	17	1.3	10	16	11	228	0.710912

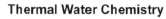
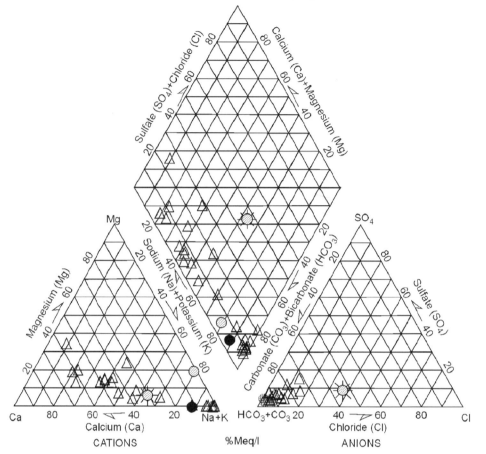

EASTERN SNAKE RIVER PLAIN
Thermal Water Chemistry

Figure 3. Water chemistries for monitoring wells and thermal springs located on eastern Snake River Plain (ESRP).

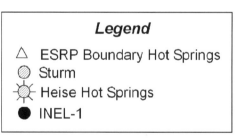

horizons are unknown. The Ca-Mg-HCO$_3$ waters characteristic of the upper portion of the aquifer are distinct from the two deep wells that are Na-K-HCO$_3$ waters. In addition, some data are intermediate between the deep wells and the upper aquifer, suggesting some mixing of water from these two sources. The lack of significant data between these two end-members may reflect a lack of water samples from intermediate depths, small amounts of mixing, or both.

Although the Sturm and INEL-1 water chemistries plot within the same region on a Piper diagram (Fig. 3), reflecting the dominance of Na and K in the cation balance, there are significant concentration differences between the two wells (Table 1). The INEL-1 water has a much higher total concentration than does the Sturm well water. The higher solute concentrations in INEL-1 water may indicate more extensive rock-water interactions than for Sturm well waters and also may indicate that the INEL-1 water has interacted with a greater diversity of host rock types. This conjecture is supported by the measured 35 ka age for INEL-1 water (Mann, 1986) and the fact that shallow aquifer water from wells near INEL-1 contains lower concentrations of dissolved minerals than deeper water. Although age dating for Sturm well water is not available, its location near the Yellowstone Plateau recharge region argues for a younger age.

EASTERN SNAKE RIVER PLAIN
Groundwater Chemistry

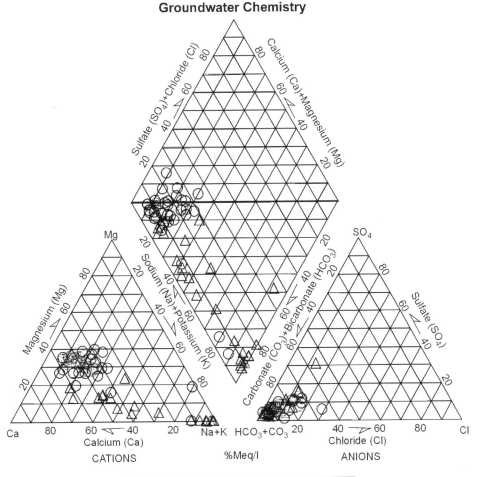

Figure 4. Piper diagram showing water chemistries from eastern Snake River Plain (ESRP) aquifer wells and thermal waters sampled for this study.

Legend

△ ESRP Boundary Hot Springs
○ ESRP Aquifer Water

Speciation calculations using MINTEQA2 indicate that water from INEL-1 is supersaturated with calcite, while water from Sturm well is undersaturated with calcite. Supersaturation with calcite is characteristic of the upper aquifer and may reflect interactions with Paleozoic and Mesozoic carbonates within the recharge areas located to the north of the eastern Snake River Plain (Robertson, 1974). Calcite undersaturation is consistent with water recharged in the silicic rocks of the Yellowstone Plateau. Other potential explanations for the observed differences in saturation states include cation exchange and/or the dissolution of plagioclase and other minerals in the basalt host rock. Although ion exchange is a process that is likely occurring in the eastern Snake River Plain aquifer, it cannot be the dominant process controlling water chemistry because anions such as chloride that are expected to be conserved during cation exchange show significant variation. In addition, previous inves-

tigators (Wood and Low, 1988; McLing, 1994) concluded that the weathering of mineral phases in basalt is of minimal importance in the aquifer.

The $^{87}Sr/^{86}Sr$ isotope ratios for Sturm well water and INEL-1 water are 0.70871 and 0.70980, respectively. The value of 0.70980 for deep INEL-1 water is much lower than the value of 0.71091 that is characteristic of the upper aquifer in the vicinity of the well (see following section). The $^{87}Sr/^{86}Sr$ isotopic ratio for INEL-1 water is intermediate between the values of the upper aquifer and the values measured for deep silicic rock on the eastern Snake River Plain (0.7067–0.7072; M. McCurry, 1998, written commun.), suggesting that significant water-rock interactions have occurred. In addition, physical observation and previous studies of deep INEL-1 core show that significant diagenesis has occurred in the deep aquifer (Mann, 1986; Morse and McCurry, this volume).

The isotopic ratio of 0.70871 for Sturm well water is lower than for INEL-1 water (0.70980) and lower than surface water derived from the Yellowstone Plateau (0.70931), suggesting that Sturm well water may have equilibrated with the silicic volcanics. In addition, the low value, coupled with low total solute concentrations, suggests that water from the Sturm well had its source entirely within silicic volcanics of the Yellowstone Plateau.

Deep water from INEL-1 and the Sturm well are Na-K-HCO_3 type and reflect a chemistry significantly different than the Ca-Mg-HCO_3 waters of the upper portion of the aquifer. Although the deep waters share some compositional similarities, their individual chemistries reflect their differences in residence time and recharge and host rock geology (see Table 1). If the hypothesis is true that the water from the Sturm well has only reacted with silicic volcanics, then by comparison INEL-1 waters have probably undergone significant interaction with aquifer rocks, including volcanic and carbonate sequences.

GEOCHEMISTRY OF THERMAL WATERS

Examination of Figure 3 shows that most of the thermal springs bordering the eastern Snake River Plain aquifer (Fig. 1) have major ion compositions intermediate between the lower aquifer system and the upper effective portion of the aquifer. This observation suggests that these springs may be mixtures of groundwater derived from the lower and the upper aquifer systems. However, the high $^{87}Sr/^{86}Sr$ isotope ratios for many of the springs cannot be explained by simple mixing, and require Sr contributions from a source such as Paleozoic carbonate and siliciclastic rocks. We propose that the most likely explanation for the data presented in Figure 4 is that upwelling of deep thermal water is important, at least locally. The contribution of deep thermal water to the upper aquifer was first hypothesized ~30 yr ago, based on geological arguments (Robertson et al., 1974). Mann (1986) inferred from INEL-1 that as much as 19 $\times$ 10^6 m^3 of thermal water might be upwelling into the effective aquifer beneath the INEEL site. Wood and Low (1988) determined that, on the basis of ^{14}C and δ^2H values, water in the deep thermal system may be older than 17 ka and may have been recharged during a colder climatic time. Doherty et al. (1979) postulated that the thick sequence of unbroken volcanics in INEL-1 indicate the presence of a caldera complex and the associated ring fractures that would provide vertical conduits for upwelling thermal water. McLing (1994) assessed the relative importance of mixing and water-rock interactions in the evolution of groundwater composition along flow paths crossing the INEEL. He concluded that with the exception of the precipitation of small amounts of calcite, there were only limited water-rock interactions, and that mixing of waters from different sources accounts for the observed changes in water composition along flow paths. One of the sources McLing (1994) proposed was deep thermal water similar to that observed at Heise Hot Springs (Fig. 3). He found that only a small (<1.0%) amount of the chloride-rich Heise water was required to obtain mass balance for chloride. However, the Heise water is atypical (Fig. 2) of the thermal springs along the margin of the eastern Snake River Plain because its dominant anion is chloride rather than bicarbonate. Larger amounts of thermal waters would have been required if McLing (1994) had used waters with compositions consistent with the main trend shown in Figure 3. These lines of evidence suggest that a significant flux of thermal water impinges on the upper aquifer.

CONCLUSION

Several lines of indirect evidence support the possibility of thermal upwelling in the eastern Snake River Plain aquifer. These include diagenetic phases in well cores and high heat flows. In addition, the complete absence of geothermal springs on the recently volcanically active Snake River Plain and the large number of hot springs located on the periphery of the plain support this conclusion. While there is little doubt concerning the existence of thermal water deep in the eastern Snake River Plain aquifer, little is known about its chemistry. Examination of water compositions from two deep wells provides some information about the chemical composition of groundwater in the lower aquifer system. This study suggests that the deep Na-K-HCO_3 waters are distinct from the water of the upper aquifer system and that the deep Na-K-HCO_3 waters are widespread (based on two wells 80 km apart). However, the composition of deep waters, similar to the shallower aquifer, reflects differences in source area recharge and aquifer host rock. Water from INEL-1 has apparently undergone significant interactions with aquifer host rocks, including basalt, Paleozoic carbonates, and siliciclastic sequences. Although the areal extent of this water is unknown, it is likely that the composition is typical of deep waters that are proximal to edges of the eastern Snake River Plain. In contrast, Sturm well water has undergone only very limited interactions with rhyolitic aquifer host rocks. The chemical composition of Sturm well water is consistent with infiltrating rain or snow through the rhyolitic terrain of the Yellowstone Plateau.

REFERENCES CITED

Blackwell, D.D., 1992, Heat flow modeling of the Snake River Plain, Idaho: Idaho Falls, Idaho, Idaho National Engineering and Environmental Laboratory, EGG-NPR-10790, 109 p.

Doherty, D.J., McBroome, L.A., and Kuntz, M.A., 1979, Preliminary geological interpretation and lithologic log of the exploratory geothermal test well (INEL-1), Idaho National Engineering Laboratory: U.S. Geological Survey Water-Resources Investigations Report, 79-1248, 17 p.

Johnson, T.M., Roback, R.C., McLing, T.L., Ballen, T.D., Depaolo, D.J., Doughty, C., Hunt R.J., Murrell, M.T., and Smith, R.W., 2000, Groundwater "Fast Paths" in the Snake River Plain Aquifer: Radiogenic isotopes ratios as natural groundwater tracers: Geology, v. 28, p. 871–874.

Lou, S., Teh-Lung Ku, Roback, R.C., Murrell, M.T., and McLing, T.L., 2000, In-situ radionuclide transport and preferential groundwater flows at Idaho National Engineering and Environmental Laboratory Decay-series disequilibrium studies: Geochimica et Cosmochimica Acta, v. 64, p. 867–881.

Mann, L.J., 1986, Hydraulic properties of rock units and chemical quality of water for INEL-1: A 10,365-foot-deep test hole drilled at the Idaho National Engineering Laboratory, Idaho: U.S. Geological Survey Water-Resources Investigations Report 86-4020, 23 p.

Mariner, R.H., Young, H.W., Evans, W.C., and Parliman, D.J., 1995, Lead and strontium isotope data for thermal waters of the Regional regional Geothermal System in the Twin Falls and Oakley areas, South-Central Idaho: Geothermal Resources Council Transactions, v. 19, p. 201–206.

McLing, T.L., 1994, The pre-anthropogenic groundwater evolution at the Idaho National Engineering Site, Idaho [M.S. thesis]: Pocatello, Idaho, Idaho State University, 62 p.

Pierce, K.L., and Morgan, L.A., 1992, The track of the Yellowstone Hot Spot: Volcanism, faulting, and uplift, *in* Link, P.K., Kuntz, M.A., and Platt, L.B., eds., Regional geology of eastern Idaho and western Wyoming: Geological Society of America Memoir 179, p. 1–54.

Robertson, J.B., Schoen, R., and Barraclough, J.T., 1974, The influence of liquid waste disposal on the geochemistry of water at the National Reactor Testing Station, Idaho, 1952–1970: U.S. Geological Survey Open-File Report ID-22053, 231 p.

Smith, R.W., and McLing, T.L., 1998, Investigation of groundwater flow paths in fractured aquifers through combined inversion of strontium isotope ratios and hydraulic head data: The role of mixing of groundwaters from different source areas: Washington, D.C., Environmental Management Science Program Workshop, CONF-980736, U.S. Department of Energy, Office of Science and Risk Policy EM-52, p. 425–426.

Wood, W.W., and Low, W.H., 1988, Aqueous geochemistry and diagenesis in the eastern Snake River Plain aquifer system, Idaho: Geological Society of America Bulletin, v. 97, p. 1456–1466.

Manuscript Accepted by the Society November 2, 2000

Geological Society of America
Special Paper 353
2002

Genesis of alteration of Quaternary basalts within a portion of the eastern Snake River Plain aquifer

Lee H. Morse
Michael McCurry*
Department of Geosciences, Idaho State University, Pocatello, Idaho 83209, USA

ABSTRACT

We characterize and interpret three-dimensional spatial patterns of alteration in basalt from core from five deep boreholes located across the Idaho National Engineering and Environmental Laboratory. The basalts range in age from ca. 0.5 to 2.5 Ma and host the eastern Snake River Plain aquifer.

Consistent patterns of alteration occur. Basalts in the upper parts of wells are remarkably fresh, aside from minor caliche or drusy calcite deposits in vesicles. At depths ranging from 320 to 508 m there are pronounced increases in the intensity of authigenic pore mineralization. Changes from largely unaltered to moderately to strongly altered basalt occur over narrow vertical intervals in at least two of the wells; in all cases these transitions correlate closely with sharp inflections in the temperature gradients in these wells. Many pores and fractures are partially or completely filled by nontronite ± saponite ± calcite; intersertal glass and olivine are partially to completely altered to nontronite ± saponite. Alteration of the basalts increases in intensity downward from the temperature-gradient inflections and appears to have been largely isochemical.

Pronounced downward transitions from unaltered to altered basalts do not correlate systematically with depth, age of the basalts, prevailing temperatures, or stratigraphic features. We propose that the alteration is produced by transient thermal inputs into the base of the aquifer from a deep-seated geothermal source and that the coincidence of alteration and the temperature-gradient inflections can be used to identify the effective base of the aquifer with a high degree of confidence.

INTRODUCTION

Purpose

Low-temperature alteration and authigenic mineralization of Quaternary olivine tholeiite basalts in the eastern Snake River Plain aquifer affect the solute balance of the aquifer, its hydrological properties, and the capacity of the aquifer system to retard the migration of many anthropogenic contaminants. In an integrated geohydrological, petrologic, and geochemical analysis of the eastern Snake River Plain, Wood and Low (1986, 1988) proposed that, despite high water-rock ratios and short residence times within the most active part of the aquifer, 20% of the aquifer solute load is derived from the reaction of groundwater with Quaternary basaltic host rocks. In contrast, Blackwell et al. (1992) suggest that alteration occurs primarily at and below the effective base of the aquifer as defined by a sharp inflection in the temperature gradient in deep wells.

The principal purpose of our work is to better define and interpret the origin of three-dimensional patterns of basalt

*E-mail: mccumich@isu.edu

Morse, L.H., and McCurry, M., 2002, Genesis of alteration of Quaternary basalts within a portion of the eastern Snake River Plain aquifer, *in* Link, P.K., and Mink, L.L., eds., Geology, Hydrogeology, and Environmental Remediation: Idaho National Engineering and Environmental Laboratory, Eastern Snake River Plain, Idaho: Boulder, Colorado, Geological Society of America Special Paper 353, p. 213–224.

alteration and authigenic mineralization in the eastern Snake River Plain aquifer, to determine the relationship between the alteration and the sharp inflections in the temperature gradients observed in deep wells, and to explore how the alteration and temperature inflections may be related to the effective base of the aquifer. To accomplish this we conducted a systematic study of core from five deep wells (Fig. 1) that penetrate the inferred base of the aquifer, as defined by sharp inflections in the temperature gradients of the wells (Smith et al., 1994; Brott et al., 1981; Blackwell et al., 1992). We evaluate hypotheses that basalt alteration and authigenic mineralization (1) correlate to the inflections in the temperature gradients of the wells, (2) correlate to depth of burial of the basalts, (3) correlate to the age of the basalts, and (4) are the product of the interaction of basalts with geothermal waters upwelling from depth.

Geohydrologic setting and previous work

The eastern Snake River Plain occupies an area of ~28 000 km² in the upper Snake River basin of Idaho. The surface of the plain is underlain primarily by thin, nearly flat-lying, Quaternary olivine tholeiite basalts (Leeman, 1982; Kuntz et al., 1992; Lanphere et al., 1994) with intercalated sedimentary interbeds of lacustrine, eolian, and alluvial origin (Olmsted, 1962; Wood and Low, 1986; Anderson and Bowers, 1995; Bartholomay, 1990; Bestland et al., this volume; Mark and Thackray, this volume). The Quaternary basalt-sediment sequence, which

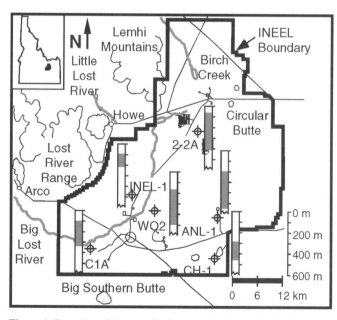

Figure 1. Location of deep wells discussed in this study. Wells are all located within boundaries of Idaho National Engineering and Environmental Laboratory (INEEL). Location of INEEL is indicated in black on map of Idaho inset in upper left of figure. Well locations are indicated by circles with crosses. Vertical bar graphs represent wells; shaded parts of bar graphs represent portion of aquifer above temperature-gradient inflection in each well (Smith et al., 1994).

contains the active part of the eastern Snake River Plain aquifer system, has accumulated to thicknesses between 700 and 1100 m (Whitehead, 1992; Hackett and Smith, 1992; Smith et al., 1994; Wood and Low, 1986; Geslin et al., this volume) and overlies thick Pliocene and upper Miocene silicic lava flows and pyroclastic deposits.

The eastern Snake River Plain aquifer is one of the largest unified aquifer systems in North America (Heath, 1984) and is largely unconfined (Wood and Low, 1986) except where local, interbedded clays and dense, unfractured basalt lava flows create semiconfined conditions (Whitehead, 1992). Regional groundwater flow is from the northeast to the southwest (Olmsted, 1962) and occurs primarily along rubble zones between basalt flows. Transmissivities range from ~400 m²/day to >100 000 m²/day (Garabedian, 1986; Whitehead, 1992).

Most of the groundwater from the upper part of the aquifer is of calcium and magnesium carbonate type; bicarbonate is the principal anion (Olmsted, 1962). Waters from deeper in the aquifer tend to have a sodium bicarbonate character (Mann, 1986). Waters are slightly alkaline, with an average pH of ~8, and have an average temperature of ~13 °C (Olmsted, 1962; McLing, 1994).

Basalts and basalt alteration. Although there are subtle geochemical variations within sequences of eastern Snake River Plain basalts (Knutson et al., 1990; Lanphere et al., 1994; Geist et al., this volume; Hughes et al., this volume) nearly all have similar bulk chemistries, textures, and mineralogical assemblages. The basalts are typically porphyritic, consisting of phenocrysts of olivine and plagioclase in a fine-grained matrix of interlocking plagioclase, augite, Fe-Ti oxide minerals, and intersertal tachylitic glass (Knutson et al., 1990; Lanphere et al., 1993, 1994). Flow margins are generally highly vesicular, and large parts of flows are characterized by abundant diktytaxitic cavities (Knutson et al., 1990). Upper and lower flow contacts often contain coarse basalt rubble with high hydraulic conductivity.

Previous studies of alteration of basalt flows in the shallow portions of wells indicate that it is generally minor, and consists of reddish oxidation of olivine (Olmsted, 1962; Lanphere et al., 1993). Olivine is locally altered to chrysotile (?), and intersertal glass to opaque minerals (Lanphere et al., 1994).

Studies of more intensely altered basalts in the eastern Snake River Plain have been of a reconnaissance nature. In a well-log report, Doherty et al. (1979) documented alteration and zeolite mineralization of basalts below 488 m depth in well INEL-1; alteration was most intense between 610 and 658 m. Fromm et al. (1994) described aspects of alteration in well WO2. Basalts above 468 m are little altered, and deeper basalts contain authigenic smectite, calcite, and phillipsite; they observed that alteration is more predominant in flow tops and bottoms than in mid-flow intervals.

There are few other pertinent published studies of low-temperature alteration of Quaternary basalts in the northwest United States. Keith and Bargar (1988) documented low-temperature isochemical alteration of Quaternary basalt in a 932-m-deep drill core near Newberry caldera, Oregon. Benson and Teague (1982) described aspects of alteration of Miocene basalts of the Columbia River Basalt Province. Other unpublished studies of altered Columbia River Basalts have documented the presence of a wide variety of alteration phases including green, yellow-green, and greenish-brown smectite, mordenite, heulandite, chalcedony, black manganese oxide, opal, quartz, and pyrite (Basalt Waste Isolation Project Staff, 1979).

Base of the aquifer. Studies of vertical temperature profiles in six deep wells, which include the five deep wells of this study, located at the Idaho National Engineering and Environmental Laboratory (INEEL) on the northern edge of the eastern Snake River Plain (Fig. 1), document changes in temperature gradient suggestive of rapid upward transitions from conductive to convective heat flow (Smith et al., 1994; Brott et al., 1981; Blackwell et al., 1992). These researchers suggested that sharp temperature-gradient inflections (Fig. 2) in these wells may correlate with the effective base of the aquifer. Above the inflections the temperature gradients are nearly isothermal. This convective regime is most likely due to the mixing of cold recharge waters with the aquifer water and short residence times in this part of the aquifer. Below the inflections, the temperature gradients more nearly approximate the regional geothermal gradient, indicating that the water in this part of the aquifer has a much greater residence time. Blackwell et al. (1992) noted that the temperature-gradient inflections are generally associated with an increase in basalt alteration, implying that the alteration may be genetically related to the inflections. Welhan and Wylie

(1997) documented a rapid downward decrease in porosity and permeability at the same depth (~335 m) as the temperature-gradient inflection in well C1A, one of the wells of this study.

Mann (1986) found that the hydraulic conductivities of basalt in the aquifer in the upper 60–245 m of well INEL-1 were two to five orders of magnitude greater than in the rocks below 460 m. He indicated that the base of the aquifer may be coincident with one of two sedimentary interbeds between 260 and 470 m depth in this well, and that these interbeds are likely to be continuous over a large area. The temperature-gradient inflection in this well occurs at a depth of 230 m.

Anderson and Bowers (1995) correlated the effective base of the aquifer in wells 2-2A and WO2 with ca. 1.8 Ma basalts correlative with the Glenns Ferry Formation of the Idaho Group (cf. Armstrong et al., 1975). In well 2-2A the ca.1.8 Ma basalts are at a depth of ~259 m and the temperature-gradient inflection occurs at a depth of 408 m, while in well WO2 the ca. 1.8 Ma basalts are at a depth of ~505 m and the temperature-gradient inflection occurs at a depth of 508 m. Blackwell et al. (1992) had previously concluded that basalt alteration and the effective base of the aquifer correlate with age or depth of burial of the basalts.

Groundwater. Robertson et al. (1974) related variations in groundwater composition to variations in lithology of recharge source areas, and discounted significant water-basalt reactions within the aquifer. McLing (1994) concluded that groundwater compositions are the product of extensive mixing of water from multiple sources rather than extensive diagenesis due to water-rock interactions. In hydrogeologic studies of well INEL-1, Mann (1986) demonstrated that waters beneath the effective base of the aquifer have a strong upward potential for flow. The paucity of hot springs within the eastern Snake River Plain suggests that any geothermal waters that are upwelling from depth are mixing with, and subdued by, the cooler waters of the effective portion of the aquifer (Robertson et al., 1974; Wood and Low, 1988; McLing, 1994; McLing et al., 1997, and this volume). Previous thermodynamic modeling of eastern Snake River Plain aquifer waters suggests that the calcium-bicarbonate waters are in equilibrium with calcium-montmorillonite and that the aquifer system is currently undergoing open-system precipitation of calcite (McLing, 1994; Knobel et al., 1997).

Methodology and scope

Selected sections of available core from the wells that penetrate deeper than the temperature-gradient inflections (2-2A, middle of Big Lost Trough; CH-1, near East Butte at the southern boundary of the INEEL; WO2, south edge of Big Lost Trough; C1A, Arco rift zone at the Radioactive Waste Management Complex; ANL-1, near Argonnne National Laboratory West) were examined at the USGS Core Library at the INEEL. We did not examine core from the sixth deep well shown in Figure 1, INEL-1; very little core is available, all from below the temperature-gradient inflection, and mostly of rhyolite.

Intervals were selected from both above and below the

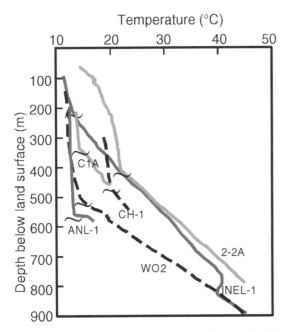

Figure 2. Temperature profiles of six deep wells discussed in this study. Inflection in temperature gradient for each well is accented by tilde. Modified from Blackwell et al. (1992) and Smith et al. (1994).

temperature-gradient inflection of each well in an effort to observe spatial variations in alteration and authigenic mineralization and possible correlations to the inflections. We logged and sampled from a total of 824 m of drill core from these wells. We collected 52 samples of core for petrographic, scanning electron microscope (SEM), energy-dispersive spectrometry (EDS), electron microprobe, X-ray powder diffraction, bulk chemical, and trace element analyses from both unaltered and altered sections of the cores as determined by the presence of authigenic minerals and by variations in color of the core from the typical gray of fresh basalts to green for altered basalts.

Petrographic analysis of standard thin sections and X-ray powder diffraction, instrumental neutron activation analysis (INAA), and inductively coupled plasma-atomic emission spectroscopy (ICP-AES) analyses were performed at Idaho State University using a Phillips X-ray powder diffractometer, HPGe detectors coupled to a multichannel analyzer system for INAA analyses (Reed et al., 1997), and a JY70C ICP-AES. Electron microprobe, SEM, and EDS analyses were performed at the University of Utah using a Cameca SX50 electron microprobe

and a Hitachi S-500 scanning electron microscope with a Kevex 5500 X-ray energy spectrometer.

CHARACTERISTICS OF CORES

In our description of unaltered and altered basalts we use "alteration" to refer to replacement of preexisting mineral phases and "authigenic" to refer to minerals that have precipitated in preexisting pore spaces (i.e., vesicles, diktytaxitic cavities, voids). Evidence for hydration alteration of intersertal glass is common, in the form of low microprobe analysis totals and anomalously low Na_2O contents, when compared to bulk rock compositions. However, we did not analyze similar glasses from all the samples; therefore in the following we do not distinguish unaltered from altered samples on this basis.

All of the basalts examined are fine to medium grained, aphyric to porphyritic, olivine bearing, and have a diktytaxitic texture. Bulk geochemical and trace element data are presented in Table 1. The $Na_2O/FeO*$ versus $CaO/FeO*$ data are plotted for altered and unaltered basalt (Fig. 3A) constituents. We con-

TABLE 1. BASALT GEOCHEMICAL DATA

Borehole	2-2A	2-2A	2-2A	2-2A	2-2A	2-2A	2-2A	WO2	WO2	WO2	WO2	WO2	WO2	WO2
Sample Depth (m)	185	351	389	484	485	682	691	401	450	522	525	601	1063	1065
Unaltered/altered (U/A)	U	U	U	A	A	A	A	U	U	A	A	A	A	A
Oxides (wt%)**														
SiO_2	48.07	47.11	47.38	46.71	47.21	48.60	46.76	47.24	46.84	43.86	46.49	47.30	49.07	49.67
TiO_2	2.66	3.29	2.94	2.29	2.30	2.20	2.26	2.01	2.68	3.39	3.35	2.53	1.46	1.32
Al_2O_3	15.11	13.95	14.78	14.91	15.10	15.59	15.49	15.17	15.14	14.11	14.25	14.96	17.32	17.12
$FeO*$	13.33	14.58	14.35	12.93	12.80	12.83	13.13	12.54	13.77	15.43	15.12	13.09	11.80	11.25
MnO	0.20	0.22	0.21	0.20	0.20	0.16	0.20	0.19	0.20	0.24	0.18	0.19	0.14	0.21
MgO	7.38	7.14	7.22	9.27	9.31	5.64	9.00	9.40	7.88	6.12	8.57	8.31	5.79	7.89
CaO	9.39	9.87	9.62	10.48	10.20	11.63	10.12	10.07	9.78	13.50	8.92	10.20	10.99	9.40
Na_2O	2.61	2.43	2.55	2.40	2.28	2.52	2.34	2.41	2.64	2.27	1.98	2.49	2.76	2.65
K_2O	0.61	0.65	0.40	0.41	0.20	0.42	0.31	0.54	0.52	0.26	0.50	0.47	0.37	0.29
P_2O_5	0.64	0.76	0.55	0.39	0.41	0.41	0.39	0.43	0.54	0.82	0.64	0.46	0.29	0.19
Trace elements (ppm)														
Sc	28.6	31.3	30.0	29.9	29.0	31.3	33.3	31.2	29.5	31.7	31.3	33.0	29.9	26.9
Cr	216	215	224	281	279	148	161	378	209	234	242	294	212	196
Co	54.0	65.9	57.2	55.8	54.8	60.7	55.7	68.4	57.7	57.1	54.0	58.0	53.5	54.4
Ni	90	115	105	135	87	25	110	113	100	84	93	90	na	na
Zn	138	195	144	109	110	106	96	149	152	184	185	142	40	102
Rb	5	8	7	4	6	22	7	10	6	4	4	8	8	9
Sr	370	370	430	385	285	280	250	243	278	267	223	230	238	230
Cs	0.17	0.39	0.24	0.30	0.08	30.2	1.51	0.11	0.10	0.11	0.14	0.03	0.21	0.16
Ba	315	450	395	226	167	229	157	334	308	404	369	304	215	215
La	31.4	30.4	21.7	15.6	15.0	15.1	15.1	18.5	18.6	24.0	20.6	18.3	17.9	15.1
Ce	71.1	69.9	51.9	33.0	35.5	36.8	37.1	42.7	42.6	52.3	50.2	43.9	38.3	31.1
Nd	28	39	30	19	19	19	19	27	31	37	27	28	21	19
Sm	7.55	8.86	7.09	4.90	4.72	4.73	4.86	5.05	5.91	7.29	7.00	5.78	4.70	4.11
Eu	2.71	3.35	2.67	1.87	1.76	1.82	1.90	1.97	2.2	2.76	2.76	2.24	1.36	1.34
Tb	1.26	1.66	1.31	0.84	0.77	0.97	0.97	0.91	1.20	1.26	1.13	1.10	0.89	0.77
Yb	3.46	4.12	3.40	2.67	2.70	2.87	2.75	2.52	2.98	4.28	3.44	2.94	3.10	2.35
Lu	0.44	0.54	0.45	0.38	0.35	0.38	0.36	0.40	0.48	0.62	0.48	0.48	0.42	0.37
Zr	240	294	241	147	149	147	148	191	201	266	260	204	143	133
Hf	5.95	7.42	6.21	3.74	3.68	3.72	3.72	3.84	4.66	5.06	6.76	6.68	5.35	4.45	3.63
Ta	1.70	1.43	1.00	0.72	0.82	0.83	0.75	0.92	0.99	1.15	1.10	0.96	0.76	0.56
Th	1.22	0.93	0.72	0.66	0.49	0.49	na	1.04	0.93	1.22	0.58	0.35	2.04	1.76
U	0.8	1.0	0.9	0.2	0.5	1.2	0.5	0.2	0.4	0.5	0.3	na	1.2	1.4

(continued)

sider Na and Ca to be mobile in typical groundwater environments, while Fe is considered immobile. We also plotted fields for what we consider high-P series (greater than 1.4 wt% P_2O_5) and low-P series (<1.4 wt% P_2O_5) basalts. Most of the analyses are within the low-P series field, except for two altered basalt analyses that are below this field. No unusual alteration features were observed in either of these samples, although the sample with higher CaO/FeO* had an unidentified zeolite-like mineral filling fractures and may have undergone minor migration of some Na due to dissolution of matrix minerals. Other altered basalts plot in the same region as unaltered basalts. Sr, Ba (two elements we consider to be mobile in typical groundwater environments), and Zr (considered immobile) are plotted on the ternary diagram in Figure 3B. Most of the altered samples plot in the same region as the unaltered samples.

Analyses of samples of altered and unaltered rocks exhibit little to no systematic differences in bulk geochemical, trace element, or isotopic compositions (Hughes et al., this volume) that could be attributed to open-system chemical alteration.

This suggests that the authigenic minerals observed were precipitated from solutions produced by the alteration and incongruent dissolution of minerals in the aquifer rocks.

Unaltered basalts and silicic volcanics

Unaltered basalts are generally medium to dark gray, and have vesicles and diktytaxitic cavities free of authigenic minerals, except near the flow boundaries on some flows. Authigenic minerals consist primarily of minor caliche (vadose zone) or drusy calcite (aquifer) deposits in vesicles and along fractures.

Basalts of well ANL-1 are unaltered throughout. Authigenic minerals consist of calcite and, possibly, clay. Clay coexists with calcite in some vesicles, but it was difficult in these instances to distinguish authigenic clay from clay infiltrated from overlying sediment interbeds. There was very little calcite or clay above 309 m. Calcite increased in abundance below 488 m. Although a temperature-gradient inflection occurs 22 m

TABLE 1. BASALT GEOCHEMICAL DATA (continued)

Borehole	C1A	C1A	C1A	C1A	ANL-1	ANL-1	ANL-1	ANL-1	ANL-1	ANL-1	ANL-1	ANL-1	ANL-1	ANL-1
Sample Depth (m)	318	363	400	493	133	147	183	190	306	478	530	541	550	582
Unaltered/altered (U/A)	U	A	U	A	U	U	U	U	U	U	U	U	U	U
Oxides (wt%)**														
SiO$_2$	47.44	47.15	47.30	46.70	46.87	46.88	46.97	47.55	46.83	46.10	46.66	46.93	46.21	47.10
TiO$_2$	2.07	2.77	2.38	2.67	3.40	3.44	3.00	3.19	3.04	3.08	2.44	2.30	3.14	2.51
Al$_2$O$_3$	15.16	15.18	14.99	14.94	14.66	14.41	15.26	14.42	14.84	14.61	15.78	15.76	14.58	15.38
FeO*	12.26	13.69	12.83	13.44	14.45	14.81	13.74	14.00	13.98	14.23	12.56	12.26	14.84	12.98
MnO	0.19	0.19	0.19	0.20	0.22	0.23	0.21	0.21	0.21	0.22	0.19	0.19	0.21	0.21
MgO	9.09	7.86	8.65	8.64	6.90	6.32	7.14	6.64	7.29	7.80	8.58	8.82	7.46	7.64
CaO	10.52	9.38	9.97	10.23	9.31	9.29	9.47	9.44	9.73	10.23	10.36	10.40	9.90	10.34
Na$_2$O	2.44	2.58	2.54	2.40	2.74	2.75	2.69	2.76	2.69	2.55	2.54	2.51	2.47	2.53
K$_2$O	0.40	0.57	0.59	0.38	0.77	1.02	0.86	0.83	0.67	0.46	0.45	0.47	0.47	0.61
P$_2$O$_5$	0.43	0.63	0.56	0.41	0.69	0.84	0.68	0.95	0.71	0.73	0.44	0.36	0.72	0.69
Trace elements (ppm)														
Sc	34.5	28.1	33.3	34.0	30.3	na	28.9	30.0	30.1	31.4	32.2	32.7	29.2	32.1
Cr	471	190	419	373	167	na	159	152	196	209	286	335	198	234
Co	68.0	51.3	63.6	66.7	51.5	na	50.8	47.5	52.5	56.5	50.4	52.0	66.5	50.6
Ni	na	88	na	na	30	na	38	60	55	75	25	65	117	89
Zn	112	178	150	135	141	na	153	151	162	153	88	122	237.7	138
Rb	19	7	15	11	17	na	18	21	23	11	11	13	2	14
Sr	238	277	257	264	329	329	323	313	306	300	266	253	324	225
Cs	0.15	0.13	0.21	0.14	0.20	na	0.46	0.20	0.27	0.16	0.22	0.28	0.23	0.25
Ba	255	370	322	265	478	702	482	505	385	327	273	240	397	425
La	17.6	27.2	24.3	14.1	29.4	na	29.7	44.0	28.4	28.4	18.6	15.8	27.2	30.3
Ce	40.1	61.7	53.1	34.9	67.8	na	66.4	96.0	63.1	60.5	42.2	37.6	62.3	64.3
Nd	25	35	29	24	34	na	33	45	34	29	22	19	37	40
Sm	5.41	6.90	6.28	5.58	8.31	na	8.07	10.30	8.00	7.53	5.74	4.97	8.04	7.81
Eu	2.04	2.34	2.22	2.07	3.15	na	2.89	3.52	2.84	2.65	2.13	1.81	2.90	2.61
Tb	0.61	1.22	1.24	0.94	1.54	na	1.28	1.52	1.52	1.37	1.09	0.86	1.29	1.40
Yb	2.88	3.24	3.08	2.84	3.90	na	3.62	4.14	3.78	3.37	2.90	2.78	3.34	4.02
Lu	0.39	0.46	0.44	0.39	0.51	na	0.49	0.51	0.51	0.42	0.43	0.36	0.48	0.56
Zr	179	251	213	178	320	444	291	350	275	250	181	159	277	332
Hf	4.81	5.71	5.36	4.48	7.80	na	7.12	8.05	6.37	5.80	4.43	4.01	6.24	7.42
Ta	1.23	1.39	1.35	0.79	1.70	na	1.51	2.06	1.60	1.29	0.99	0.95	1.31	1.29
Th	0.98	0.51	1.09	0.40	1.21	na	1.31	2.61	1.59	0.62	0.91	1.05	0.35	1.31
U	0.7	0.2	1.2	0.9	0.7	na	1.0	1.5	0.6	1.3	0.8	0.6	0.5	na

Note: na—not analyzed for this constituent.
*Denotes all Fe as FeO
**Oxides normalized to 100%

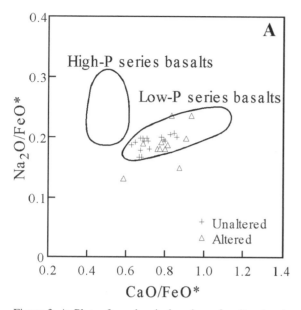
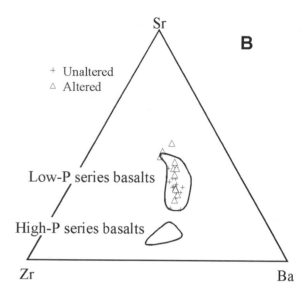

Figure 3. A: Plots of geochemical analyses for altered and unaltered basalt samples from wells 2-2A, C1A, WO2, and ANL-1. Na_2O/FeO^* vs. CaO/FeO^* data are plotted. High-P series basalt and low-P series basalt fields are based on analyses of Snake River Plain basalts that are either high in P (>1.4 wt%; P_2O_5) or low in P (<1.4 wt%; P_2O_5). B: Ternary diagram shows relation of Sr and Ba to Zr. Most samples on both graphs plot within low-P series basalt fields.

above the bottom of this well, we did not observe any evidence of alteration in the interval.

Authigenic mineralization in unaltered basalts from wells 2-2A and C1A is limited to minor precipitation of calcite and smectite clays, while in well WO2 it is limited to minor precipitation of calcite. We observed no alteration of basalts above the temperature-gradient inflection in wells WO2 and C1A, and only minor alteration in a 4 m section of core above the temperature-gradient inflection in well 2-2A.

We observed no alteration in the core we examined from well CH-1. No authigenic mineralization was observed above the temperature-gradient inflection. Petrographic and microprobe analyses reveal that below the temperature-gradient inflection, authigenic minerals consist of smectite clay, calcite, Fe-Mg-rich calcite, calcite, and minor siderite.

Altered basalts

Basalts from deep portions of wells 2-2A, WO2, and C1A are generally green. This change in color with depth was used as a guide in distinguishing between altered and unaltered basalts for sampling purposes. Most vesicles and diktytaxitic cavities are filled with authigenic minerals. Figure 4 is a photo of core samples collected from well WO2 and illustrates the change in color of the core and increased vesicle filling with depth. Note also that vesicles and diktytaxitic cavities in the two deepest samples are completely filled with white calcite or green clay. Only in well 2-2A was alteration of basalts above the temperature gradient inflection observed. The alteration oc-

curred at a depth interval of 184–188 m and was limited to minor alteration of matrix glass to clay.

The only zeolite observed is authigenic chabazite in vesicles and diktytaxitic cavities of core from well 2-2A, from depths between 478 and 488 m. Large chabazite crystals in a vesicle of a sample from 485 m have a rhombohedral habit; globular clusters of prismatic crystals were also observed in this section of core.

In addition to chabazite, fluorapophyllite and a zeolite-like tectosilicate were identified in vesicles and diktytaxitic cavities from 484–485 m in well 2-2A (Fig. 5). The zeolite-like mineral has a radiating habit and coexists with chabazite and/or fluorapophyllite. A ternary diagram of microprobe analyses for chabazite and the zeolite-like mineral is presented in Figure 6. A Sr-rich crystal was observed in a sample from 485 m. Sr-rich and Sr-poor chabazites are similar in appearance, are distinguished only by their compositions, and are present, in places, side by side, suggesting that they are coexisting phases or, texturally speaking, exhibit an unusual growth sequence. Compositions of these phases based on microprobe analyses are as follows: an Sr-poor chabazite was $(Na_{0.70}K_{1.01}Ca_{1.08})$ $(Si_{7.88}Al_{4.20})O_{24} \cdot 12H_2O$; an Sr-rich chabazite was $(Na_{0.74}K_{0.36}$ $Ca_{1.40}Sr_{0.16})(Si_{7.95}Al_{3.99})O_{24} \cdot 12H_2O$; and a zeolite-like mineral was $(Na_{1.05}K_{0.61}Ca_{0.40})(Si_{9.10}Al_{3.04})O_{24}$.

Smectite clays were identified by X-ray powder diffraction in 15 of 16 clay samples from wells 2-2A, C1A, and WO2. Kaolinite may have been present in a sample from well 2-2A.

SEM analysis of a light green clay from a sample from 351 m in well 2-2A showed a botryoidal habit. Caterpillar-like aggregates of clays are present with a globular shaped clay lining

Figure 4. Photograph of core samples from well WO2. Core is placed in order by depth, with core from most shallow depth in upper right and core from greatest depth in lower left. Note change in color of core and increase in cavity filling with authigenic minerals with depth. Core changes from gray to dark green with depth due to alteration of primary mineral phases to clays and filling of cavities with clays. Bar scale at left of photo is 10 cm long.

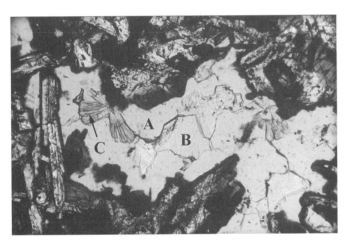

Figure 5. Microphotograph of diktytaxitic cavity in basalt sample from well 2-2A, from depth of 484 m, taken with plane light. Cavity walls are lined with brown clay and cavity contains three other minerals. Fluorapophyllite (A) is mostly near cavity walls. Chabazite (B) is located in central part of cavity. The prismatic crystals (C) are unidentified zeolite-like mineral. Horizontal field of view is 1.3 mm.

a vesicle of a sample from 485 m. The clay morphologies observed are consistent with smectite group minerals.

In thin section, we commonly observe two types of clays intergrown with each other, indicating that they are in textural equilibrium. Electron microprobe analyses of clays from well 2-2A span a range of compositions, suggesting that the smectite clays identified by X-ray powder diffraction methods are probably mixtures of dioctahedral and trioctahedral types. Structural formulas inferred from representative analyses from opposing parts of the compositional spectrum are shown below, and most closely resemble Fe^{3+}-rich nontronite and Mg-rich saponite: nontronite $(Ca_{0.36}Na_{0.05}K_{0.06})(Fe^{3+}_{2.53}Mg_{1.91}Al_{0.22})(Si_{7.11}Al_{0.89})O_{20}(OH)_4 \cdot nH_2O$ and saponite $(Ca_{0.09}Na_{0.02}K_{0.01})$ $(Fe^{3+}_{2.19}Mg_{4.01})$ $(Si_{5.08}Al_{2.12}Fe^{3+}_{0.80})$ $O_{20}(OH)_4 \cdot nH_2O$. However, because very fine Fe-Ti minerals may be present within the clays, the mineral compositions we have proposed are only estimates. Based upon the similar appearance of clays, and similar alteration features of matrix glass and minerals, we believe that the smectite clays identified in all the wells are similar in composition to the clays in well 2-2A.

Averages of microprobe analyses for chabazite, fluorapophyllite, and the unidentified zeolite-like mineral and for selected smectite samples are presented in Table 2. The H_2O^+ data were calculated by allocating measured oxygen to each of the cation proportions, and then allocating all remaining oxygen to H_2O^+. Due to an apparent problem with measured oxygen for some suites of analyses, not all average analyses have H_2O^+ data listed.

Basalts from below the temperature gradient inflection in wells 2-2A, WO2, and C1A exhibit increased filling with depth of vesicles and diktytaxitic cavities with calcite and smectite clays, and alteration of matrix glass and minerals to smectite clays. Alteration of olivine and plagioclase in wells 2-2A and WO2 to smectite clays increases in intensity with depth. In well WO2 alteration of olivine is more pervasive than in well 2-2A, but alteration of matrix glass is less pervasive. Authigenic mineralization and alteration is much less intense in well C1A than in wells 2-2A and WO2. Pyroxenes are distinguished by their lack of any apparent alteration.

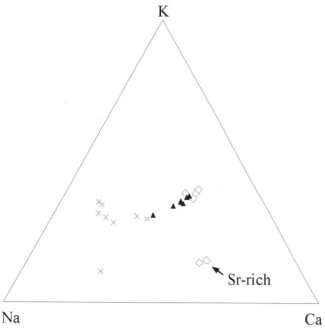

Figure 6. Ternary diagram of microprobe analyses for chabazite and zeolite-like mineral that we have not identified. Samples are from well 2-2A, from depths of 494 m and 495 m. Multiple analyses were performed on three crystals in diktytaxitic cavities of sample from 494 m and four crystals in vesicle of sample from 495 m.

PARAGENESIS

Textures of the authigenic minerals indicate that calcite and smectite-group clays coprecipitated at depths less than the temperature gradient inflection in wells 2-2A, WO2, and C1A. Although we cannot rule out the possibility that the calcite originally precipitated within the vadose zone, the coarse grain size and euhedral form of many of the calcite grains suggests that some of the calcite was in equilibrium with groundwater. The botryoidal and spherulitic habits of clays mantling some apparently unaltered vesicle walls suggest they precipitated congruently from aqueous solution.

Below the temperature-gradient inflections clays are much more abundant and are accompanied by a range of other authigenic mineral phases. In well 2-2A, several samples exhibit a time-growth sequence of calcite ± clay followed by chabazite ± fluorapophyllite ± zeolite-like mineral. Chabazite and fluorapophyllite that completely fill vesicles appear to have been the last authigenic minerals to form. When chabazite occurs in combination with other authigenic minerals, it typically occurs as overgrowths. In other wells the paragenetic sequence is not as clear. Clay and calcite often occur alone in vesicles and dik-

tytaxitic cavities. Where they occur together, either may have precipitated before the other or they may be intimately intergrown.

DISCUSSION

Figure 7 depicts alteration and authigenic mineralization and shows basalt ages based on our interpolation of paleomagnetic ages of basalt flows and sedimentary interbeds (Anderson and Bowers, 1995; Anderson and Bartholomay, 1995; Anderson et al., 1997; Champion and Lanphere, 1997; Bestland et al., this volume; S.R. Anderson, 1999, personal commun.). Paleomagnetic reversals in basalts and sediments are used to delineate the Brunhes-Matuyama boundary (0.78 Ma) and the Olduvai normal polarity subchron (1.77 Ma) (Berggren et al., 1995). The Brunhes-Matuyama boundary and the upper limit of the Olduvai subchron are indicated in Figure 7 for four of the wells.

Alteration and authigenic mineralization of basalts correlate well with the temperature-gradient inflection in four of the five wells examined. The temperature gradient inflection for each well occurs at the following depth below land surface (BLS): 2-2A, 408 m; CH-1, 455 m; WO2, 508 m; C1A, 337 m; ANL-1, 560 m. In well ANL-1 the inflection occurs 22 m above the bottom of the well; no alteration or increase in authigenic mineralization was observed in the core below the inflection. In wells 2-2A and WO2 both the degree of alteration and authigenic mineralization increase with depth below the inflections. The CH-1 cores we examined, all from below 118 m, consisted entirely of rocks ranging in composition from basaltic andesite to rhyolite. Authigenic mineralization in this well correlates well with the inflection, but no alteration was observed, perhaps due to the silicic composition and lower apparent permeability of the rocks in this well. Only minor alteration was observed in well C1A. However, as in other wells, authigenic mineralization increases with depth below the inflections.

The lack of correlation of alteration and authigenic mineralization to depth of burial of the basalts is apparent in Figure 7. In well 2-2A precipitation of clays and alteration of mineral phases is observed first at depths of ~360–460 m. In well WO2 this occurs at a depth of ~510 m, while in well ANL-1 it is not observed at all, even at a depth of 582 m.

Anderson and Bowers (1995) suggested that the base of the aquifer is associated with alteration and is primarily constrained to ca. 1.8 Ma or older basalts. At well WO2 Anderson and Bowers picked the ca. 1.8 Ma basalts at a depth of ~500 m, which is approximately the depth at which we first observe alteration and most authigenic mineralization. However, there is no apparent correlation between alteration and the age of basalts in well 2-2A. At that well Anderson and Bowers picked the ca. 1.8 Ma basalts at ~259 m BLS (but see Geslin et al., this volume, for an updated correlation). However, we do not observe alteration of the basalts above 351 m BLS. Substantial

TABLE 2. AVERAGE MICROPROBE ANALYSES OF AUTHIGENIC MINERALS

	Chabazite (1) Sr-rich	Chabazite (2) Sr-poor	Saponite (3)	Nontronite (4)	Fluorapophyllite (5)	Unidentified Mineral (6)
SiO_2	51.71	52.64	33.42	50.74	52.39	67.72
Al_2O_3	22.01	23.77	11.80	6.71	1.08	19.16
TiO_2	na	na	0.01	0.01	na	na
Fe_2O_3	nd	0.04	26.16	23.98	0.14	0.10
Cr_2O_3	na	na	0.01	0.00	na	na
MgO	nd	nd	17.69	9.13	nd	0.01
MnO	na	na	0.34	0.05	na	na
CaO	8.50	6.70	0.55	2.38	23.48	2.80
Na_2O	2.50	2.40	0.08	0.19	0.28	4.01
K_2O	1.86	5.31	0.03	0.34	4.02	3.52
SrO	1.85	0.01	na	na	0.01	0.01
BaO	0.06	0.32	na	na	0.04	0.04
F	na	na	na	na	2.06	na
H_2O+		7.40			15.93	2.91
Total	88.48	98.58	90.10	93.53	99.43	100.29
Oxygens in structural formula	24	24	22	22	21 (O, OH,F)	24
Si	7.95	7.88	5.08	7.11	7.96	9.10
Al	3.99	4.20	2.12	1.11	0.19	3.04
Ti	na	na	0.00	0.00	na	na
Fe_3+	nd	0.00	2.99	2.53	0.02	0.01
Cr	na	na	0.00	0.00	na	na
Mg	nd	nd	4.01	1.91	0	0.00
Mn	na	na	0.04	0.01	na	na
Ca	1.40	1.08	0.09	0.36	3.83	0.40
Na	0.74	0.70	0.02	0.05	0.08	1.05
K	0.36	1.01	0.01	0.06	0.78	0.61
Sr	0.16	0.00	na	na	0.00	0.00
Ba	0.00	0.02	na	na	0.00	0.00
F	na	na	na	na	0.99	na
Total Cat	14.61	14.88	14.37	13.13	12.87	14.21

Note: H_2O+ calculated from oxygen analyses by allocating required oxygens to oxides and assigning remaining oxygens to H_2O+. H_2O+ data not used in calculating structural formulas; na—this constituent not analyzed for; nd—not detected
(1) average of 2 analyses; (2) average of 11 analyses; (3) analysis of one selected clay that represents the saponite end-member of the smectite clays; (4) analysis of one selected clay that represents the nontronite end-member of the smectite clays; (5) average of 7 analyses; (6) average of 8 analyses.

alteration was only observed at depths >477 m BLS, which corresponds to an estimated age of ca. 2.6 Ma.

The high degree of correlation between alteration and authigenic mineralization with the temperature-gradient inflections in the wells studied makes us confident that we can use the coincidence of these two properties to identify the base of the aquifer in this part of the eastern Snake River Plain aquifer. The temperature-gradient inflections represent the change from a conductive regime deeper in the aquifer to a convective regime in the shallower parts of the aquifer. Average residence times in the convective regime are 200–250 yr (Wood and Low, 1986, 1988). The ^{14}C age dating of geothermal waters of the conductive regime indicates they have an age of ca. 35 ka (J.T. Barraclough, *in* Mann, 1986). It is the rapid movement of water cooled by recharge in the upper part of the aquifer relative to the slow movement of much older water at depth that produces the sharp temperature-gradient inflection observed in the wells we studied. The slower, deeper waters have time to equilibrate with the temperature of the host rocks, while the cooler, shallower waters carry away and subdue the heat produced by the host rocks.

The geometry of the base of the aquifer and therefore the effective thickness of the aquifer varies across the area studied. Rather than being a planar feature, the base of the aquifer, as inferred from temperature-gradient inflections, has a very irregular shape, as illustrated in Figures 1 and 2.

Robertson et al. (1974) and Mann (1986) suggested that the base of the aquifer correlates to sedimentary interbeds. While the temperature gradient inflections for wells 2-2A and WO2 occur within thick (10–70 m) sedimentary interbeds and the temperature gradient inflection for well ANL-1 occurs ~1 m above a 4-m-thick sedimentary interbed, there are no similar interbeds near the temperature gradient inflections in wells C1A and CH-1. Therefore, the base of the aquifer doesn't correlate well with sedimentary interbeds.

It thus appears that there is no systematic correlation between the depth at which basalt alteration becomes pronounced and the prevailing characteristics of the basalts or sediment interbeds. We therefore evaluate the possibility that a transient event, specifically the inputs of warmer waters from greater depths into the base of the active portion of the aquifer, may play a role in the alteration of the basalts. The plausibility of

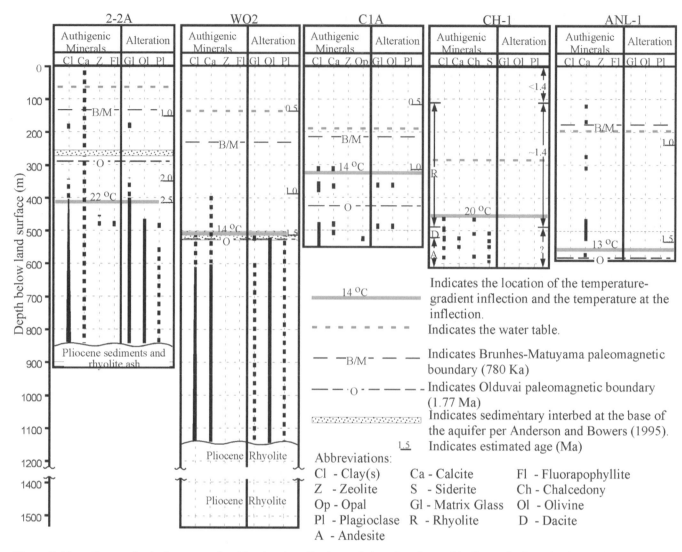

Figure 7. These five graphs depict extent of authigenic mineralization and alteration observed in five wells. Length of graph for each well is proportional to actual depth of well. Vertical lines on graphs are used to depict vertical ranges of respective authigenic minerals and alteration and widths of lines depict abundance. Dashed lines indicate lower abundance than solid lines. Shaded solid horizontal lines indicate location of temperature-gradient inflection in each well. Shaded dashed horizontal lines indicate location of water table in each well. An age of ca. 1.42 Ma has been assigned to rhyolitic unnamed dome penetrated by well CH-1 (Kuntz and Dalrymple, 1979), but ages have not been determined for rocks above and below rhyolites. Basalts overlie sequence of rhyolite, dacite, and andesite indicated for well CH-1.

this idea is supported by the occurrence of hot springs in areas marginal to the eastern Snake River Plain (Robertson et al., 1974; McLing et al., 1997), and continuing volcanism in the area (Kuntz et al., 1992; Hackett and Smith, 1992).

The computer program MINTEQA2 (Allison et al., 1991) was used to calculate solute activities over a range of plausible compositions (Mann, 1986; McLing, 1994; Fromm, 1995) and temperatures ranging from 12 to 100 °C. Thermodynamic relations for pertinent authigenic mineral phases were modeled using The Geochemist's Workbench (Bethke, 1998). Thermodynamic data for chabazite are not available, so laumontite was used as a surrogate. Calcium-beidellite, a calcium-smectite clay, was used as a surrogate for saponite and nontronite. Results are

shown in Figure 8: 12°C corresponds to typical water temperatures in the upper part of the aquifer, and 150°C corresponds to warm water (146°C) from a depth of 3043 m in well INEL-1 (Mann, 1986). The water compositions plotted on diagrams are for waters from the upper part of the aquifer (McLing, 1994; Fromm, 1995) and for geothermal waters from depths ranging from 461 to 3159 m (Mann, 1986). Note that water chemistry does not significantly influence the mineral stability relationships, but that they are significantly dependent upon temperature. An increase of ~50°C shifts the water chemistry in the direction of laumontite stability. Thus an increase in water temperature alone could produce important authigenic mineral components of some of the most altered basalts we observed.

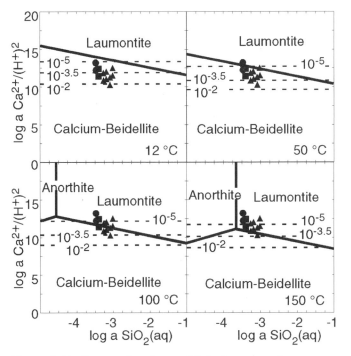

Figure 8. Zeolite-smectite clay equilibria for waters from eastern Snake River Plain aquifer. Closed circles and squares represent waters chemistries for waters from upper part of aquifer (McLing, 1994; Fromm, 1995). Triangles represent analyses of geothermal waters from deep in aquifer (Mann, 1986). Horizontal dashed lines represent calcite solubility at fugacity of CO_2 listed for each line.

SUMMARY AND CONCLUSIONS

Our petrographic, mineralogical, and geochemical analyses of basalts from deep boreholes from a part of the eastern Snake River Plain indicate that there is a regionally significant, pronounced, and rather sudden downward change from largely unaltered to moderately to strongly altered basalts. Our observations are in agreement with assertions by McLing (1994) and Robertson et al. (1974), that very little diagenesis is occurring in the active portion of the aquifer. However, they appear to conflict with the conclusion of Wood and Low (1986, 1988) that diagenesis is a significant process within the active part of the aquifer. Because the work of Wood and Low encompassed the entire eastern Snake River Plain aquifer, it is possible that diagenesis is occurring in the active part of the aquifer in other parts of the eastern Snake River Plain.

Dominant alteration phases include smectite clays (nontronite and saponite) and calcite ± chabazite. The change appears over a narrow depth interval, defining a spatially highly irregular surface having the form of an alteration front. It is spatially correlated with pronounced changes in temperature gradient profiles and rock permeability. However, it does not correlate systematically with any other obvious prevailing aspects of the rocks (e.g., host rock age, depth of burial, or ratio of basalt to sediment interbeds).

Authigenic mineralization and alteration associated with the transient upward influx and mixing of hydrothermal waters with the Snake River Plain aquifer may account best for the assemblages and space-time patterns of alteration observed in the deep wells we have investigated. Variations in the degree of authigenic mineralization and alteration may be due to varying amounts of hydrothermal water that enters the system at different locations and moves upward along fractures or other structural control mechanisms.

ACKNOWLEDGMENTS

We gratefully acknowledge David Blackwell for permitting us to use his temperature log data. Linda Davis provided valuable assistance in our logging of drill core. Ray Lambert assisted in the operation of the electron microprobe and scanning electron microscope equipment. Steve Anderson provided his insights into correlations of basalt flows and sedimentary interbeds. Scott Hughes provided assistance in instrumental neutron activation analysis analyses. Chris Martin was instrumental in our work on ANL-1 core. This paper was reviewed by Richard P. Smith, Robert W. Smith, Thomas H. Giordano, and LeRoy L. Knobel. This research is supported by Department of Energy Remediation Technologies at the Idaho National Engineering and Environmental Laboratory, grant DE-FG07-96ID13420, through the Idaho Water Resources Research Institute–Idaho Universities Consortium.

REFERENCES CITED

Allison, J.D., Brown, D.S., and Novo-Gradac, K.J., 1991, MINTEQA2/PRODEFA2, A geochemical assessment model for environmental systems 3.0: Athens, Georgia, U.S. Environmental Protection Agency, EPA-/600/3-91/021.

Anderson, S.R., and Bartholomay, R.C., 1995, Use of natural-gamma logs and cores for determining stratigraphic relations of basalt and sediment at the Radioactive Waste Management Complex, Idaho National Engineering Laboratory, Idaho: Journal of the Idaho Academy of Science, v. 31, 10 p.

Anderson, S.R., and Bowers, B., 1995, Stratigraphy of the unsaturated zone and uppermost part of the Snake River Plain aquifer at Test Area North, Idaho National Engineering Laboratory, Idaho: U.S. Geological Survey Water-Resources Investigations Report 95-4130, 47 p.

Anderson, S.R., Liszewski, M.J., and Cecil, L.D., 1997, Geologic ages and accumulation rates of basalt-flow groups and sedimentary interbeds in selected wells at the Idaho National Engineering Laboratory, Idaho: U.S. Geological Survey Water-Resources Report 97-4010, 39 p.

Armstrong, R.L., Leeman, W.P., and Malde, H.E., 1975, K-Ar dating, Quaternary and Neogene volcanic rocks of the Snake River Plain, Idaho: American Journal of Science, v. 275, no. 3, p. 225–251.

Bartholomay, R.C., 1990, Mineralogical correlation of surficial sediment from area drainages with selected sedimentary interbeds at the Idaho National Engineering Laboratory, Idaho: U.S. Geological Survey Water-Resources Investigations Report 90-4147, 18 p.

Benson, L.V., and Teague, L.S., 1982, Diagenesis of basalts from the Pasco Basin, Washington. 1. Distribution and composition of secondary mineral phases: Journal of Sedimentary Petrology, v. 52, no. 2, p. 595–613.

Berggren, W.A., Hilgren, F.J., Langereis, C.G., Kent, D.V., Obradovich, J.D., Raffi, I., Raymo, M.E., and Shackleton, N.J., 1995, Late Neogene chronology: New perspectives in high-resolution stratigraphy: Geological Society of America Bulletin, v. 107, no. 11, p. 1272–1287.

Bethke, C.M., 1998, The geochemist's workbench: Urbana, Illinois, University of Illinois, 184 p.

Blackwell, D.D., Kelly, S., and Steele, J.L., 1992, Heat flow modeling of the Snake River Plain, Idaho: EG&G Idaho Report EGG-C91-103450, 109 p.

Brott, C.A., Blackwell, D.D., and Ziagos, J.P., 1981, Thermal and tectonic implications of heat flow in the eastern Snake River Plain, Idaho: Journal of Geophysical Research, v. 86, no. B12, p. 11709–11734.

Champion, D.E., and Lanphere, M.A., 1997, Age and paleomagnetism of basaltic lava flows in corehole ANL-OBS-AQ-014 at Argonne National Laboratory-West, Idaho National Engineering and Environmental Laboratory: U.S. Geological Survey Open-File Report 97-700, 34 p.

Doherty, D.J., McBroome, L.A., and Kuntz, M.A., 1979, Preliminary geological interpretation and lithologic log of the exploratory geothermal test well (INEL-1), Idaho National Engineering Laboratory, eastern Snake River Plain, Idaho: U.S. Geological Survey Open-File Report 79-1248, 9 p.

Fromm, J.M., 1995, Characterizing aquifer hydrogeology and anthropogenic chemical influences on groundwater near the Idaho Chemical Processing Plant, Idaho National Engineering Laboratory, Idaho [M.S. thesis]: Pocatello, Idaho, Idaho State University, 312 p.

Fromm, J.M., Hackett, W.R., and Stephens, J.D., 1994, Primary mineralogy and alteration of basalts and sediments in drillcores from the Idaho National Engineering Laboratory, eastern Snake River Plain, *in* Abstracts of the 7th International Symposium on the Observation of the Continental Crust Through Drilling: Santa Fe, New Mexico,

Garabedian, S.P., 1986, Application of a parameter-estimation technique to modeling the regional aquifer underlying the eastern Snake River Plain, Idaho: U.S. Geological Survey Water-Supply Paper 2278, 60 p.

Hackett, W.R., and Smith, R.P., 1992, Quaternary volcanism, tectonics, and sedimentation in the Idaho National Engineering Laboratory Area, *in* Wilson, J.R., ed., Field guide to geologic excursions in Utah and adjacent areas of Nevada, Idaho, and Wyoming: Boulder, Colorado, Geological Society of America, p. 1–18.

Heath, R.C., 1984, Ground-water regions of the United States: U.S. Geological Survey Water-Supply Paper 2242, 78 p.

Keith, T.E.C., and Bargar, K.E., 1988, Petrology and hydrothermal mineralogy of U.S. Geological Survey Newberry 2 drill core from Newberry Caldera, Oregon: Journal of Geophysical Research, v. 93, no. B9, p. 10174–10190.

Knobel, L.L., Bartholomay, R.C., and Orr, B.R., 1997, Preliminary delineation of natural geochemical reactions, Snake River Plain aquifer system, Idaho National Engineering Laboratory and vicinity, Idaho: U.S. Geological Survey Water-Resources Investigations Report 97-4093, 52 p.

Knutson, C.F., McCormick, K.A., Smith, R.P., Hackett, W.R., O'Brien, J.P., and Crocker, J.C., 1990, FY 89 report Radioactive Waste Management Complex vadose zone basalt characterization: EG&G Idaho Report EGG-WM-8949, 412 p.

Kuntz, M.A., Covington, H.R., and Schorr, L.J., 1992, An overview of basaltic volcanism of the eastern Snake River Plain, Idaho, *in* Link, P.K., Kuntz, M.A., and Platt, L.B., eds., Regional geology of eastern Idaho and western Wyoming: Geological Society of America Memoir 179, p. 227–267.

Kuntz, M.A., and Dalrymple, G.B, 1979, Geology, geochronology, and potential volcanic hazards in the Lava Ridge-Hells Half Acre area, eastern Snake River Plain, Idaho: U.S. Geological Survey Open-File Report 79-1657, 66 p.

Lanphere, M.A., Champion, D.E., and Kuntz, M.A., 1993, Petrography, age, and paleomagnetism of basalt lava flows in coreholes Well 80, NRF 89-04, NRF 89-05, and ICPP 123, Idaho National Engineering Laboratory: U.S. Geological Survey Open-File Report 93-327, 40 p.

Lanphere, M.A., Kuntz, M.A., and Champion, D.E., 1994, Petrography, age, and paleomagnetism of basaltic lava flows in coreholes at Test Area North (TAN), Idaho National Engineering Laboratory: U.S. Geological Survey Open-File Report 94-686, 49 p.

Leeman, W.P., 1982, Olivine tholeiitic basalts of the Snake River Plain, Idaho, *in* Bonnichsen, B., and Breckenridge, R.M., eds., Cenozoic geology of Idaho: Idaho Bureau of Mines and Geology Bulletin 26, p. 181–191.

Mann, L.J., 1986, Hydraulic properties of rock units and chemical quality of water for INEL-1—a 10,365-foot deep test hole drilled at the Idaho National Engineering Laboratory, Idaho: U.S. Geological Survey Water-Resources Investigations Report 86-4020, 23 p.

McLing, T.L., 1994, The pre-anthropogenic groundwater evolution at the Idaho National Engineering Laboratory, Idaho [M.S. thesis]: Pocatello, Idaho, Idaho State University, 62 p.

McLing, T.L., Smith, R.W., and Johnson, T.M., 1997, The effect of hydrothermal water on the Snake River Plain aquifer: Geological Society of America Abstracts with Programs, v. 29, no. 6, p. A323.

Olmsted, F.H., 1962, Chemical and physical character of ground water in the National Reactor Testing Station, Idaho: U.S. Geological Survey Report IDO-22043-USGS, 124 p.

Reed, M.F., Bartholomay, R.C., and Hughes, S.S., 1997, Geochemistry and stratigraphic correlation of basalt lavas beneath the Idaho Chemical Processing Plant, Idaho National Engineering Laboratory: Environmental Geology, v. 30, p. 108–118.

Robertson, J.B., Shoen, R., and Barraclough, J.T., 1974, The influence of liquid waste disposal on the geochemistry of water at the National Reactor Testing Station, Idaho: 1952–1970: U.S. Geological Survey Open-File Report IDO-22053, 231 p.

Smith, R.P., Hackett, W.R., Josten, N.E., Knutson, C.F., Jackson, S.M., Barton, C.A., Moos, D., Blackwell, D.D., and Kelley, S., 1994, Synthesis of deep drill hole information at the Idaho National Engineering Laboratory (INEL): Upper crustal environment in the continental track of a mantle hotspot, *in* Proceedings of the 7th International Symposium on the Observation of the Continental Crust Through Drilling: Santa Fe, New Mexico, p. 89–92.

Staff, Basalt Waste Isolation Project, 1979, Geological characterization of drill holes DC-6, DC-8, and DC-4, final report, RHO-BWI-C-69: Richland, Washington, Rockwell Hanford Operations, 24 p.

Welhan, J., and Wylie, A., 1997, Stochastic modeling of hydraulic conductivity in the Snake River Plain aquifer. 2. Evaluation of lithologic controls at the corehole and borehole scales, *in* Sharma, S., and Hardcastle, J.H., eds., Proceedings of the 32nd Symposium on Engineering Geology and Geotechnical Engineering: Boise, Idaho, p. 93–107.

Whitehead, R.L., 1992, Geohydrologic framework of the Snake River Plain aquifer system, Idaho and eastern Oregon: U.S. Geological Survey Professional Paper 1408-B, 32 p.

Wood, W.W., and Low, W.H., 1986, Aqueous geochemistry and diagenesis in the eastern Snake River Plain aquifer system, Idaho: Geological Society of America Bulletin, v. 97, p. 1456–1466.

Wood, W.W., and Low, W.H., 1988, Solute geochemistry of the Snake River Plain regional aquifer system, Idaho and eastern Oregon: U.S. Geological Survey Professional Paper 1408-D, 79 p.

Manuscript Accepted by the Society November 2, 2000

Geological Society of America
Special Paper 353
2002

Stochastic simulation of aquifer heterogeneity in a layered basalt aquifer system, eastern Snake River Plain, Idaho

John A. Welhan
Idaho Geological Survey, Department of Geosciences, Idaho State University, Pocatello, Idaho 83209, USA
Thomas M. Clemo
*Center for Geophysical Investigations of the Shallow Subsurface, Boise State University,
Boise, Idaho 83725, USA*
Edith L. Gégo
Idaho Water Resources Research Institute, University of Idaho, Idaho Falls, Idaho 83402, USA

ABSTRACT

A stochastic approach to modeling aquifer heterogeneity in basalt lava flows was evaluated using different methods. Direct simulation of permeability was not justified with the limited amount of permeability data available. The lithology was interpreted from borehole geophysical logs, and three lithologic categories were used as permeability surrogates: massive basalt and fine-grained sediment, both of low permeability, and high-permeability interflow zones between lava flows.

Sequential multiple indicator simulation of lithology was the first modeling method evaluated. Indicator semivariograms of lithology were described with a nested model. The horizontal ranges of the sediment and massive basalt categories were 140 m and 150 m, respectively, and isotropic. The maximum horizontal range of the interflow zones was 110 m, with a horizontal range anisotropy of 7:1, elongated east-west. A 1.74 km^2 by 60-m-thick volume of aquifer was discretized on a 6 × 6 × 0.6 m grid and simulations were conditioned to borehole lithology. A major limitation of this approach is that permeabilities assigned to the simulated lithologies are not conditional to available well-test permeability data.

The second approach was simulated annealing of well-scale permeability, conditioned to limited well-test data and to previously simulated lithology, with a vertical resolution matching the average 6 m measurement support of well-test data. A relationship was demonstrated between well-test permeability (K_b) and number of interflow zones, N_Z, in the well-test interval, thereby allowing both lithologic data and simulated lithologic structure to be used as additional conditioning information. The K_b-N_Z relationship indicated that interflow zone permeability is lognormal with a relatively small variance.

A preliminary structure-imitating model was developed to investigate the feasibility of a variant of an object-based simulation approach. The model mimics the physical geologic structure created by processes such as eruption, inflation, ponding, and sedimentation, by optimizing specified rules and parameters that produce stochastic realizations which reproduce the morphologic and geometric characteristics of lava flows.

Note: Additional information for this chapter can be found in GSA Data Repository item 2002041. See footnote 1 on page 231.

Welhan, J.A., Clemo, T.M., and Gégo, E.L., 2002, Stochastic simulation of aquifer heterogeneity in a layered basalt aquifer system, eastern Snake River Plain, Idaho, *in* Link, P.K., and Mink, L.L., eds., Geology, Hydrogeology, and Environmental Remediation: Idaho National Engineering and Environmental Laboratory, Eastern Snake River Plain, Idaho: Boulder, Colorado, Geological Society of America Special Paper 353, p. 225–247.

INTRODUCTION

The eastern Snake River Plain aquifer is one of the most prolific aquifers in the western United States. Hydraulic conductivity varies more than six orders of magnitude (Ackerman, 1991) and preferential flow paths likely control much of the mass flux (Sorenson et al., 1996) within a complex sequence of interlayered basalt and minor intercalated sedimentary interbeds. Accurate predictive models of groundwater flow and transport in this aquifer will require a detailed model of this heterogeneous distribution of permeability (for convenience, the term permeability is used herein to denote hydraulic conductivity). However, relatively few permeability measurements are available in even the most densely drilled and instrumented portions of the system such as shown in Figure 1, at the Idaho National Engineering and Environmental Laboratory (INEEL). Furthermore, almost all subsurface data represent measurements from large open well intervals (3–30 m, or more), so that well-test permeabilities represent bulk averages over the open well intervals rather than the actual permeability of localized permeable zones.

Stochastic methods of simulating hydraulic heterogeneity in aquifers and petroleum reservoirs have been used to quantify uncertainty in geologic interpretations of subsurface heterogeneity and in fluid flow and transport predictions where limited subsurface information precludes accurate deterministic predictions. A large body of work on stochastic aquifer simulation has developed since Freeze (1975) published his landmark paper on the impact of statistically uncertain permeability fields on groundwater flow. Much of the stochastic groundwater literature since has focused on simulating heterogeneity in porous intergranular media in sedimentary aquifers and is not directly applicable to fractured rocks. The emphasis has been on capturing more realism and incorporating soft geologic information to constrain simulations of heterogeneity so as to be as faithful as possible to the structure of the geologic environment (Phillips and Wilson, 1989; Fogg, 1989; Koltermann and Gorelick, 1996; Carle et al., 1998).

Permeable interflow zones between lava flows are believed to be the most important hydraulic features of the eastern Snake River Plain aquifer because of their high permeability and ability to localize large groundwater fluxes (Lindholm and Vaccaro, 1988; Knutson et al., 1990; Welhan et al., this volume, Chapter 9). Preferential flow and mass transport in this aquifer are attributed to interflow zones (Sorenson et al., 1996). Figure 2 portrays a conceptual model of interflow zones (Welhan et al., 2002; this volume, Chapter 9), as subhorizontal features hosted within inflated pahoehoe lava flows, the spatial geometry of which is controlled by a complex interleaving of different scales of lava flows and interflow zones.

Accurate simulation of the geometry and spatial distribution of interflow zones is a key to predicting contaminant movement in the eastern Snake River Plain aquifer. Knutson et al. (1990, 1992) collected extensive data on basalt lavas for defin-

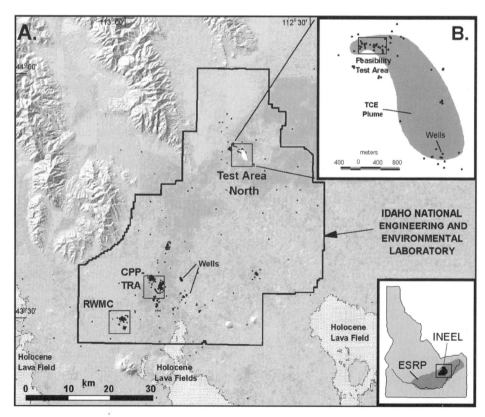

Figure 1. A: Location map of Test Area North (TAN) facility area at Idaho National Engineering and Environmental Laboratory (INEEL), showing density of wells site-wide and in other test facilities: Chemical Processing Plant–Test Reactor Area (CPP-TRA); Radioactive Waste Management Complex (RWMC). B: Location map of feasibility test subarea at TAN (rectangle), showing wells within feasibility test subarea and density of wells facility-wide. TCE is trichloroethylene, ESRP is eastern Snake River Plain.

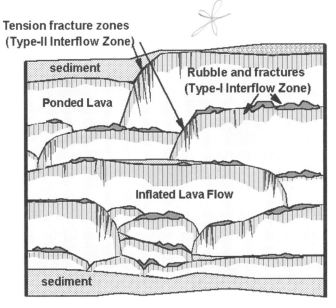

Figure 2. Conceptual model of eastern Snake River Plain aquifer stratigraphy, showing stacked pahoehoe lava flows, interflow zones, and sedimentary interbeds. Based on Welhan et al. (this volume).

ing stochastic simulations and developed a complex expert system to generate stochastic geologic structure simulations (Knutson and Lee, 1991). Welhan et al. (1997) and Welhan and Wylie (1997) suggested that the large-scale features of eastern Snake River Plain aquifer heterogeneity could be analyzed and modeled stochastically with an adaptation of approaches developed for sedimentary environments; they developed the conceptual groundwork for a geostatistical analysis and stochastic modeling approach in this geologic environment.

Scope and goals

The goal of this study was to evaluate different stochastic approaches for simulating basalt heterogeneity and to evaluate the feasibility of such an approach in this aquifer. A related goal has been to determine whether the spatial architecture of interflow zones can be deduced from available subsurface information and, if so, whether the interpreted architecture corroborates observations of the geometry and scale of interflow zones found in outcrop (Welhan et al., this volume, Chapter 9). The scale of the simulation model domain was defined by the spatial scale in flow and transport modeling: hundreds to thousands of meters laterally, 100 m vertically. However, to be useful in characterizing the uncertainty of preferential flow paths, simulations must reflect the architecture of interflow zones on a submeter vertical scale. Hence, both data collection and model discretization will have to reflect this fine-scale vertical resolution (Gego et al., this volume).

Study area

This work focused on a small (1.74 km^2) test subarea of the INEEL's Test Area North (TAN) reactor testing facility (Fig.

1A). The feasibility study subarea is located in a densely drilled portion of the aquifer in which a trichloroethylene (TCE) plume has developed downgradient of a deep waste injection well (Fig. 1B). The subarea contains numerous monitoring wells comprising mostly open boreholes with interval lengths of between 2.7 and 70 m (6 m median length) and open to the aquifer between 60 and 130 m below land surface. The regional-scale stratigraphic dip of basalt lava flows is less than 1° to the south (Anderson and Bowers, 1995). Much steeper dips may occur locally on the flanks of large, inflated lava flow structures, but at the scale of individual lava flow units, inflationary lava growth would not be expected to initiate on slopes greater than ~1°–2° (Hon et al., 1994).

Slug test measurements of permeability (as hydraulic conductivity) were available from open borehole intervals and straddle packer intervals within this 70-m-thick rectangular volume. In addition, numerous wells in and near the subarea had adequate geophysical log information with which to assess lithologic heterogeneity. The areal density of wells was sufficient to justify three-dimensional spatial correlation analysis of lithology, but was insufficient to evaluate the three-dimensional spatial structure of well-scale permeability. The analysis and modeling work was restricted to the basalt portion of the aquifer, which is bounded above and below by major sedimentary interbed units (known as the PQ and QR interbeds; Sorenson et al., 1996).

Stochastic methods compared

Two conventional stochastic simulation methods were developed and evaluated in the course of this study. In addition, a third stochastic approach, based on the concept of geologic structure-imitating models (Koltermann and Gorelick, 1996), was developed to demonstrate a potentially more powerful alternative simulation approach. This communication focuses on the first two methods: sequential indicator simulation of fine-scale borehole lithology and simulated annealing of well-scale permeability. The structure-imitating model is summarized only briefly; it is described in detail elsewhere (Clemo and Welhan, 2000).

The present evaluation of stochastic approaches considered several points: (1) suitability of existing data and data availability; (2) appropriate simulation scale; (3) choice of simulation method; and (4) limitations, alternatives, and recommendations. Welhan et al. (1997) and Welhan and Wylie (1997) discussed much of the rationale for data availability and scale. The following subsection summarizes the salient aspects of these topics, with subsequent sections focusing primarily on 3 and 4, and on specific scale- and data-related issues that bear directly on the simulation approaches chosen.

Stochastic simulation approach. Two generally different approaches to stochastic simulation of lithologic heterogeneity have been developed: object based and cell based (Kupfersberger and Deutsch, 1999). Object-based methods (Deutsch and

Wang, 1996) rely on describing the shape and dimensions of geometric objects representing any number of lithofacies categories (such as gravel bars, sand lenses, and deltaic fans) and positioning these shapes within an aquifer volume to satisfy specified limitations (e.g., volumetric proportion, size distribution, orientation, location of the lithofacies encountered in wells). The traditional object-based approach has not been pursued in this work because the characteristic shape of an interflow zone has yet to be defined. Simulations based on assumed shapes would be arbitrary at best. Furthermore, because the areal shape of interflow zones appears to be fractal (Welhan et al., this volume, Chapter 9), considerable computational effort could be involved in generating and conditioning such shapes plus satisfying collateral limitations. The structure-imitating approach (Clemo and Welhan, 2000) is a type of object-based modeling approach that generates the geometry of interflow zones as a consequence of the emplacement of simulated lava flows and sediments. This approach is being considered to circumvent the limitations imposed by linear (two point) statistical descriptions of spatial geometry used in conventional cell-based simulation.

Cell-based simulation methods build the spatial architecture of continuous or categorical variables stochastically, using geostatistical modeling techniques based on semivariograms or other two-point geostatistical measures (Deutsch and Journel, 1997). The primary limitation imposed by this method is in situations where the spatial architecture of geologic formations cannot be represented accurately with two-point geostatistical measures such as the semivariogram. Multipoint statistical methods are necessary to represent complex geometric shapes but are difficult to implement (Guardino and Srivastava, 1993). A number of different approaches within the class of cell-based methods have been developed to incorporate a variety of constraints other than variogram statistics to simulate realistic heterogeneous systems; these include the incorporation of anisotropy, juxtapositional relationships, transition probabilities, and conditioning to a variety of soft information (Fogg, 1989; Poeter and McKenna, 1995; Carle et al., 1998).

In contrast to hard data, for which there is no (or negligible) uncertainty in the data values representing the primary attribute of interest, soft information does not represent precise, single-valued measurements of the attribute. Rather, information on the probability distribution of a continuous variable or on threshold-classified intervals can be considered (Goovaerts, 1997). For example, a continuous variable, z, at location x_j, has a conditional cumulative distribution function, F, defined by an indicator transform, I, at one or more thresholds, z_c, that subdivide the range of z values into $k + 1$ intervals (Journel, 1989):

$$F(z_c; x_j) = I(z_c; x_j) \in [0, 1]. \tag{1}$$

At each threshold, the indicator variable at location x_j is defined by

$$I(z_c; x_j) = 0 \text{ for } z > z_c \tag{2}$$

and

$$I(z_c; x_j) = 1 \text{ for } z \leq z_c \tag{3}$$

For example, if permeability is the primary variable and hard data are unavailable at location x_j, soft information such as porosity class might be used to estimate the likelihood that permeability at x_j is high or low valued.

Categorical information also can be treated as hard or soft data, utilizing an indicator formalism. For example, a single lithologic category, z_c, within a class of any number of categories, Z, can be represented with an indicator transform:

$$I(z_c; x_j) = 1 \text{ if } Z(x_j) = z_c \tag{4}$$

and

$$I(z_c; x_j) = 0 \text{ if } Z(x_j) \neq z_c \tag{5}$$

Welhan et al. (1997) proposed that categorical information on lithology be used to simulate aquifer heterogeneity, and Welhan and Wylie (1997) provided a rationale for identifying key lithologic categories as surrogates for high- and low-permeability classes in the eastern Snake River Plain aquifer. Indirect information on lithologic variability interpreted from geophysical logs inherently contains uncertainty and is best dealt with in an indicator formalism.

Once the spatial distribution of lithologic categories has been simulated, representative permeability values can be assigned to them to create a stochastic realization of aquifer permeability. As an alternative to this approach, we present a method for directly simulating well-scale permeability, while utilizing simulated lithology as a type of soft information to delimit the permeability simulations.

Limitations and advantages. Direct simulation of permeability is considered infeasible at this time primarily because of the paucity and measurement scale (support) of available permeability data (Welhan et al., 1997; Welhan and Wylie, 1997). The scale of support of well-test permeabilities from large open intervals does not adequately reflect the spatial architecture of high-permeability interflow zones (interflow zones) within the aquifer, and simulated well-scale permeability cannot be appropriately conditioned to reflect the spatial architecture of these zones.

The advantages of modeling lithologic heterogeneity as a surrogate for permeability are two fold: (1) the stochastic representation of lithologic heterogeneity can be modeled at a fine spatial scale (<1 m vertical discretization) due to the scale at which borehole lithologic information is available; and (2) outcrop information on lithologic surrogates at the surface can be used to define the spatial structure of these surrogates in the subsurface, thereby overcoming some of the limitations imposed by large interwell spacings (Kupfersberger and Deutsch, 1999; Welhan et al., this volume, Chapter 9).

The principal disadvantage of modeling lithologic heterogeneity to represent the spatial distribution of permeability is that the assignment of representative permeability values to the simulated lithologic categories must be based primarily on core-scale permeability information. Hence, the permeability assignments cannot be conditioned readily to observed borehole-scale permeability because of the difference in measurement scales.

HYDROGEOLOGIC BACKGROUND

The primary hydrogeologic control on fluid flow in the eastern Snake River Plain aquifer is exerted by high-permeability interflow zones distributed between stacked lava flow lobes that have mostly massive interiors of relatively low permeability (Fig. 2). Although this is primarily a volcanic aquifer system, the spatial characteristics and architecture of its highly permeable interflow zones bear a resemblance to the distribution of coarse fluvial facies in sedimentary aquifers (Fogg, 1989; Carle et al., 1998). Similarities include abrupt transitions from high to low permeability over short distances when moving from one lithologic type into another; marked spatial anisotropy in the correlation ranges of lithologic classes, with much larger ranges horizontally than vertically; and characteristically high transition probabilities between certain lithologic classes (e.g., massive basalt and interflow zones) relative to other lithologic classes (e.g., infrequent sedimentary interbeds).

The geometry of lava flow lobes that support interflow zones is described by Welhan et al. (this volume, Chapter 9). They propose that interflow zones are of two types: type-I interflow zones are thin, lobate and areally extensive "pancakes" of high- to extremely high permeability material draped over the massive interior elements of lava flow lobes (Fig. 2). Type-II zones are supported in the intensely fractured margins of flow lobes and flow units, and may be interconnected over comparatively very large horizontal distances. Although type-II interflow zones may be important at large scales for conducting fluid and contaminant mass over long distances, they cannot be distinguished from type-I interflow zones with the type of borehole data currently available. Therefore, occurrences of type-I and type-II porosity are lumped together in this work.

A geostatistical description of the eastern Snake River Plain aquifer permeability was first proposed by Welhan and Reed (1997). Their analysis of well-scale permeability data collected from open boreholes across the entire INEEL demonstrated the existence of regional-scale correlation structure. Their analysis was necessarily first order because of the scale and quantity of permeability data they analyzed. However, the similarity between the orientations and lateral scale of lava flow groups and the observed anisotropy and correlation scale of borehole permeability suggested that regional-scale geologic controls were responsible for the observed spatial structure of permeability. They suggested that it might be possible to model aquifer permeability stochastically, based on a knowledge of underlying geologic controls at the appropriate scale. Welhan et al. (1998) provided corroboration of this conjecture by showing that the local-scale correlation structure and anisotropy of interflow zones deduced from borehole lithologic data was similar to (1) the magnitude and orientation of permeability anisotropy observed in a multiwell pumping test, (2) the local direction of contaminant transport, and (3) the inferred orientation of lava flows in the aquifer.

DATA

Availability and scale of permeability data

Borehole permeability information in the eastern Snake River Plain aquifer is extremely limited; mean borehole spacing (number of wells / $\sqrt{area}$) over the entire INEEL being on the order of kilometers (Fig. 1A). Within individual test facilities, where most drilling and monitoring has been conducted, mean borehole spacing can be of the order of tens to hundreds of meters. Thus, our efforts to evaluate the feasibility of stochastic modeling approaches in this aquifer were necessarily limited to those areas in which sufficient borehole information was available, i.e., where mean spacing between boreholes is of the order of tens of meters (such as parts of the TAN facility shown in Fig. 1B).

Furthermore, the great majority of INEEL boreholes are cased to the top of the basalt section and open over intervals of several meters to hundreds of meters. For the wells shown in Figure 1B, the mean vertical open interval length is 6 m. Therefore, except for a few wells in which short-interval packer or flowmeter profiles have been obtained, almost all available borehole data represent vertical averages of not only head and permeability, but also of contaminant concentration.

Figure 3 summarizes the ranges and means of measured, core-scale permeabilities for massive basalt, vesicular basalt, and fine-grained sediment, as well as available measured and estimated values for basalt rubble and open fractures. The most detailed set of basalt permeability data available for the eastern Snake River Plain is from permeameter measurements on small plugs collected from basalt core at the INEEL (Knutson et al., 1990). However, these data reflect the permeability of the rock matrix, and are not representative of the magnitude and spatial variability of permeability observed in field-scale tests arising from large-scale heterogeneities.

Estimates of well-scale permeability (as bulk average hydraulic conductivity, K_b) in the TAN feasibility test subarea (Fig. 1B) were available for 60 individual open intervals and straddle packer intervals in 33 wells. A compilation of all available well-test information from existing INEEL sources to October 1996 (A. Wylie, 1996, written commun.) is summarized in Table 1. The data represent K_b values for individual hydraulic tests in the indicated well intervals. In all but two cases, the

J.A. Welhan, T.M. Clemo, and E.L. Gégo

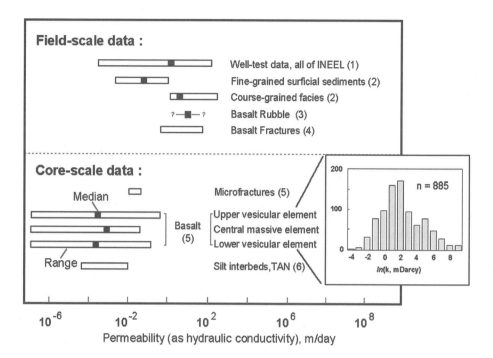

Figure 3. Ranges of field-scale and core-scale permeability measurements in Idaho National Engineering and Environmental Laboratory (INEEL) sediments and basalts. Data sources indicated in parentheses: 1, A. Wylie (1996, written commun.); 2, Mark (1999); 3, Knutson et al. (1992); 4, Knutson et al. (1993); 5, Knutson et al. (1990); 6, Kaminsky et al. (1993). TAN is Test Area North.

tests were performed by slug injection or withdrawal. In these two cases, the reported K_b value represents the transmissivity computed from pumping test data divided by the length of the well's test interval.

The permeability data of Figure 3 and Table 1 can be used to classify lithologic elements into permeability categories. As pointed out by Welhan and Wylie (1997), the distribution of K_b is lognormal and the 90th percentile of the TAN borehole $\ln(K_b)$ frequency distribution corresponds closely to the hydraulic conductivity of interflow zones estimated by Knutson et al. (1992) and of basalt fractures (Knutson et al., 1993). Welhan and Wylie (1997) suggested this cutoff as a convenient interim value that represents the interflow zone lithologic category. They also proposed the 10th percentile of $\ln(K_b)$ values as an interim cutoff representing both massive basalt and fine-grained sedimentary material. This threshold also corresponds to the upper permeability range of these lithologic categories in Figure 3.

The choice of the 10th and 90th percentile cutoffs is not based on analysis of permeability data for the proposed lithologic categories because direct measurements of interflow zone permeability (either core scale or borehole scale) are unavailable, and the available core-scale permeability data for basalts (Fig. 3) cannot be used to assign appropriate cutoffs for borehole-scale data. Therefore, as Welhan and Wylie (1997) did, we considered these as interim cutoffs, subject to revision when more and better information becomes available. However, herein we present an alternative analysis that supports the interim choice of cutoffs.

Well-scale permeabilities between the 10th and 90th percentile cutoffs may represent variable degrees of fracturing, variable proportions of coarse-grained, channel facies sedi-

ments (Mark, 1999), and/or spatial averaging of high- and low-permeability materials within the tested well intervals. In support of spatial averaging, Welhan et al. (1997) first pointed out that a weak relationship exists between the bulk average permeability measured in an open well bore interval and the number of interflow zones encountered in the interval at the TAN facility. If the number of interflow zones influencing a well-test interval has a direct bearing on the borehole-scale permeability measured in the interval, then it may be possible to exploit the relationship between interflow zone spatial density and bulk permeability for modeling purposes. This relationship is explored later in this paper for modeling purposes.

Perhaps the most problematic simulation issue is the scale at which the available permeability measurements were collected. Each of the data values listed in Table 1 represents the bulk average K_b value for the entire length of a well-test interval or a packer test interval. Welhan et al. (this volume, Chapter 9) estimated that the average effective thickness of an interflow zone may be as thin as the average rubble zone (0.2 m) or more likely a combination of rubble and the most densely jointed upper crust of a lava flow (~1–1.5 m). Either estimate is far less than the average well-test interval length of 6 m at TAN. Similarly, the average thickness of the massive basalt element in lava flow lobes is 2.3 m and that of sedimentary interbeds is 1.5 m (Welhan et al., this volume, Chapter 9). Thus, the scale at which most borehole permeability data were collected cannot resolve the permeability variations arising from individual lithologic elements in the aquifer that control the vertical permeability structure on a spatial scale much smaller than 6 m. This problem is most acute from the perspective of interflow zones; they are not thick, but their permeability is believed to control

TABLE 1. WELL-TEST PERMEABILITY DATA IN THE AREA OF THE TAN PLUME

Well name	Type of well test	Test interval length (m)	Test interval center depth (m below surface)	Measured permeability (K_b, m/day)
TAN-24a	Open well	3.35	137.35	194.01
TAN-24a	Packer	6.10	69.51	60.98
TAN-24	Packer	10.98	93.29	25.41
TAN-24	Packer	16.46	110.37	1.13
Lptf-disp	Packer	31.40	78.51	10.36
TAN-06	Open well	6.10	74.70	4.27
TAN-07	Open well	6.10	93.81	14.63
TAN-22a	Open well	9.30	155.56	69.97
TAN-22a	Packer	6.10	158.84	60.98
TAN-23	Packer	4.73	113.95	137.69
TAN-23	Packer	7.62	140.09	85.37
TAN-23	Packer	3.05	132.32	213.41
TAN-15	Open well	6.10	73.78	19.51
TAN-16	Open well	6.10	95.12	37.20
TAN-23a	Open well	6.10	135.67	30.49
TAN-01	Open well*	46.04	85.21	58.55
TAN-21	Packer	5.49	120.12	8.47
TAN-21	Packer	16.16	71.49	1.15
TAN-21	Packer	14.63	83.54	0.13
TAN-21	Packer	12.20	99.09	2.29
TAN-21	Packer	14.18	135.14	2.62
TAN-03	Open well	10.00	75.55	55.49
TAN-02	Open well*	30.49	86.89	48.78
USGS-024	Packer	21.34	88.41	60.98
TAN-04	Open well	9.60	69.89	14.94
TAN-05	Open well	6.55	89.10	102.13
TAN-18	Open well	6.10	154.27	91.46
TAN-18	Packer	12.80	142.99	50.81
TAN-19	Packer	7.32	105.79	88.92
TAN-19	Packer	12.80	129.88	1.45
TAN-19	Open well	6.10	123.78	0.09
TAN-08	Open well	6.52	72.96	3.05
TAN-29	Open well	12.20	73.78	16.84
TAN-30a	Open well	6.10	94.51	10.72
TAN-28	Open well	12.20	73.17	11.20
TAN-27	Open well	9.15	71.65	41.16
TAN-25	Open well	24.39	78.35	11.43
Tsf05	Open well#	25.91	76.22	0.11
TAN-09	Open well	6.68	94.92	8.54
TAN-10	Open well	9.22	70.16	153.05
TAN-11	Open well	6.10	91.46	4.88
TAN-10a	Open well	10.24	71.10	9.45
TAN-12	Open well	6.10	113.41	1.13
TAN-CH-1	Packer	3.66	81.10	1.08
TAN-CH-1	Packer	3.66	62.80	0.48
TAN-CH-1	Packer	3.66	65.85	0.10
TAN-CH-1	Packer	3.66	75.00	1.53
TAN-CH-1	Packer	3.66	68.90	0.64
TAN-CH-1	Packer	3.66	71.95	0.79
TAN-CH-1	Packer	3.66	102.44	0.10
TAN-CH-1	Packer	3.66	78.05	1.69
TAN-CH-1	Packer	3.66	96.34	0.69
TAN-CH-1	Packer	3.66	99.39	1.03
TAN-CH-1	Packer	3.66	93.29	0.44
TAN-CH-1	Packer	3.66	116.46	33.73
TAN-CH-1	Packer	3.66	84.15	3.50
TAN-CH-1	Packer	3.66	90.85	7.81
TAN-CH-1	Packer	3.66	106.10	4.57
TAN-14	Open well	6.10	117.68	0.07
TAN-13a	Open well	6.10	68.90	4.27
TAN-20	Packer	3.96	80.64	0.09
TAN-20	Packer	2.74	120.58	0.14
TAN-20	Packer	6.10	108.54	0.61
Loft-2	Open well	72.87	100.15	14.03
Loft-1	Open well	30.49	85.37	94.51
let-disp	Open well	30.49	82.01	0.49
Fet-disp	Packer	29.27	75.30	47.64

Note: Permeability expressed as bulk average hydraulic conductivity (K_b), averaged over the well-test interval length. From a compilation by A. Wylie (1996, written commun.).
* Open well test conducted as a pumping test.
\# Well-test represents two test intervals: 59.5–74.4 m and 82–93 m below surface.

much of the mass flux through the aquifer (Lindholm and Vacarro, 1988; Sorenson et al., 1996).

Interpreted lithology from geophysical logs

For other than the few boreholes that have been cored at the INEEL, available lithologic drilling logs lack the necessary degree of specificity to accurately define lava flow contacts, positions of interflow zones, or other high-porosity features (lava tubes, major fractures) in the basalts.

Several borehole logging methods are useful in identifying interflow zones in the eastern Snake River Plain aquifer, including borehole video, acoustic televiewer, seismic and radar tomography, and thermal flowmeter logging. However, these logging tools have not been widely applied on the eastern Snake River Plain. A relatively large amount of information is available from conventional borehole geophysical logs at the INEEL, and these have been shown to be useful in locating basalt interflow zones (Knutson, 1993; Bennecke, 1996; Glover et al., GSA Data Repository[1]). The approach of Anderson (1991), based on natural gamma logs, has been successful in mapping sediment occurrences within the basalt stratigraphy and was used in this work.

A methodology proposed by Knutson (1993) and refined by Glover et al. (GSA Data Repository[1]) utilizes a combination of neutron porosity, induced-gamma density, and caliper diameter logs to identify the most prominent zones of enhanced porosity within the basalt. This approach exploits the fact that interflow zones are localized zones of low density and high porosity, often associated with the physical breakout of incompetent, rubbly, or fractured material. A coincidence of large signal deflections in neutron-porosity and induced gamma-density sondes, plus caliper log deflections, was found to provide a reasonably accurate representation of the most significant interflow zones. This interpretational protocol, calibrated against 2700 m of basalt core, was necessarily qualitative because the variable quality of most of the geophysical logs and because quantitative comparisons between wells could not be made due to lack of calibration of the sondes.

Figure 4 shows examples of the low-density-high-porosity zones identified with this method for two wells, TAN-34 and TAN-35, which are 15 m apart. Interpreted zones are compared with zones of preferential flow and the locations of lava flow lobe contacts identified in the TAN-34 core. Interpreted zones were identified from density-porosity logs as shown in Figure 4 and represent the highest confidence picks; these are referred to as zones of enhanced permeability (interflow zones). Note that in TAN-34, two extraneous picks were made at 77 and 80 m below surface and one lava flow contact at 104 m was not identified. However, all three major groundwater flow zones (10–12, 93–95, and 117 m) were identified. In TAN-35, seven of nine flowing zones were identified correctly using this inter-

[1]GSA Data Repository item 2002041, Identification of basalt interflow zones with borehole geophysical and videologs at the Idaho National Engineering and Environmental Laboratory, Idaho, is available on the Web at http://www.geosociety.org/pubs/ft2002.htm. Requests may also be sent to editing@geosociety.org.

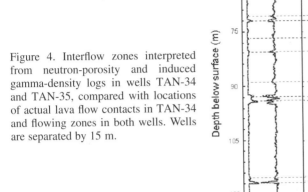

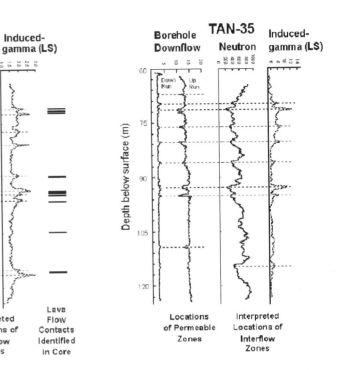

Figure 4. Interflow zones interpreted from neutron-porosity and induced gamma-density logs in wells TAN-34 and TAN-35, compared with locations of actual lava flow contacts in TAN-34 and flowing zones in both wells. Wells are separated by 15 m.

pretational protocol. Comparisons between induced gamma-neutron-caliper picks and visual examination of video and acoustic televiewer logs also show good correspondence (M. Hankins, 1998, oral commun.), lending further credibility to the interpreted lithology data.

The results of this interpretational effort compose a database (Glover et al., GSA Data Repository*) of the locations of the three lithologic categories (interflow zones, massive basalt, and sediment-filled zones) in 36 boreholes within the entire TAN facility area, 20 of which occur within the feasibility test subarea (Fig. 1B).

The method used to interpret density-porosity logs in this work did not produce estimates of the actual thicknesses of individual interflow zones, only their point locations. Because lithology was discretized and represented in the database on a 0.6 m interval, interflow zones in the database were effectively assigned a constant, 0.6 m thickness. Estimates of the actual thickness of interflow zones (mean thickness 0.5 m) in these boreholes were compiled from an examination of density and porosity log deviations and visual inspection of borehole video and acoustic televiewer logs (M. Hankins, 1998, written commun.) and support this choice of the data discretization scale. This value also falls within the range of effective interflow zone thickness inferred from basalt core data (Welhan et al., this volume, Chapter 9). However, such estimates may be low because of their reliance on borehole data, which underrepresent the density of vertical cooling joint fracturing and hence underestimate the vertical extent of fractured zones in the upper crust of lava flow lobes.

In addition to subsurface lithologic information, detailed mapping of basalt lava morphology on two of the youngest and

best-exposed Holocene basalt lava fields on the eastern Snake River Plain has provided considerable soft geologic information to guide stochastic simulation (Welhan et al., this volume, Chapter 9). As first suggested by Knutson et al. (1990) and corroborated by Welhan et al. (1997; 2002), Holocene basalts can be considered direct analogs of their counterparts in the aquifer. The work of Johannesen (2001), summarized in Welhan et al. (this volume, Chapter 9), has provided quantitative measures of lava flow geometry, constraints on the spatial geometry of type-I interflow zones, and the inferred potential importance of type-II interflow zones.

Data sources utilized

Table 1 summarizes all well-test permeability (K_b) data for the TAN area, as compiled by A. Wylie (1996, written commun.). These data are plotted as a histogram and a cumulative distribution of log-transformed K_b values in Figure 5.

Table 2 summarizes the interpreted information on lithologic variability derived from geophysical logs in the TAN facility (Glover et al., GSA Data Repository*), with the available permeability data. A total of 36 well-test intervals from 20 wells in the feasibility test subarea have both permeability data and lithologic information from interpreted borehole geophysics.

SIMULATION METHODS

Indicator simulation of lithologic categories

Indicator semivariogram analysis. An indicator formalism was used to represent the lithologic categories. The lithologic category, z_c, representing the presence of a particular lithologic category, *c,* in the set of all lithologic categories, $Z(x)$, at lo-

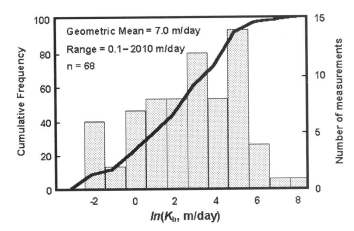

Figure 5. Observed well-test hydraulic conductivity data in all TAN wells for which slug-test data are available. Data are from compilation by A. Wylie (1996, written commun.).

cation x_j was transformed into a binary indicator variable, $I(x_j; z_c)$, according to equations 4 and 5. The indicator transform I is a random function with a defined mean and variance (Deutsch and Journel, 1997), in which the mean is equal to the global proportion of category z_c and the variance is related to the sill of the indicator semivariogram.

The experimental indicator semivariogram of I is defined as

$$\gamma(h; z_c) = \frac{1}{2N(h)} \sum_{i=1}^{N(h)} [I(x_{j+h}; z_c) - I(x_j; z_c)]^2, \quad (6)$$

where h is the lag or separation vector. The indicator semivariogram takes on values between 0 and 1. Other measures of spatial correlation also can be defined (Isaaks and Srivastava, 1989) and were used in this work to verify conclusions based on the semivariograms. Indicator semivariograms provide a useful means for defining the spatial dependence of categorical variables such as rock type and can be modeled in the same manner as conventional semivariograms. For example, an exponential semivariogram model has the form

$$\gamma(h; z_c) = c_o + c\left(1 - \exp\left[\frac{-3h}{a}\right]\right), \quad (7)$$

where the parameter c_o is the nugget variance, c is the sill, and a is the practical range. If the spatial structure is anisotropic, the sill and/or range vary with the orientation of the lag vector h. In three dimensions the principal orthogonal ranges define a range anisotropy ellipsoid, the orientation of which characterizes the spatial orientation of subsurface heterogeneity (Johnson and Dreiss, 1979).

A horizontal lag spacing of 30 m and a vertical lag spacing of 0.6 m were used to calculate experimental semivariograms, with a horizontal angular tolerance of 45°, no horizontal maxi-

imum band width, and a 5° vertical tolerance with 3 m maximum vertical band width. In addition, a 50% overlap between lag intervals was employed to further smooth the computed experimental semivariogram statistics in order to help identify spatial structure.

Small-scale spatial structure may not be evident in the experimental semivariograms because of the lack of closely spaced wells to define spatial statistics at small lags. Because lithology is expected to be continuous at small spatial scales, semivariograms that appeared to comprise mostly uncorrelated variance were modeled with an arbitrary, small nugget value and a nonzero range. This follows accepted practice for simulation purposes (Deutsch and Journel, 1997) and is particularly appropriate for semivariograms of lithologic indicators where constraints on semivariogram model structure can be imposed by soft geologic information (e.g., Kupfersberger and Deutsch, 1999).

Figures 6 and 7 show directional indicator semivariograms and fitted models for the sediment-filled and massive basalt categories, respectively, in the horizontal and vertical directions. An exponential semivariogram model was fitted to the semivariograms of the sediment-filled zones. For the massive basalt category, a nested, exponential-spherical model was used. In the horizontal plane, no compelling evidence for anisotropy was detected for either lithologic category (e.g., Fig. 6).

The very slight ($\sim 1°$) regional southward stratigraphic dip in the TAN area was ignored in this analysis and in simulations. In the 300 m north-south width of the test subarea this would amount to only 5 m of vertical deviation in the model domain, well within the vertical search tolerances used for the semivariogram analysis. Only that part of the basalt section between the PQ and QR sedimentary interbeds was analyzed and modeled in this work.

Like the sediment-filled and massive basalt categories, modeling of the vertical indicator semivariogram for the interflow zone category was straightforward. As shown in Figure 8C, a zero nugget was also chosen for the semivariogram model, on the basis of the preceding rationale, even though the vertical experimental semivariogram implies a pure nugget effect at all lags down to the 0.6 m resolution of the lithologic database. However, the modeled vertical range is smaller than that of either sediment-filled zones or massive basalt, which is consistent with the inference that the average thickness of interflow zones is less than the average thicknesses of either the massive basalt or sedimentary interbeds (see Table 1 in Welhan et al., this volume, Chapter 9).

The horizontal model semivariogram ranges and anisotropy of the interflow zone category were more difficult to determine. The irregular geometry of lava flow lobes, their variable thickness, and the interleaving of different scales of lava flow lobes tends to produce considerable noise in directional indicator semivariograms of the interflow zone category in the horizontal plane, even with the generous search windows and tolerances used in these semivariogram calculations.

TABLE 2. WELLS IN THE TAN AREA HAVING INTERPRETED LITHOLOGIC INFORMATION AND WELL-TEST DATA

Well name	In feasibility subarea	Land surface elevation (m)	Well location* X	Well location* Y	Lithologic information: Number of IFZs# in test interval	Lithologic information: Number of sediment-filled IFZs	Permeability information: Test interval center elevation (m)	Permeability information: Interval length (m)	Permeability information: Kb (m/day)
GIN 01		1459.43	360675.04	788854.08					
GIN 02		1459.27	361169.41	788930.14					
GIN 04		1459.34	361119.58	788922.62					
GIN 05		1459.37	361357.07	789392.93					
TAN 05	Yes	1463.98	358354.04	795647.38	2	0	1374.88	6.55	102.13
TAN 06		1459.41	361772.37	793962.34	3	0	1384.72	6.10	4.27
TAN 07		1459.38	361768.58	793915.64	1	0	1365.57	6.10	14.63
TAN 08		1460.48	358066.16	793500.87	2	0	1387.52	6.52	3.05
TAN 09	Yes	1457.54	356986.83	795489.42	1	0	1362.62	6.68	8.54
TAN 10	Yes	1457.41	356986.83	795489.42	3	0	1387.26	9.24	153.05
TAN 11	Yes	1457.47	356932.27	795159.75			1366.01	6.10	4.88
TAN 18	Yes	1464.32	358254.68	795251.22	2	0	1321.33	12.80	50.82
TAN 18		1464.32	358254.68	795251.22	2	0	1310.05	6.10	91.46
TAN 19	Yes	1464.45	358236.25	795265.72	4	0	1358.66	7.32	88.93
TAN 19		1464.45	358236.25	795265.72	0	0	1340.67	6.10	0.09
TAN 20		1457.70	355660.72	794753.04	2	0	1349.16	6.10	0.61
TAN 20		1457.70	355660.72	794753.04			1377.05	3.96	0.09
TAN 20		1457.70	355660.72	794753.04			1337.12	2.74	0.12
TAN 21		1459.55	359254.62	791009.26	2	0	1388.05	16.16	1.16
TAN 21		1459.55	359254.62	791009.26	2	0	1376.01	14.63	0.12
TAN 21		1459.55	359254.62	791009.26	2	0	1339.42	5.49	8.48
TAN 21		1459.55	359254.62	791009.26	3	0	1360.46	12.20	2.29
TAN 21		1459.55	359254.62	791009.26	3	0	1324.41	14.18	2.62
TAN 22a		1459.48	361724.74	792012.37	1	0	1303.91	9.30	69.97
TAN 22a		1459.48	361724.74	792012.37			1300.63	6.10	60.98
TAN 23a		1459.50	361668.55	792050.42	1	0	1323.83	6.10	30.49
TAN 23		1459.50	361722.44	792054.51			1319.41	7.62	85.37
TAN 23		1459.50	361722.44	792054.51			1327.19	3.05	213.41
TAN 23		1459.50	361722.44	792054.51			1345.55	4.73	137.80
TAN 24		1459.98	362861.09	788240.19			1366.69	10.98	25.40
TAN 24		1459.98	362861.09	788240.19			1349.62	16.46	1.13
TAN 24a		1459.95	362886.29	788264.45			1390.44	6.10	60.98
TAN 24a		1459.95	362886.29	788264.45			1322.60	3.35	194.02
TAN 25	Yes	1457.99	357019.34	795386.07	4	2	1379.64	24.39	11.43
TAN 26	Yes	1457.99	357040.58	795372.26					
TAN 27		1457.45	357207.35	795158.39			1385.80	9.15	41.16
TAN 28		1457.75	357261	795380.65			1384.58	12.20	11.22
TAN 29		1457.68	357508.06	795330.84			1383.90	12.20	16.86
TAN 30a		1457.73	357269.79	795363.63			1363.22	6.10	10.73
TAN 32	Yes	1459.02	357706.26	795024.83					
TAN 34	Yes	1458.17	357749.12	795197.74					
TAN 35	Yes	1458.01	357707.11	795225.24					
TAN 36	Yes	1461.54	358257.73	794843.19					
TAN 38	Yes	1462.85	358234.29	795047.44					[†]
TAN 39	Yes	1463.48	358064.67	795155.77					[†]
TAN 40	Yes	1458.63	357863.13	795288.44					[†]
TAN 41	Yes	1458.57	357841.23	795281.77					[†]
TAN 42	Yes	1463.55	357988.22	795200.1					
TAN 44	Yes	1462.97	358214.93	795039.6					
TAN 46	Yes	1461.73	358240.94	794840.58					
TAN dd1	Yes	1460.02	358628.8	794349.32					
TAN dd3		1457.10	354966.9	797827.06					
TCH 1		1457.02	356797.62	795930.51	1	0	1394.21	3.66	0.49
TCH 1		1457.02	356797.62	795930.51	0.5	0	1391.16	3.66	0.09
TCH 1		1457.02	356797.62	795930.51	1	0	1388.12	3.66	0.64
TCH 1		1457.02	356797.62	795930.51	0.5	0	1385.07	3.66	0.79
TCH 1		1457.02	356797.62	795930.51	1	0	1382.02	3.66	1.52
TCH 1		1457.02	356797.62	795930.51	1	0	1378.97	3.66	1.71
TCH 1		1457.02	356797.62	795930.51	0.5	0	1375.92	3.66	1.07
TCH 1		1457.02	356797.62	795930.51	1	0	1372.87	3.66	3.51
TCH 1		1457.02	356797.62	795930.51	0	0	1369.82	3.66	10.52
TCH 1		1457.02	356797.62	795930.51	0	0	1366.16	3.66	7.80
TCH 1		1457.02	356797.62	795930.51	1	0	1363.73	3.66	0.43
TCH 1		1457.02	356797.62	795930.51	0	0	1360.68	3.66	0.70
TCH 1		1457.02	356797.62	795930.51	0	0	1357.63	3.66	1.04
TCH 1		1457.02	356797.62	795930.51	0	0	1354.58	3.66	0.09
TCH 1		1457.02	356797.62	795930.51	1	0	1350.92	3.66	4.57
TCH 1		1457.02	356797.62	795930.51	1.5	0	1340.55	3.66	33.72
TCH 2		1460.63	358050.07	793457.42					
USGS 24	Yes	1462.14	358398.92	795215.02	4.5	2	1373.73	21.34	60.98

Note: Location data are from A. Wylie (1996, written commun.), augmented with data from J. Bukowski (1998, written commun.).
Lithologic information from interpreted geophysical logs (J. Glover, 1998, written commun.). IFZs—interflow zones.
* Well locations referenced to 1983 North American Datum, state plane feet.
# Interflow zones (IFZs) falling outside of a test interval, but within 0.6 m, were assigned a weight of 0.5.
† Borehole flow meter permeability data are available, but of unknown reliability.

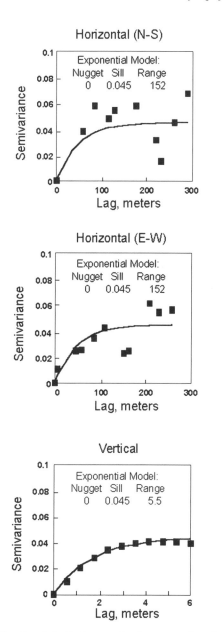

Figure 6. Directional indicator semivariograms of sediment-filled zones between PQ and QR sedimentary interbeds of feasibility test subarea, based on well database of interpreted lithology. Upper two panels are in north-south and east-west horizontal directions; lower panel is in vertical direction.

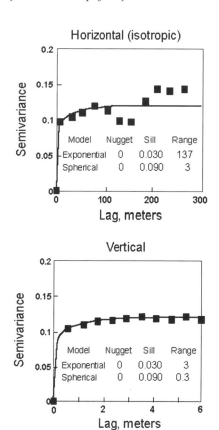

Figure 7. Directional indicator semivariograms of massive basalt category, based on well lithologic database. Upper panel is for horizontal plane; lower panel is in vertical direction.

Figure 8A shows the north-south and east-west directional indicator semivariograms of the interflow zone category assuming that interflow zones have a constant 0.6 m thickness; no anisotropy is apparent in the horizontal plane and the variogram range is ~45 m. These estimates of spatial structure are considered poor at best and misleading at worst because of the geometry-based problems.

As Welhan et al. (this volume, Chapter 9) point out, the effective average thickness of interflow zones is not known. They estimate it to be >~0.2 m (rubble zone thickness only), likely in the range of 1–1.5 m (rubble plus the most densely fractured upper crust), and possibly as great as 2.5 m (rubble plus entire average upper vesicular zone thickness). Because we cannot decide a priori on the appropriate thickness at which to represent interflow zones, derivative data sets were created from the 0.6 m discretized data, representing interflow zones with a constant thickness of 1.2 m and 2.4 m, respectively (i.e., by assigning multiple grid blocks to an interflow zone occurrence).

Indicator semivariograms using these data sets showed improved definition of lateral spatial correlation structure and clearer evidence of anisotropy in the horizontal plane. Figure 8B shows the result for the 2.4-m-thick interflow zone case. Semivariograms based on the 1.2 m data set are noisier but show similar structure.

Note that anisotropy in Figure 8B is very clear. This remains so regardless of search parameters and lag spacings used to compute the variograms. This is also manifested in the systematic variation of correlation range with horizontal orientation of the directional semivariogram. Figure 9 shows a plot of the maximum model semivariogram ranges for the exponential structure that was fitted to the directional indicator semivariograms for various azimuths. Both the 1.2 m and 2.4 m assumption for interflow zone thickness show a maximum semivario-

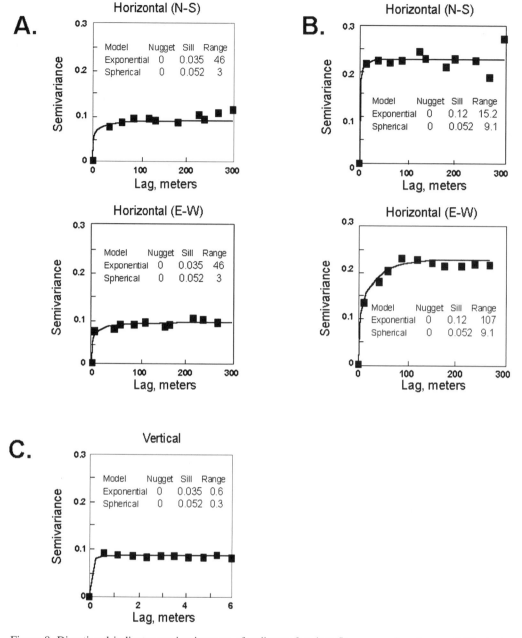

Figure 8. Directional indicator semivariograms of sediment-free interflow zones, based on well lithologic database. A: North-south and east-west horizontal semivariograms, assuming interflow zones have constant 0.6 m thickness. B: Horizontal semivariograms based on assumption that interflow zones are 2.4 m thick; note marked correlation range anisotropy. C: Vertical semivariogram, assuming interflow zones are 0.6 m thick.

gram range in approximately the 80°–90° direction (east-west). We have used the 2.4 m assumption to estimate the range anisotropy of interflow zone spatial structure.

In essence, the assumption of a constant 2.4 m interflow zone thickness improved the definition of semivariogram structure by effectively increasing the degree of vertical averaging in the computation of horizontal directional semivariograms. However, the global proportion of interflow zone material in the aquifer is overestimated by the 2.4 m data set; the indicator mean (proportion) of the 2.4 m data set is 0.35 compared to 0.09 for the original 0.6 m interflow zone data set.

Indicator semivariograms for the 0.6 m, constant-thickness interflow zone data set and M. Hankins' (1998, written commun.) data set of estimated interflow zone thicknesses show very similar semivariogram sills, ranges, and indicator means. For subsequent modeling, therefore, the 0.6 m, constant-

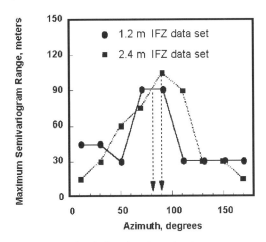

Figure 9. Maximum fitted directional semivariogram ranges as function of horizontal direction. Results for both 1.2 and 2.4 m interflow zone (IFZ) data sets are shown.

thickness interflow zone data set was used to define the volumetric proportion of interflow zone material (indicator means), as well as to model the semivariograms of all but the horizontal range of the interflow zone category, which was estimated from the 2.4 m interflow zone data set.

Table 3 summarizes all fitted semivariogram parameters used in this study. Also shown are the indicator means and variances of the three lithologic categories. The vertical range for the interflow zone category is a direct function of the effective interflow zone thickness assumed; therefore, the fitted range based on the 0.6 m data set is likely a minimum estimate. However, to avoid subjectivity in the choice of effective interflow zone thickness, we used the vertical range for the 0.6 m interflow zone data set shown in Figure 8C in all simulations.

Although range anisotropy in the horizontal plane was not evident for either the sediment-filled or massive basalt categories, a pronounced correlation range anisotropy is present in the interflow zone category. The anisotropy ratio is of the order of 7:1 (Fig. 9) but, because of the generous search tolerances used, this is likely a minimum estimate.

Welhan et al. (this volume, Chapter 9) showed that pahoehoe lava flow lobes (which support type-I interflow zones) in Holocene lavas of the eastern Snake River Plain have a mean length of 85 m, and a mean length:width aspect ratio of ~3:1, and that type-I interflow zones have a characteristic thickness of ~1 m. On this basis, type-I interflow zones supported on lava lobes would have mean dimensions of 85 m:26 m:1 m. Note that the maximum horizontal range of the interflow zone category listed in Table 3 (107 m) is similar to the median length of lava flow lobes (85 m) and the vertical correlation range (0.5 m) is also similar to the thickness of type-I interflow zones (~1 m). However, the minimum horizontal correlation range anisotropy identified from the variogram analysis (7:1) is much larger than the mean length:width aspect ratio of lava flow lobes (3:1). This may be indirect evidence that type-II interflow zone material, with an expected large length:width anisotropy (Welhan et al., this volume, Chapter 9), is relatively abundant in the TAN subsurface.

Sequential indicator simulation. Sequential indicator simulation was used to produce various realizations of the spatial arrangement of the three lithologic categories. The realizations honor the available lithologic data in the borehole data set, their global proportions, and their semivariogram models. The simulation grid was a rectangular volume 543 m long (oriented east-west), 299 m wide, and 85 m deep comprising 6.1 × 6.1 × 0.6 m grid cells. This provided the maximum vertical resolution consistent with the discretization scale of the lithologic data set. The grid is positioned between the PQ and QR sedimentary interbeds.

Simulations were created with SISIMPDF, a program for multiple categorical variables (Deutsch and Journel, 1992). All realizations reproduce the observed lithologic categories at available data points, honor specified global indicator means, and honor individual model semivariograms for each category.

Figure 10 shows two realizations of a simulation run, showing the nature of the variability of the stochastically generated results. These realizations were created with the semivariogram models summarized in Table 3, conditioned to the 0.6 m interflow zone data set. Because this is a conditional

TABLE 3. SUMMARY OF SEMIVARIOGRAM MODELS AND INDICATOR STATISTICS FOR LITHOLOGIC CATEGORIES

Lithologic category	Data set used	Indicator mean	Indicator variance	Modeled total sill	Modeled nugget	Type of correlation structure	Modeled sill component	Max. correlation range, m, E-W	Correlation structure anisotropy (x:y:z)
Sediment	0.6-m	0.063*	0.059	0.045*	0*	Exp#	0.045*	152*	1:1:0.036*
Massive basalt	0.6-m	0.844*	0.132	0.13*	0*	Sph	0.09*	3*	1:1:0.1*
						Exp	0.03*	137*	1:1:0.022*
Interflow zones	0.6-m	0.094*	0.085		0*	Sph	0.052*	0.3	1:1:10
				0.085*		Exp	0.035*	30	1:1:0.05
Interflow zones	2.4-m	0.351	0.228	0.23	0	Sph	0.11	9.1*	1:1:0.03†
						Exp	0.12	107*	1:0:14:0.006†

Note: All variogram modeling was performed on the 0.6-meter lithologic data set, except where indicated.
* These values were used in simulations.
# Exp = exponential structure model; Sph = spherical structure model.
† The vertical range was determined from 0.6 meter-thick IFZs.

simulation, the lithologic categories encountered at borehole locations are reproduced exactly in all realizations.

Gego et al. (this volume) provide an example of the application of categorical indicator simulation using this algorithm and these borehole conditioning data, to model permeability heterogeneity in the TAN aquifer for evaluating the feasibility of stochastically representing flow and transport. As discussed previously, the principal limitation of lithologic simulation is that the assignment of appropriate (core scale) permeability values to lithologic categories cannot readily be conditioned to observed well-scale permeability data because of the difference in permeability measurement support. However, fine-scale simulations of lithology provide an appropriate means to estimate bulk porosity from core-scale porosity assignments at any coarser discretization scale by simple averaging.

Direct simulation of well-scale permeability

We have explored an alternative approach to simulate permeability variations directly. In this, we sought to accomplish several objectives: (1) utilize available borehole-scale permeability information at the appropriate measurement support; (2) condition permeability realizations to this information; and (3) force the permeability realizations to honor available borehole lithologic data and the fine-scale spatial structure of lithologic heterogeneity. In this approach, the scale of the simulated permeability structure is dictated by the scale of available permeability information (i.e., vertically averaged over a mean test interval length of 6 m); this is significantly coarser than the scale at which lithologic heterogeneity was simulated (0.6 m).

In the following we analyze available well-test permeability to substantiate a quantitative model of the effective permeability of interflow zones. This model is based on a linear average of the permeabilities of interflow zone material and massive basalt and the number of interflow zones within a well-test interval. The relationship was used to effectively scale realizations of lithologic heterogeneity created by categorical indicator simulations up to the scale of well tests, in order to condition simulations of borehole-scale permeability to information on lithologic heterogeneity. In effect, we integrate two scales of information: coarse-scale, borehole permeability (hard) data with lithologic variability previously simulated on a fine vertical scale, which is treated as soft data for the purpose of conditioning simulated permeability.

Correlation structure of well-scale permeability. Figure 11 shows the horizontal isotropic semivariogram of normal-score transformed K_b values calculated from the data in Table 1. A horizontal lag spacing of 60 m was used, and a 3 m vertical band width. An analysis of structural anisotropy was precluded by the paucity of data; even an isotropic semivariogram model could not be identified with great confidence. In the approach we develop here, the limitation of a poorly defined permeability semivariogram was ameliorated because it is not the only source of information with which the spatial variability of K_b can be defined. In the following, information on the spatial architecture of interflow zones derived from previous lithologic simulations is the primary limitation on K_b simulations.

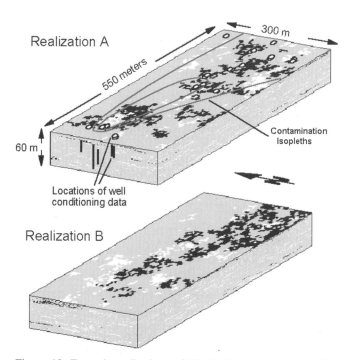

Figure 10. Example realizations of lithologic categories created by sequential indicator simulation, conditioned to 0.6 m interflow zone data set. White, interflow zones; gray, massive basalt; black, sediment-filled interflow zones; white circles, locations of well-conditioning data. Trichloroethylene isopleths indicate contaminant plume location for reference.

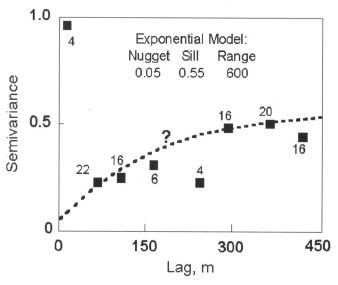

Figure 11. Experimental, nondirectional semivariogram of normally transformed well-test permeability, K_b, in TAN wells, in horizontal plane. Numbers of data pairs per lag interval are shown adjacent to experimental semivariogram points. Well pairs beyond bounds of feasibility test subarea were included in analysis. Note poor definition of correlation structure.

Dependence of well-scale permeability on interflow zones. Figure 12 shows a weak, semilogarithmic relationship between bulk permeability (K_b) measured in large open interval slug tests and the number of interflow zones encountered within a well-test interval (N_z), in the 36 slug test intervals at TAN for which data on both are available (Table 2). A similar relationship was observed in three wells in the Chemical Processing Plant (CPP) facility for which packer slug test data are available from multiple intervals (Welhan et al., 1999; Glover et al., GSA Data Repository*). Figure 13 shows additional evidence of a correlation in the TAN facility, where the means of conditional distributions of normal-transformed K_b values (not lnK_b) are correlated with N_z. Because too few data points exist in the highest interflow zone class, the data in Figure 13 are classified into four groups based on the number of interflow zones in the well-test interval: zero, one, two, and three or more.

Statistical evaluations of these conditional distributions were performed to evaluate the statistical significance of the implied correlation. A Student's t-test of the significance of the regression slope was performed on the log-transformed K_b data shown in Figure 12 as well as on the conditional classes of normally transformed K_b values in Figure 13B. Additional tests

of significance between the conditional means of the individual N_z classes in Figure 13B also showed that a statistically significant difference exists in the mean transformed K_b among all N_z classes, at the 95% confidence level.

From these results, we conclude that the number of interflow zones (N_z) communicating with a well during a slug test exerts a primary control on the bulk permeability sensed by the well. Because a similar relationship was observed in the CPP area, we propose that this represents a fundamental relationship between K_b and N_z in the eastern Snake River Plain aquifer.

Permeability model for interflow zones. It is proposed that K_b reflects a spatial average of different permeabilities encountered within the test interval. The scatter of lnK_b values in Figure 12 suggests that several factors may affect K_b: test interval length, permeability of interflow zone material, the number of interflow zones communicating with the well-test interval, effective interflow zone thicknesses, vertical interconnectedness of interflow zones, and/or a combination of these factors. The large within-class scatter of K_b in Figure 13 also suggests that the effective permeability of an individual interflow zone is highly variable and dependent on its thickness, lateral extent, and interconnectedness.

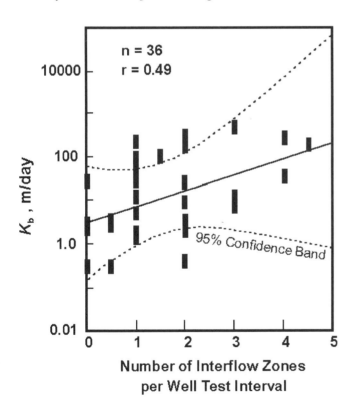

Figure 12. Empirical dependence of well-test permeability on number of interflow zones encountered in TAN well-test intervals from data in Table 2. Number of interflow zones per well-test interval was determined from interpreted lithologic database. Interflow zones outside of test interval but within 0.6 m were assigned weight of 0.5.

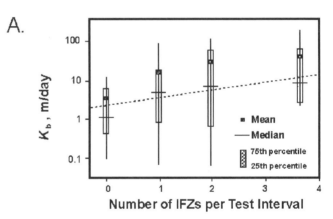

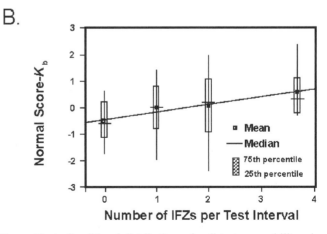

Figure 13. A: Conditional distributions of well-test permeability relative to number of interflow zones (IFZ) per test interval. B: Conditional distributions of normally transformed well-test permeabilities.

*See footnote 1, p. 231.

Direct measurements of interflow zone permeability variations are unavailable to test this hypothesis. However, the statistical distribution of K_b shown in Figure 5 and its variability in Figure 12 can be reconciled with a simple model of how interflow zone permeability may control K_b within a stochastic framework, which sheds light on the distribution of K_b.

Because of the high transmissivity of the basalt section in general, flow to or from the aquifer in communication with a well-test interval is considered to be substantially horizontal, i.e., partial penetration effects are negligible (Ackerman, 1991). Therefore, the bulk permeability (K_b) sensed by a well in a test interval of length b, in which flow to the test interval is only horizontal and parallel to layered heterogeneity, can be expressed as a weighted arithmetic mean (Freeze and Cherry, 1979):

$$K_b = \frac{\Sigma K_i b_i}{\Sigma b_i} = \frac{\Sigma K_i b_i}{b} \qquad (8)$$

where the K_i and b_i are, respectively, the permeabilities and thicknesses of the various contributing layers within the interval b. Assuming that sediment-filled zones are of minor importance within the aquifer, equation 8 can be rearranged to yield the following expression:

$$K_b = (K_Z - K_M) \frac{N_Z b_Z}{b} + K_M \qquad (9)$$

or

$$K_b = (K_Z - K_M)F_Z + K_M \qquad (10)$$

where the mean permeabilities of massive, low-permeability basalt and high-permeability interflow zones are represented by K_M and K_z, respectively, N_z is the number of interflow zones encountered in the well-test interval, b_z is the mean thickness of interflow zones in the interval, and F_z is the fraction of the test interval length composed of interflow zone material. This model predicts that K_b will approach K_M as either F_z or the contrast between K_z and K_M approaches zero. It views K_b as a linear, weighted average of the permeabilities of interflow zones and massive basalt in the well-test interval, that is directly dependent on the number of and thicknesses of interflow zones intersected by the test interval.

Because b_Z, K_Z, and K_M are expected to be variable, the dependence of K_b on N_z could be expected to display considerable scatter (cf. Fig. 12). To test this hypothesis, values of $\ln K_b$ were stochastically generated by randomly sampling a lognormal distribution of numbers of interflow zones of random thicknesses in each of the TAN well-test intervals in Table 1 and computing a possible K_b value with equation 10. The distribution of TAN rubble zone thicknesses observed in INEEL cores (Welhan et al., this volume, Chapter 9) was used as a surrogate for the distribution of b_z. An estimate of N_z was ob-

tained by randomly generating a lobe thickness from the distribution of lava flow lobe thicknesses in INEEL cores and dividing the well-test interval length, b, by this value. These values of N_z, b_z, and b were used in equation 10 to generate stochastic estimates of K_b for each well-test interval.

For a constant K_M value of 0.3 m/day, the value of K_z was adjusted so that the means of all realizations of synthesized K_b values approximated the mean of the TAN $\ln K_b$ distribution in Figure 5. The value of K_z determined in this manner (100 m/day) is similar to the 125 m/day value estimated by Knutson et al. (1992).

Figure 14 shows a suite of 10 realizations of K_b values calculated from equation 10 for the TAN well-test intervals, using an assumed constant value of K_z equal to 100 m/day.

The realizations define an envelope of cumulative distribution functions (cdfs) of K_b for the 68 well-test intervals for which we have data. As is evident in Figure 14, the variance of any of these synthesized $\ln K_b$ cdfs is significantly less than that of the $\ln K_b$ distribution observed at TAN (Fig. 5). Furthermore, the modeled variance cannot be increased substantially by increasing the variances of either the b_z or N_z populations used in equation 10.

This suggests that most of the variability in K_b is due to the variance of K_z.

To test this hypothesis and to better define the probability distribution of K_z, random values of K_z were drawn from a lognormal distribution with a specified variance of K_z and used in equation 10 along with randomly drawn values of b_z and N_z.

The variance of the K_z distribution was adjusted until a satisfactory match to the modeled $\ln K_b$ variance was obtained. An example of the match is shown in Figure 15. Note that the variance of $\ln K_z$ relative to the mean of $\ln K_z$ is much smaller ($s^2[\ln K_z]/m[\ln K_z] = 3.0/5.8$) than the relative variance of $\ln K_b$ ($s^2[\ln K_b]/m[\ln K_z] = 5.6/3.1$). Thus, it appears that K_z values may be lognormally distributed, with a relatively small lognormal variance.

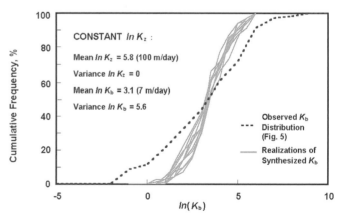

Figure 14. Cumulative frequency distributions in 10 realizations of synthesized well-test permeabilities, K_b, using equation 10 and interpreted lithologic database, with constant permeability value (K_z = 100 m/day) assigned to interflow zones.

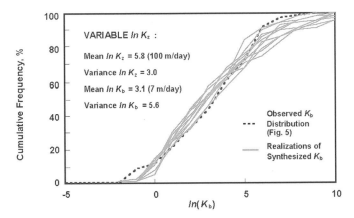

Figure 15. Cumulative frequency distributions in ten realizations of synthesized well-test permeabilities, K_b, using equation 10 and interpreted lithologic database, with statistically variable value of interflow zone permeability (lognormal distribution, mean of $\ln K_z = 5.8$, variance of $\ln K_z = 3.0$). Note that relative variance (variance/mean) of $\ln K_z$ is small (3.0/5.8) compared to that of $\ln K_b$ (5.6/3.1).

Because the simulations produced in this manner assume that the variables N_z, b_z, and K_z are independent, the scatter of the simulated K_b values relative to N_z is likely larger than if these variables displayed some degree of multivariate dependence (e.g., K_z correlated with b_z). However, realizations of K_b plotted semilogarithmically against N_z display a similar type of scatter and hint of a correlation, as seen in Figure 12 (Welhan et al., 1999).

On the basis of this analysis, we conclude that the observed distribution of K_b values in TAN wells can be approximated to first order by a simple linear model in which the permeability of interflow zones exerts the dominant control on well permeability. That is, the permeability sensed by a well is primarily controlled by the relative proportion of interflow zone material in a well-test interval and the variance of interflow zone permeability. Furthermore, our analysis suggests that the permeability of interflow zones should be viewed as lognormally distributed with a relatively modest $\ln K_z$ variance.

Simulation of well-scale permeability with simulated annealing. On the basis of the interflow zone permeability model developed here and the empirical relationship between the number of interflow zones communicating with a well-test interval and the bulk average K_b value measured in the test interval, we attempt to overcome a principal shortcoming of the indicator lithologic simulation approach. That is, maps of permeability heterogeneity inferred from lithologic heterogeneity are not constrained by borehole permeability data and therefore are not conditional simulations of permeability. However, direct conditional simulation of permeability is considered infeasible because of the relatively small amount of borehole K_b data available and because the spatial structure of borehole permeability is poorly characterized by available data (Fig. 11).

In any approach to conditioning simulated permeability to borehole data we must recognize that the simulations will have much coarser vertical resolution than the lithologic simulations presented in Figure 10, because the K_b data represent a large vertical measurement support or averaging volume. This may or may not be a serious limitation for the intended use of the simulations. Although the resulting permeability maps will have less vertical resolution, a coarser scale of discretization may still be appropriate for predictive modeling and remediation assessment if well monitoring, sampling, and hydraulic testing and validation are based on vertical test intervals of this length. That is, both the modeled and observed flow fields represent vertical spatial averages of preferential flow within the monitoring interval length.

We seek to exploit the conditional K_b-N_z relationship shown in Figure 13, so as to incorporate lithologic information (in the form of the simulated spatial architecture of interflow zones) in conditional permeability simulations: we effectively condition the resulting realizations to both permeability and lithology data. A simulated annealing algorithm was utilized, allowing greater flexibility in the types of information to which simulations can be conditioned. It is an algorithmic approach to stochastic simulation of a continuous variable that relies on minimization of one or more objective criteria of goodness of fit in order to reproduce one or more desired spatial or statistical characteristics in the resulting realizations. The objective criteria can be as simple as requiring realizations to reproduce the semivariogram of the simulated variable or as complex as satisfying a combination of criteria: a specified correlation with a secondary variable or soft data, reproduction of a data histogram, or conformance with the structural characteristics of a training image such as anisotropy, or complex spatial heterogeneity.

The SASIM algorithm in GSLIB (Deutsch and Journel, 1997) is a simulated annealing routine with considerable flexibility to satisfy multiple objective criteria. The algorithm begins by assigning a starting distribution of values randomly in the simulation grid then swapping pairs of data at random locations in the grid. After each swap, an objective function representing a quantitative measure of goodness of fit is recomputed. If the swap improves the goodness of fit it is retained; if not the swap is rejected and another initiated. To avoid trapping within local minima of the objective function, not all bad swaps are rejected. The process ceases after either a specified maximum number of swaps or no further improvement in the objective function.

Realizations of the spatial distribution of interflow zones in the TAN aquifer were created with indicator simulation, as described previously, providing maps of the spatial architecture of interflow zones at a relatively fine vertical resolution of 0.6 m (Fig. 10). This provided a simulated spatial distribution that was used as a secondary variable to define the permeability simulation by making use of the empirical conditional distributions shown in Figure 13. The number of simulated interflow zones (N_z) was summed within 6-m-thick vertical blocks (corresponding to the median well-test interval length in Table 1 and equal to 10 of the indicator simulation grid blocks). Figure 16 shows an example of a lithology realization from the indicator simulation and its corresponding N_z map on this 10 times coarser vertical grid.

SASIM was used to produce realizations of K_b that were conditioned to borehole K_b data and to the N_z map with the conditional expectation shown in Figure 13. Various combinations of other constraints could also be used, such as the K_b semivariogram and the K_b histogram. For the purposes of demonstrating the approach for this feasibility analysis, all K_b data used for conditioning were located at the mid-points of well-test intervals and were assumed to represent constant, 6-m-long intervals. For actual modeling applications, the K_b data from intervals shorter than 6 m could be normalized, and those representing longer test intervals could be assigned over multiple grid cells.

In Figure 17A, measured K_b values and the conditional expectation of K_b on N_z (defined in Fig. 13) were used as conditioning information. In Figure 17B, the semivariogram of K_b (Fig. 11) was included as an additional constraint, and the histogram of K_b was included as a fourth constraint in the realization shown in Figure 17C.

The value of incorporating additional simulation constraints is that the realizations can be forced to honor a variety of soft and hard information. Realizations produced in this manner can be compared in a flow and transport model to evaluate the impact of different combinations of conditioning data to identify the optimum set of constraints with which to describe the spatial architecture of preferential flow paths.

Geologic structure-imitating model

The major limitation of the previous cell-based modeling approaches is that they rely on a spatial structure interpretation, based on limited subsurface data, that presupposes that type-I interflow zones are the only significant high-permeability zones in the aquifer. As Welhan et al. (this volume, Chapter 9) have pointed out, there is documented evidence suggesting that type-II interflow zones may also be important high-permeability zones with quite different spatial structure that cannot be represented with two-point geostatistical measures. Object-based simulation may be an effective alternative approach.

We have developed a prototype simulator with the goal of generating representations of aquifer heterogeneity that capture the spatial architecture of eastern Snake River Plain basalt and sediment. The approach is a generalization of object-based simulation, which generates three-dimensional shapes of basalt and sediment that reproduce the architecture of high and low-

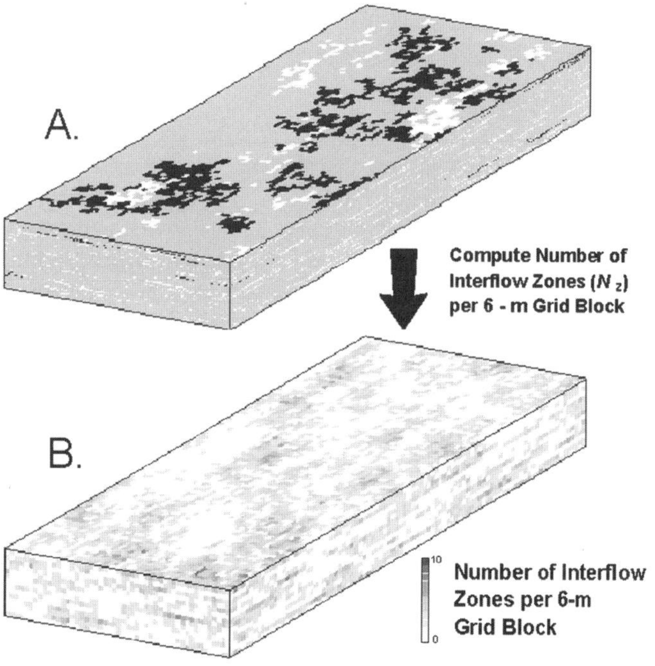

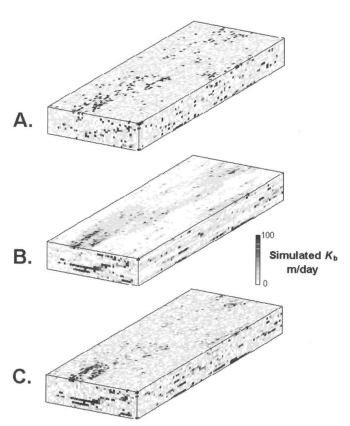

Figure 16. Conversion of simulated interflow zones from 0.6 m vertical support (A) to 6.0 m vertical support (B) commensurate with measurement scale of well-test permeability data, showing number, N_z, of simulated interflow zones per grid block.

Figure 17. Example realizations of well-test permeability, K_b, produced by simulated annealing, conditioned to different combinations of information. A: K_b conditioned to distribution of N_z produced by sequential indicator simulation (see Fig. 16). B: K_b conditioned to N_z and semivariogram of K_b (Fig. 11). C: K_b conditioned to N_z, model semivariogram of K_b (Fig. 11), and histogram of K_b values.

permeability zones. Here we briefly describe the first stage of development of this approach; a detailed description of the simulation rules is beyond the scope of this chapter. Furthermore, we have not taken the next step of assigning permeability and porosity to the simulated geometry, nor is the process of constraining the simulations with field measurements fully developed. Clemo and Welhan (2000) described the details of the prototype model and evaluated its performance.

This approach is within the class known as structure-imitating models (Koltermann and Gorelick, 1996), where the imitation of morphology and structure is defined by quantitative measures obtained through field mapping and core logging. Structure imitation creates geometric realizations of the internal stratigraphy of a layered sequence of basalt flows without modeling the physics of the flow emplacement process. Structure-imitation uses a set of ad hoc rules that are judged to be successful if the resulting geometric configurations match those observed in the field.

The simulation of basalt structure is divided into three stages: (1) specifying the locations, sequencing, and volume of eruptions; (2) simulating individual lava flows in each eruption and any sediment accumulation between eruptions; and (3) subdividing each lava flow into internal lithologic elements (massive, vesicular, rubbly).

Constraining simulated geometry to measured geometry has been set up as an optimization problem. Simulated geometry is compared to data from field mapping and core logs. A simulation rule base was created that allows for a wide spectrum of possible geometries. The best rules are selected from the rule base using a genetic algorithm to pick rules and set parameters (Goldberg, 1987; Whitley, 1998). A genetic algorithm was chosen because of its ability to handle the noncontinuous sensitivity of the simulated measurements to rule changes.

Figure 18 shows examples of aerial patterns of individual simulated lava flows. The flow patterns are created one grid cell at a time until the volume of flow material for an eruptive event is exhausted. Rules for selecting the direction of lava flow growth include closeness to the source of the flow and the length of time previously filled cells have been occupied. Panels A, B, and C depict the effects of different rule choices on the simulated geometry. Inflated lava flow units in the distal portions of Holocene lava flow groups on the eastern Snake River Plain (panel D) resemble the models produced in panels B and C. This indicates that flow patterns that are similar in geometric character to those of the Snake River Plain basalt flows can be generated with appropriate rule selection criteria.

In the genetic algorithm approach, the parameters and rule selection are encoded into a numerical structure called a gene. A population of genes is created by randomly selecting rules and parameters. The structure imitation code creates a three-dimensional realization of geologic structure using the rules and parameters described by each gene. The algorithm tests each realization by comparing the realization to field measurements. The best genes from the current population and the previous

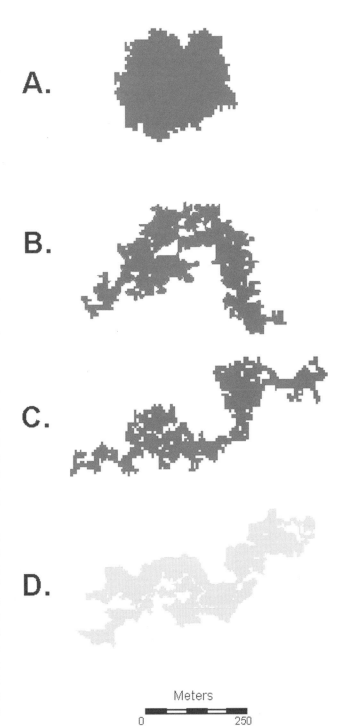

Figure 18. Effect of cell selection rules on areal geometry of lava flows produced by structure-imitating model. A: Preferentially selecting cells near eruptive source to be filled with lava during eruptive event. B: Weighting selection of cells to be filled by square of distance from eruptive source. C: Weighting by distance from source and preferentially selecting cells near most recently filled cells. D: Example of lava flow unit on Wapi Flow Group (Welhan et al., this volume, Chapter 9), for comparison.

population are selected to be the parents of a subsequent new population and a new three-dimensional realization is created. Various combinations of gene swaps and mutations are allowed to introduce variability to ensure convergence to a truly optimal geologic structure. We have only used the genetic algorithm in an exploratory manner to develop a structure-imitating model that reproduces observed geometry of lava flows, including shape, fractal dimension, and internal lava flow structure.

The genetic algorithm was used to optimize the match between simulated proportions of vesicular, massive, and rubbly material generated by the structure-imitating model and from core-derived measurements. The combined upper vesicular zone and rubble layer thickness, as well as the thickness of massive interior and lower vesicular layer, were matched against data from Knutson et al. (1992) and summarized in Figure 19. The light bars are data from Knutson et al. (1992) and the dark bars are simulation results. No attempt has been made at this stage of model development to improve the initial rule base for simulating the internal stratigraphy of flow lobe elements.

Further development of a functional model will require continuing integration of field measurements and the modeling effort. This includes characterization and modeling of the sedimentary interbeds as well as refinement of the lava flow models. As with most modeling efforts, development of the structure-imitation code will be an iterative, evolutionary process. Comparing the simulated realizations conditioned with available geologic data will lead to the identification of additional constraints that may be necessary. In cases where the model fails to provide reasonable matches to the conditioning data, new rules or submodels may be needed. In line with the goal of providing detailed three-dimensional maps of heterogeneity for flow and transport simulations, emphasis will be placed on characterizing and modeling the rubble zones and the fracture networks (type-I interflow zones) within the margins of inflated lava flows.

SUMMARY AND CONCLUSIONS

A study was conducted to determine whether stochastic modeling of aquifer heterogeneity beneath the eastern Snake River Plain is feasible, and to evaluate the advantages and disadvantages of different stochastic modeling approaches in view of the unique geology of this aquifer, the availability and type of subsurface data, and the scale of simulation.

Sufficient information was available from borehole geophysical logs to identify three principal lithologic categories as surrogates for permeability within the basalt aquifer: (1) low-permeability, massive basalt; (2) low-permeability zones filled with fine-grained sediment; and (3) high-permeability interflow zones. Because the different types of interflow zones identified by Welhan et al. (this volume, Chapter 9) currently cannot be distinguished in borehole data, all interflow zone material was assumed to be on the surfaces of lava flow lobes (i.e., type-I

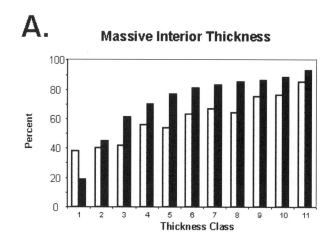

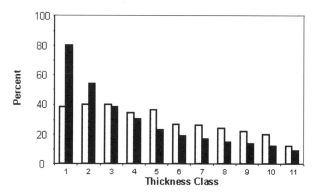

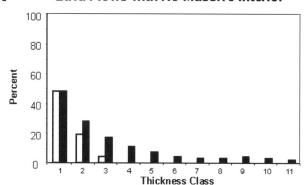

Figure 19. Internal lava flow geometry produced by genetic algorithm (dark bars), compared to Knutson et al.'s (1992) measured geometry (light bars). Data are plotted against lava flow lobe thickness classes. A: Simulated and measured thicknesses of massive lava flow lobe interiors. B: Combined thickness of upper vesicular zone and rubble, as measure of interflow zone thickness. C: Number of simulated lobes without central massive element as fraction of total number of lobes simulated.

interflow zones). The correlation ranges of interflow zone material deduced from variograms of borehole data are similar to the dimensions expected for interflow zones supported on the surfaces of eastern Snake River Plain lava flow lobes, although the horizontal anisotropy is much larger than expected. This may be an indication that a significant amount of type-II interflow zone material is present in the TAN subsurface, with much larger horizontal anisotropy.

Basalt heterogeneity between the PQ and QR sedimentary interbeds was simulated stochastically in a 543 × 299 × 85 m test subarea (1.74 km²) of the TAN facility at INEEL, on a simulation grid that was discretized horizontally into 6.1 × 6.1 m blocks. This model domain encompassed 20 wells, the lithology of which was interpreted from geophysical logs and 36 well-test intervals for which well-test permeability data were available.

Two cell-based simulation approaches were developed and tested: (1) sequential indicator simulation of lithologic heterogeneity, on a fine (0.6 m) vertically discretized grid, conditioned to borehole lithologic data; and (2) simulated annealing of well-test permeability, on a coarser (6.1 m) vertical grid, conditional to both well-test permeability data and to the spatial arrangement of interflow zones previously simulated by sequential indicator simulation. A third modeling alternative was developed to investigate the feasibility of an object-based simulation approach: a prototype structure-imitating model of basaltic lava flows.

Sequential indicator simulation of lithology

The major advantages of lithologic simulation are that it is straightforward and readily implemented. In addition, lithologic heterogeneity can be simulated at any arbitrarily fine vertical spatial resolution that is commensurate with the need to represent the spatial architecture of relatively thin, hydraulically important interflow zones.

A major disadvantage to modeling lithologic heterogeneity is that the results rely entirely on the semivariogram models deduced from borehole data and presuppose that the spatial architecture of the lithologic categories can be represented with two-point geostatistical measures. This assumption may be invalid if type-II interflow zone material composes a significant proportion of the aquifer. A second serious limitation is that appropriate values of permeability have to be assigned to the simulated lithologic categories. These permeability values most likely would be derived from core-scale data sets. Values could be assigned as constant throughout each category, as random values drawn from appropriate distribution functions for each category, or with a combination of these approaches in different categories. Our analysis of TAN well-scale permeability has shown that interflow zone permeability should be assigned from a lognormal distribution with a mean log-transformed permeability corresponding to 100 m/day and a relatively modest lognormal variance. Regardless of the permeability assignment

method used, conditioning assigned permeabilities to well-scale permeability data will be difficult because of the difference in support volumes between the simulation grid and the borehole permeability data.

It is recommended that additional research be focused toward identifying the effective thickness of type-I interflow zones to define the volumetric proportion of interflow zone material in the aquifer. It is also recommended that methods to discriminate between type-I and type-II interflow zones in boreholes be developed. The geometry of the two types of interflow zones is quite different (Welhan et al., this volume, Chapter 9), and the presence of type-II interflow zones could substantially alter the spatial architecture of simulated high-permeability zones in the aquifer.

Simulated annealing of well-scale permeability

The amount of well-scale permeability data in and near the model domain was barely adequate to estimate an isotropic semivariogram model, and was totally inadequate to define anisotropy in the semivariogram structure. However, we found that well-scale permeability (K_b) could be related to the number of interflow zones (N_z) in well-test intervals, so that realizations of interflow zone spatial architecture generated by indicator simulation could be used as soft information to condition realizations of K_b. Simulated annealing was chosen for its flexibility to incorporate various types of conditioning information.

A model to predict K_b from the permeability and abundance of interflow zones was found to account for much of the statistical variability of well-scale permeability observed at the TAN facility. The model is based on a simple linear average of the permeability of massive basalt and interflow zone material. It was used to show that interflow zone permeability must be lognormally distributed with a modest variance.

In the manner that simulated annealing was implemented in this work, its chief disadvantage was the sacrifice in vertical resolution because of the measurement support of K_b data. In contrast, the approach offers several advantages. First, it avoids the problem of correctly assigning representative permeabilities to simulated lithologic categories, and simplifies the conditioning of simulated permeabilities to well-test data. Second, realizations can be conditioned to the spatial architecture of interflow zones. This implicitly conditions realizations to borehole lithologic information, geostatistical structural limitations, and any additional soft geologic limitations that may have been imposed in the lithology simulations (e.g., mean aspect ratio of interflow zones; lateral spatial continuity of sediment-filled zones). The simulated permeability therefore honors many more different types and scales of available subsurface information.

The method integrates two different scales of information: well-scale permeability and fine-scale vertical lithologic variability. The stochastic realizations of K_b generated with this approach reflect not only the fine-scale spatial distribution of interflow zone and massive basalt permeability, but also their volume-averaged hydraulic response. Thus, future development of the approach could incorporate conditioning of realizations to the hydraulic anisotropy observed in large-scale, multiwell pumping tests (Welhan et al., 1998).

Basalt structure-imitating model approach

Elongated fracture networks along the margins of inflated lava flows (type-II interflow zones) may exert a substantial control on the spatial architecture of high-permeability zones (Welhan et al., this volume, Chapter 9). If so, linear geostatistics will be inadequate to simulate their complex, meandering spatial patterns, and an object-based simulation approach will be necessary. Such an approach may ultimately supersede conventional simulation methods in the eastern Snake River Plain because of its ability to capture the spatial architecture of geometrically complex heterogeneity.

The structure-imitating approach is a variant of object-based simulation. A structure-imitating model was developed to recreate the complex geometries of lava flows, interflow zones, flow groups, and sedimentary interbeds through sets of rules and parametric equations that mimic the structure generated by physical processes of eruption, inflation, ponding, and sediment deposition. A genetic algorithm was used to optimize the rules and parameters of the model in order to produce stochastic realizations that honor observed structural constraints (e.g., lava flow thickness distributions, fractal dimension of lava flows, vent locations).

Work to date has demonstrated the feasibility and potential power of the structure-imitating approach. Future work will focus on: (1) development of refined rule bases for lava emplacement and sedimentation; (2) development of additional matching constraints based on lava flow measurements obtained by Welhan et al. (this volume, Chapter 9) and others; and (3) mapping permeability to geologic structures in order to condition the realizations to borehole data.

ACKNOWLEDGMENTS

We thank Allan Wylie for use of his database of Idaho National Engineering and Environmental Laboratory well-test information, and Matt Hankins for use of his estimates of interflow zone thicknesses. We also thank John Glover and Linda Lee Davis for their work in interpreting borehole geophysical data from which the lithologic database was produced. John Bukowski, Lockheed-Martin Idaho Technologies Co., and Allan Wylie provided much pertinent information about Test Area North area hydrology, as well as valuable discussion and ideas on the geologic controls on lithologic and hydraulic heterogeneity in the aquifer. Cheryl Whittaker, LMIT Co, provided valuable assistance in accessing digital data files. Steve Anderson and Mel Kuntz, U.S. Geological Survey, Scott Hughes, Idaho State University, and Denny Geist, University of Idaho, provided fruitful discussion and insights into the geology of pahoehoe lava flows. We benefited tremendously from thoughtful reviews by Stanley Miller, University of Idaho, and Gary Weissmann, Michigan State University. This work was funded by the U.S. Department of Energy through grant DE-FG07-96ID13420 to the Idaho Water Resources Research Institute–Idaho Universities Consortium.

REFERENCES CITED

Ackerman, D.J., 1991, Transmissivity of the Snake River Plain aquifer at the Idaho National Engineering Laboratory, Idaho: U.S. Geological Survey Water-Resources Investigations Report 91-4058, 35 p.

Anderson, S.R., 1991, Stratigraphy of the unsaturated zone and uppermost part of the Snake River Plain aquifer at the ICPP, Idaho National Engineering Laboratory, Idaho: U.S. Geological Survey Water-Resources Investigations Report 91-4010, 71 p.

Anderson, S.R., and Bowers, B., 1995, Stratigraphy of the unsaturated zone and uppermost part of the Snake River Plain aquifer at Test Area North, Idaho National Engineering Laboratory, Idaho: U.S. Geological Survey Water-Resources Investigations Report 95-4130, 47 p.

Bennecke, W., 1994, The identification of basalt flow features from borehole television logs: Hydrogeology, waste disposal, science and politics, *in* Proceedings of the 30th Symposium on Engineering Geology and Geotechnical Engineering: Boise, Idaho, Idaho State University, p. 371–383.

Carle, S.F., Labolle, E.M., Weissmann, G.S., Van Brocklin, D., and Fogg, G.E., 1998, Conditional simulation of hydrofacies architecture: A transition probability/Markov approach, *in* Fraser, G.S., and Davis, J.M., eds., Hydrogeologic models of sedimentary aquifers: Concepts in hydrogeology and environmental geology Number 1: SEPM (Society for Sedimentary Geology) Special Publication, p. 147–179.

Clemo, T. and Welhan, J., 2000, Simulating basalt lava flows using a structure imitation approach, *in* Bentley, L.R., Sykes, J.F., Brebbia, C.A., Gray, W.G., and Pinder, G.F., eds., Computational Methods in Water Resources, Proceedings, 13th International Conference, Calgary: Rotterdam, A.A. Balkema, p. 841–848.

Deutsch, C.V., and Journel, A.G., 1997, GSLIB: Geostatistical Software Library and User's Guide (second edition): New York, Oxford University Press, 340 p.

Deutsch, C.V., and Wang, I., 1996, Hierarchical object-based stochastic modeling of fluvial reservoirs: Mathematical Geology, v. 28, p. 857–880.

Freeze, R.A., 1975, A stochastic-conceptual analysis of one-dimensional groundwater flow in non-uniform, homogeneous porous media: Water Resources Research, v. 11, p. 725–741.

Freeze, R.A., and Cherry, J.A., 1979, Groundwater: New York, Prentice-Hall, 604 p.

Fogg, G.E., 1989, Emergence of geologic and stochastic approaches for characterization of heterogeneous aquifers, *in* Proceedings of Conference on New Field Techniques for Quantifying the Physical and Chemical Properties of Heterogeneous Aquifers: Dallas, Texas, National Water Well Association, p. 1–17.

Goldberg, D., 1987. Genetic algorithms in search, optimization, and machine learning: Reading, Massachusetts, Addison-Wesley, 412 p.

Goovaerts, P., 1997, Geostatistics for natural resources evaluation: New York, Oxford University Press, 483 p.

Guardino, F., and Srivastava, R.M., 1993, Multivariate geostatistics: Beyond bivariate moments, *in* Soares, A., ed., Geostatistics troia '92: Norwell, Massachusetts, Kluwer Academic Publishers, v. 1, p. 133–144.

Hon, K., Kauahikaua, J., Denlinger, R., and Mackay, K., 1994, Emplacement and inflation of pahoehoe sheet flows: Observations and measurements of active lava flows on Kilauea Volcano, Hawaii: Geological Society of America Bulletin, v. 106, p. 351–370.

Isaaks, E.H., and Srivastava, R.M., 1989, An introduction to applied geostatistics: New York, Oxford University Press 561 p.

Johnson, N.M., and Dreiss, S.J., 1989, Hydrostratigraphic interpretation using indicator geostatistics: Water Resources Research, v. 25, p. 2501–2510.

Johanneson, C., Funderberg, T., and Welhan, J., 1998, Basalt surface morphology as a constraint in stochastic simulations of high-conductivity zones in the Snake River Plain aquifer, Idaho: Eos (Transactions of the American Geophysical Union), v. 79, p. F243.

Johannesen, C.M., 2001, Basalt surface morphology on the eastern Snake River Plain, Idaho: Implications for the emplacement of high-porosity elements in the Snake River Plain Aquifer [M.S. thesis]: Pocatello, Idaho, Idaho State University, 140 p.

Journel, A.G., 1989, Fundamentals of geostatistics in five lessons: Washington, D.C., American Geophysical Union, Short Course in Geology, v. 8, 40 p.

Kaminsky, J.F., Keck, K.N., Schafer-Perini, A.L., Hersley, C.F., Smith, R.P., Stormberg, G.J., and Wylie, A.H., 1993, Remedial investigation final report with addenda for the Test Area North Groundwater Operable Unit 1-07B at the Idaho National Engineering Laboratory: Idaho Falls, Idaho, EG&G-Idaho Report EGG-ER-10643, 98 p.

Knutson, C.F., 1993, Geophysical well logging and formation characterization, Radioactive Waste Management Complex area: EG&G, Idaho, unpublished report, 57 p.

Knutson, C.F., and Lee, C.B., 1991, Vadose-zone modeling using geostatistical inferences: Extended abstract presented at Denver GeoTech/Geochautauqua '91: Lakewood, Colorado, 10 p.

Knutson, C.F., Cox, D.O., Dooley, K.J., and Sisson, J.B., 1993, Characterization of low-permeability media using outcrop measurements, *in* Proceedings of the 68th Annual Technical Conference and Exhibition of the Society of Petroleum Engineers: Houston, Texas, p. 729–739.

Knutson, C.F., McCormick, K.A., Smith, R.P., Hackett, W.R., O'Brien, J.P., and Crocker, J.C., 1990, Radioactive Waste Management Complex vadose zone basalt characterization: Idaho Falls, Idaho, EG&G-Idaho Report EGG-WM-8949, 126 p.

Knutson, C., McCormick, K.A., Crocker, J.C., Glenn, M.A., and Fishel, M.L., 1992, 3D Radioactive Waste Management Complex vadose zone modeling: Idaho Falls, Idaho, EG&G-Idaho Report EGG-ERD-10246, 78 p. plus appendices.

Knutson, C.F., Dooley, K.J., and Sullivan, W.H., 1994, Geotechnical logging evaluation of the eastern Snake River Plain basalts, *in* Proceedings of the 35th Annual Logging Symposium: Tulsa, Oklahoma, Society of Professional Well Log Analysts, p. 27.

Koltermann, C.E., and Gorelick, S.M., 1996, Heterogeneity in sedimentary deposits: A review of structure-imitating, process-imitating and descriptive approaches: Water Resources Research, v. 32, p. 2617–2658.

Kupfersberger, H., and Deutsch, C.V., 1999, Methodology for integrating analog geologic data in 3-D variogram modeling: American Association of Petroleum Geologists Bulletin, v. 83, p. 1262–1278.

Lindholm, G.F., and Vaccaro, J.J., 1988, Region 2, Columbia Lava Plateau, *in*

Back, W., Rosenshein, J.S., and Seabar, P.R., eds., Hydrogeology: Boulder, Colorado, Geological Society of America, Geology of North America, v. O-2, Decade of North American Geology, p. 37–50.

Mark, L.E., 1999, Hydrologic and sedimentary characterization of surficial sedimentary facies, northern Idaho National Engineering and Environmental Laboratory, eastern Idaho [M.S. thesis]: Pocatello, Idaho, Idaho State University, 186 p.

Phillips, F.M., and Wilson, J.L., 1989, An approach to estimating hydraulic conductivity spatial correlation scales using geological characteristics: Water Resources Research, v. 25, p. 141–143.

Poeter, E.P., and McKenna, S.A., 1995, Reducing uncertainty associated with ground-water flow and transport predictions: Ground Water, v. 33, p. 899–905.

Sorenson, K.S., Wylie, A.H., and Wood, T.R., 1996, Test Area North site conceptual model and proposed hydrogeological studies, Operable Unit 1-07B: Lockheed-Martin Idaho Technologies Company Report INEL-96/0105, 140 p.

Welhan, J.A., Johannesen, C.M., Glover, J.A., Davis, L.L., and Reeves, K.S., 2002, Overview and synthesis of lithologic controls on aquifer heterogeneity in the eastern Snake River Plain, Idaho, *in* Bonnichsen, W., White, C., and McCurry, M., eds., Tectonic and magmatic development of the Snake River Plain: Idaho Geological Survey Bulletin 30 (in press).

Welhan, J.A., and Reed, M.F., 1997, Geostatistical analysis of regional hydraulic conductivity variations in the Snake River Plain aquifer, eastern Idaho: Geological Society of America Bulletin, v. 109, p. 855–868.

Welhan, J.A., and Wylie, A., 1997, Stochastic modeling of hydraulic conductivity in the Snake River Plain aquifer. 2. Evaluation of lithologic controls at the core and borehole scales, *in* Proceedings of the 32nd Engineering Geology and Geotechnical Engineering Symposium: Boise, Idaho, p. 93–108.

Welhan, J.A., Funderberg, T., Smith, R.P., and Wylie, A., 1997, Stochastic modeling of hydraulic conductivity in the Snake River Plain aquifer. 1. Hydrogeologic constraints and conceptual approach, *in* Proceedings of the 32nd Engineering Geology and Geological Engineering Symposium: Boise, Idaho, p. 75–92.

Welhan, J.A., Bukowski, J.M., Wylie, A.H., and Hankins, M.R., 1998, Directional transmissivity and spatial correlation structure of high-porosity zones in a fractured basalt aquifer: Geological Society of America Abstracts with Programs, v. 30, p. A69.

Welhan, J.A., Wylie, A.H., Hankins, M.R., Gego, E., and Johnson, G., 1999, Permeability of preferential ground water flow zones in basalt and implications for stochastic modeling: Geological Society of America Abstracts with Programs, v. 31, p. A60.

Whitley, D., 1998. A genetic algorithm tutorial: http://samizdat.mines.edu/ga_tutorial.

MANUSCRIPT ACCEPTED BY THE SOCIETY NOVEMBER 2, 2000

Geological Society of America
Special Paper 353
2002

Modeling groundwater flow and contaminant transport in the Snake River Plain aquifer: A stochastic approach

Edith L. Gégo*
Idaho Water Resources Research Institute, University of Idaho, 1776 Science Center Drive,
Idaho Falls, Idaho 83402, USA
Gary S. Johnson
Matthew R. Hankin
Department of Geology and Geological Engineering, University of Idaho, 1776 Science Center Drive,
Idaho Falls, Idaho 83402, USA
Allan H. Wylie
Idaho Water Resources Research Institute, University of Idaho, 1776 Science Center Drive,
Idaho Falls, Idaho 83402, USA
John A. Welhan
Idaho Geological Survey, Idaho State University, Pocatello, Idaho 83209, USA

ABSTRACT

The numerical framework used for solving equations in models of groundwater flow and contaminant transport requires the discretization of the flow domain into numerous cells that can be characterized by different properties, thereby enabling the modeler to simulate a heterogeneous medium. However, precisely defining the hydrological properties of each cell in extremely heterogeneous media such as the Snake River Plain aquifer in eastern Idaho is simply impossible. In this context, the use of a stochastic approach that allows quantification of the uncertainty of the model outputs resulting from our inability to accurately characterize the aquifer may be more pertinent than the use of conventional deterministic techniques. In this chapter we detail the conception of a Monte Carlo approach for the modeling of contaminant transport in the Snake River Plain aquifer. Assuming that the material forming the Snake River plain aquifer can be classified into three distinct lithologic classes and that the starkest contrasts in hydraulic conductivity, K, correspond to the boundaries between lithologies, we propose to generate equiprobable K maps by first producing alternate lithologic random fields and then converting the lithologic fields into K fields by assignment of proper K values within each lithology. An example demonstrating the use of the Monte Carlo approach to assess the uncertainty of contaminant migration in the aquifer is presented.

INTRODUCTION

The Snake River Plain aquifer is present beneath most of the eastern Snake River Plain in southeast Idaho (Fig. 1); it originates near Island Park and Ashton and terminates in the Thousand Springs area near Twin Falls, Idaho. Water flows in the aquifer from the northeast to the southwest, with occasional local deviations from this direction. The aquifer is bounded by the Basin and Range Province on the north and west and by the Snake River on the east and south.

*E-mail: psp@srv.net

Gégo, E.L., Johnson, G.S., Hankin, M.R., Wylie, A.H., and Welhan, J.A., 2002, Modeling groundwater flow and contaminant transport in the Snake River Plain aquifer: A stochastic approach, *in* Link, P.K., and Mink, L.L., eds., Geology, Hydrogeology, and Environmental Remediation: Idaho National Engineering and Environmental Laboratory, Eastern Snake River Plain, Idaho: Boulder, Colorado, Geological Society of America Special Paper 353, p. 249–261.

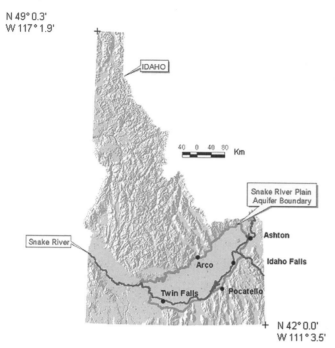

Figure 1. Location of Snake River Plain aquifer.

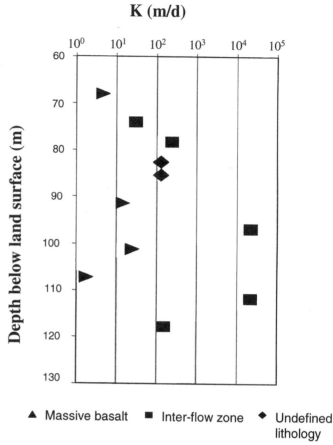

Figure 2. Vertical profile of hydraulic conductivity, K in well TAN-35.

The Snake River Plain aquifer is hosted in a complex series of thin, overlapping, lava flows, sometimes separated by lenses of sediments. According to Knutson et al. (1990), each basalt lava flow can be thought of as an assemblage of lobes extending out from a source vent, overlapping and building on the undulating topography of previous flows. The source vents responsible for the basaltic eruptions were concentrated along various volcanic fields that were active at different times during the genesis of the eastern Snake River Plain (Pierce and Morgan, 1992). Knutson et al. (1990) suggested that the average length and thickness of a lava lobe are ~265 m and 4.5 m, respectively. The ratio between a lobe length and its width varies from 2:1 to 5:1. Interpretation of borehole cores showed that, from bottom to top, a lava flow consists of a thin vesicular zone, a massive nonvesicular center, and an upper vesicular zone, densely jointed and fractured, generally thicker than the lower one. In ~50% of the cases, lava flows were deposited over a substratum of rubble, scattered in patches in the subsurface.

Due to this particular geological setting, hydraulic conductivity (K) in the aquifer can vary by several orders of magnitude over distances of a few tens or hundreds of meters horizontally and less than 1 m vertically (Lindholm and Vaccaro, 1988; Ackerman, 1991). This heterogeneity is not random. The contacts between lava flows, formed of rubble and highly vesicular basalt, often referred to as interflow zones, are thought to constitute preferential pathways for contaminant migration (Sorenson et al., 1996). Hydraulic conductivity in the interflow zones is typically >100 m/day and occasionally exceeds 10 000 m/day (Fig. 2). By opposition, K in the massive lava flow centers and

the sediment lenses inserted between basaltic masses is substantially less. K in the massive basalt typically varies between 5 and 50 m/day; K in the sediments is usually <5 m/day (Mark, 1998). The spatial organization of K, most specifically the continuity and connectivity of the interflow zones, is of extreme importance for realistic groundwater flow and contaminant transport simulations. If long and/or sufficiently interconnected, the interflow zones could convey contaminant away from their source at unsuspected high rates.

The groundwater and contaminant transport models used today rely on the numerical solution of partial differential equations quantifying the advection-dispersion processes. Some models also include diverse expressions representing other processes such as adsorption and biodegradation. Numerical solution techniques such as the finite difference and finite element techniques require discretization of the domain into numerous cells that can be characterized by different properties, permitting the user to model heterogeneous aquifers.

The Idaho National Engineering and Environmental Laboratory (INEEL), a 2300 km² site operated by the U.S. Department of Energy, is located in the northern portion of the eastern Snake River Plain, therefore above the aquifer. Groundwater contamination has been observed at different locations at the INEEL, such as beneath the Test Area North (TAN) facility, for which a CERCLA Record of Decision was issued in 1995. The U.S. Department of Energy has since encouraged the development of innovative technologies that would accelerate and improve the effectiveness of their remediation activities and would allow more accurate characterization of the subsurface.

Despite the relatively high density of wells at the INEEL, the most intensively studied area of the Snake River Plain aquifer, the complex architecture of the lava flows prevents precise spatial delineation of the high- and low-K zones. For this reason, the use of simulation models in the traditional deterministic framework in which the aquifer properties are assumed to be perfectly known may not be adequate. Experience has shown that properties of highly heterogeneous media, such as the Snake River Plain aquifer, are better represented in a model as random variables (Neuman, 1982; Freeze, 1975). When some model inputs are random variables, the model is said to be stochastic. The goal of stochastic techniques is to propagate the effect of an input uncertainty through a model and quantify the consequences of this uncertainty on the model outputs. The outputs of stochastic models are also random variables. Because random variables are best described by their probability distribution function (PDF) or their cumulative distribution function (CDF), most statements concerning the behavior of the modeled aquifer are expressed in probabilistic terms. The outputs or so-called answers of a stochastic model are therefore fundamentally different from those of a deterministic one. While the latter consists in a single number, the traditional output in the stochastic framework is a CDF that presents the whole range of possible values for the answer.

The most widely used stochastic approach relies on Monte Carlo simulations. Monte Carlo simulations consist of performing numerous runs of the model, with the value of the stochastic input, K in this case, modified in each run. The results of all the runs are compiled to form estimates of the CDF of the output variable. One of the reasons explaining the popularity of Monte Carlo simulations is their good performance with highly variable stochastic inputs, in contrast to other techniques, such as the moment techniques (Dettinger and Wilson, 1981; Graham and McLaughlin, 1989). Monte Carlo simulations can very practically be performed with traditional deterministic simulation models and do not necessitate any major modification of these models.

Despite their convenience, Monte Carlo simulations have been reproached for their computational burden for a long time, because they rely on multitudinous model runs. Modern computers are now remarkably faster than their predecessors were and, in most cases, computation time is not a major concern. Most important, improvements to Monte Carlo simulation techniques permit realistic estimates of output variability with fewer simulations. These improvements coincide with the development of geostatistics, the branch of statistics that allows characterization of the spatial organization structure of an attribute. Meaningful results can now be obtained with considerably

fewer alternate K fields if these K fields respect the spatial arrangement of K in the aquifer.

In this chapter we detail each of the major steps involved in the conception and development of a Monte Carlo approach applicable to the modeling of contaminant transport in the Snake River Plain aquifer. Using the groundwater flow model Modflow (McDonald and Harbaugh, 1988) and the particle tracker Modpath (Pollock, 1994), we show that, even in the most intensively studied area in the Snake River Plain aquifer, our current knowledge of the aquifer does not allow precise estimation of solute transport times, proving therefore the value of a stochastic approach when modeling transport in the Snake River Plain aquifer.

CONCEPTION OF A STOCHASTIC APPROACH APPLICABLE TO THE SNAKE RIVER PLAIN AQUIFER

Monte Carlo simulations require the generation of multiple equally acceptable K fields, also called realizations, and the import of these fields in the chosen flow and transport algorithms. By sorting of the model results produced by all K fields, one can estimate the variability of the model outputs resulting from our imprecise knowledge of K structure in the aquifer.

Variations of K in the aquifer are not random, but rather highly structured. While the massive basalts that constitute the bulk of the aquifer and the sediment lenses are moderately conductive, the interflow zones present between flows constitute strips of highly permeable material (Welhan et al., this volume). In such a structured medium, composed of various lithologies with very distinctive properties, simulating possible spatial organizations of the lithologies before assigning them K values has been shown to be the best technique available for preservation of the preferential pathways (Scheibe and Murray, 1998; Langsholt et al., 1998).

Welhan (1997) showed that proper interpretation of various geophysical signals allows identification of the three lithologies of interest, i.e., of fine-grained sediments, massive basalt, and interflow zones, in well boreholes. Welhan (1997) also showed that at the TAN facility of the INEEL, the borehole density was sufficient to allow inference of the spatial structure of these three lithologies. There are 31 wells spread over an area of <6 km^2 at TAN: 12 of them, located along the path of the contamination plume, are confined in an area <0.2 km^2.

Assuming that the starkest contrasts of K are located at the boundaries between lithologies and that relative uniformity exists within each lithology, we propose to generate alternate K fields in two steps: major heterogeneities are first reproduced through the generation of lithologic fields, and secondary heterogeneities are then reproduced by assigning proper K values to each lithologic category.

The following sections provide a detailed description of each step involved in the generation of stochastic lithologic fields and their conversion into K fields. The data available to implement the proposed approach, most of which originate from the TAN facility of the INEEL, are also described.

GENERATION OF STOCHASTIC LITHOLOGIC FIELDS

Data available to assess the lithologic spatial architecture of the Snake River Plain aquifer

Various geophysical logs, such as density, neutron, borehole diameter, and natural gamma, are available to evaluate the lithologic composition of boreholes at TAN. Following the methodology suggested by Welhan (1997), the occurrence and location of each lithology in well boreholes can be inferred as follows.

1. Sediments are first identified as peaks on the natural gamma radiation logs. The naturally emitting isotope ^{40}K, which is more abundant in the sediments than in the surrounding basaltic material, causes these peaks (Fig. 3).

2. The interflow zones are then identified on the neutron and density logs. Because they are typically less compact and contain more water than the adjacent massive basalt, the inter-

flow zones appear as troughs on these logs. Borehole diameter logs allow confirmation of the presence of a rubble zone between flows (Fig. 4).

3. The material not classified as either interflow zone or fined-grained sediments is assumed to be massive basalt.

After interpretation of thousands of feet of logs pertaining to the zone between 60 m (water-table level) and 125 m below the land surface, the volumetric proportions of fine-grained sediments, interflow zones and massive basalt were estimated as 4%, 12%, and 84%, respectively. The sediments in this portion of the aquifer form an extensive bed at about ~120 m below the surface and are quite scarce above that level. The thickness of massive basalt blocks appears to vary between <2 m and ~20 m with a mean and a median of 6.3 m and 4.5 m, respectively. Figure 5 shows the distribution of the massive basalt and interflow zone thickness in the studied area.

Assessment of the lithologic spatial architecture in the Snake River Plain aquifer

Following interpretation of geophysical logs, indicator transforms are used for conversion of borehole lithologic classifications into a quantitative framework, better suited for assessing the spatial characteristics of each lithology. Indicator data sets containing the spatial coordinates of all locations at which information is available (well locations) and a categorical indicator value are created for each lithology. The indicator value is set to 1 if the lithology observed at a location corresponds to the lithology considered in the data set, or to 0 otherwise. The main data set created for this study utilizes a discretization interval of 0.3 m and summarizes the lithologic composition of 22 boreholes at TAN. In parallel, Welhan et al. (this volume) created other data sets that pertain to more wells (36 wells) but use a coarser discretization interval (0.6 m). For creation of their data sets, Welhan et al. (this volume) assumed that the thickness of the interflow zones is constant and equal to 1.2 m, 1.8 m, or 2.4 m. This artificial augmentation of the actual interflow zones thickness was performed in order to improve the horizontal correlation structure of the interflow zones. Interflow zones are thin structures, having irregular topography that follows that of the lava lobe hosting them. A given interflow zone present at different depths in different boreholes could be understood as nonconnected elements. By enhancing the thickness of the interflow zone, one may increase the likelihood of correctly assessing their true continuity.

The geostatistical function used to measure the spatial correlation of each lithology is the experimental semivariogram (de Marsily, 1986; Journel and Huijbregts, 1978), defined as:

$$\gamma^*(h) = \frac{1}{2N} \sum_{i=1}^{N} \left(i(x_i) - i(x_i + h) \right)^2$$

where * is the semivariogram value at lag h, h is the lag or the

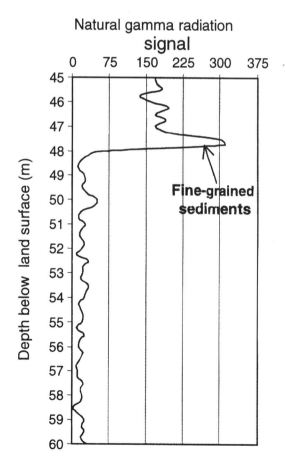

Figure 3. Location of fine-grained sediments based on natural gamma radiation probe signal.

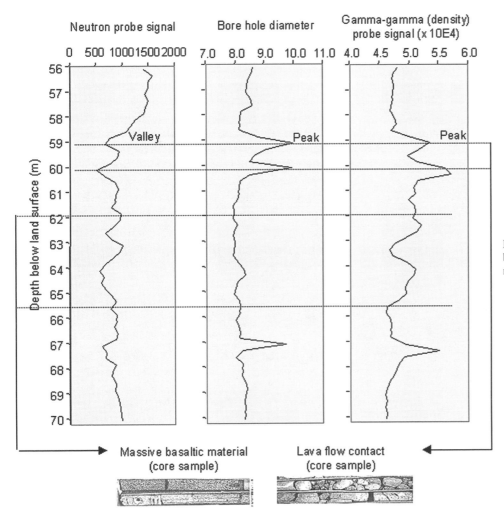

Figure 4. Location of interflow zones based on density and neutron probe signals and caliper logs.

separation distance between measured values, $i(x_i)$ and $i(x_i + h)$ are indicator values at locations x_i and $x_i + h$, and N is the number of pairs separated by the lag distance h.

Semivariograms relate the variability between points as a function of the distance or lag that separates them. Semivariogram values usually increase as the lag distance increases until they reach a maximum corresponding to the variance of the population. The lag at which the semivariogram equals the variance is called the range of influence. Beyond the range of influence, spatial correlation between points is no longer discernable.

The algorithm Gamv3 (Deutsch and Journel, 1992) is used to calculate experimental semivariograms. By allowing the user to specify data search options (such as the horizontal and vertical lag spacing and tolerance, the direction of search and tolerance, and the maximum band width of the search window), Gamv3 can compute semivariograms corresponding to specific directions (within some tolerance values), therefore allowing characterization of anisotropic processes. After trial and error, the best search strategy was identified as a horizontal lag spacing of 30.5 m (with a tolerance of 15.8 m), a dip of 0°, and a

directional tolerance of 45° for definition of the horizontal semivariograms. A lag spacing of 0.6 m with a tolerance of 0.3 m and an angular tolerance of 10° allowed best definition of the vertical semivariograms. At the TAN of the INEEL, it appears that the direction of maximum continuity for the massive basalt and the interflow zones corresponds to an azimuth of 90° and an inclination of 0° (west-east direction in a horizontal plane). An anisotropy of 1–3 was identified between the west-east and the north-south directions in the horizontal plane.

The ranges of influence in the west-east direction were estimated at about 100.6 m for the interflow zones and 121.9 m for the massive basalt (Fig. 6), values similar to those identified by Welhan et al. (this volume) working with the 0.6 m increment data sets and artificially increased interflow zones thickness. These ranges of influence are often interpreted as the average length and width of each continuous lithologic block (Langsholt et al., 1998). If so, the average lengths estimated for the blocks of massive basalt (121.9 m) along the major axis of continuity are substantially less than those reported by Knutson et al. (1990), who suggested that the average length of a lava lobe is ~265 m. The difference between the observations of

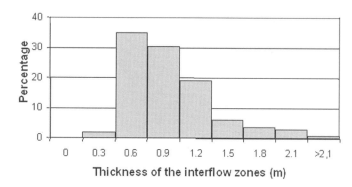

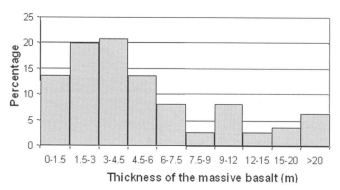

Figure 5. Distribution of thickness of interflow zones.

Knutson et al. (1990) and ours may reflect differences between subsurface morphologies at the locations studied. It should also be acknowledged that this difference might also result from our inability to accurately locate the different lithologies in the subsurface. Another possible explanation for the difference is that the algorithm that computes the semivariograms is unable to detect and reflect the continuity of thin structures that are not strictly linear, such as the interflow zones.

The vertical range of influence estimated for the massive basalt is 4.5 m, value identical to the average lava flow thickness suggested by Knutson et al. (1990). The vertical ranges of influence estimated for the interflow zones is 1.5 m (5 ft) (Fig. 7). Table 1 specifies the equation of the continuous lines adjusted through the observed semivariograms values.

Production of stochastic lithologic maps

Once the semivariograms defined for each lithology, the sequential indicator simulation algorithm Sisimpdf (Deutsch and Journel, 1992) can be used to create multiple equally possible arrangements of the lithologies in the subsurface. Each of the fields created by Sisimpdf respect the semivariograms identified for each lithology and therefore respect our understanding of the lithologic spatial architecture. In addition, the alternate lithologic fields generated by Sisimpdf also honor data that may data exist within the model domain. When using Sisimpdf, the domain simulated is divided into a finite number of cells form-

ing a regular mesh and the lithology is estimated at the center of each cell only. Therefore, fine meshes allow reproduction of more details than coarse ones. In this study, we propose a cell volume of 10.0 (L) $\times$ 3.3 (W) $\times$ 0.6(H) m³. The cell length and width correspond to 1/10 of the interflow zones ranges of influence along the two horizontal major axes of anisotropy. The cell height corresponds to 4/10 of the interflow zones range of influence in the vertical direction. The benefit of a smaller height, i.e. a height corresponding to a lesser fraction of the vertical range of influence, seems minor since most of the interflow zones on the digitized data set are at least 0.6 m thick (Fig. 5). Figure 8 shows two possible alternative interpretations of the subsurface at TAN generated by Sisimpdf.

CONVERSION OF LITHOLOGIC FIELDS INTO *K* FIELDS

Assessment of the conductive properties of each lithologic class

Using infiltrometers, Mark (1998) measured the conductive properties of surficial fine-grained sediments in the vicinity of the TAN of the INEEL. These sediments are thought to be similar to those found in the aquifer. The *K* estimates of Mark (1998) range from 0.7 m/day to 14.7 m/day with a mean and standard deviation 2.2 and 3.1 m/day, respectively.

Both laboratory and field measurements are available to characterize *K* in the massive basalt. The major source of laboratory data is probably Knutson et al. (1990): according to their findings, the median conductivity of the massive basalt is 10^{-3} m/day. However, Wylie (1993), Ackerman (1991), and Wylie et al. (1995) measured *K* in the massive basalt at the field scale. All these field-scale measurements are 1–1000 m/day. The substantial difference between laboratory and field estimates may be explained by the difference in the support volumes to which these estimates pertain; laboratory measurements are typically less (occasionally several orders of magnitude less) than the field observations obtained in the same material (Rovey and Cherkauer, 1995).

No laboratory estimate of *K* is available for the interflow zones due to the inability to collect core samples. Field estimates of *K* in the Snake River Plain aquifer are relatively abundant, but most of these estimates are not usable to characterize the interflow zones because they pertain to wells open over very large intervals, in which case the estimated *K* values characterize an amalgam of several lithologic elements.

A total of 28 field estimates of *K* only were found suitable to characterize the interflow zones: 12 of them were obtained by straddle-packer slug and pumping tests, the equipment being lowered in the well to isolate an interflow zone. The remaining 16 results (Kaminski et al., 1994) were obtained from traditional pumping tests and pertain to wells having screens <7.6 m long, and were shown, after interpretation of the corresponding

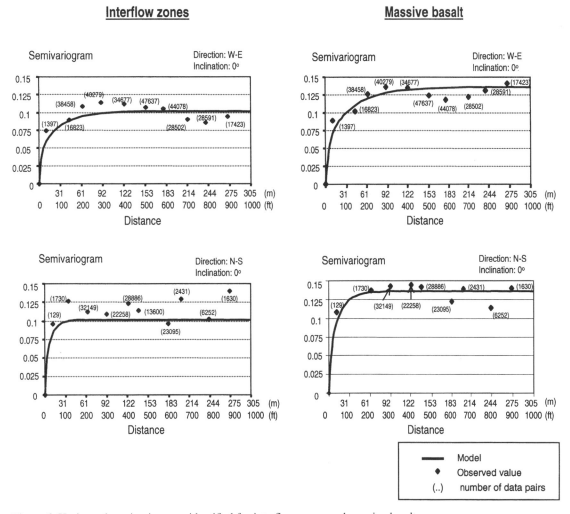

Figure 6. Horizontal semivariograms identified for interflow zones and massive basalt.

geophysical logs, to intercept a single interflow zone. For the latter measurements, it was assumed that the contribution of the massive basalt in the estimated transmissivity was negligible and that the measurement reflects a property of the interflow zone only.

The estimates of K in interflow zones range from 6 to 37 961 m/day; the mean and standard deviations are 3994 m/day and 9848 m/day, respectively (Fig. 9). The probability distribution of these estimates does not fit the traditional lognormal curve, often used to characterize K distribution (Smith and Freeze, 1979; Hess et al., 1992; Gelhar, 1986). Rather, the histogram of Figure 9 exhibits a pronounced bump on the right side of the distribution. This bump may suggest the presence of two subpopulations within the interflow zones. It may also be linked to the very limited number of data available to characterize K distribution: 28 data points may not be sufficient for this purpose. In addition, field technicians indicated that changes in the testing equipment may contribute to this abnormality.

Despite the limited number of data (28), the K estimates in interflow zones were submitted to preliminary geostatistical analysis in order to identify their spatial structure. Our results suggest that the variations of K between interflow zones are purely random. This apparent lack of spatial structure contradicts several studies, including those of Hoeksema and Kitanidis (1985), Delhomme (1979), and Welhan and Reed (1997).

A possible explanation for the absence of spatial correlation is the following. The interflow zones in the Snake River Plain aquifer do not share the same geometric characteristics (their lengths and thickness are highly variable) or the same structure (some are made of rubble whereas others appear like hollow gaps between basaltic masses). Therefore, the interflow zones do not constitute a lithology per se, but should be seen as the result of accidents that occurred during the volcanic eruptions. These accidents could have been caused by the autobrecciation of the solidified outer surface of the erupted lava flows (Geiss, D.J., 1997, personal commun.) or by new lava breaking down and transporting the material present at the surface before

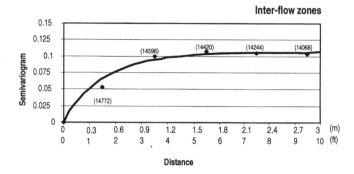

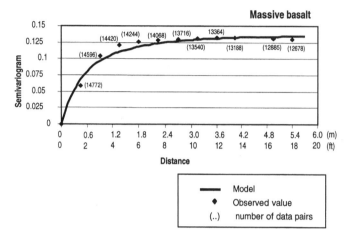

Figure 7. Vertical semivariogram identified for interflow zones and massive basalt.

the volcanic eruptions. Whatever the case, if the interflow zones are perceived as random accidents, one may conceive that their conductive properties do not conform to any spatial organization pattern.

Tomographic radar experiments between two wells at TAN allowed the estimation of the porosity of the interflow zones and the massive basalt at 0.22 and 0.09, respectively (Knoll, M., 1998, written commun.). The latter value is very compatible with the porosity estimates that Knutson et al. (1990) obtained on core samples of massive basalt. The porosity of the fine-grained sediments, the least common lithology in the aquifer, was set at 0.45, compatible with silt to clay material (Freeze and Cherry, 1979).

Conversion of the lithologic fields into K fields

Davis et al. (1993), Desbarats (1987), and Poeter and Townsend (1994) converted lithologic fields into K fields by simply assigning a uniform K value to each lithology. This simple procedure is applied for the conversion of the two least conductive lithologies. K in the sediments is set at 3 m/day, value slightly higher than the average K reported by Mark (1998). K in the massive basalt is set at 11 m/day, the average of K measurements performed in the massive basalt of well TAN-35.

Gégo et al. (2001) showed that, for small domains, characterizing the interflow zones by a single K value leads to results (in terms of volume of water transiting through the domain and particle migration times) drastically less variable than those obtained with multiple K values. Based on Gégo et al. (2001), we propose to assign varying K values to the different interflow zones in the simulated domain. The following assumptions are utilized to generate these K values: (1) K distribution in the interflow zones is assumed to be lognormal (even if the data available do not closely fit this type of distribution); (2) K variations between interflow zones are supposed random; and (3) K is assumed to be uniform within each interflow zone. The latter assumption is employed to ensure that K values are simulated over volumes similar to the support volume of the K data available, thought to correspond to a whole interflow zone or a large portion of it.

In agreement with these assumptions, the following procedure is believed to be the simplest for conversion of cells identified as interflow zones in the lithologic field into K values. Intermediate K fields pertaining to hypothetical media composed solely of interflow zones are first generated. The cell length and width in the intermediate K fields are set equal to 100.6 m and 33.5 m, respectively, i.e., to the interflow zone ranges of influence along the two horizontal principal axes of anisotropy. The cell height is set at 6.1 m or the average vertical distance between successive interflow zones. This cell size guarantees that K is almost uniform within each interflow zone. The statistics of actual K measurements in interflow zones (mean $\log_K = 2.42$, $\sigma_{\log K} = 0.98$) and the assumption of randomness are used to generate K in each cell of the intermediate K fields. K values originating from the intermediate K fields are then transferred to cells identified as interflow zones on the lithologic image or discarded otherwise.

EXAMPLE

Generating alternate equally probable maps of K in the subsurface is usually not the end point of a Monte Carlo experiment. Rather, these K maps are used as inputs in simulation models, such as groundwater flow and contaminant transport models, so that the effects of K variability on the model results can be quantified. In this context, the following example demonstrates the use of the technique described here for the generation of stochastic K fields in an exercise designed to assess the uncertainty of the time necessary for particles to migrate in the aquifer from a given location to another.

In this example we consider a 603.5 long × 201.2 wide × 36.6 high m³ domain located underneath the TAN of the INEEL, oriented in a west-east direction, i.e., approximately parallel to the lithology alignment. In order to obtain substantial conditioning of the images generated, the domain is located to include 18 boreholes for which lithologic information is available. The length and width of the domain encompass six times the interflow zone ranges of influence along the two principal

TABLE 1. PARAMETERS OF INDICATOR SEMIVARIOGRAMS

Lithology	Prop	Nugget	Sill	Horizontal west-east	Horizontal north-south	Vertical
Interflow zones*	0.12	0.0	0.045	9.1	6.1	0.6
			0.057	100.6	34.5	1.5
Massive basalt	0.84	0.0	0.060	13.7	9.1	1.5
			0.076	121.9	45.7	4.5
Fine-grained sediments†	0.04	0.0	0.044	274.3	274.3	2.7

Note: Prop is volumetric proportion of the lithology in the aquifer.
* Two nested exponential models.
† One exponential model.

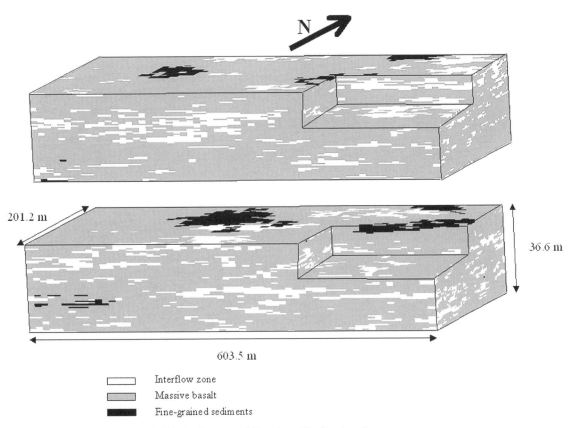

Figure 8. Example of stochastic lithologic maps of Test Area North subsurface.

axes of continuity. The domain height corresponds to six times the average thickness of a lava flow. Flow conditions are defined to generate predominantly unidirectional flow in the west-east direction. An imaginary contaminant source, shaped as a rectangular box (30.2 long × 30.2 wide × 3 high m³), is immersed in the upper portion of the aquifer, 91.4 m downgradient from the west boundary of the domain, equidistant from the north and south faces. The leakage of a conservative contaminant from the source commences at day 1 ($t = 1$ day). The question is to estimate the variability of the time at which the first particle reaches the east face of the domain.

The foremost step necessary for completion of this exercise is generating stochastic equally acceptable K fields. The ap-

proach presented in the previous section is used for this purpose. Each cell in the simulated domain is 10.0 long × 3.3 wide × 0.6 high m³. The domain is therefore composed of 216 000 cells (60 × 60 × 60).

In addition to the geostatistical software GSLib used for the generation of the K maps, the groundwater flow model Modflow (McDonald and Harbaugh, 1988) and the particle tracker Modpath (Pollock, 1994) are necessary for completion of this example. Modflow allows identification of the flow field corresponding to each K map. Modpath allows tracking of imaginary particles migrating by advection through the domain, as a function of the flow field computed by Modflow and porosity data. Modflow and Modpath require discretization of the

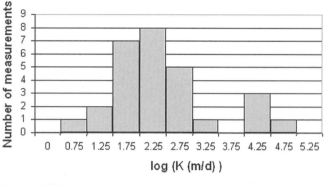

Mean *K*: 3,994 m/day   Mean log (*K*): 2.42
Standard deviation *K*: 9,848 m/day   Standard deviation log (*K*): 0.98

Figure 9. Histogram of hydraulic conductivity, *K,* in interflow zones.

domain into a finite number of regular cells. To simplify the importation of the results of GSLib into Modflow and Modpath, the same cell size is used in the three pieces of software.

Modflow is executed in steady-state conditions. Heads are fixed on the inflow (west) and outflow (east) faces of the domain in order to respect a macroscopic gradient of 0.0005. This gradient is representative of hydrological conditions at the TAN of the INEEL. A no flow condition is imposed on the four remaining faces. Porosity is supposed to be uniform within each lithology and is set at 0.45, 0.09, and 0.22 in the fine-grained sediments, the massive basalt and the interflow zone, respectively. In Modpath, a total of 10 800 particles are distributed uniformly within the source area.

If the domain were modeled as uniform (a single *K* and porosity value throughout the domain), the imaginary plume caused by the release of particles would be shaped as a rectangular box. The width and thickness of this box would correspond exactly to those of the source, because Modpath only models the advection process (diffusion and dispersion are ignored). The plume length would be proportional to the time elapsed since the particle release. When high- and low-*K* zones are differentiated through the domain, particles do not migrate at a uniform pace. Particles released in or reaching an interflow zone have the potential to migrate faster than those moving within the massive basalt or the sediments. Consequently, the plume appears as the juxtaposition of fingers of various lengths (the longest fingers corresponding to the interflow zones) rather than as a rectangular box. The time necessary for particles to reach the east face of the domain is extremely variable, depending on the lithologies encountered along their path. In this example, the output of interest is the minimum transit time from the source to the outflow face (east face) of the domain. This result (minimum arrival time) is saved after each simulation. The results of all simulations are sorted at the end of the experiment.

Estimating the number of realizations necessary to obtain accurate estimates of the CDF of the output of interest is always a critical point in a Monte Carlo experiment. The first endeavors

using this type of approach for the modeling of transport in heterogeneous aquifers suggest that hundred of model runs may be needed to achieve this goal. When data exist within the simulated domain to condition simulations, the number of model runs can be substantially reduced. In a study aimed at characterizing the uncertainty in the pathways followed by particles released from a waste disposal area in a heterogeneous aquifer, Mackay et al. (1996) indicated that 30 model runs may be sufficient. The criteria adopted by Mackay et al. (1996) to stop simulations are the steadiness of the mean and standard deviation of the output of interest. These criteria are not very strict. The contribution of a new value bears less and less weight in the computation of the output mean and standard deviation, and so the new value is an outlier.

Modifying methodology of Mackay et al. (1996) to impose stricter criteria, we propose the following strategy. At set of 30 model runs is first executed. Following these 30 simulations, we consider the number of model runs sufficient when adding the result of five new simulations to previously obtained results does not modify substantially the mean, the median, or the range of the first arrival times. Because the range of first arrival times would increase if a new extreme were encountered, our criteria for stopping simulations are tighter than those of Mackay et al. (1996).

There is no widely acceptable technique available to determine when to terminate a Monte Carlo experiment. Based on our criteria, it appears that 35 realizations may be sufficient for our problem (Fig. 10). One may argue that more model runs would lead to a more accurate estimation of the output CDF. This is certainly true. Yet the incremental value of a new result decreases as the number of model runs increases, and so does the likelihood of encountering a new extreme value. Figure 11 shows the CDF of the first arrival time estimated after 35 simulations. As a control of the methodology adopted for stopping simulations, Figure 11 also shows the CDF established after 100 simulations. Both curves are very similar between their 5th and 95th percentiles, showing that 35 simulations may be a reasonable option if the objective of the experiment is to estimate the central tendency and spreading of the first arrival time. If the extremes of the CDF are the focus on the Monte Carlo experiment, than 100 simulations are recommended.

The minimum and maximum values obtained after 35 simulations are 200 days and 2000 days, respectively. The spread of this CDF is already quite large, even in this simplified example where the only source of uncertainty is aquifer heterogeneity. In reality, other significant sources of uncertainty, such as the location of the contamination source, the direction of the hydraulic head gradient, the nature of the contaminant release, and the location of the wells where contamination is monitored, can only worsen the situation. This example shows that our current knowledge of *K* distribution in the Snake River Plain aquifer does not allow precise estimation of contaminant migration in the aquifer, and therefore stochastic simulations are a very valuable option when modeling contaminant transport.

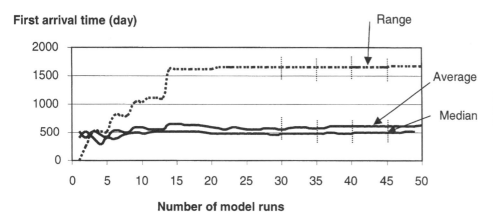

First arrival time (day)

Number of model runs

Figure 10. Evolution of mean, median, and range of first arrival time as function of number of simulations performed.

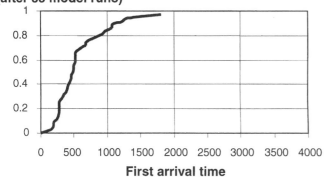

**Estimated CDF
(after 35 model runs)**

First arrival time

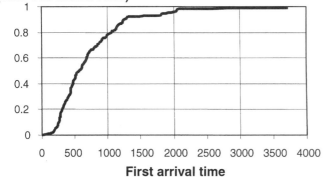

**Estimated CDF
(after 100 model runs)**

First arrival time

Figure 11. Histograms of first arrival time after 35 model runs and after 100 model runs. CDF is cumulative distribution function.

SUMMARY AND DISCUSSION

Hydraulic conductivity (K) in the Snake River Plain aquifer can differ by as much as five orders of magnitude over distances of less than 1 m vertically and tens or hundreds of meters horizontally. K variations are not random, but depend on the spatial arrangement of three distinct lithologies. From least to most conductive, these lithologies are fine-grained sediments, massive basalt, and interflow zones. Hydraulic conductivity in the interflow zones largely exceeds those in the massive basalt and the fine-grained sediments. Interflow zones constitute preferential pathways for contaminant migration.

In numerous circumstances, transport in the aquifer was enigmatic. At the TAN of the INEEL, for example, the contamination plume first extends in the east direction, then abruptly turns toward the south. At another facility of the INEEL where pump and treat operations are deployed to remediate a CCl_4 plume, a previously contaminated monitoring well is now clean, while the CCl_4 concentration in the surrounding wells is still increasing. All these unconventional plume shapes might be explained by the spatial organization of the interflow zones in the aquifer. Due to the complex overlapping nature of the lava flows, precisely delineating these interflow zones and assessing their connectivity is an impossible task, suggesting the appropriateness of a stochastic approach for modeling transport in the aquifer.

In this study we propose a technique for the generation of multiple equally acceptable maps of K in the aquifer. Integrated in a Monte Carlo experiment, these K maps allow quantification of the uncertainty of transport model outputs resulting from K variability. The conception and development of the proposed methodology rely on existing information (e.g., geologic and hydrologic) and on new data gathered throughout this study. However, we were unable to identify transport data, such as breakthrough curves, that would allow us to check the plausibility of our results and validate our technique. The preliminary results presented here should therefore be considered with caution.

Because lithologic boundaries are the major factor causing K variability, for production of stochastic K fields we propose first simulating the spatial lithologic architecture in the aquifer and then assigning appropriate K values within each lithology. This two-step approach has been shown to be the most appropriate for preservation of strips of high conductivity (Scheibe and Murray, 1998).

Geophysical signals are interpreted in order to assess the occurrence and location of each lithologic class in bore holes. The radius of investigation of the geophysical probes used may be as much as 0.6 m (Keys and MacCary, 1971), in which case some thin, pinched interflow zones may not be detected, therefore biasing the results of this study. The geometry and continuity of each lithology is assessed by identification of semivariograms. Because the interflow zones are thin and their topography is very irregular, semivariograms may not allow correct assessment of their physical characteristics and their connectivity. Tracer tests are probably the only means available to assess the interflow zones connectivity. However, the aquifer is very productive and dilution often prevents detection of the tracer at modest distance (>50 m) from the source.

A sequential indicator simulation program (Sisimpdf in GSLib; Deutsch and Journel, 1992) is used for the generation of multiple lithologic maps. The algorithm Sisimpdf ensures that each map created respects the spatial characteristics of each lithology and retains existing data in the simulated domain. This algorithm, however, does not preserve juxtapositional relationships (Carle and Fogg, 1996), i.e., the natural lithologic sequence in the aquifer. When many data are available for conditioning the images generated, the natural lithologic sequence in these images is probably faithfully reproduced. Respecting juxtapositional relationships could greatly improve the quality of the random fields generated when less conditioning information is available.

The conversion of the lithologic fields into K fields is performed by assigning random K estimates to the interflow zones while considering K to be constant within the other two lithologies (massive basalt and fine-grained sediments), which are substantially less conductive. The constant values chosen to represent the massive basalt and the fine-grained sediments are compatible with existing data in the corresponding lithology but may not be optimal, in part because the support volume of these data differs from the cell size in the lithologic fields. K is assumed to be uniform within each interflow zone, but randomly distributed (log-normal distribution) between interflow zones. The randomness assumption is used because no spatial correlation could be detected in the data available. This assumption is plausible if the interflow zones are seen as the result of random accidents that occurred during the lava deposition process, yet contradicts various studies. Additional data are needed for each lithology to refine the approach proposed.

This chapter contains an example presenting the result of a Monte Carlo experiment designed to assess the uncertainty of the time necessary for particles to migrate from a given location in the aquifer to another. This example uses the approach proposed for the generation of K fields in conjunction with the groundwater flow model Modflow (McDonald and Harbaugh, 1988) and the particle tracker Modpath (Pollock, 1994). The results of this example suggest that, when studying local contamination, our knowledge of the aquifer hydraulic properties is still too limited to allow precise estimation of the evolution of a contamination plume. Stochastic simulations therefore look like a reasonable option for fine-scale modeling of contaminant transport in the aquifer.

ACKNOWLEDGMENTS

This research was funded by a grant from the U.S. Department of Energy to the Idaho Universities Consortium, through the Idaho Water Resources Research (award DE-FG07-96ID13420).

REFERENCES CITED

Ackerman, D.J., 1991, Transmissivity of the Snake River Plain aquifer at the Idaho National Engineering Laboratory, Idaho, U.S. Geological Survey Water-Resources Investigations, Report 91–4058, DOE/ID-22097, 35 p.

Carle, S.E., and Fogg, G.E., 1996, Transition probability-based indicator geostatistics: Mathematical Geology, v. 28, no. 4, p. 453–466.

Davis, J.M., Lohmann, R.C., Philips, F.M., Wilson J.L., and David, W.L.,1993, Architecture of the Sierra Ladrones formation, central New Mexico: Depositional controls on the permeability correlation structure: Geological Society of America Bulletin, v. 105, p. 998–1007.

de Marsily, G., 1986, Quantitative hydrology: Groundwater hydrology for engineers: Orlando, Florida, Academic Press, 440 p.

Delhomme, J.P., 1979, Spatial variability and uncertainty in groundwater flow parameters: A geostatistical approach: Water Resources Research, v. 15, no. 2, p. 269–280.

Desbarats, A.J., 1987, Numerical estimation of effective permeability in sandshale formations: Water Resources Research, v. 23, no. 2, p. 273–286.

Dettinger, M.D., and Wilson, J.L., 1981, First order analysis of uncertainty in numerical models of groundwater flow. 1. Mathematical development: Water Resources Research, v. 17, no. 1, p.149–161.

Deutsch, C.V., and Journel, A.G., eds., 1992 GSLIB: Geostatistical software library and user's guide: Oxford, Oxford University Press, 340 p.

Freeze, R.A., 1975, Stochastic-conceptual analysis of one-dimensional groundwater flow in non-uniform homogeneous media: Water Resources Research, v. 11, no. 5, p. 725–741.

Freeze, R.A., and Cherry, J.A., 1979, Groundwater: Englewood Cliffs, New Jersey, Prentice-Hall, 604 p.

Gégo, E.L., Johnson, G.S., and Hankin, M.R., 2001, An evaluation of methodologies for the generation of stochastic hydraulic conductivity fields in highly heterogeneous aquifers: Stochastic Environmental Research and Risk Assessment, v. 15, p. 47–64.

Gelhar, L.W., 1986, Stochastic subsurface hydrology from theory to applications: Water Resources Research, v. 22, no. 9, p.135S–145S.

Graham, G., and McLaughlin, D., 1989, Stochastic analysis of nonstationary subsurface solute transport. 1. Unconditional moments: Water Resources Research, v. 25, no. 2, p. 215–232.

Hess, K.M., Wolf, S.H., and Celia, M.A., 1992, Large-scale natural gradient tracer test in sand and gravel, Cape Cod, Massachusetts. 3. Hydraulic conductivity variability and calculated macrodispersivities: Water Resources Research, v. 28, no. 8, p. 2011–2027.

Hoeksema, R.J., and Kitanidis, P.K., 1985, Analysis of the spatial structure of properties of selected aquifers: Water Resources Research, v. 21, no. 4, p. 563–572.

Journel, A.G., and Huijbregts, J., 1978, Mining geostatistics: New York, Academic Press, 600 p.

Kaminski, J.F., Keck, K.N., Schafer-Perini, A.L., Hersley, C.F., Smith, R.P., Stomberg, G.J. and Wylie, A.H., 1994, Remedial investigation final report with addenda for the Test Area North groundwater operable unit 1-07B at the Idaho National Engineering Laboratory: Idaho Falls, Idaho, Department of Energy, EGG-ER-10643-V2, variously paged.

Keys, W.S., and MacCary, L.M., 1971, Application of borehole geophysics to water-resources investigations: U.S. Geological Survey Techniques of Water Resources Investigations, Book 2, Chapter E1: Denver, Colorado, U.S. Geological Survey, 126 p.

Knutson, C.F., McCormick, K.A., Smith, R.P., Hackett, W.R., O'Brien, J.P., and Crocker, J.A., 1990, FY 89 Report: Radioactive Waste Management Complex vadose zone basalt characterization: Idaho Falls, Idaho, EG&G Idaho, Informal Report No. EGG-WM-8949, 426 p.

Langsholt, E., Kitterrod, N.-O., and Gottschalk, L., 1998, Development of three-dimensional hydrostratigraphical architecture of the unsaturated zone based on soft and hard data: Ground Water, v. 36, no. 1, p. 104–111.

Lindholm, G. F., 1993, Summary of the Snake River Plain aquifer-system analysis in Idaho and eastern Oregon: U.S. Geological Survey Open-File Report 91-98, 62 p.

Lindholm, G.F., and Vaccaro, J.J., 1988, Region 2, Columbia lava plateau, *in* Back, W., Rosenhein, J.S., and Seabar, P.R., eds., Hydrogeology: Boulder, Colorado, Geological Society of America, Geology of North America, v. O-2, Decade of North America Geology, p. 37–50.

Mackay, R., Cooper, T.A., Metcalfe, A.V., and O'Connell, P.E., 1996, Contaminant transport in heterogeneous porous media: A case study. 2. Stochastic modeling: Journal of Hydrology, v. 175, p. 429–452.

Mark, L., 1998, Hydrologic and sedimentologic characterization of surficial sedimentary facies, Idaho National Engineering and Environmental Laboratory, eastern Idaho [M.S. thesis]: Pocatello, Idaho, Idaho State University, 186 p.

McDonald, M.G., and Harbaugh, A.W., 1988, A modular three-dimensional finite-difference ground-water flow model: U.S. Geological Survey Techniques of Water-Resources Investigations, Book 6, Chapter A1, Report number TWI 06081: Denver, Colorado, U.S. Geological Survey, p. 586.

Neuman, S.P., 1982, Statistical characterization of aquifer heterogeneities: An overview, Geological Society of America Special Paper 189, p. 89–102.

Pierce, K.L., and Morgan, L.A., 1992, The track of the Yellowstone hotspot: Volcanism, faulting and uplift: Geological Society of America Memoir 179, p. 1–53.

Poeter, E., and Townsend, P., 1994, Assessment of critical flow path for improved remediation management: Ground Water, v. 32, no. 3, p. 439–447.

Pollock, D.W., 1994, Source code and ancillary data files for the MODPATH particle tracking package of the ground-water flow model MODFLOW, Version 3, Release 1: U.S. Geological Survey Open-File Report 94-463, on disk.

Rovey, C.W., and Cherkauer, D.S., 1995, Scale dependency of hydraulic conductivity measurements: Ground Water, v. 33, no. 5, p. 769–779.

Schiebe T.D., and Murray, C.J., 1998, Simulation of geological patterns: A comparison of stochastic simulation techniques for ground water transport modeling, *in* Fraser, G.S., and Davis, J.M., eds., Hydrogeologic models of sedimentary aquifers: SEPM (Society for Sedimentary Geology), Concepts in Hydrogeology and Environmental Geology and Environmental Geology, v. 1, p. 107–117.

Smith, L., and Freeze, R.A., 1979, Stochastic analysis of steady state groundwater flow in a bounded domain. 2. Two-dimensional simulations: Water Resources Research, v. 15, no. 6, p. 1543–1557.

Sorenson, K.S., Wylie, A.H., and Wood, T.R., 1996, Test Area North site conceptual model and proposed hydrogeologic studies, operable unit 1-07B: Lockeed-Martin Idaho Technologies Company Report INEL-96/0105, variously paged.

Welhan, J.A., 1997, Stochastic modeling of hydraulic conductivity in the Snake River Plain aquifer. 1. Hydrogeologic constraints and conceptual approach, *in* Sharma, S., and Hardcastle, J.H., eds., Proceedings of the 32nd Engineering Geology and Geotechnical Engineering Symposium, Moscow, Idaho: Pocotello, Idaho, Idaho State University College of Engineering, p. 75–92.

Welhan, J.A., and Reed, M.F., 1997, Geostatistical analysis of the regional hydraulic conductivity variations in the Snake River Plain aquifer, eastern Idaho: Geological Society of America Bulletin, v. 109, p. 855–868.

Wylie, A.H., 1993, Appendix H: Test Area North RI aquifer properties testing results, *in* Remedial investigation final report with addenda for the Test Area North groundwater operable unit 1-o7B at the Idaho National Engineering and Environmental Laboratory: Report EGG-ER-10643, EG&G Idaho, variously paged.

Wylie, A.H., McCarthy, J.M., Neher, E., and Higgs, B.D., 1995, Large-scale aquifer pumping tests results: Report INEEL-95/012, EG&G Idaho, variously paged.

MANUSCRIPT ACCEPTED BY THE SOCIETY NOVEMBER 2, 2000

Geological Society of America
Special Paper 353
2002

Recirculating tracer test in fractured basalt

Robin E. Nimmer
Dale R. Ralston
Department of Geosciences, University of Idaho, Moscow, Idaho, USA
Allan H. Wylie
Gary S. Johnson
Department of Geosciences, University of Idaho, Idaho Falls, Idaho, USA

ABSTRACT

Groundwater contamination in a fractured rock aquifer is very difficult to remediate. An understanding of the transport mechanisms that occur in the fractures and the rock matrix is important in order to implement cleanup strategies such as in situ bioremediation. The Snake River Plain aquifer underlying the Test Area North site at the Idaho National Engineering and Environmental Laboratory is hosted in fractured basalt and is contaminated with trichloroethylene, its degradation products, and other compounds. The objective of this study was to examine and compare responses of various tracers to gain an understanding of transport characteristics in the aquifer and assist in the cleanup of the TAN site.

A multiple-well, recirculating, radially convergent tracer test was conducted at the University of Idaho Ground Water Research site, on Columbia River basalt, in Moscow, Idaho. The test focused on a horizontal fracture zone (E fracture) formed as the result of cooling lava. Conservative or near-conservative tracers and particulate tracers included fluorescein, iodide, veratryl alcohol (VA), polystyrene noncarboxylated fluorescent microbeads (6 µm) and *Bacillus thermoruber* spores. The tracer test results indicate that transport is governed by preferential pathways and is affected by aquifer heterogeneity, advection, dispersion, molecular diffusion, channeling, transport in different channels, and borehole storage. In this short duration experiment, transport was not significantly affected by sorption or matrix diffusion. The particulate tracers may also be affected by density and filtration, as seen by the very low tracer recovery.

There are many implications from the outcome of this tracer test regarding the use of various tracers. Fluorescein and VA exhibited a breakthrough curve similar to iodide, and consequently are considered conservative tracers in this environment. The early arrival of polystyrene particle tracer relative to dissolved tracers provides an insight into the preferential flow paths of the system. *Bacillus thermoruber* spores were not found in any of the wells used in the test; this could be due to filtration, sorption, density affects, predation, or the result of laboratory analysis problems. Without further testing, these spores are not recommended as groundwater tracers.

INTRODUCTION

Groundwater contamination is a significant problem in the United States and around the world. More than 300 000 sites are contaminated in the United States, some of which overlie fractured rock (National Research Council, 1994). Examples include the Oak Ridge National Laboratory in Tennessee that overlies fractured shale (Stafford et al., 1998), the Hanford Nuclear Reservation in southeastern Washington State (DeJue, 1979), and the Idaho National Engineering and Environmental

Nimmer, R.E., Ralston, D.R., Wylie, A.H., and Johnson, G.S., 2002, Recirculating tracer test in fractured basalt, *in* Link, P.K., and Mink, L.L., eds., Geology, Hydrogeology, and Environmental Remediation: Idaho National Engineering and Environmental Laboratory, Eastern Snake River Plain, Idaho: Boulder, Colorado, Geological Society of America Special Paper 353, p. 263–277.

Laboratory (INEEL) in southeastern Idaho (Sorenson et al., 1996). The latter two overlie fractured basalt. In layered basalts, contaminants are transported not only through the fractures, but also between lava flow contacts in the sediment-filled rubble zones or interbeds (Welhan et al., this volume, chapter 9).

In 1987, trichloroethylene (TCE) and other compounds were discovered in the groundwater at the Test Area North (TAN) site at the INEEL (Sorenson et al., 1996). These contaminants were disposed of in an injection well in the Snake River Plain aquifer several decades earlier. Site characterization and cleanup efforts have been underway and continue to better characterize the plume and the site geology as well as to develop remediation techniques.

Groundwater cleanup at fractured rock sites is particularly difficult due to the dual porosity nature of the material. Dual porosity encompasses the porosity of the fractures and the porosity of the rock matrix. Block or matrix porosity includes both primary porosity and secondary porosity related to microfractures and nature of the rock (i.e., a weakly cemented shale will have a high primary porosity). Most of the flow and transport occur in the fractures, yet most of the storage exists in the rock matrix; this is true of fractured limestones, siltstone, shale, basalt, some clays, and clayey tills (Freeze and Cherry, 1979). This significantly hampers cleanup efforts and reduces the success rate of treatment methods. However, during brief contaminant exposure, matrix storage is less of a concern.

Cleanup of these sites is dependent on understanding the fracture network, bedding characteristics, and interchange of the matrix and fractures. A variety of cleanup methods exist today, among them in situ bioremediation that is an increasingly popular technology receiving much attention (National Research Council, 1994). This method may be a cost-effective approach for the removal of a high percentage of the contaminant (National Research Council, 1994).

Tracer experiments are a useful tool in acquiring information on the migration of contaminants in fractured rock systems. The aquifer porosity, groundwater velocity, and dispersion and diffusion values can be determined from conservative dissolved tracers. In fractured rock flow systems, processes such as channeling and matrix diffusion can also be investigated. The knowledge gained from these tests can be used in concert with other investigations, such as microbial analysis and geophysical methods, to determine the optimal remediation strategy.

A tracer test was conducted at the University of Idaho Groundwater Research Site (UIGRS) in Moscow, Idaho, in four wells located within the same horizontal fracture zone in basalt. The fracture is the result of cooling lava and is found at a depth of approximately one-third the total basalt thickness. The interflow zones differ from cooling fractures in the massive flow interior by the rubbly and vesicular flow top mixed with sediment filled interbeds (Welhan et al., this volume, chapter 9). The tracer test was conducted using fluorescein, iodide, veratryl alcohol (VA), spores, and microbeads as tracers. The Columbia

River Basalt, in which the experiment was conducted, has structural features (i.e., vesicular flow top, massive lava flow interior, and vesicular flow bottom) similar to those of the Snake River Plain basalt, which underlies the INEEL.

The focus of this study was twofold. The first objective was to develop a methodology for tracer experiments that can be conducted at the INEEL and to evaluate a variety of conservative tracers that can be used in the same experiment. The second objective was to gain an understanding of how dissolved and particulate tracers migrate in fractured basalt.

BACKGROUND

Transport processes

Most fractures do not have smooth parallel surfaces but rather have rough surfaces, causing the fracture walls to have partial contact, which physically and chemically affect solute and particulate transport (Neretnieks, 1993). The degree of fracture interconnectedness partially controls the extent to which transport occurs. Thompson and Brown (1991) suggested that the orientation of fracture roughness is more important than the degree of roughness when determining transport characteristics. The average solute transport rates increase significantly if the long axis of the roughness is oriented parallel to the flow direction. This becomes more apparent as the surface contact area increases. Solute transport slows if the surface roughness is orientated normal to the direction of flow.

Zones of mobile and stagnant fluid are formed within a single fracture due to tortuous flow over rough surfaces (Turner, 1958; Aris, 1959; Coats and Smith, 1964; Raven et al., 1988). The mobile zone contains an inertial core that carries most of the tracer at higher flow velocities. Stagnant zones are located along irregularities in the fracture walls where vortices and eddies resulting from nonlaminar flow form. Solute in the mobile zone diffuses into the stagnant zone and is stored there until the concentration of solute in the mobile zone decreases and is then released back into the mobile zone.

Channeling or channelized transport is a major mechanism controlling solute and particulate transport in fractured media as well as rubble zones. Fluid in a single fracture moving through preferential tortuous channels causes channeling (Tsang and Tsang, 1987; Tsang et al., 1988). A channel is a narrow pathway of least resistance; there may be multiple channels within a single fracture that vary their position in space and time based upon the direction and rate of flow. Moreno et al. (1988), Tsang et al. (1991), and the National Research Council (1996) attributed channeling to nonuniform velocity distributions caused by highly variable apertures in a fracture. Raven et al. (1988) and Abelin et al. (1991b) suggested that this process is caused by fracture roughness in addition to the contact area effects of the apertures. Channeling effects grow for a fracture with <10 percent contact area (Thompson, 1991). Tsang and Tsang (1989) stated that channeling occurs in fractures with highly variable apertures and originates due to variable permeability within the porous medium. Dverstorp et al. (1992) determined that the degree of channeling is controlled by the permeability distribution and the geometry of the fractures.

Channeling causes regions of higher velocities within a fracture (Moreno et al., 1988), most of the flow occurring on only a portion of the fracture surface (Cliffe et al., 1993). Channelized solute velocities are two to three times greater than the average velocity. These higher velocity zones may limit the contact between the solute and rock surface, thus limiting many chemical and physical reactions (Moreno et al., 1988; Tsang and Tsang, 1989).

Retardation results in the solute moving at a slower rate through an aquifer system than the average groundwater velocity. The process is sensitive to fracture width, fracture porosity, matrix porosity, diffusion, and sorption (Novak, 1993). The two largest retardation processes associated with transport in fractured rock are matrix diffusion and sorption (Moreno and Neretnieks, 1993; Moreno et al., 1985).

Both molecular and matrix diffusion retard transport movement in fractured rock. Diffusion is the process by which "ionic or molecular constituents move under the influence of their kinetic activity in the direction of their concentration gradient" (Freeze and Cheery, 1979, p. 103). Molecular diffusion thus will transport solute in the mobile zone into water stored in stagnant zones in rough-walled fractures (Raven et al., 1988; Abelin et al., 1991a) and into dead-end fractures (National Research Council, 1996) when the concentrations are highest in the mobile zone. Matrix diffusion can transfer the solute in the macrofractures to microfractures, pore spaces, and joints in the rock matrix (Neretnieks, 1980). When the concentration in the pathway becomes less than the surrounding area, the process is reversed, causing significant tailing (i.e., a higher than normal concentration on the downside of the breakthrough curve) of breakthrough curves during tracer tests (Novakowski et al., 1985).

Matrix diffusion may be a major retardation factor for solute movement in fractured rock (Grisak and Pickens, 1980; Neretnieks, 1980, 1983; Moreno and Neretnieks, 1993; Novak, 1993; Novakowski and Lapcevic, 1994; National Research Council, 1996). It may be orders of magnitude more effective in slowing plume migration when compared to retardation by fracture surface reactions alone (Neretnieks, 1980). However, the magnitude depends upon the amount of rock matrix available to the solute (Neretnieks, 1980) and the length of contact time between the solute and matrix blocks (National Research Council, 1996). Abelin et al. (1991a) concluded that fractures with large surface areas have a greater rate of matrix diffusion and induce a larger matrix diffusion value than smaller surface areas. Even at low block porosities, Abelin et al. (1991b) and Neretnieks et al. (1982) found that matrix diffusion retards solute transport; however, Malowzewski and Zuber (1992) stated that matrix diffusion is insignificant at high transport velocities when low block porosities exist.

Transport characteristics of particulates

Knowledge of particulate (microbial and particle) transport is significant for use in enhanced in situ bioremediation. Bacteria and spores may be used to degrade contaminants and specific particles may be used to deliver organisms and nutrients to the contaminant plume in the subsurface (Brown, 1998). In addition, chemicals may piggyback on particles, enhancing the mobility of the contaminant (McCarthy and Zachara, 1989). Particulate transport is influenced by similar processes affecting dissolved constituents (i.e., advection, dispersion, and channeling), but is governed more specifically by preferential pathways, filtration, sorption, and density effects.

In forced gradient tests, microorganisms and particles travel along high velocity preferential pathways (i.e., zones of relatively high hydraulic conductivity). Particulates are transported only through fractures that can pass their size. These may constitute only a small percentage of the total fracture network; therefore, the effective porosity is less, which in turn increases the average velocity for a given flux. This phenomenon is referred to as the porosity exclusion effect and was described by Enfield and Bengtsson (1988). Dissolved tracers also follow these preferential pathways but follow other pathways with slower velocities as well. Consequently, the first arrival for the particulate tracers will be prior to the center of mass of the conservative tracer (Petrich, 1995). The initial rise of the breakthrough curve for the dissolved tracer and the particulate tracer should occur at approximately the same time (Petrich, 1995). Table 1 lists several forced gradient tracer experiments conducted using microbial and/or particle tracers in conjunction with conservative tracers. In each experiment the breakthrough peak for the particulate tracer arrived prior to that for the conservative tracer.

Particulate transport is affected by filtration, which can decrease the aquifer permeability and may severely alter or block the transport pathway. Most microorganisms and particles are filtered as they are transported through the subsurface due to the heterogeneities of the aquifer. The smaller the pore space or fracture aperture, the fewer large-sized particles are transported. This was demonstrated in forced gradient tests by Harvey et al. (1989), Petrich (1998), and Brown (1998), who employed different sized particles during their experiments. They discovered that the particle recovery varied inversely with particle size. Cumbie and McKay (1999) found similar findings at particle sizes >0.5 μm.

Reversible and irreversible sorption causes retardation or loss of particulates during transport (Harvey and Garabedian, 1991). Reversible sorption is the equilibrium partitioning of bacteria (or other particles) between the liquid phase and the fracture walls (Hendry et al., 1997). Some bacteria produce adhesive substances on their cell walls. These bacteria anchor themselves to a porous medium and over time become difficult to remove, causing irreversible sorption (Hendry et al., 1997).

Particulate sorption and desorption are affected by several

TABLE 1. SUMMARY OF PARTICULATE TRACER WORK IN FORCED GRADIENT TESTS

Source	Conservative tracers	Microbial tracers	Particle tracers	Test environment
Wood and Ehrlich (1978)	Bromide and iodide	Yeast—*Saccharomyces cerevisiae*		Sand and gravel aquifer
Pyle and Thorpe (1981) in Davis et al. (1985)	Rhodamine WT	Bacteria—*E. coli*		?
Bales et al. (1989)	?	Virus—*bacteriophage*		Sandy soil and fractured tuff
Harvey et al. (1989)	Bromide	Bacteria	Microspheres	Sandy aquifer
Gannon et al. (1991)	Chloride	Bacteria—*Pseudomonas, Achromobacter, Bacillus, Enterobacter*		Soil
McKay et al. (1993)	Bromide	Virus—*bacteriophage*		Fractured till
Reimus et al. (1994)	Iodide		Microspheres	Fractured tuff
Petrich et al. (1998)	Bromide		Microspheres	Sandy aquifer
Pang et al. (1998)	Chloride and rhodamine WT	Spores—*Bacillus subtilis*		Alluvial gravel aquifer
Brown (1998)	Bromide	Spores—*Clostridium bifermentans*	Microspheres	Sandy aquifer

mechanisms. These mechanisms include the surface roughness of the fracture, fluid dynamics, substratum chemistry, and solute chemistry (Harvey et al., 1993). Fontes et al. (1991) discovered that sorption is affected by the ionic strength and pH of the solution on the charge density and electrostatic repulsion. Hendry et al. (1997) found that desorption may be a function of bacterial residence time on the media.

In column studies, Hendry et al. (1997) discovered that the breakthrough curve for the vegetative bacterium, *Klebsiella oxytoca,* had an attenuated peak and a substantial tail with respect to the chloride breakthrough curve. Hendry et al. (1997) attempted to exclude all other transport variables during the experiments and concluded that the attenuated peak and tail were caused by sorption.

Particulates that are denser than water tend to fall out of suspension, resulting in a large tracer loss. This is more pronounced at slower flow velocities, particularly under a natural gradient. A significant breakthrough curve tail may result from particulates that fall out of suspension but return to the flow stream by an increase in velocity. Tracer results in some natural gradient field tests suggest that the peak breakthroughs for microorganisms and particles are attenuated with respect to dissolved tracers. Harvey et al. (1989) conducted a natural gradient field tracer test using chloride and different types (i.e., noncarboxylated latex, polyacrolein, and carboxylated) and sizes of microspheres. They discovered that the bacteria-sized microsphere breakthrough peaks were retarded with respect to the chloride peak in a sandy aquifer. This result was likely related to particle settling at lower groundwater velocities and, to a smaller degree, sorption. The noncarboxylated latex spheres were the first to arrive, followed by the polyacrolein spheres (carbonyl surface groups are attached) and finally by the carboxylated latex spheres. The carboxylated latex spheres have a net charge that attracts them to other charged substances, thereby retarding the sphere transport. The authors concluded that particle size, and more important, surface characteristics of the particle, affect transport. Harvey et al. (1993) also found that bacteria and bacteria-sized carboxylated latex microspheres were retarded with respect to bromide in a natural-gradient test in a sandy aquifer.

Some authors have found that diffusion retards solute transport to a greater extent than particulate transport, despite the type of tracer test. McKay et al. (2000) conducted a natural gradient experiment in fractured shale saprolite. They found that four colloidal tracers were transported much faster than the solute tracers (rhodamine-WT, He, and Ne). This was attributed to the greater diffusion of the solute tracers. Hinsby et al. (1996) found similar results in an experiment in a column of fractured clay rich till in which they injected chloride, colloid-sized bacteriophage, and uncharged latex microspheres. They found it took a much longer time for chloride to reach steady state than it did for the particulate tracers.

Microorganism transport may be affected by a number of more microbial-specific processes; these include microbial growth, death, starvation, predation, motility, and chemotaxis. Microbial growth may increase the problems associated with filtration (i.e., reduction in permeability) and result in an increased recovery. Death, starvation, and predation of the microbes result in a tracer loss. Bacteria are either motile or nonmotile. Motility results from the flagella (long, hair-like tail) attached to the bacterium (Chapelle, 1993). Breakthrough may occur much faster for the motile bacteria caused by the forward swimming motion of the flagella. Chemotaxis may also affect microbial transport, which is the movement of an organism in relation to chemicals.

MATERIALS AND METHODS

Selection of tracers

Several different tracers were needed for these experiments because multiple tests were to be run in paired wells. Velocity and dispersivity were determined from conservative tracers and contrasted with similar determinations of particulate tracers. Biological tracers and particulate tracers were used to study how microorganisms are transported in the subsurface.

There were several criteria for selecting the tracers used in the experiments. These included the ease of use, ease of tracer solution preparation, cost of the tracer, detection limit, ease and cost of the analysis, and the level of information from available previous studies. Fluorescein, iodide, veratryl alcohol, *Bacillus thermoruber* spores, and polystyrene fluorescent microbeads (6 μm) were chosen as the tracers for the experiments at the UIGRS.

Fluorescein, a fluorescent dye, was chosen as a tracer for many reasons. Fluorescein is considered to be a fairly conservative tracer. The dye does not occur naturally in the groundwater at the UIGRS and could be used as a qualitative tracer numerous times as long as the residual concentrations remained at a low level when used in multiple tests. It is easily detectable in the field; e.g., a yellowish tint can be seen with the unaided eye above 0.1 mg/L. Fluorescein is also inexpensive.

Iodide (KI form) was chosen as the ionic tracer for several reasons. It is a conservative tracer and had a low background concentration at the site. Iodide has properties similar to commonly used bromide and thus much is known about its behavior. Bromide was not used in this experiment because it had been used in a previous experiment at the site and had an elevated background concentration. Iodide is easily prepared but has been shown to be biologically unstable and may sorb more than bromide. It is inexpensive and fairly easy to analyze.

Veratryl alcohol (3,4-dimethoxybenzlalcohol or $[CH_3O]_2C_6H_3CH_2OH$), or VA, was chosen as an organic tracer because of its desirable qualities. It is fairly soluble, should not sorb, and is inexpensive. However, veratryl alcohol oxidizes to veratraldehyde within hours and is biodegradable under normal conditions (nonacidic conditions). Analysis can be costly, time consuming, and require solvents in the analyses using a gas chromatograph; however, it is easily analyzed by the Hewlett-Packard liquid chromatograph.

Spores of *Bacillus thermoruber*, a red-pigmented thermophilic bacterium, were chosen as the biological tracer for several reasons. This spore is easily identified as dark red and/or brown colonies. They do not sporelate at the temperature of the groundwater at the UIGRS; in order for their rejuvenation they need to be heat shocked at 80°C for 10 min. Spores are easily grown in a laboratory and fairly simple to analyze; however, filtration is necessary when their numbers are small, and this process is very time consuming.

Fluoresbrite plain microspheres (fluorescent labeled polystyrene, 2.5 percent solids, latex microbeads; 6 μm) were chosen as the particle tracer. They are relatively new, and previous studies have shown them to be successful groundwater tracers. They are easily identifiable under an electron microscope by their spherical shape and bright fluorescent color. The 6-μm-diameter size was selected to compliment the spores (~3 μm) in the particle transport experiments. (Polysciences in Warrington, Pennsylvania, supplied the microbeads.)

METHOD OF ANALYSIS

Each tracer was analyzed using a separate method and all had different detection limits. Fluorescein samples were analyzed by the F4500 Hitachi fluorescence spectrophotometer

2504060-04, which had a detection limit of 0.5 μg/L. Iodide analysis was conducted using an Orion combination iodide selective electrode. The detection limit for iodide was 0.5 mg/L. Veratryl alcohol was analyzed by the Hewlett Packard liquid chromatograph. The VA detection limit was ~1.0 mg/L. The spores were analyzed by plate counts in a laboratory. It is difficult to determine a minimum detection limit for spores because of their particulate nature. The microbeads were filtered from 100 mL of sample and the microbeads on the entire filter were counted under a microscope with a magnification of 10×. In this experiment the detection limit is one bead per 100 mL of sample. A more detailed discussion of the analyses can be found in Nimmer (1998).

DESCRIPTION OF TEST FIELD SITE

The field work for this study was conducted at the University of Idaho Groundwater Research Site located on the western edge of campus in Moscow, Idaho. This site was chosen because of the ease of access to the university and the setup of the wells along a transect within a single fracture zone. The location and plan view map of the wells at the UIGRS is shown in Figure 1. The UIGRS is situated in the Moscow-Pullman basin, which is on the eastern margin of the Columbia River Basalt system (Li, 1991; Provant, 1995). Miocene basalt flows originating from fissures located in southeastern Washington and northeastern Oregon cover the irregular surface of the Precambrian bedrock. The flows are interbedded with sedimentary deposits (Li, 1991; Lum et al., 1990).

SITE GEOLOGY

There are three main stratigraphic units underlying the UIGRS. From the bottom, these are the crystalline basement rock, Columbia River Basalt with associated sediments, and surficial sediments. The crystalline basement rock consists of Precambrian orthoquartzite and Cretaceous granite intrusions (Provant, 1995). The overlying Miocene Columbia River Basalt Group comprises two basalt formations, the Grande Ronde and the Wanapum, with Latah Formation sediments deposited between or overlying some of the flows. Pleistocene Palouse loess and alluvium make up the uppermost stratigraphic unit.

The Wanapum Formation is the upper basalt unit; only the Priest Rapids Member is present in the basin, and is the focus of this chapter. Lava flows dated as 14.5 Ma created the Wanapum Formation. The Lolo flow of the Priest Rapids Member is the uppermost basalt unit in the basin; it is between ~4.5 and 61 m under the UIGRS (Li, 1991). This lava flow consists of mostly dense basalt with subhorizontal fractures, numerous microfractures, vertical joints, and vesicles that formed as the result of cooling patterns. The UIGRS wells in the basalt are completed in subhorizontal fractures in the upper third of the flow at depths ranging from 19 to 27 m. Less continuous horizontal fractures, vertical joints that may connect horizontal

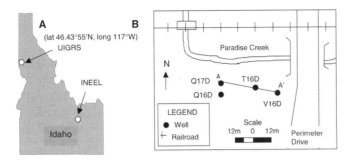

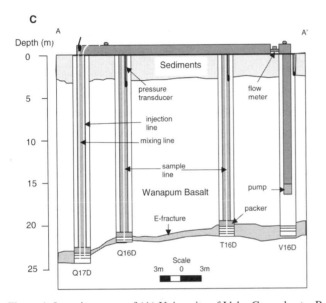

Figure 1. Location maps of (A) University of Idaho Groundwater Research Site (UIGRS; INEEL is Idaho National Engineering and Environmental Laboratory), (B) map view of field site (lat 46.43°55′N, long 117°W), and (C) cross section of recirculating tracer test experiment.

fractures, and at a smaller scale, microfractures and vesicles are found at a variety of depths.

STRUCTURE OF A COLUMBIA RIVER BASALT FLOW

The structural characteristics of a basalt unit are formed by the cooling and emplacement of the flow (Bush and Seward, 1992). A typical Columbia River Basalt flow consists of (from top to bottom) a vesicular flow top, entablature, colonnade, and pillows (Fig. 2). Static conditions during the cooling of a flow are necessary to form this structure. Vesicles are found in several portions of a flow, but are most abundant in the flow top. They are formed from trapped gas bubbles within the cooling magma. The colonnade is composed of large columns formed in the basal section of a flow. These columns have diameters from ~30 cm to 5 m with an average of ~1 m and a length of up to 75 m, with an average between 15 m and 30 m. The entablature is composed of smaller columns in the upper section

of a flow and makes up most of the flow thickness. Columns in the entablature are usually <1 m in diameter and are less consistent in orientation. They are often bundled together to form fans, synforms, antiforms, or other odd-shaped structures. A distinct contact is formed where the colonnade and entablature meet, creating an extensive horizontal fracture zone that can be several kilometers long. Multiple sequences of colonnade and entablature can occur in a single flow.

The structure of the Lolo flow is somewhat different, and also varies laterally (Li, 1991). The top of the flow is oxidized. The upper section of the flow consists of large blocks with subhorizontal fractures. A hackly entablature then grades in the center portion of the flow to alternating entablature and colonnade structures. Large diameter columns of the colonnade make up the bottom of the flow.

A fracture zone located in the middle of the upper basalt unit, known as the E fracture zone, was tested in this study. It is called the E fracture because it is in the eastern portion of the site. There is also a W fracture zone located in the western portion of the site that was not used in this experiment. Subhorizontal fracture zones occur at numerous sites within the basalt. They vary in lateral continuity and thickness. Wells Q17D, Q16D, T16D, and V16D penetrate the E fracture zone at the UIGRS (Fig. 1). The fracture zone is 0.153–0.914 m thick at a depth of ~19.5–24.1 m below land surface and dips <10° to the west (Li, 1991). The depth to water in the wells varies between 1.52 and 3.05 m. The E fracture zone acts as a confined, heterogeneous, and anisotropic aquifer with well yields from <227 to 6810 L/s. The average transmissivity value from previous aquifer tests of the E fracture is 8.6×10^{-5} m²/s; storativity ranges between 2×10^{-5} and 5×10^{-4} (Li, 1991).

DESIGN AND DESCRIPTION OF EXPERIMENT

The subject of this chapter is a radially convergent, two-well recirculation test that was conducted for the hydrogeologic analysis of the E fracture zone in the Wanapum basalt. This experiment used a pumping-injection well pair, the tracer injection occurring in the injection well. Figure 1 shows the location of the test site and a cross section of the test setup. For a more detailed discussion of the test design see Nimmer (1998).

The design of the recirculation tracer test experiment was controlled by the unique characteristics of the selected wells. Wells V16D and Q17D were chosen as the recirculation well pair and wells Q16D and T16D were used as monitoring wells; T16D is located between the recirculation well pair and Q16D is just off this transect. This setup was chosen because a transect with multiple wells was desired to better interpret the heterogeneous environment of the E fracture. All of these wells are 4 in (~10.16 cm) in diameter, cased throughout, and are completed with short perforated intervals (0.15–1.22 m). Well V16D was pumped and the water was injected into well Q17D below a packer. The pumping well was pumped at a rate of

~0.31 L/s for 3 h before the tracer injection to establish a steady gradient. Recirculation continued for the duration of the experiment (314 min). An Aardvark packer was placed directly above the screened interval in the injection well (Q17D) to restrict circulation to the screened interval.

Both dissolved and particulate tracers were injected into Q17D. In the order of injection they included: fluorescein, iodide, VA, *Bacillus thermoruber* spores, and YG polystyrene microbeads (6 μm). The dissolved tracers were injected prior to the spores and microbeads because the particulates could plug the fractures and alter the flow paths. The microbeads were injected last because they have a larger size than the spores. A time lag between 10 and 30 min followed each injection before the next tracer was injected in order to allow for a more complete dispersal of each.

Samples were collected from the two monitoring wells Q16D and T16D (below the packers) and the main pumping well (V16D). Packers were placed above the screened intervals in wells Q16D and T16D to prevent tracer from migrating up the well column. The sample lines were placed below the packers. Peristaltic pumps were used in wells Q16D and T16D for sample collection.

Water levels were measured by two methods; an electric sounder and pressure transducers. A transducer was placed in each well and connected to a data logger, which was programmed to record measurements on a 15 min frequency. The electrical sounder measurements were used to check for drift in the transducer data.

RESULTS

Breakthrough curve analysis by well

The tracer results for well V16D in the test are presented as relative concentration (C/C_0) and relative particulate counts versus time (Fig. 3). The concentration of the sample is represented by C. The initial concentration of the water below the packer calculated immediately after injection and assuming homogeneous mixing is represented by C_0. For the microbeads C_0 is the total number of beads injected. The breakthrough curves shown are fluorescein, iodide, veratryl alcohol (VA), and microbeads. There is a background concentration of fluorescein because it was used in previous tracer experiments. The residual concentration of fluorescein is subtracted from each sample and is represented by C-res (concentration minus residual). The VA relative concentrations are one order of magnitude higher than those of fluorescein and iodide. Spores were not detected in any of the samples.

The dissolved tracer breakthrough curves for well V16D resemble each other well; each has at least two main peaks, one peak clearly greater than the other (Fig. 3). The second peak is greater for fluorescein and veratryl alcohol (VA); both peaks occur at very similar times. The first peak is greater for iodide and occurs at nearly the same time as the first peak for the other

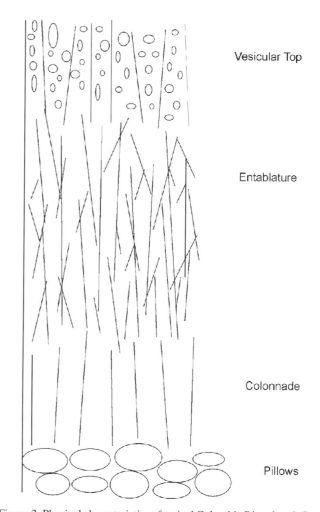

Figure 2. Physical characteristics of typical Columbia River basalt flow (Bush and Seward, 1992).

two dissolved tracers. The second peak for iodide matches very closely with the second peak for fluorescein. All three tracer breakthrough curves have long tails with similar slopes. Because the breakthrough curves for fluorescein and VA follow a similar pattern with iodide, a conservative tracer, it is concluded they also act as conservative tracers.

The microbead breakthrough curve for well V16D has two main peaks, although there are multiple smaller peaks (Fig. 3). The initial peak breakthrough occurred prior to the conservative tracer peaks in well V16D. This was expected due to the existence of preferential flow paths and porosity exclusion effects. The initial peak at a relative count of 8.88×10^{-10} occurred at $t = 44$ min ($t = 0$ at time of injection). The first peak for the conservative tracers arrived ~22 min later. The second bead peak of 6.51×10^{-9} occurred at $t = 169$ min. The second peak was lagged for several possible reasons. The discharge rate was decreased by ~0.032 L/s preceding the microbead injection due to a problem with the water injection, possibly giving rise to later-than-expected particle peak when compared

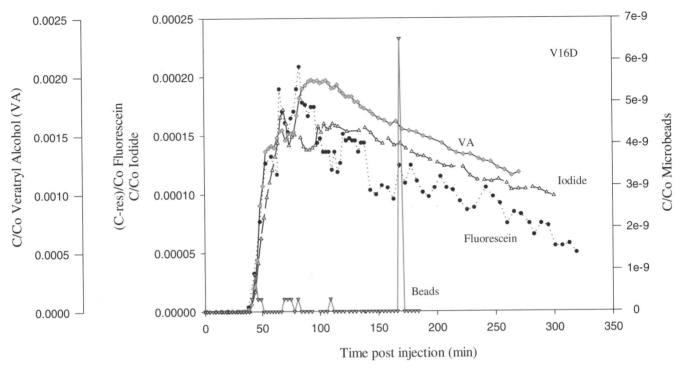

Figure 3. Relative concentration (C/C_0 or $C\text{-res}/C_0$; res is residual) vs. time (min) for tracers in well V16D (C is number of beads per 100 mL).

to the dissolved tracer peaks because the hydraulic gradient was slightly less for the beads. The second peak may also be the result of recirculated beads or possibly the result of a longer pathway.

The relative concentration data and particulate counts for well T16D are graphed versus time in Figure 4. No spores were detected. The initial rise of the breakthrough curves begins with the beads, followed by VA, iodide, and finally fluorescein with the latest rise. This is the same sequence with respect to relative concentration and particulate count peaks. The peak of the beads occurs more than 140 min later than any dissolved tracer peak. The early rises of the iodide and fluorescein breakthrough curves are very similar; after they deviate, the iodide and VA curves resemble each other. All three of the dissolved tracers have broad highs, and therefore the times of peak concentrations are difficult to identify. Had the test been conducted for a longer period of time it may have been easier to identify the peak and the tail. It is interesting to note higher concentrations of iodide and fluorescein are found in well T16D as compared to V16D. The beads peaked at a relative count (C/C_0) of 8.88×10^{-10} at $t = 34$ min, VA with a C/C_0 of 5.27×10^{-4} at $t = 178$ min, iodide with a C/C_0 of 4.76×10^{-4} at $t = 192$ min, and fluorescein with a C/C_0 of 3.20×10^{-4} at $t = 211$ min.

The relative concentrations of VA and iodide are graphed versus time for well Q16D (Fig. 4). Analysis was conducted only for iodide and VA. Analysis of fluorescein and spores was not conducted because they were injected into this well during a previous experiment and their residual concentrations were very high. VA was also previously injected, but believed to have been degraded by the time this test was run. There was a significant amount of unidentified precipitate in each sample that clogged the membranes used to filter the microbeads; therefore, no samples could be analyzed for the microbeads. There was uncertainty in the residual level of VA subtracted from the sample results, causing some of the data values to be below zero. The breakthrough curves begin to ascend at similar times, but the VA curve rises much faster than iodide. The highest measured C/C_0 for iodide is 2.55×10^{-4} and for VA C/C_0 is 5.10×10^{-4}.

Breakthrough curve analysis by tracer

The breakthrough curves for the different tracers are compared among the various wells. Figure 5 is a graph of relative fluorescein concentration versus time for wells V16D and T16D. The first appearance of fluorescein is in well T16D. This well also has a higher concentration throughout most of the test. The peak concentration in well V16D occurs before the T16D peak. Figure 5 shows a graph of relative iodide concentration versus time. The first arrival for iodide occurred in well V16D, the second in T16D, and the last in Q16D. Well T16D has the higher concentration. Figure 6A is a graph of relative VA concentration versus time. The first arrival and peak concentration for VA begin with V16D, then T16D, and finally Q16D. The difference is that well V16D has the highest concentration, followed by Q16D (at the peak), ending with T16D. In each graph of the dissolved tracers the breakthrough peak occurs first in

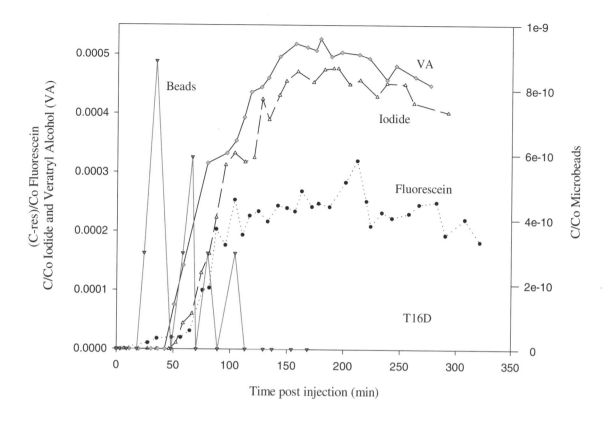

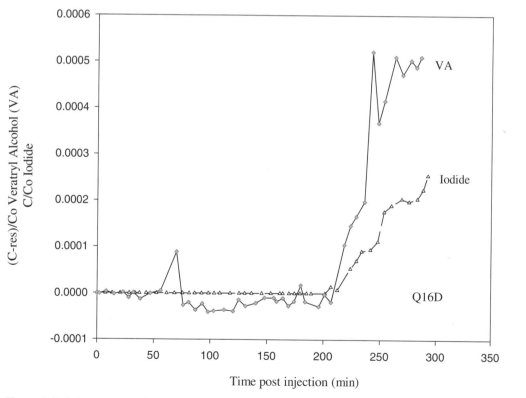

Figure 4. Relative concentration (C/C$_0$ or C-res/C$_0$; res is residual) vs. time (min) for tracers in well T16D and well Q16D (C is number of beads per 100 mL).

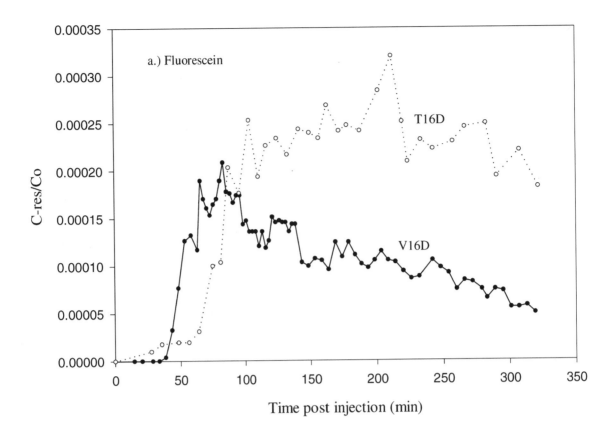

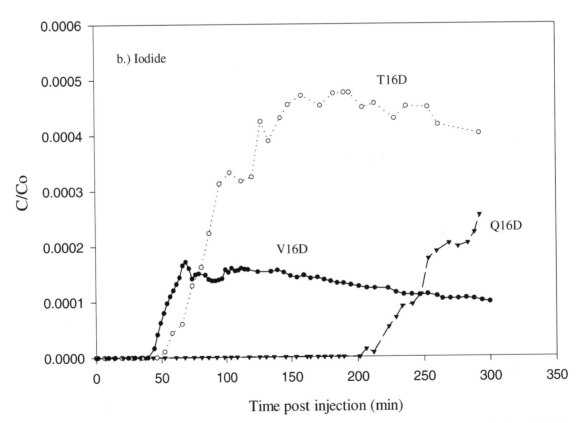

Figure 5. A: Relative fluorescein concentrations (C-res/C_0; res is residual) vs. time (min) for wells V16D and T16D. B: Relative iodide concentrations (C/C_0) vs. time (min) for wells V16D, T16D, and Q16D.

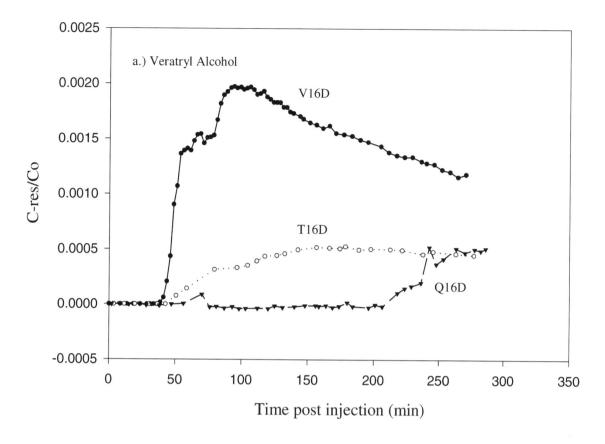

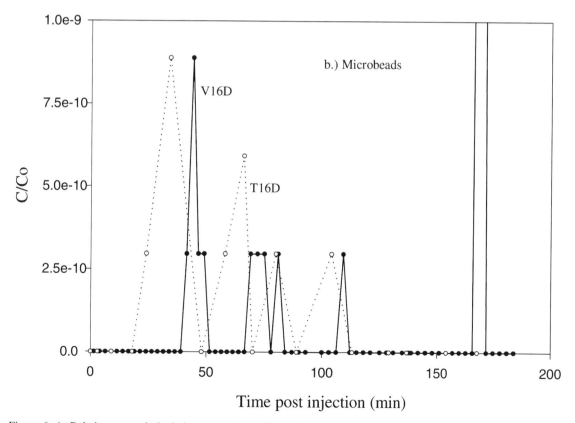

Figure 6. A: Relative veratryl alcohol concentrations (C-res/C_0; res is residual) vs. time (min) for wells V16D, T16D, and Q16D. B: Relative microbead counts (C/C_0) vs. time (min) for wells V16D and T16D.

well V16D, second in T16D, and last in Q16D. However, the particulate tracer breakthrough peak arrives first in well T16D and last in V16D, as seen in Figure 6B. The initial rise of the breakthrough curve is first in well T16D and last in V16D. The same amount of beads for the peak concentration is found in both wells V16D and T16D.

PARAMETER CALCULATIONS

The average hydraulic gradient (*dh/dl*), peak breakthrough times, and average tracer velocity for each tracer in the three monitoring wells are presented in Table 2. Based on the hydraulic gradients and distances between wells, the tracer peak breakthrough times should be smallest for well Q16D, followed by T16D, and last by V16D; just the opposite is seen. In addition, different patterns of tracer velocities in the various wells are apparent. In well V16D the beads are the first to arrive, followed by iodide, fluorescein, and VA. In well T16D the beads are the first to arrive, but are followed by VA, iodide, and finally fluorescein. The velocities for well T16D are more alike than those for well V16D.

GENERAL DISCUSSION OF TRACER TEST RESULTS

The tracer test results provide information on the physical aspects of the E fracture. The results indicate this fracture zone is hydraulically continuous within the area of the four tested wells and is able to transport particles of 6.0 μm size. The fracture zone is heterogeneous, based on the fact that none of the breakthrough curves have a bell shape, as would be expected in a homogeneous environment, and several curves have multiple peaks.

The breakthrough peak for the microbeads occurred prior to that of the dissolved tracers. The size of the particulates restricts movement to the larger fractures. Particles follow the preferential pathways, which have the highest velocities. All the tracers should arrive at the monitoring well at the same time if there was no dispersion, but because the dissolved tracers follow both slower and faster pathways, it takes longer for their peak concentrations to emerge. However, in well T16D the bead peak arrived before any other tracer was detected. Another explanation for the dissolved tracer peaks occurring after the mi-

crobead peak may be the greater diffusion coefficients of the dissolved tracers, as was seen by McKay et al. (2000).

The tracer recovery was low, especially for the particulate tracers, <1 percent for the beads, and no recovery of the spores. The low retrieval of the tracers may be attributed to loss within dead-end fractures and pore spaces, borehole storage, sorption, biodegradation, density effects, dilution, dispersion, and diffusion (Cumbie and McKay, 1999). The duration of the experiment was too short for biodegradation to occur. The particulates were probably lost to density effects and filtration. Spores may also have been lost by sorption and predation. Unlike the spores, the microbeads are probably not affected by sorption because they do not have a surface charge and will not biodegrade. The loss of spores may also be attributed to problems with the analysis methods and the analyses themselves. The dissolved tracers are considered to act conservatively based on the findings in this experiment; therefore, sorption probably does not affect their transport.

The dissolved tracer breakthrough curves have a fairly rapid rise, shoulders, and multiple peaks. The rapid increase in concentration may be the result of channeling. Channeling is caused by areas of faster movement within a fracture resulting from tortuous flow through rough-walled fractures. The multiple peaks may result from flow through different channels or fractures (Pang et al., 1998). Each peak may represent breakthrough within a single fracture. The peaks may be the result of sampling or analysis errors.

Many of the breakthrough curves have significant tails. Tailing is caused by borehole storage, dispersion, matrix and molecular diffusion (Novakowski et al., 1985), transport in different channels (National Research Council, 1996), and desorption (Hendry et al., 1997). Tracer remaining in the injection borehole may have acted as a continuous source. In the early portion of the experiment the tracers diffuse from their flow channel into the eddies, dead-end fractures, pore spaces, and the block matrix. The long tail is a result of the reverse of this process. Matrix diffusion is probably not a major mechanism because the experiments were short in duration. If the experiment was conducted for several days, weeks or months, matrix diffusion would likely have had a significant effect in retarding the migration of the plume. Flow through different channels causes enhanced tailing of the breakthrough curve by a com-

TABLE 2. PARAMETER CALCULATIONS

	Distance from injection well (m) [ft] - Q17D	Hydraulic gradient (avg)	Fluorescein PBT (min)	Iodide PBT (min)	Veratryl alcohol PBT (min)	Microbead PBT (min)	Fluorescein velocity (avg)		Iodide velocity (avg)		VA velocity (avg)		Microbead velocity (avg)	
							m/min	ft/min	m/min	ft/min	m/min	ft/min	m/min	ft/min
V16D	30.5 [100]	0.16	83	69	106	44	0.366	1.20	0.442	1.45	0.287	0.94	0.692	2.27
T16D	21.3 [70]	0.19	211	192	178	34	0.101	0.33	0.110	0.36	1.119	0.39	0.628	2.06
Q16D	9.1 [20]	0.48	N.A.	>292	>287	N.A.	N.A.		<0.69	0.210	<0.70	0.213	N.A.	

N.A.—not applicable
PBT—peak breakthrough time
VA—veratryl alcohol

bination of lower velocities in some channels and faster velocities in other channels (National Research Council, 1996).

COMPARISON OF BREAKTHROUGH CURVES AMONG DIFFERENT WELLS

Based on the hydraulic gradients, peak breakthrough of all the tracers should first arrive in well Q16D, followed by T16D, and then by V16D; yet the reverse is seen for the dissolved tracers. Features possibly affecting peak breakthrough times include the flow line regime generated by the recirculation of the pumped water, heterogeneities of the aquifer, fracture orientation, and preferential pathways. More flow lines converge at the pumping well (V16D), and therefore the peak arrives first in this well and with lower concentrations. A longer time period is necessary for the breakthrough peak to occur in well T16D. Tracers are detected in well Q16D much later than the other two wells. Based on the data from a previous experiment, breakthrough in well Q16D should have occurred much sooner. However, this well is located within a very low hydraulic conductivity area. Another explanation for the late arrival of the tracers in well Q16D is the blockage of certain flow paths caused by biofouling or the filtration of microbeads and spores from previous tests. The new flow paths may be longer, thus causing later breakthrough peaks. It should be noted that the first arrivals are a function of the average hydraulic conductivity.

The peak breakthrough of the microbeads occurs first in well T16D, followed by well V16D, which follows the order of the hydraulic gradients. There are no data for well Q16D. Transport of the dissolved tracers is controlled by the pumping-recirculation array, which should also control transport of the microbeads. The arrival of the earlier peak in well T16D may be explained by the longer flow paths leading to the pumping well increasing the opportunity for filtration leading to a later peak in well V16D.

The initial detection of the tracers should occur at about the same time in the same well, at least for the dissolved tracers. However, fluorescein and the microbeads are detected first in well T16D. VA arrives at approximately the same time in wells V16D and T16D, but iodide is first detected in well V16D. A possible explanation for the differences is the different tracer detection limits.

The sequence of the peak concentrations is not the same for all the wells, possibly indicating a heterogeneous environment or variances in chemical properties and velocities. As expected, slightly higher fluorescein and much higher iodide concentrations are found in the closer down gradient well, T16D, compared to well V16D. The pumping well (V16D) has lower concentrations caused by a greater dilution from radial flow. However, concentrations of VA are highest in well V16D, which is unexpected. VA has the greatest recovery, almost 10 times higher than the other conservative tracers in V16D. It is unknown why this occurred. The number of microbeads collected from well V16D and T16D is nearly the same. Well V16D should have a greater breakthrough peak because more flow lines converge at the well; however, the flow paths may also be much longer, in which case the particles have a greater chance of encountering filtration.

CONCLUSIONS

The general conclusions of this study are: (1) the E fracture zone is heterogeneous and (2) conducting a recirculating test with multiple tracers (dissolved and particulate tracers) provides an insight into the aquifer, transport in fractured basalt, and individual tracer characteristics. More specific conclusions from conducting the tracer experiments are as follows.

Fluorescein and veratryl alcohol (VA) act as conservative tracers denoted by their similar breakthrough curve shape and peak timing with iodide, a known conservative tracer.

Particles and conservative tracers in the same tracer experiment help to identify the existence of preferential flow paths in a fractured rock environment as seen by the early arrival of the microbeads relative to the conservative tracers.

The recovery of the tracers was quite low. Possible reasons for this include borehole storage in the injection well and flow paths that are sufficiently long that tracer was not recovered in the given test duration.

Because no spores were found in any of the samples, the use of *Bacillus thermoruber* spores is questionable in future experiments. More than 99.999 percent of the particulate tracers were not recovered and therefore were relatively unsuccessful in this type of environment at these well spacings. Studies were not found in a similar geologic setting with a similar well field.

The particulates were lost mainly by filtration and density effects. Spores may also have been lost by sorption and predation.

Processes that affect dissolved tracer transport in fractured rock include flow through different channels, dispersion, channeling, molecular diffusion, flow into dead-end fractures, borehole storage, and density effects.

ACKNOWLEDGMENTS

We thank Ronald Crawford for his help with the microbial and particle analysis, and John Bush for his insight into the area geology. We also thank Stacy Guess for her work with the spore analysis, Susan Smart for her assistance in the microbead analysis, Lisa Allenbach for her work with the veratryl alcohol analysis, and Paul Brown, Roy Mink, Roger Jensen, and Craig Tesch for their field help. This work was funded by U.S. Department of Energy grant DE FG07-96ID13420 and by Environmental Protection Agency grant R-821284012.

REFERENCES CITED

Abelin, H., Birgersson, L., Gidlund, J., and Neretnieks, I., 1991a, A large-scale flow and tracer experiment in granite. 1. Experimental design and flow distribution: Water Resources Research, v. 27, no. 12, p. 3107–3117.

Abelin, H., Birgersson, L., Moreno, L., Widén, H., Ågren, T., and Neretnieks, I., 1991b, A large-scale flow and tracer experiment in granite. 2. Results and interpretation: Water Resources Research, v. 27, no. 12, p. 3119–3135.

Aris, R., 1959, The longitudinal diffusion coefficient in flow through a tube with stagnant pockets: Chemical Engineering Science, v. 10, p. 194–198.

Bales, R.C., Gerba, C.P., Grondin, G.H., and Jensen, S.L., 1989, Bacteriophage transport in sandy soil and fractured tuff: Applied Environmental Microbiology, v. 55, p. 2861–2967.

Brown, P.A., 1998, Spore, microsphere and encapsulated cell transport in a heterogeneous subsurface environment [M.S. thesis]: Moscow, Idaho, University of Idaho, 137 p.

Bush, J.H., and Seward, W.P., 1992, Geologic field guide to the Columbia River Basalt, northern Idaho and southeastern Washington: Idaho Geological Survey Information Circular 49, 35 p.

Chapelle, F.H., 1993, Ground-water microbiology and geochemistry: New York, New York, John Wiley and Sons, 424 p.

Cliffe, K.A., Gilling, D., Jefferies, N.L., and Lineham, T.R., 1993, An experimental study of flow and transport in fractured slate: Journal of Contaminant Hydrology, v. 13, p. 73–90.

Coats, K.H., and Smith, B.D., 1964, Dead end pore volume and dispersion in porous media: Society of Petroleum Engineers Journal, v. 4, p. 73–84.

Cumbie, D.H., and McKay, L.D., 1999, Influence of diameter on particle transport in a fractured shale saprolite: Journal of Contaminant Hydrology, v. 37, no. 1-2, p. 139–157.

Davis, S.N., Campbell, D.J., Bentley, H.W., and Flynn, T.H., 1985, Groundwater tracers: National Water Well Association, Robert S. Kerr Environmental Research Laboratory, U.S. Environmental Protection Agency, 200 p.

DeJue, P., 1979, Evaluation of basalt flows as a waste isolation medium, in Proceedings from the 19th Annual Symposium, Geological Disposal of Nuclear Waste: Albuquerque, New Mexico, American Society of Mechanical Engineers, New Mexico Section [and the] University of New Mexico, College of Engineering, v. 19, pp. 55–57.

Dverstorp, B., Andersson, J., and Nordqvist, W., 1992, Discrete fracture network interpretation of field tracer migration in sparsely fractured rock: Water Resources Research, v. 28, no. 9, p. 2327–2343.

Enfield, C.G., and Bengtsson, G., 1988, Macromolecular transport of hydrophobic contaminants in aqueous environments: Ground Water, v. 26, no. 1, p. 64–70.

Fontes, D.E., Mills, A.L., Hornberger, G.M., and Herman, J.S., 1991, Physical and chemical factors influencing transport of microorganisms through porous media: Applied and Environmental Microbiology, v. 57, no. 9, p. 2473–2481.

Freeze, R.A., and Cherry, J.A., 1979, Groundwater: Englewood Cliffs, New Jersey, Prentice Hall, 604 p.

Gannon, J.T., Mingelgrin, U., Alexander, M., and Wagenet, R.J., 1991, Bacterial transport through homogeneous soil: Soil Biology and Biochemistry, v. 23, n. 12, p. 1155–1160.

Grisak, G.E., and Pickens, J.F., 1980, Solute transport through fractured media. 1. The effect of matrix diffusion: Water Resources Research, v. 16, no. 4, p. 719–730.

Harvey, R.W., and Garabedian, S.P., 1991, Use of colloid filtration theory in modeling movement of bacteria through a contaminated sandy aquifer: Environmental Science Technology, v. 25, no. 1, p. 178–185.

Harvey, R.W., George, L.H., Smith, R.L., and LeBlanc, D.R., 1989, Transport of microspheres and indigenous bacteria through a sandy aquifer: Results of natural- and forced-gradient tracer experiments: Environmental Science Technology, v. 23, no. 1, p. 51–56.

Harvey, R.W., Kinner, N.E., MacDonald, D., Metge, D.W., and Bunn, A., 1993, Role of physical heterogeneity in the interpretation of small-scale laboratory and field observations of bacteria, microbial-sized microsphere, and bromide transport through aquifer sediments: Water Resources Research, v. 29, no. 8, p. 2713–2721.

Hendry, M.J., Lawrence, J.R., and Maloszewski, P., 1997, The role of sorption in the transport of klebsiella oxytoca through saturated silica sand: Ground Water, v. 35, no. 4, p. 574–584.

Hinsby, K., McKay, L.D., Jørgensen, P., Lenczewski, M., and Gerba, C.P., 1996, Fracture aperture measurements and migration of solutes, viruses, and immiscible creosote in a column of clay-rich till: Ground Water, v. 34, no. 6, p. 1065–1075.

Li, T., 1991, Hydrogeologic characterization of a multiple aquifer fractured basalt system [Ph.D. thesis]: Moscow, Idaho, University of Idaho, 307 p.

Lum, W.E., Smoot, J.L., and Ralston, D.R., 1990, Geohydrology and numerical model analysis of ground-water flow in the Pullman-Moscow area, Washington and Idaho: U.S. Geological Survey Water-Resources Investigations Report 89-4103, 73 p.

Maloszewski, P., and Zuber, A., 1992, On the calibration and validation of mathematical models for the interpretation of tracer experiments in groundwater: Advances in Water Resources, v. 15, p. 47–62.

McCarthy, J.F., and Zachara, J.M., 1989, Subsurface transport of contaminants: Environmental Science and Technology, v. 23, no. 5, p. 496–502.

McKay, L.D., Sanford, W.E., and Strong, J.M., 2000, Field-scale migration of colloidal tracers in a fractured shale saprolite: Ground Water, v. 38, no. 1, p. 139–147.

McKay, L.D., Cherry, J.A., Bales, R.C., Yahya, M.T., and Gerba, C.P., 1993, A field example of bacteriophage as tracers of fracture flow: Environmental Science and Technology, v. 27, p. 1075–1079.

Moreno, L., and Neretnieks, I., 1993, Flow and nuclide transport in fractured media: The importance of the flow-wedded surface for radionuclide migration: Journal of Contaminant Hydrology, v. 13, p. 49–71.

Moreno, L., Neretnieks, I., and Eriksen, T., 1985, Analysis of some laboratory tracer runs in natural fissures: Water Resources Research, v. 21, no. 7, p. 951–958.

Moreno, L., Tsang, Y.W., Tsang, C.F., Hale, F.V., and Neretnieks, I., 1988, Flow and tracer transport in a single fracture: A stochastic model and its relation to some field observations: Water Resources Research, v. 24, no. 12, p. 2033–2048.

National Research Council (NRC), 1994, Alternative for ground water cleanup: Washington D.C., National Academy Press, 314 p.

National Research Council (NRC), 1996, Rock fractures and fluid flow: Washington D.C., National Academy Press, 314 p.

Neretnieks, I., 1980, Diffusion in the rock matrix: An important factor in radionuclide retardation: Journal of Geophysical Research, v. 85, no. B8, p. 4379–4397.

Neretnieks, I., 1983, A note on fracture flow dispersion mechanisms in the ground: Water Resources Research, v. 19, no. 2, p. 364–370.

Neretnieks, I., 1993, Solute transport in fracture rock: Applications to radionuclide waste repositories, flow and contaminant transport in fractured rock: New York, Academic Press, p. 39–127.

Neretnieks, I., Erikson, T., and Tähtinen, P., 1982, Tracer movement in a single fissure in granitic rock: Some experimental results and their interpretation: Water Resources Research, v. 18, no. 4, p. 849–858.

Nimmer, R.E., 1998, Ground water tracer studies in Columbia River Basalt [M.S. thesis]: Moscow, Idaho, University of Idaho, 148 p.

Novak, C.F., 1993, Modeling mineral dissolution and precipitation in dual-porosity fracture-matrix systems: Journal of Contaminant Hydrology, v. 13, p. 91–115.

Novakowski, K.S., and Lapcevic, P.A., 1994, Field measurement of radial solute transport in fractured rock: Water Resources Research, v. 30, no. 1, p. 37–44.

Novakowski, K.S., Evans, G.V., Lever, D.A., and Raven, K.G., 1985, A field example of measuring hydrodynamic dispersion in a single fracture: Water Resources Research, v. 21, no. 8, p. 1165–1174.

Pang, L., Close, M., and Noonan, M., 1998, Rhodamine WT and *Bacillus subtilis* transport through an alluvial gravel aquifer: Ground Water, v. 36, no. 1, p.112–122.

Petrich, C.R., 1995, Microsphere and encapsulated cell transport in a heterogeneous subsurface environment [Ph.D. thesis]: Moscow, Idaho, University of Idaho, 390 p.

Petrich, C.R., Stormo, K.E., Ralston, D.R., and Crawford, R.L., 1998, Encapsulated cell bioremediation: Evaluation on the basis of particle tracer tests: Ground Water, v. 36 no. 5, p. 771–778.

Provant, A.P., 1995, Geology and hydrogeology of the Viola and Moscow west quandrangles; Latah County, Idaho and Whitman County, Washington [M.S. thesis]: Moscow, Idaho, University of Idaho, 116 p.

Pyle, B.H., and Thorpe, H.R., 1981, Evaluation of the potential for microbiological contamination of an aquifer using a bacterial tracer: Proceedings of the Groundwater Pollution Conference, Australian Water Resources Council Conference Series (1), p. 213–233, Meeting: Groundwater pollution conference, Perth, Western Australia, February 19–23, 1979.

Rasmuson, A., and Neretnieks, I., 1986, Radionuclide transport in fast channels in crystalline rock: Water Resources Research, v. 22, p. 1247–1256.

Raven, K.G., Novakowske, K.S., and Lapcevic, P.A., 1988, Interpretation of field tracer tests of a single fracture using a transient solute storage model: Water Resources Research, v. 24, no. 10, p. 2019–2032.

Reimus, P.W., Robinson, B.A., Nuttall, H.E., and Kale, R., 1994, Simultaneous transport of synthetic colloids and a nonsorbing solute through single saturated natural fractures: Los Alamos National Laboratory, U.S. Department of Energy, LA-UR-94-TSA-11-94-R104.

Sorenson, K.S., Jr., Wylie, A.H., and Wood, T.R., 1996, Test Area North site conceptual model and proposed hydrogeologic studies Operable Unit 1-07B: INEL-96/0105, Parsons ES-25.9.9.31, 164 p.

Stafford, P., Toran, L., and McKay, L., 1998, Influence of fracture truncation on dispersion: A dual permeability model: Journal of Contaminant Hydrology, v. 30, no. 3, p. 79–100.

Thompson, M.E., 1991, Numerical simulation of solute transport in rough fractures: Journal of Geophysical Research, v. 96, no. B3, p. 4157–4166.

Thompson, M.E., and Brown, S.R., 1991, The effect of anisotropic surface roughness on flow and transport in fractures: Journal of Geophysical Research, v. 96, no. B3, p. 21923–21932.

Tsang, Y.W., and Tsang, C.F., 1987, Channel model of flow through fractured media: Water Resources Research, v. 23, no. 3, p. 467–479.

Tsang, Y.W., and Tsang, C.F., 1989. Flow channeling in a single fracture as a two-dimensional strongly heterogeneous permeable medium: Water Resources Research, v. 25, no. 9, p. 2076–2080.

Tsang, C.F., Tsang, Y.W., and Hale, F.V., 1991, Tracer transport in fractures: Analysis of field data based on a variable-aperture channel model: Water Resources Research, v. 27, n. 12, p. 3095–3106.

Tsang, Y.W., Tsang, C.F., Neretnieks, I., and Moreno, L., 1988, Flow and tracer transport in fractured media: a variable aperture channel model and its properties: Water Resources Research, v. 24, no. 12, p. 2049–2060.

Turner, G.A., 1958, The flow structure in packed beds: Chemical Engineering Science, v. 7, p. 156–165.

Wood, W.W., and Ehrlich, G.G., 1978, The use of baker's yeast to trace microbial movement in ground water: Ground Water, v. 16, n. 6, p. 398–403.

MANUSCRIPT ACCEPTED BY THE SOCIETY NOVEMBER 2, 2000

Geological Society of America
Special Paper 353
2002

Characterization of microbial isolates from the Idaho National Engineering and Environmental Laboratory Test Area North aquifer: Identifying potential enzymatic pathways for toluene oxidation

Mary E. Watwood*
Department of Biological Sciences, Idaho State University, Campus Box 8007, Pocatello, Idaho 83209, USA
William K. Keener
*Biotechnology Department, Idaho National Engineering and Environmental Laboratory, P.O. Box 1625,
Idaho Falls, Idaho 83415, USA*
William A. Smith
*Department of Biological Sciences, Idaho State University, Campus Box 8007, Pocatello, Idaho 83209,
and Biotechnology Department, Idaho National Engineering and Environmental Laboratory,
P.O. Box 1625, Idaho Falls, Idaho 83415, USA*

ABSTRACT

We obtained 45 microbial isolates from the Test Area North (TAN) aquifer of the Idaho National Engineering and Environmental Laboratory by sterile filtration of groundwater from well 35 within the trichloroethylene (TCE) contamination plume. In addition to standard morphological and biochemical screening, isolates were analyzed for the expression of various enzyme pathways responsible for toluene oxidation. Many of the known toluene oxidation enzyme systems are capable of cometabolically degrading TCE, and preliminary enrichment experiments have demonstrated the presence of toluene-degrading organisms within this aquifer. Isolates were tested for reaction with phenylacetylene, an enzyme-activity-dependent probe that, when used as a chromogenic substrate (yields colored product), can identify toluene-degradative pathways that include toluene 2,3-dioxygenase or toluene 2-monooxygenase. Isolates were also screened for growth and pigment production on indole plates.

Another indicator test was inhibition of growth on toluene during exposure to 1-pentyne, a selective growth inhibitor for bacteria that express toluene 2-monooxygenase or toluene 3-monooxygenase. Results for the TAN isolates were compared with those obtained for control organisms known to exhibit various toluene oxidation pathways. Isolates were also subjected to analysis with Biolog GN plates. Principal components analysis was used to distinguish groups of isolates and to compare them with various control species based on their ability to oxidize individual carbon sources. Results of this study suggest the presence of known toluene oxidation pathways within the aerobic microbial community at this site. Results also indicate that the 45 isolates represent fewer than 10 distinct bacterial species.

---

*E-mail: watwmari@isu.edu.

Watwood, M.E., Keener, W.K., and Smith, W.A., 2002, Characterization of microbial isolates from the Idaho National Engineering and Environmental Laboratory Test Area North aquifer: Identifying potential enzymatic pathways for toluene oxidation, *in* Link, P.K., and Mink, L.L., eds., Geology, Hydrogeology, and Environmental Remediation: Idaho National Engineering and Environmental Laboratory, Eastern Snake River Plain, Idaho: Boulder, Colorado, Geological Society of America Special Paper 353, p. 279–285.

INTRODUCTION

Bioremediation, the use of biological systems to remediate contaminated environments, has been an important component of numerous remediation campaigns in the United States and in other countries. Implementation of any bioremediation system must be preceded by a comprehensive and focused microbial characterization of the contaminated subsurface area. The importance of this characterization cannot be overestimated, although such data must be examined within the overall context of a thorough hydrogeological investigation (Troy, 1994).

Groundwater contaminants present at the Idaho National Engineering and Environmental Laboratory (INEEL) Test Area North (TAN) site include trichloroethylene (TCE), tetrachloroethylene (PCE), dichloroethylene (DCE), radionuclides, and sewage sludge (Sorenson et al., 1996). This type of complex subsurface contamination is not unique; myriad sites across the United States have unfortunately been contaminated with a broad suite of petroleum-based chemicals, chlorinated solvents, radionuclides, and other waste forms (Kovalick, 1991). This complex mixture of contaminants is often present within extremely heterogeneous subsurface strata, as is the case for the TAN site; hydrogeological complexity further confounds the situation. It is imperative at this type of site that data derived from numerous disciplines, including hydrology, geology, chemistry, engineering, and microbiology, be incorporated into the preliminary site characterization. Likewise, a strong interdisciplinary approach should be used to develop design plans for any remedial action (Troy, 1994).

One parameter, critical to microbial characterization, is assessment of specific microbial populations capable of degrading the contaminants of interest. This is often a sizable undertaking, because numerous potential biodegradation pathways must be considered for certain chemical pollutants, aerobic and anaerobic communities must be assessed, and potential methods of manipulation and stimulation must be tested. Once this work is completed the remediation team will have information regarding microbial populations in the contaminated zones, types of degradation rates they exhibit for the contaminants of interest, and how these microbial populations can be stimulated to enhance the degradative process in situ (Baker and Herson, 1994). At this point it is again necessary to consider the hydrogeological environment, especially when engineering the intended remediation system. This continuous exchange between components of the multidisciplinary remediation team is imperative to maximize the likelihood of a successful outcome.

In this multidisciplinary context, specific investigations were conducted in the TAN contaminant plume to characterize resident aerobic microbial populations and to determine their ability to cometabolically degrade TCE, one of the primary TAN contaminants. In this study, several distinct, yet complementary, methods were used to characterize microbial isolates obtained from the aquifer, to assess key enzymatic activities indicative of TCE biodegradation, and to compare these isolates with standard cultures known to exhibit these enzymatic activities.

Cometabolism refers to a biodegradation process that does not provide the microorganism with carbon or energy, and often results from enzyme nonspecificity. Several different bacterial genera can cometabolically degrade TCE via oxygenase-catalyzed reactions, including bacteria that use toluene as a natural growth substrate (Ensley, 1991). Bioremediation efforts can be expedited and fine-tuned by using methods that monitor microbial populations based on their biodegradative activity.

Several pathways for toluene degradation are expressed in bacteria. *Burkholderia cepacia* G4, *Burkholderia pickettii* PKO1, *Pseudomonas mendocina* KR1, and *Pseudomonas putida* mt-2 initiate toluene catabolism by hydroxylation at the ortho, meta, para, and methyl positions of toluene, respectively (Worsey and Williams, 1975; Shields et al., 1989; Whited and Gibson, 1991; Olsen et al., 1994; Leahy et al., 1996). *Pseudomonas putida* F1, *P. fluorescens* CFS215, and strain JS 150 produce toluene 2,3-dioxygenase, which converts toluene to cis-toluene dihydrodiol, a dehydrogenase for production of 3-methylcatechol, and a meta-ring-fission enzyme that cleaves the aromatic ring (Gibson et al., 1990; Haigler et al., 1992; Mikesell et al., 1993).

Toluene dioxygenase in *P. putida* F1 has a broad substrate range and is responsible for degradation of TCE (Wackett and Gibson, 1988). *P. fluorescens* CFS215, strain JS 150, and toluene-degrading bacteria using other pathways have also been shown to degrade TCE (Haigler et al., 1992; Leahy et al., 1996). In a mutant derived from *P. putida* F1 (*P. putida* 39/D), toluene dioxygenase catalyzes the formation of cis-2,3-dihydrodiols from many monoaromatic substrates, including phenylacetylene (Gibson et al., 1990; Williams et al., 1990). In addition, toluene dioxygenase activity yields several colored products from indole, providing a convenient selective indicator for bacterial colonies on semisolid media (Luu et al., 1995). Bacterial colonies on indole-containing media may reproducibly develop pigments associated with specific strains that express specific enzyme systems. However, color formation on indole medium is incompletely understood and may result partly from enzymatic processes unrelated to toluene catabolism. Color formation should be viewed as indicative of particular enzyme pathways, rather than confirmatory.

Methods currently used for monitoring microbial populations include trapping CO_2 derived from radiolabeled substrates (Fan and Scow, 1993), colony counting on semisolid media (Graham et al., 1992), most-probable-number analysis using liquid media, phospholipid analysis (Bowman et al., 1993), and direct counts with fluorescent DNA stains (Rodriguez et al., 1992). Other methods are based on nucleic acids (Brockman, 1995), enzyme-linked immunosorbent assays (Archer, 1984), luminescent genetically engineered microorganisms (Prosser et al., 1996), and molecular probes consisting of fluorescent molecules linked to oligodeoxynucleotides (Tsien et al., 1990) or antibodies (Völsch et al., 1990). However, most of these methods are not direct measures of enzymatic activity, but rather indicate the potential for enzymatic activity. In order to maximize the likelihood of success of any planned bioremediation campaign, direct enzymatic activity of indigenous subsurface microbial populations should be assessed.

We reported the use of phenylacetylene as a fluorogenic alternative substrate (yields fluorescent product) for detecting a meta fission pathway, involving toluene 2,3-dioxygenase, for toluene degradation in single bacterial cells (Keener et al., 1998). We have also used phenylacetylene as a chromogenic substrate that identifies meta fission pathways in bacterial colonies expressing toluene 2,3-dioxygenase, toluene 2-monooxygenase, or toluene 3-monooxygenase (Keener et al., 1998). The transformation of phenylacetylene yields a fluorescent product, which enables direct visualization of biodegradative activity within bacteria expressing this pathway for toluene catabolism. Phenylacetylene also acts as a selective inhibitor of the toluene 4-monooxygenase of *P. mendocina* KR1, preventing growth on toluene of strain KR1.

We have utilized α,α,α-trifluoro-*m*-toluic acid as a chromogenic substrate for toluene side chain-monooxygenase and additional enzymes of the toluene oxidation (TOL) pathway that collectively produce a yellow, meta fission product. We have also used 1-pentyne as a selective growth inhibitor of bacteria expressing toluene 2-monooxygenase or toluene 3-monooxygenase. These methods, along with indole plating and Biolog identification, were used to characterize isolates obtained from the TAN aquifer.

MATERIALS AND METHODS

Chemicals and control cultures

All chemicals were obtained from Aldrich Chemical Company (Milwaukee, Wisconsin), and media components except $(NH_4)_2SO_4$ (Aldrich) were obtained from Fisher Scientific Company (Fairlawn, New Jersey. All chemicals were of analytical reagent grade.

Pseudomonas putida F1 and *Burkholderia* (*Pseudomonas*) strain JS150 were provided by D.T. Gibson and J.C. Spain, respectively. *Pseudomonas fluorescens* CFS215 and *Burkholderia pickettii* PKO1 were provided by R.H. Olsen. *Pseudomonas mendocina* KR1 was obtained from D.T. Gibson via T.E. Ward. *Pseudomonas putida* mt-2 (ATCC 33015) and *Burkholderia cepacia* G4 (ATCC 53617) were obtained commercially.

Isolation of bacteria from TAN groundwater

Groundwater was collected in May 1997 from well 35, located within the trichloroethylene (TCE) plume at the TAN site at the INEEL (Fig. 1). Aliquots of groundwater (2 L) were passed via suction filtration through sterile filters (0.45 μm Nylon acrodisc, Gelman, Ann Arbor, Michigan). The filtrate for each aliquot was collected in a separate sterile container. Each

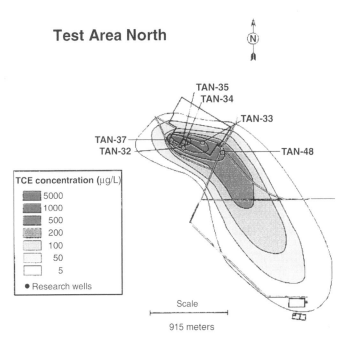

Test Area North

Figure 1. Aerial depiction of trichloroethylene (TCE) plume and several existing monitoring wells at Idaho National Engineering and Environmental Laboratory Test Area North (TAN) site (from Lehman et al., 1997). This study focused on samples from well 35.

filter was placed cell side up onto a 100×15 mm petri plate filled with semisolid minimal medium consisting of $(NH_4)_2SO_4$ (0.1% w/v), modified Hutner's mineral solution (2% v/v), Na_2HPO_4/KH_2PO_4 buffer (pH 7.4), Nobel agar (1.5% w/v), and yeast extract (0.01% w/v). The filtrate was passed through a sterile 0.2 μm filter (Nylon acrodisc). Each 0.2 μm filter was placed cell side up onto the same semisolid medium described here, and the resulting filtrate was collected and discarded as a listed waste.

All petri plates were placed upside down in a glass desiccator (2.25 L, total volume) at 25 °C. Toluene (10 μL per plate; certified ACD) was added to an open amber vial placed on its side at the bottom of the desiccator. The toluene was replenished every 2–3 days during incubation. After 5 days of incubation, plates were examined with a $10 \times$ dissecting microscope for colonies growing on the filters. Colonial morphologies were recorded, and each colony was labeled and transferred onto a gridded plate containing minimal medium (prepared as above), which was placed into a separate desiccator with toluene (10 μL per plate). Three successive transfers onto fresh, gridded plates were performed. The final transfer was made onto plates containing minimal medium with no yeast extract. We obtained 45 isolates in this manner; these isolates were characterized with respect to Gram reaction, indicator reactions (indole, phenylacetylene, 1-pentyne and α,α,α-trifluor-*m*-toluic acid), and Biolog GN identification. Each isolate was also grown on tryptic soy agar (TSA; ATCC Medium 77; Difco Laboratories), without toluene or specific indicator compounds, to assess its baseline morphological characteristics. For each indicator test, control tests were performed with cultures of known bacterial strains exhibiting specific metabolic pathways.

Growth of isolates on indole medium

All isolates were streaked onto semisolid medium consisting of $(NH_4)_2SO_4$ (0.1% w/v), modified Hutner's mineral solution (2% v/v), Na_2HPO_4/KH_2PO_4 buffer (pH 7.4), Nobel agar (1.5% w/v), yeast extract (0.01% w/v), NaCl (0.1% w/v), L-arginine·HCl (0.1% w/v), disodium succinate (0.03% w/v), and indole (50 ppm). Colony morphology on the indole medium was recorded for each isolate. Control cultures were similarly examined for growth on indole-containing medium.

Exposure of isolates to phenylacetylene during growth on toluene

Isolates were streaked onto semisolid medium consisting of minimal medium with no yeast extract added. Plates were inverted and placed into a desiccator with a vial containing toluene (10 µL per plate) and phenylacetylene (phenylacetylene:toluene = 3.5:100 v/v). Plates were examined at 2, 4, 7, and 10 days to assess microbial growth and color development. Phenylacetylene inhibits growth of a standard organism expressing the toluene 4-monooxygenase system (*P. mendocina* KR1), and results in a yellow pigment if meta fission pathways involving toluene 2,3-dioxygenase or toluene 2-monooxygenase are active (when applied during growth on toluene). Control cultures were similarly examined for growth on toluene in the presence of phenylacetylene.

Exposure of isolates to 1-pentyne

Isolates were streaked onto semisolid agar plates consisting of minimal medium (as described above) with no yeast extract added. Plates were inverted and placed into a dessicator with a vial containing toluene (10 µL per plate) and two different concentrations of 1-pentyne in toluene (1-pentyne:toluene = 1:20 or 1:10 v/v). Plates were examined at 2, 4, 7, and 10 days to assess microbial growth and colony morphology. 1-pentyne inhibits growth of standard cultures expressing toluene 2-monooxygenase or toluene 3-monooxygenase. Control cultures were similarly examined for growth on toluene in the presence of 1-pentyne.

Exposure of isolates to α,α,α-trifluoro-m-toluic acid (TFTA)

Colonies, which had grown for 7 days on minimal medium plates (no yeast extract) with toluene as sole carbon source, were exposed to TFTA (1 drop of 10 mM in phosphate buffer). Colonies were examined for a development of a yellow pigment. TFTA results in the formation of a yellow pigment when transformed by standard organisms expressing toluene side-chain monooxygenase. Subsequent enzymatic steps result in a yellow, meta ring fission product. Control cultures were similarly examined for color production with TFTA after 7 days of growth on toluene.

Biolog GN analysis of isolates

Each of the 45 isolates was subjected to analysis using Biolog GN microtiter plates, which contained 96 wells: 95 of the wells were impregnated with distinct carbon sources, mineral nutrients, and a redox indicator that is reduced and turns purple proportionate to oxidation of the carbon sources. The 96th well of each plate contained only the redox indicator with no added carbon source. Oxidation levels within each well were read by an automated plate reader.

Each isolate was grown on tryptic soy agar (TSA; ATCC Medium 77; Difco Laboratories), then transferred to fresh TSA plates. Colonies from the second TSA plates were diluted to a standard density and inoculated into Biolog GN plates. Biolog GN plates were incubated for 60–72 h at 25°C. Absorbance readings generated by the automated plate reader were recorded every 2 h. Readings were automatically adjusted for the level of color development within the control well. Factor scores for each isolate were derived from principal components analysis (PCA) of the adjusted absorbance readings.

RESULTS AND DISCUSSION

All of the 45 isolates recovered from TAN well 35 were Gram negative rods, 2–5 µm in length. Colonies formed from these isolates on tryptic soy agar were small (1–4 mm), round, and beige or yellowish with smooth margins. Two main colony morphologies, filmy and mist-like, were observed on minimal medium with toluene. The filmy colonies were large, 1–10 mm, transparent, flat, and spreading with irregular margins. The mist-like colonies were very small (<1 mm), transparent, slightly raised, and round with smooth margins. Table 1 lists the 45 isolates, divided into 9 distinct groups, along with the corresponding morphologies and results of the indicator tests. Most of the isolates produced a pale blue-green pigment when grown on the indole-containing medium. This unusual color production may be the result of several colors being produced. One isolate (#1) produced a pale yellow pigment on this medium, which has not been described in the literature. Another isolate (#10) produced no color on this medium. The control organisms exhibited expected patterns on indole medium, consistent with previous observations for specific toluene-degrading enzymes as expressed by these organisms (Table 2), with one partial exception. *Pseudomonas putida* mt-2 exhibited beige color at low indole concentrations, but produced the expected light blue color at higher indole concentrations.

Independent of exposure to phenylacetylene, colonies of isolate #1 exhibited a yellow color, thus indicating that this isolate is distinct from all of the other isolates (Table 1). All of

TABLE 1. MORPHOLOGICAL CHARACTERISTICS AND INDICATOR TEST RESULTS FOR DISTINCT ISOLATE GROUPS RECOVERED FROM TAN WELL #35

Isolate group (individual isolate numbers composing group)	Morphology on minimal medium with toluene	Color of growth on indole medium	Color of growth with phenylacetylene (3.5% v/v in toluene mixture)	Growth with 1-pentyne (10% v/v in toluene mixture)
Group 1 (#1)	Yellow; translucent	Pale yellow	Yellow; no inhibition	Beige; no inhibition
Group 2 (#2, #3, #4)	Mist-like	Blue-green	Filmy growth; partial inhibition*	Beige; partial inhibition
Group 3 (#5, #6, #7, #8, #17, #18, #19, #25, #26, #27, #28, #30, #31, #32, #34, #35, #36, #41, #42, #43, #44, #45)	Filmy	Blue-green	Complete inhibition	Substantial inhibition
Group 4 (#9, #11, #12, #13, #14, #21, #22, #23, #24)	Mist-like	Blue-green	Complete inhibition	Substantial inhibition
Group 5 (#10)	Mist-like	Clear	Complete inhibition	Substantial inhibition
Group 6 (#15)	Mist-like	Light brown/Green	Complete inhibition	Substantial inhibition
Group 7 (#20)	Filmy	Light brown	Complete inhibition	Substantial inhibition
Group 8 (#29, #33)	Filmy	Growth inhibited; green	Complete inhibition	Substantial inhibition
Group 9 (#37, #38, #39, #40)	Mist-like →filmy	Blue-green	Complete inhibition	Substantial inhibition

* Partial inhibition = ~10%–50% inhibition; substantial inhibition = >90% inhibition; complete inhibition = 100% inhibition.

TABLE 2. MORPHOLOGICAL CHARACTERISTICS AND INDICATOR TEST RESULTS FOR CONTROL CULTURES

Control culture identity	Morphology on minimum medium with toluene	Color of growth on indole medium	Color of growth with phenylacetylene (3.5% v/v in toluene mixture)	Growth with 1-pentyne (10% v/v in toluene mixture)
Pseudomonas fluorescens CFS215	Beige; translucent	Green	Yellow; no inhibition	Beige; no inhibition
Pseudomonas putida F1	Brown; opaque	Green	Brown; no inhibition	Brown; no inhibition
Burkholderia cepacia G4	Beige; opaque	Brown	Pale yellow; partial inhibition*	Beige; substantial inhibition
Burkholderia sp. JS150	Beige; opaque	Green	Pale yellow; partial inhibition	Beige; no inhibition
Pseudomonas mendocina KR1	Beige; translucent	Blue	Complete inhibition	Beige; no inhibition
Pseudomonas putida mt-2	Beige; translucent	Beige/light blue	Beige, translucent; no inhibition	Beige; no inhibition
Burkholderia pickettii PKO1	White; translucent	Pale purple	Substantial inhibition	Beige; substantial inhibition

* Partial inhibition = ~10%–50% inhibition; substantial inhibition = >90% inhibition; complete inhibition = 100% inhibition.

the other isolates exhibited growth inhibition by phenylacetylene, suggesting that the groundwater isolates may express a toluene 4-monooxygenase system that is similar to that expressed by *Pseudomonas mendocina* KR1. The control organisms, including *Pseudomonas mendocina* KR1, exhibited responses (Table 2) identical to those previously observed.

Although none of the isolates exhibited inhibition at relative pentyne concentrations of 1:20, pentyne:toluene, most of the isolates were substantially inhibited (>90% inhibition) when exposed to 1-pentyne at a relative concentration of 1:10, pentyne:toluene (Table 1). This result may suggest the presence of the toluene 2-monooxygenase or toluene 3-monooxygenase. Three isolates (group 2; #2, #3, #4) exhibited partial inhibition upon exposure to 1-pentyne, and a single isolate (#1) exhibited no inhibition. The two control organisms, which should have been inhibited by 1-pentyne, *Burkholderia pickettii* PKO1 and *Burkholderia cepacia* G4, were substantially inhibited at both 1-pentyne concentrations tested (Table 2). In fact, all control cultures gave the same results at both 1-pentyne concentrations tests. Therefore, the pentyne-sensitive groundwater isolates were less sensitive to 1-pentyne than strains G4 or PK01, but were more sensitive than the remaining control cultures.

None of the isolates, grown on toluene as sole carbon source, produced yellow pigment upon exposure to TFTA, in-

dicating the likely absence of the toluene side-chain monooxygenase pathway (data not shown). The control organism, *Pseudomonas putida* mt-2, produced a yellow color upon exposure to TFTA, as previously shown (Engesser et al., 1988).

Taken together these results suggest that known aerobic toluene oxidation enzyme systems, as previously characterized for standard culture microorganisms, may be present in the isolates. However, isolate #1 was clearly unique. It exhibited yellow color when grown on toluene regardless of exposure to phenylacetylene. The failure of 1-pentyne to inhibit the growth of this organism on toluene suggests the lack of toluene 2-monooxygenase activity, as expressed in standard culture bacteria; this result will need to be further substantiated. A green color reaction on indole-containing media, noted for some standard organisms that express toluene 2,3-dioxygenase activity, was not noted for this isolate. As mentioned in the Introduction, indole reactions are merely indicative of a particular pathway. The inhibition by 1-pentyne of isolates #15 and #20, combined with the light brown color produced on indole-containing media, is indicative of the presence of toluene 2-monooxygenase or toluene 3-monooxygenase. In addition, isolates #15 and #20 were strongly inhibited in their growth in the presence of phenylacetylene, as was strain PKO1 but not G4.

The results of analyses using Biolog GN microtiter plates

are shown in Figure 2. Most of the isolates, inoculated into the individual wells of the plates, grew to late log phase within 24 h; this is the growth phase at which results were compared. Principal components analysis showed that most of the data plotted fairly close together, in a rather well defined cluster. However, data derived from nearly all of the standard organisms plotted separately from the groundwater organisms. The exception to this was *Burkholderia pickettii* PKO1, which plotted within the same cluster of most of the isolates. Thus, the collective data suggest that at least some of the groundwater organisms are quite similar to strain PKO1, although the data are insufficient to assign identities. Two of the groundwater isolates (#1 and #17) exhibited slower growth in the Biolog plates; #21 exhibited more rapid growth in the plates. These organisms, along with one other (#33, exhibiting typical growth rate), plotted separately from the main data cluster. Of these, only #1 appears to be unique, based on physiological analysis and indicator tests. The other isolates that plotted outside of the main cluster are all members of groups with similar characteristics. It may be possible that these represent distinct strains, but additional determinations would be required to verify this observation.

Based on these data, the following conclusions can be made. It is very likely that the 45 isolates, which were obtained from TAN groundwater, represent fewer than 10 separate bacterial species. Furthermore, there is some evidence to indicate the presence of the toluene-degradative pathway of strain PKO1 in at least some of the isolates. Those isolates inhibited by phen-

ylacetylene may express an enzyme similar to toluene-4-monooxygenase of strain KR1. Clearly, because these 45 organisms grew on toluene as the sole carbon source, they are expressing some type of toluene metabolizing enzyme system(s). It is also interesting that the Biolog-PCA analysis grouped the majority of the isolates as separate from all but one of the standard organisms. Taken together, these observations appear to reflect inherent differences between groundwater isolates and the standard, laboratory cultured bacterial strains. There may be other physiological parameters, unique to subsurface bacteria, which result in altered probe responses and carbon utilization patterns, as determined by the Biolog method.

FUTURE WORK

Further inquiry regarding the physiological differences between these sets of organisms could strengthen the interpretation of probe results with subsurface organisms. As a hypothetical example, there could be a meta-fission pathway expressed by some of the isolates, which would produce a yellow product from phenylacetylene. However, if other enzymes that hydrolyze this product are very active in these organisms, then the yellow product would not accumulate. This possibility will be tested by generating the product with *Pseudomonas putida* F1, then determining if individual isolates can degrade the product more rapidly than does F1. Numerous additional scenarios could explain why the TAN groundwater isolates appear to respond differently to the probes than do the standard cultures. The isolates may simply express novel toluene degradative enzymes, which could be isolated and characterized. Regardless, these results clearly indicate that additional probes, and a better understanding of the physiological basis for probe response, are required.

ACKNOWLEDGMENTS

This study was administered through the Idaho Water Resources and Research Institute under U.S. Department of Energy grant DE-FG07-96ID13420. We thank Lorena Turick for isolation work and preliminary laboratory analyses, Ron Crawford for groundwater screening of potential enzyme inducer compounds, and Marjorie Nelson for technical assistance. We also thank Bill Apel and Kay Austin for thorough reviews of the manuscript.

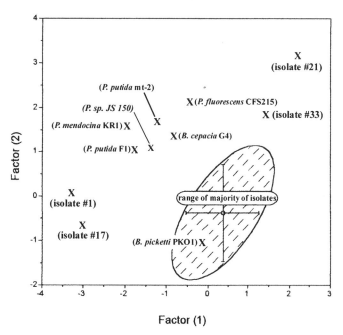

Figure 2. Principal components analysis of Biolog GN analysis of Test Area North (TAN) groundwater isolates and standard bacterial cultures. Nondimensional factors 1 and 2 represent 30% and 20%, respectively, of total variance in data.

REFERENCES CITED

Archer, D.B., 1984, Detection and quantitation of methanogens by enzyme-linked immunosorbent assay: Applied and Environmental Microbiology, v. 48, p.797–801.
Baker, K.H., and Herson, D.S., 1994, Microbiology and biodegradation, *in* Baker, K.H., and Herson, D.S. eds., Bioremediation: New York, McGraw Hill, p. 9–60.
Bowman, J.P., Jiménez, L., Rosario, I., Hazen, T.C., and Sayler, G.S., 1993,

Characterization of the methanotrophic bacterial community present in a trichloroethylene-contaminated subsurface groundwater site: Applied and Environmental Microbiology, v. 59, p. 2380–2387.

Brockman, F.J., 1995, Nucleic-acid-based methods for monitoring the performance of in situ bioremediation: Molecular Ecology, v. 4, p. 567–578.

Engesser, K.H., Cain, R.B., and Knackmuss, H.J., 1988, Bacterial metabolism of side chain fluorinated aromatics: cometabolism of 3-trifluoromethyl (TFM)-benzoate by *Pseudomonas putida (arvilla)* mt-2 and *Rhodococcus rubropertinctus* N657: Archives of Microbiology, v. 149, p. 188–197.

Ensley, B.D., 1991, Biochemical diversity of trichloroethylene metabolism: Annual Reviews in Microbiology, v. 45, p. 283–299.

Fan, S., and Scow, K.M., 1993, Biodegradation of trichloroethylene and toluene by indigenous microbial populations in soil: Applied and Environmental Microbiology, v. 59, p. 1911–1918.

Gibson, D.T., Zylstra, G.J., and Chauhan, S., 1990, Biotransformations catalyzed by toluene dioxygenase from *Pseudomonas putida* F1, *in* Silver, S., Chakrabarty, A.M., Iglewski, B., and Kaplan, S., eds., *Pseudomonas*: Biotransformations, pathogenesis, and evolving biotechnology: Washington, D.C., American Society for Microbiology, p. 121–132.

Graham, D.W., Korich, D.G., LeBlanc, R.P., 1992, Applications of a colorimetric plate assay for soluble methane monooxygenase activity: Applied and Environmental Microbiology, v. 58, p. 2231–2236.

Haigler, B.E., Pettigrew, C.A., and Spain, J.C., 1992, Biodegradation of mixtures of substitued benzenes by *Pseudomonas* sp. strain JS150: Applied and Environmental Microbiology, v. 58, p. 2237–2244.

Keener, W.K., Watwood, M.E., and Apel, W.A., 1998, Activity-dependent fluorescent labeling of bacteria that degrade toluene via toluene 2,3-dioxygenase: Applied Microbiology and Biotechnology, v. 49, p. 455–462.

Kovalick, W.W., Jr., 1991, Removing impediments to the use of bioremediation and other innovative technologies, *in* Saylor, G.S., Fox, R., and Blackburn, J.W., eds., Environmental Biotechnology for Waste Treatment: New York, Plenum Press, p. 53–60.

Leahy, J.G., Byrne, A.M., and Olsen, R.H., 1996, Comparison of factors influencing trichloroethylene degradation by toluene-oxidizing bacteria: Applied and Environmental Microbiology, v. 62, p. 825–833.

Lehman, R.M., O'Connell, S.P., Garland, J.L., and Colwell, F.S., 1997, Evaluation of remediation by community-level physiological profiles *in* Insam, H., and Rangger, A., eds., Microbial communities: Functional versus structural approaches: Berlin, Springer Verlag, p. 94–108.

Luu, P.P., Yung, C.W., Sun, A.K., and Wood, T.K., 1995, Monitoring trichloroethylene mineralization by *Pseudomonas cepacia* G4 PR1: Applied Microbiology Biotechnology, v. 44, p. 259–264.

Mikesell, M.D., Kukor, J.J., and Olsen, R.H., 1993, Metabolic diversity of aromatic hydrocarbon-degrading bacteria from a petroleum contaminated aquifer: Biodegradation, v. 4, p. 249–259.

Olsen, R., Kukor, J.J., and Kaphammer, B., 1994, A novel toluene-3 monooxygenase pathway cloned from *Pseudomonas pickettii* PKO1: Journal of Bacteriology, v. 176, p. 3749–3756.

Prosser, J.I., Killham, K., Glover, L.A., and Rattray, E.A.S., 1996, Luminescence-based systems for detection of bacteria in the environment: Critical Reviews in Biotechnology, v. 16, p. 157–183.

Rodriguez, G.G., Phipps, D., Ishiguro, K., and Ridgway, H.F., 1992, Use of a fluorescent redox probe for direct visualization of actively respiring bacteria: Applied and Environmental Microbiology, v. 58, p. 1801–1808.

Shields, M.S., Montgomery S.O., Chapman, P.J., Cuskey, S.M., Prichard, P.H., 1989, Novel pathway of toluene catabolism in the trichloroethylene-degrading bacterium G4: Applied Environmental Microbiology, v. 55, p. 1624–1629.

Sorenson, Jr., K.S., Wylie, A.H., and Wood, T.R., 1996, Test Area North Site Conceptual Model and Proposed Hydrogeological Studies Operable Unit 1-07B: Prepared for US DOE, DOE Idaho Operations Office, INEL-96/0105.

Troy, M.A., 1994, Bioengineering of soils and ground waters, *in* Baker, K.H., and Herson, D.S., eds., Bioremediation: New York, McGraw Hill, p. 173–201.

Tsien, H.C., Bratina, B.J., Tsuji, K., and Hanson, R.S., 1990, Use of oligodeoxynucleotide signature probes for identification of physiological groups of methylotrophic bacteria: Applied and Environmental Microbiology, v. 56, p. 2858–2865.

Völsch, A., Nader, W.F., Geiss, H.K., Nebe, G., and Birr, C., 1990, Detection and analysis of two serotypes of ammonia-oxidizing bacteria in sewage plants by flow cytometry: Applied and Environmental Microbiology, v. 56, p. 2430–2435.

Wackett, L.P., and Gibson, D.T., 1988, Degradation of trichloroethylene by toluene dioxygenase in whole-cell studies with *Pseudomonas putida* F1: Applied and Environmental Microbiology, v. 54, p. 1703–1708.

Whited, G.M., and Gibson, D.T., 1991, Separation and partial characterization of the enzymes of the toluene-4-monooxygenase catabolic pathway in *Pseudomonas mendocina* KR1: Journal of Bacteriology, v. 173, p. 3017–3020.

Williams, M.G., Olson, P.E., Tautvydas, K.J., Bitner, R.M., Mader, R.A., and Wackett, L.P., 1990, The application of toluene dioxygenase in the synthesis of acetylene-terminated resins: Applied Microbiology and Biotechnology, v. 34, p. 316–321.

Worsey, M.J., and Williams, P.A., 1975, Metabolism of toluene and the xylenes by *Pseudomonas putida* (arvilla) mt-2: Evidence for a new function of the TOL Plasmid: Journal of Bacteriology, v. 124, p. 7–13.

MANUSCRIPT ACCEPTED BY THE SOCIETY NOVEMBER 2, 2000

Geological Society of America
Special Paper 353
2002

Effect of basalt heterogeneity on intrinsic bioremediation processes in groundwater

Allan H. Wylie*
Dale R. Ralston
Gary S. Johnson
Idaho Water Resources Research Institute, University of Idaho, Idaho Falls, Idaho 83402, USA

ABSTRACT

Investigations indicate that volatile organic compounds have contaminated groundwater at Test Area North on the Idaho National Engineering and Environmental Laboratory. Contaminants include tetrachloroethene (PCE), trichloroethene, and dichloroethene (DCE). The source is an injection well that was in operation from 1953 to 1972. Two boreholes have been drilled 213 m down gradient of the injection well. A straddle packer system was used to test aquifer properties and collect samples from discrete intervals. Packer tests revealed stratification of both aquifer properties and contaminants. PCE concentrations are much lower in the upper, more permeable portion of the aquifer, suggesting that it may be degraded within an anaerobic zone near the injection well. Concentrations of cis-DCE are between two and four times greater than trans-DCE, providing further evidence for microbial activity. Kriged aquifer temperature data imply that the injection well is within a thermal high, perhaps a result of microbial respiration. The mole ratio of Cl^-/Na^+ is elevated downgradient of the injection well, implying that the plume is more enriched in chloride than sodium. The source of chloride may be degradation of the chlorinated ethenes. Although none of these factors individually are proof of intrinsic bioremediation, each represents a piece of evidence that contributes support for the concept.

INTRODUCTION

The Snake River Plain aquifer underlies the eastern Snake River Plain east of Bliss, Idaho (Fig. 1). The Snake River Plain aquifer consists of interlayered basalt flows, pyroclastics, and sedimentary material. The aquifer extends from Bliss on the west to Ashton on the northeast. Its lateral boundaries are formed by contacts with less permeable rocks at the margins of the plain (Mundorff et al., 1964). The Idaho National Engineering and Environmental Laboratory (INEEL) and Test Area North (TAN) overlie the north-central portion of the aquifer.

Aquifer permeability is largely controlled by the distribution of basalt flow contacts (interflow zones) with some additional permeability contributed by fractures, vesicles, and intergranular pore spaces (Mundorff et al., 1964; Welhan et al., this volume, chapter 15). On a large scale, the dense basalt flow interiors act as "grains" while the "intergranular porosity" is reflected in the interflow zones (Whitehead, 1992). These grains are formed as basalt flow sequences are deposited in an overlapping and coalescing manner, where younger flows build on the complex undulating topography of previous flows. Flow through the aquifer follows a tortuous path around, through, and between large particles (dense units of basalt) in the general direction of the regional hydraulic gradient.

*E-mail: awylie@uidaho.edu

Wylie, A.H., Ralston, D.R., and Johnson, G.S., 2002, Effect of basalt heterogeneity on intrinsic bioremediation processes in groundwater, *in* Link, P.K., and Mink, L.L., eds., Geology, Hydrogeology, and Environmental Remediation: Idaho National Engineering and Environmental Laboratory, Eastern Snake River Plain, Idaho: Boulder, Colorado, Geological Society of America Special Paper 353, p. 287–296.

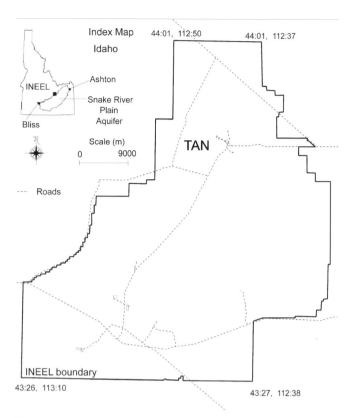

Figure 1. Location of eastern Snake River Plain aquifer, Idaho National Engineering and Environmental Laboratory (INEEL), and Test Area North (TAN).

Trichloroethene (TCE) was discovered in the groundwater at TAN in 1987 during routine sampling of water supply wells. Ensuing investigations revealed the presence of a substantial plume of dissolved chlorinated organic compounds in the aquifer beneath TAN (Fig. 2). A decommissioned injection well (TSF-05) was identified as the source of the contamination (Sorenson et al., 1996). The TSF-05 injection well was drilled to a depth of 93 m below land surface (bls) in 1953 and used for waste disposal, including sewage, until 1972. In addition to TCE, six other contaminants of concern were identified for the site: tetrachloroethene (PCE), 1,2-dichloroethene (DCE), ^{90}Sr, tritium, ^{137}Cs, and ^{234}U (Kaminsky et al., 1994). The subject of this document is biodegradation of the chlorinated ethenes within an anaerobic zone adjacent to the TSF-05 injection well. Radionuclide remediation is not addressed.

The hypothesis being tested is that geologic heterogeneity at the scale of individual basalt flows affects the distribution of biological substrates and thereby affects the intrinsic bioremediation of contaminants. This hypothesis can be tested by analyzing: (1) basalt stratigraphy and vertical distribution of hydrologic properties; (2) electron acceptor distribution; (3) contaminant and daughter product distribution; (4) microorganism distribution; and (5) aquifer temperature distribution.

This chapter is organized into four sections: methods, results, analysis, and summary and conclusions. The methods section describes how the data were collected. The results section briefly presents the data. The analysis section offers integrated interpretations of the data.

METHODS

Hypothesis evaluation involved an interdisciplinary research team from three Idaho universities, the University of Idaho, Idaho State University, and Boise State University. Research was focused on wells TAN-34 and TAN-35 (Fig. 2). The wells were drilled in the fall of 1996 to ~126 m bls. Both wells have a 20 cm intermediate casing string to 59 m bls and are open from 59 to ~126 m bls. The depth to water in wells TAN-34 and TAN-35 is ~64 m bls. Core was collected using aseptic precautions at several intervals between 60 and 127 m bls in TAN-34 and analyzed for indigenous microbial populations. Colwell et al. (1992) described aseptic coring techniques in detail. Investigations conducted in both wells included hydraulic testing, groundwater sampling, geochemical analysis of the basalt core, and borehole to borehole radar tomography.

Basalt stratigraphy and vertical distribution of hydrologic properties

Basalt stratigraphy evaluation and location of the primary advective flow paths within the aquifer at wells TAN-34 and TAN-35 involved a three-pronged approach. One approach utilized radar tomography. Geophysical tomography involves propagating waves through a medium. Average material properties can be computed from one wave; however, if numerous wave paths intersect, material properties can be computed at each intersection (Witten and King, 1990). Knoll and Lane (1997) collected and analyzed the radar tomographic data. This involved conducting a level run survey, cross-hole tomography, a vertical radar profile, and a single-hole reflection survey in wells TAN-34 and TAN-35. Data were inverted and slowness values in a petrophysical model were used to infer porosity variations between the wells.

A second technique involved basalt geochemical correlations. Geist et al. (this volume, chapter 4) and Hughes et al. (this volume, chapter 10) discuss this technique in more detail. This technique involves analyzing basalt samples to determine the geochemical makeup and then correlating the results to determine which samples are from the same basalt flow.

Another approach involved using a straddle packer system to isolate and test specific zones within the aquifer. The packer system consists of two inflatable sliding head packers separated by a pump and pump shroud. The straddle packer system has three pressure transducers, one above the upper packer, one below the lower packer and one between the packers. The upper and lower transducers are used to determine if the packers are seating properly. If the packers are seating properly, the upper and lower transducers should not detect head changes during a

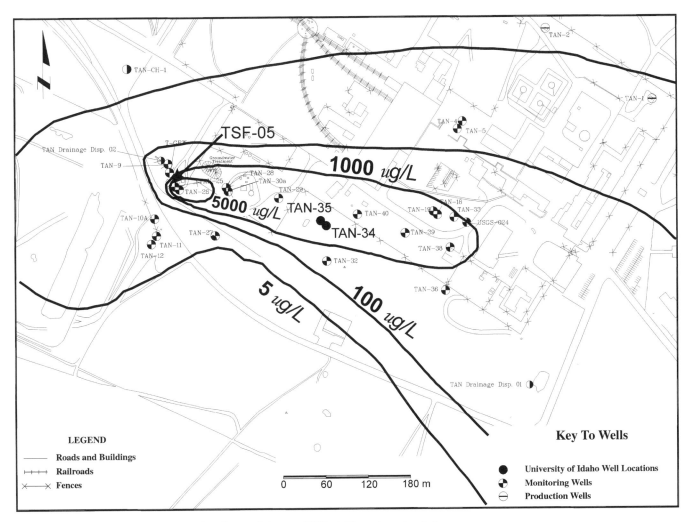

Figure 2. Test Area North (TAN) well locations and August 1996 trichloroethene isopleths.

slug test. If leakage is detected, steps are taken to minimize the leakage by changing the inflation pressure in the packers, relocating the assembly either up or down the hole, and/or changing the separation distance between the packers.

The general procedure involved: (1) selecting zones of interest and packer seats by analyzing geophysical and video logs; (2) lowering the packers down the well to the zone of interest; (3) inflating the packers to isolate the test zone; (4) conducting pneumatic slug tests to determine whether the packers seats are adequate; and (5) once the packer seats are adequate, determine hydraulic conductivity.

Slug tests were analyzed by employing one or more of several techniques described by Hvorslev (1951), Bouwer and Rice (1976), and van der Kamp (1976).

Electron acceptor distribution and contaminant and degradation product distribution

After conducting the slug tests, groundwater samples were collected from the isolated zone. After pumping three well vol-

umes (~284 L), purge parameters, including temperature, pH, specific conductance, and dissolved oxygen (DO) were allowed to stabilize and then samples were collected.

Samples were analyzed using a variety of techniques. Purge parameters were monitored in a flow-through cell. Chloride, fluoride, and nitrate were analyzed in the field with ion-specific electrodes. Sulfate concentrations were determined using a colorimetric assay. Alkalinity was determined by titration. Volatile organics compounds were determined in a laboratory at the University of Idaho using purge and trap gas chromatography coupled with a mass spectrometer (Environmental Protection Agency, 1995, EPA method 524.2). Metals were analyzed in a laboratory at Idaho State University with inductively coupled plasma.

Microorganism distribution

Griffiths et al. (1999) and Pyle (1999) describe the microbial analysis conducted and results from the TAN-34 core. Only a cursory overview is provided in this chapter. Core was collected using clean equipment and a series of tracers to assess potential contamination introduced during the drilling process using techniques modified from Colwell et al. (1992). Upon retrieval, the core was quickly transferred from the core barrel to sterile PVC tubes. The tubes were inserted into a glove bag with an anaerobic atmosphere. While within the glove bag the core was sampled for later microbial and tracer analysis. The samples were chilled to 4°C and shipped to a laboratory within 24 h. Cultures were grown using core for inoculum on several different culture media and mixtures of electron donors and acceptors in both aerobic and anaerobic environments. TCE was then added to these cultures to evaluate their ability to degrade this contaminant.

RESULTS

Basalt stratigraphy and vertical distribution of hydrologic properties

The results of the hydraulic testing and the borehole radar tomography allow identification of primary advective flow paths near the well pair. Figure 3 contains hydraulic testing results and an acoustic televiewer log depicting the borehole lithology. The vertical lines in the graph to the right of the televiewer log indicate the tested intervals and the calculated hydraulic conductivity.

The acoustic televiewer log can be viewed as a picture of the borehole wall. The logging tool transmits an acoustic signal that is then reflected back to the tool by the borehole wall. The borehole wall tends to be smooth within the dense basalt interiors and thus reflects a high percentage of the signal back to the tool, resulting in a brighter segment on the log. The vesicular sections tend to have a rougher borehole wall and more of the signal is scattered, resulting in a darker segment on the log. Interflow zones lacking a cylindrical borehole wall do not reflect the signal back to the tool, resulting in a black segment on the log. The televiewer log has a strangely shaped object in the test interval from 66 to 70 m bls. This has been identified as a pipe vesicle. The interval from 109 to 115 m bls has been identified as a columnar jointed section.

Note that the tests adjacent to brighter segments of the televiewer log tend to have lower hydraulic conductivities. For example, the tests from 66–70 m bls, 101–105 m bls, and 83–87 m bls are all adjacent to sections of dense basalt, and they all resulted in low hydraulic conductivities. Compare these results with tests conducted adjacent to darker segments of the televiewer log, for example, the tests conducted from 81–84 m bls, 87–90 m bls, and 94–98 m bls. The tests conducted adjacent to the interflow zones all have much higher hydraulic conductivies than tests conducted in dense basalt. These data suggest that the more transmissive zones tend to be located in interflow zones while the basalt flow interiors tend to be much less transmissive.

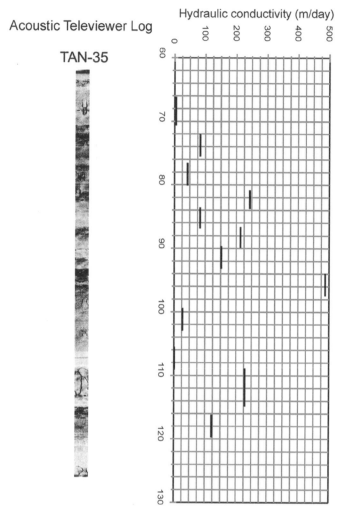

Figure 3. Hydraulic testing results and acoustic televiewer log for Test Area North well TAN-35. Bars indicate tested interval and calculated hydraulic conductivity.

Borehole radar tomography data were used to help characterize the hydrostratigraphy at the site. Several tests were run, including a level run survey, cross-hole tomography, vertical radar profiles, and a single-hole reflection survey. The level run, tomography, and vertical radar profile data show significant variations in the traveltimes and amplitudes of radar direct arrivals. Knoll and Lane (1997) used the computed slowness values in a petrophysical model to infer porosity variations between the wells (Fig. 4). Several zones having high estimated porosities are identifiable in the radar data. The two zones that are believed to have the greatest impact on flow and transport are a high-porosity zone in the depth interval of 91–96 m bls and a low-porosity dense basalt zone in the depth interval of 98–115 m bls.

The basalt geochemical correlations suggest that the basalt flow at 98–115 m bls in wells TAN-34 and TAN-35 is laterally extensive in the TAN area (Geist et al., this volume, chapter 4).

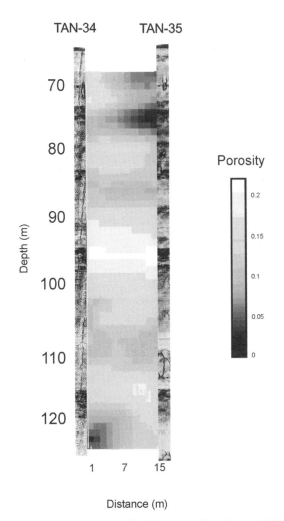

Figure 4. Radar tomogram modified from Knoll and Lane (1997).

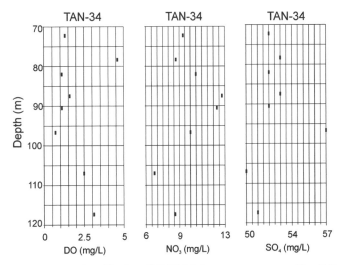

Figure 5. Test Area North well TAN-34 electron acceptor profiles. DO is dissolved oxygen.

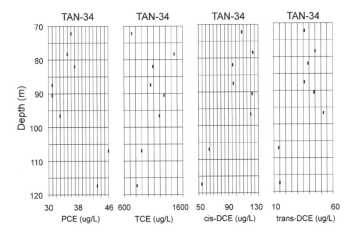

Figure 6. Test Area North well TAN-34 volatile organic compound profiles. PCE is tetrachloroethene, TCE is trichloroethene, and DCE is dichloroethene.

Electron acceptors

The potential electron acceptor profiles, DO, nitrate, and sulfate indicate the dominant type of biological respiration taking place in the aquifer (Fig. 5). Note that above 98 m bls DO concentrations are lower while nitrate and sulfate are elevated.

Contaminant and daughter product distribution

Figure 6 presents PCE, TCE, and DCE profiles from well TAN-34. Note how the TCE and DCE profiles have higher concentrations above 98 m bls and lower concentrations below that depth. These data make sense given that the injection well is open to the aquifer from 56 to 93 m bls and there appears to be a laterally extensive and thick basalt unit below 98 m bls. The PCE profile, however, shows lower concentrations above 98 m bls and higher concentrations below that depth.

Microbial analysis

Interesting trends are evident in the data from the bacterial isolates from the TAN-34 core. Aerobic bacteria are readily isolated from basalt from all depths (Pyle, 1999). Some of them (e.g., phenol utilizers) are able to degrade TCE by cometabolism. Griffiths et al. (1999) indicated that anaerobic bacteria were found throughout the core. More anaerobic bacteria were cultured from core collected in interflow zones than from dense basalt core. Denitrifiers were more easily isolated than bacteria utilizing other electron acceptors. Numerous glucose fermenters were isolated from core collected within interflow zones. Some of the anaerobes are capable of reductive dehalogenation of PCE and TCE.

TAN-34

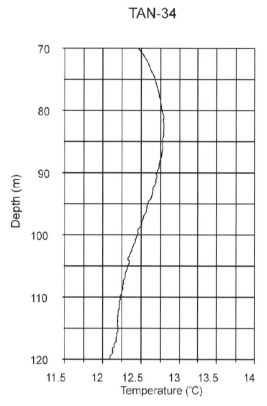

Figure 7. Temperature log from Test Area North well TAN-34.

Aquifer temperature

Geophysical logs from wells TAN-34 and TAN-35 indicate a reversed geothermal gradient; the upper portion of both wells was nearly 0.75°C warmer than the bottom (Fig. 7). Aquifer temperature data are collected during sampling. Temperature is one of the purge parameters monitored prior to sample collection (Idaho National Engineering Laboratory, 1997). Data from INEEL environmental monitoring logbooks were used to help refine the spatial distribution of the heat signal: Table 1 presents the raw data obtained from these logbooks.

ANALYSIS

More culturable microorganisms were recovered from interflow zones than from within the basalt flows (Griffiths et al., 1999). Atlas and Bartha (1998) indicated that <1% of the microorganisms typically recovered are culturable, so this observation alone does not directly imply more biomass, activity, or biodiversity within the flow contacts. However, Welhan et al. (this volume, chapter 9) and hydraulic testing (Fig. 3) suggest that the primary advective flow paths are also within the interflow zones. This inference was supported by the radar tomography data indicating that these zones are more porous (Fig. 4). Thus the observation that more culturable microorganisms are concentrated within and adjacent to the flow contacts lends

support to the hypothesis that the flow contacts are the primary advective flow paths within the aquifer.

Biotransformation products such as DCE, vinylchloride, ethane, formate, dichloroacetate, and/or glyoxylate should be in evidence if microorganisms are metabolizing contaminants. Figure 8 contains the primary aerobic and anaerobic degradation pathways for TCE. DCE, vinylchloride, or ethene should be present if reductive dechlorination is taking place. TCE epoxide is formed if aerobic degradation is taking place. TCE epoxide spontaneously hydrolyzes to formate, dichloroacetate, and/or glyoxylate, which are utilizable by microorganisms and, therefore, are generally not observed in the field (Gibson et al., 1995). Figure 6 contains profiles of the organic compounds detected.

Cis-DCE and trans-DCE appear to be the only biotransformation products present. The cis-DCE concentrations are much higher than the trans-DCE concentrations. Trans-DCE, being the more stable isomer, is a common commercial product and was probably introduced as a contaminant in the TCE. Cis-DCE is the most common biotransformation product of TCE (Vogel et al., 1987). The cis-DCE isomer is often favored microbiologically because of the specificity of the enzyme that dehalogenates TCE. Therefore the fact that more cis-DCE is present in the aquifer than trans-DCE is evidence that anaerobic biotransformation of TCE is taking place (Weidemeier et al., 1994; Ellis et al., 1997).

The electron acceptor profile (Fig. 5) suggests that there is oxygen present and negligible quantities of other electron acceptors have been consumed, and that nitrate and sulfate concentrations appear to be elevated relative to concentrations below 98 m. This is not consistent with anaerobic activity in the upper horizon, unless you consider that the source of nitrate and sulfate is probably the waste stream disposed of in the injection well. These contaminants were probably injected into the well with the other waste products. These observations are reconcilable with the observation that all detectable biotransformation products are products of anaerobic decay if it is assumed most of the anaerobic decay is taking place up gradient from TAN-34 and TAN-35 near injection well TSF-05, where raw sewage was injected into the aquifer. TCE would be degraded using the anaerobic pathway shown in Figure 8. Oxygen would diffuse back into the aquifer as it flows down gradient, increasing the oxygen concentration, leaving the anaerobic biotransformation products present in the aquifer. The concentrations of TCE and DCE may be high enough that detection of other organics at lower concentrations is problematic; or perhaps vinylchloride and ethylene are quickly consumed and the steady-state concentration is below the detection limit. Another possibility is that vinylchloride and ethylene are not there. It is not uncommon, especially under carbon-limiting conditions, for reductive dechlorination to stop at DCE.

The PCE profile (Fig. 6) also provides evidence of anaerobic biodegradation within the aquifer. There are two points to keep in mind while evaluating this profile: (1) PCE degrades

TABLE 1. INEEL ENVIRONMENTAL MONITORING TEMPERATURE DATA THIRD QUARTER, 1997

Well name	Temperature (°C)	Location Easting (m)	Northing (m)	Elevation (m)
TAN-1	11.23	109534.85	242619.76	1351.00
TAN-2	10.98	109386.98	242712.39	1358.53
TAN-4	12.57	109261.70	242586.25	1395.54
TAN-10A	11.67	108818.02	242451.06	1387.14
TAN-11	11.93	108820.81	242426.75	1367.20
TAN-25	14.92	108847.36	242495.75	1362.59
TAN-26	13.76	108853.84	242491.54	1340.87
TAN-28	15.57	108921.04	242494.10	1384.47
TAN-30A	13.64	108923.72	242488.91	1361.47
TAN-33	11.23	109245.27	242450.96	1362.24
TAN-34	13.1	109069.85	242438.34	1362.26
TAN-35	13.2	109057.05	242446.72	1362.97
TAN-36	12.28	109224.92	242330.24	1360.88
TAN-38	11.74	109217.77	242392.51	1361.59
TAN-39	11.72	109166.06	242425.54	1362.05
TAN-40	12.02	109104.61	242465.99	1362.41
TAN-41	12.66	109097.94	242463.95	1362.70
TAN-42	11.8	109142.75	242439.05	1362.56
TAN-43	11.57	109162.18	242420.55	1362.71
TAN-44	11.24	109211.87	242390.12	1361.91
TAN Drainage				
Disp 01	12.6	109338.05	242179.67	1376.27
TSF-05	18.16	108841.40	242500.50	1386.96
USGS-024	11.06	109267.96	242443.60	1378.66

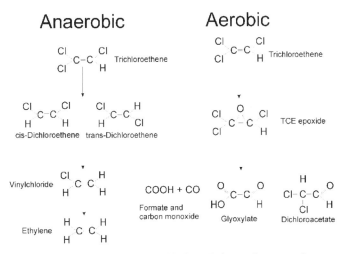

Figure 8. Anaerobic and aerobic biodegradation pathways under normal physiologic pH (after Vogel et al., 1987). TCE is trichloroethene.

readily anaerobically and is not known to degrade aerobically (Vogel et al., 1987), and (2) the injection well is completed at 93 m bls and was used to dispose raw sewage among other waste products. The particulate fraction of sewage can't diffuse through the interior of a basalt flow, so it must migrate horizontally along interflow zones. The chlorinated ethenes, however, being molecular in size, can migrate by both advection and molecular diffusion. The chlorinated volatile organic compounds migrating horizontally through the sewage are within the anaerobic zone longer than those that migrate below the

completion depth of the injection well, which is below the sewage and the anaerobic zone. Figure 5 contains a DO profile indicating an elevated oxygen concentration below 98 m bls.

Once PCE is in the aerobic zone, it is recalcitrant. As a result, the fraction of water migrating horizontally through the sewage stays in the anaerobic zone longer and loses a larger percentage of PCE than the fraction migrating below the sewage. PCE is reductively dechlorinated to TCE, which is in turn reductively dechlorinated to cis-DCE; at TAN the process may stop. Thus, the waste products that migrate horizontally within the anaerobic zone lose PCE and gain cis-DCE. This is consistent with the volatile organic compounds profiles in Figure 6.

Chloride is another biotransformation product because degradation of TCE releases chloride ions. Every mole of completely dehalogenated or cometabolically oxidized TCE releases 3 moles of chloride, so chloride concentrations should increase relative to sodium concentrations (Fig. 9). Figure 9 shows low background concentrations of Na^+ and Cl^-. Both concentrations increase sharply at the injection well. Down gradient from the injection well the molar concentration of Na^+ decreases because of dispersion, while the molar concentration of Cl^- increases or is constant due to the addition of Cl^- from degradation of chlorinated organic compounds. The background Cl^-/Na^+ mole ratio at TAN is ~1 (0.98 in well ANP-6). Although the mole ratio of the injectate when the injection well was in use from 1953 to 1972 is uncertain, an educated guess for the mole ratio is ~1. The primary sources of sodium and chloride probably consisted of sewage and water-softener discharge. The mole ratio at TAN-34 and TAN-35 is ~1.4 and

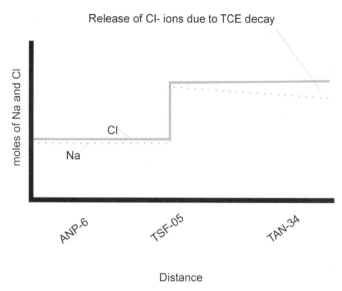

Figure 9. Theoretical molar concentration of chloride and sodium used to infer trichloroethene (TCE) degradation.

1.2, respectively. This enrichment in chloride suggests that chlorinated ethenes are being degraded.

Aquifer temperature data were compiled from numerous monitoring wells to determine whether the heat signal detected during geophysical logging was a result of biological respiration. Wells near and within the anaerobic zone should have the highest aquifer temperatures while the up gradient wells should have the lowest temperatures, if biologic respiration were responsible. Figure 10 is a kriged map of aquifer temperature.

Heat-flow modeling using TETRAD (Vinsome and Shook, 1993) indicates that a 10 kW heat source would be required to generate the heat signature observed in Figure 10. In this model the aquifer was simulated as a 61-m-thick basalt having a porosity of 10% fully saturated with groundwater. The modeled groundwater had a velocity of 0.76 m/day to simulate groundwater velocities observed in a tracer test at TAN (K. Sorenson, INEEL, 1998, personal commun.). The heat source was simulated as a point source.

Assuming that a carbohydrate ($C_6H_{12}O_6$) is being degraded anaerobically and nitrate (NO_3^-) is the electron acceptor, the heat generated by respiration can be calculated using heats of formation for the chemical reaction $5\ C_6H_{12}O_6 + 24\ NO_3^- + 24\ H^+ \rightarrow 30\ CO_2 + 24\ N_2 + 42\ H_2O$.

The heat of formation for this reaction is about 3.1×10^3 kcal per mole of carbohydrate. Typically bacteria use ~50% of the energy and the rest escapes as heat (Metcalf & Eddy, Inc., 1991). Assuming a 50% yield, ~1.6×10^3 kcal of heat is generated per mole of carbohydrate consumed. To match the modeled source of 10 kW, ~24 kg of carbohydrate must be consumed per day.

Although not conclusive, it is plausible that biodegradation of sewage is causing an elevated temperature plume. If this is the case it supports the concept of an anaerobic zone within the major basalt interflow zones near the point of injection.

SUMMARY AND CONCLUSIONS

Investigations completed to date indicate that volatile organics compounds have contaminated groundwater beneath TAN on the INEEL. The primary organic contaminants identified include PCE, TCE, cis-DCE, and trans-DCE. The source of the contamination is a wastewater injection well that operated from ca. 1953 to 1972. Two boreholes have been drilled 15 m apart within the contaminant plume, penetrating ~61 m of aquifer. A straddle packer system was used to test aquifer properties and collect samples from discrete intervals. Packer tests and radar tomography revealed stratification of both aquifer properties and contaminants. There are numerous thin basalt flows in the upper portion of the aquifer, from the water table to ~98 m bls. Below ~98 m bls a 21–24-m-thick basalt flow appears to be limiting downward vertical contaminant migration. The culturable heterotrophic anaerobic bacteria population appears to be concentrated within the high hydraulic conductivity interflow zones. This suggests that the bacteria are metabolizing organic carbon transported with the groundwater. PCE concentrations are much lower above 98 m bls, suggesting it may be degraded within the anaerobic zone near the injection well. Concentrations of cis-DCE are between two and four times greater than trans-DCE, providing further evidence for anaerobic microbial activity. Geophysical logging identified a thermal anomaly. The aquifer above ~98 m bls is ~0.75 °C warmer than below 98 m bls. Kriged aquifer temperature data imply that the injection well is within a thermal high, possibly due to microbial respiration. The mole ratio of Cl^-/Na^+ is elevated down gradient of the injection well, implying that the plume is more enriched in chloride than sodium. The source of chloride may be degradation of the chlorinated solvents. Although no single factor conclusively demonstrates that intrinsic bioremediation is taking place, collectively the evidence supports the hypothesis.

The combination of microbial research results with the field-testing results reported in this paper may provide a basis for assessing the role that intrinsic microbial activity can play in remediation of the dissolved groundwater plume at the TAN site after the contaminants in the source area have been remediated or removed. Key factors in this assessment will be rates and locations of microbial degradation relative to movement of contaminated groundwater, both in the high hydraulic conductivity flow contact zones and in the less fractured basalt matrix.

ACKNOWLEDGMENTS

We gratefully acknowledge financial support from U.S. Department of Energy grant DE-FG07-96ID12420, administered through the Idaho Water Resources Research Institute. We thank the many people at the Idaho National Engineering and

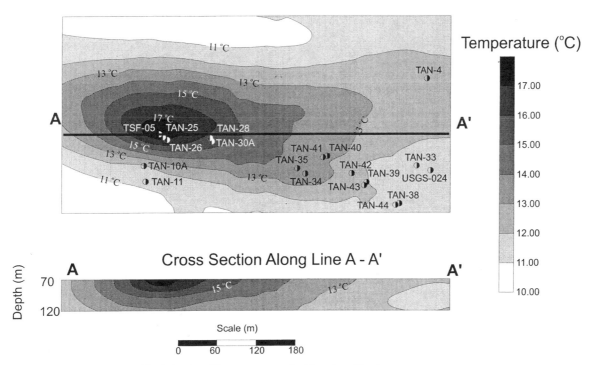

Figure 10. Third quarter, 1997 kriged aquifer temperature for Test Area North area.

Environmental Laboratory who helped us, including Patricia St. Clair, Robert Creed Jr., Kenneth Gilbert, Kent Sorenson Jr., Lance Peterson, Paul Bills, Marty Bartholomei, and Heidi Bullock.

REFERENCES CITED

Atlas, R.M., Bartha, R., 1998, Microbial ecology, fundamentals and applications: Menlo Park, California, Benjamin/Cummings Science Publishing, 694 p.

Bouwer, H., and Rice, R.C., 1976, A slug test for determining hydraulic conductivity of unconfined aquifers with completely or partially penetrating wells: Water Resources Research, v. 12, no. 3, p. 423–428.

Colwell, F., Stormberg, G., Phelps, T., Birnbaum, T., McKinley, J., Rawson, S., Veverka, C., Goodwin, S., Long, P., Russell, B., Barland, T., Thompson, D., Skinner, P., and Grover, S., 1992, Innovative techniques for collection of saturated and unsaturated subsurface basalts and sediments for microbiological characterization: Journal of Microbiological Methods, v. 15, p. 279–292.

Ellis, D.E., Lutz, E.J., Klecka, B.J., Pardieck, D.L, Salvo, J.J., Heitkamp, M.A., Gannon, D.J., Mikula, C.C., Vogel, C.M., Sayles, G.D., Kampbell, D.H., Wilson, J.T., and Maiers, D.T, 1997, Remediation technology development forum intrinsic remediation project at Dover Air Force Base, Delaware, *in* Proceedings, Symposium on Natural Attenuation of Chlorinated Organics in Ground Water: Dallas, Texas, EPA/540/R-97/504, p. 95–99.

Environmental Protection Agency (EPA), 1995, Measurement of purgable organic compounds in water by capillary column gas chromatography/mass spectrometry: EPA Method 524.2, Rev 1.4.

Gibson, D.T., Resnick, S.M., Lee, K., Brand, J.M., Torok, D.S., Wackett, L.P., Schocken, M.J., and Haigler, B.E., 1995, Desaturation, dioxygenation, and monooxygenation reactions catalyzed by naphthalene dioxygenase from *pseudomonas* sp. strain 9816-4: Journal of Bacteriology, v 177, p. 2615–2621.

Griffiths, E.C., Mobarry, B.K., Crawford, R.L., and Ely, R.L., 1999. Anaerobic bacteria capable of reductive dechlorination of tricholoroethylene obtained from Idaho National Engineering and Environmental Laboratory, Test Area North, *in* Geological Society of America Abstracts with Programs, v. 31, no. 4, p. A14.

Hvorslev, M.J., 1951, Time lag and soil permeability in ground-water observations. U.S. Army Corps of Engineers Bulletin Number 36, 50 p.

Idaho National Engineering Laboratory (INEL), 1995, Record of Decision: Declaration for the technical support facility injection well (TSF-05) and surrounding groundwater contamination (TSF-23) and miscellaneous no action sites final remedial action, operable Unit 1-07B, Waste Area Group 1, Idaho National Engineering Laboratory: Idaho Falls, Idaho, INEL-10139, variously paged.

Idaho National Engineering Laboratory (INEL), 1997, Groundwater monitoring plan, Test Area North, Operable Unit 1-07B: Idaho Falls, Idaho, INEL-96/0015, Revision 2, variously paged.

Kaminsky, J.F., Keck, K.N., Schafer-Perini, A.L., Hersley, C.F., Smith, R.P., Stormberg, G.J., and Wylie, A.H., 1994, Remedial investigation final report with addenda for the Test Area North Groundwater Operable Unit 1-07B at the Idaho National Engineering Laboratory: Idaho Falls, Idaho, EGG-ER-10643v2, variously paged.

Knoll, M.D., and Lane, J.W., 1997, Use of borehole-radar methods to characterize hydrostratigraphy in basalt, Eos (Transactions, American Geophysical Union), v. 17, no. 46, p. 319.

Metcalf & Eddy, Incorporated, 1991, Wastewater engineering treatment, disposal, and reuse: New York, McGraw-Hill, 1334 p.

Mundorff, M.J., Crosthwaite, E.G., and Kilburn, C., 1964, Groundwater for irrigation in the Snake River Basin in Idaho: U.S. Geological Survey Water-Supply Paper 1654, 224 p.

Pyle, T., 1999, Characterizations of subsurface bacteria isolated from Idaho National Engineering and Environmental Laboratory-Test Area North co-metabolism of trichloroethylene by an environmental, phenol-oxidizing bacterium [MS thesis]: Moscow, Idaho, University of Idaho, 144 p.

Sorenson, K.S., Jr., Wylie, A.H., and Wood, T.R., 1996, Test Area North site conceptual model and proposed hydrogeologic studies Operable Unit 1-07B: INEL-96/0105, variously paged.

van der Kamp, G., 1976, Determining aquifer transmissivity by means of well response test: The underdamped case: Water Resources Research, v. 12, no. 1, p. 71–77.

Vinsome, P.K.W., and Shook, G.M., 1993, Multi-purpose simulation: Journal of Petroleum Science and Engineering, v. 9, no. 1, p. 29–38.

Vogel, T.M., Criddle, C.S., and McCarty, P.L., 1987, Transformation of halogenated aliphatic compounds: Environmental Science and Technology, v. 21, no. 8, p. 722–736.

Whitehead, R.L., 1992, Geohydrologic framework of the Snake River Plain regional aquifer system, Idaho and eastern Oregon: U.S. Geological Survey Professional Paper 1409-B, 32 p.

Wiedemeier, T.H., Wilson, J.T., Miller, R.N., and Kampbell, D.H., 1994, United States Air Force guidelines for successfully supporting intrinsic remediation with an example from Hill Air Force Base, *in* The conference on petroleum hydrocarbons and organic chemicals in ground water: Prevention, detection, and restoration: Houston, Texas, National Ground Water Association, p. 317–334.

Witten, A.J., and King, W.C., 1990, Acoustic imaging of subsurface features: Journal of Environmental Engineering, v. 166, no. 1, p. 166–181.

Wylie, A.H., McCarthy, J.M., Neher, E., and Higgs, B.D., 1995, Large-scale aquifer pumping test results: Idaho Falls, Idaho, INEL-95/012, variously paged.

MANUSCRIPT ACCEPTED BY THE SOCIETY NOVEMBER 2, 2000

Index